Particle Technology Forum 2019

Held at the 2019 AIChE Annual Meeting

Orlando, Florida, USA
10 - 15 November 2019

ISBN: 978-1-7138-0527-4

Printed from e-media with permission by:

Curran Associates, Inc.
57 Morehouse Lane
Red Hook, NY 12571

CURRAN ASSOCIATES INC. proceedings .com

Some format issues inherent in the e-media version may also appear in this print version.

Printed with permission by Curran Associates, Inc. (2020)

For permission requests, please contact AIChE
at the address below.

AIChE
120 Wall Street, FL 23
New York, NY 10005-4020

Phone: (800) 242-4363
Fax: (203) 775-5177

www.aiche.org

Additional copies of this publication are available from:

Curran Associates, Inc.
57 Morehouse Lane
Red Hook, NY 12571 USA
Phone: 845-758-0400
Fax: 845-758-2633
Email: curran@proceedings.com
Web: www.proceedings.com

TABLE OF CONTENTS

(49A) USE OF BENTONITE BINDER TO OPTIMIZE GRANULATION PROCESS DURING FERTILIZER MANUFACTURE 1
Natsai Mushonga, Nkosikhona Hlabangana

(49B) COMPARTMENTAL MODELLING OF WET GRANULATION IN TWIN SCREW USING POPULATION BALANCE 2
Hamza Ismail, Mehakpreet Singh, Saeed Shirazian, Darren Whitaker, Ahmad B. Albadarin, Gavin Walker

(49D) COAGULATION OF POLYDISPERSE PRIMARY PARTICLES IN THE TRANSITION REGIME 3
Eirini Goudeli, Georgios A. Kelesidis

(77A) FLUIDIZATION OF GRAPHENE NANOPLATELETS: FROM MICROSTRUCTURE TO HYDRODYNAMICS 4
Damiano La Zara, Fabio Grillo, J. Ruud Van Ommen

(77B) BUBBLE SELF-ARRANGEMENT IN ANNULAR GAS-SOLID FLUIDIZED BEDS 6
Kaiqiao Wu, Victor Francia, Marc-Olivier Coppens

(77C) ELECTROSTATIC CHARGING AND FLUIDIZATION BEHAVIORS OF DIELECTRIC PARTICLES IN THE PRESENCE OF NOBLE GASES 8
Guanhe Rim, Dongyi Wang, Sankaran Sundaresan, Ah-Hyung Alissa Park

(77D) AN IMMISCIBLE FLUIDS APPROACH FOR CORRECTLY PREDICTING AGGLOMERATE DYNAMICS DURING PARTICLE ALD 9
Julia Hartig, Alan W. Weimer

(77E) MODELING COHESIVE GAS-PARTICLE FLOWS WITH A PARCEL-BASED LAGRANGIAN APPROACH 10
Stefan Radl, Josef Tausendschön

(77F) RAYLEIGH-TAYLOR INSTABILITY IN VIBRATED GAS-FLUIDIZED PARTICLES 12
Christopher McLaren, Thomas M. Kovar, Alexander Penn, Christoph R. Müller, Christopher M. Boyce

(96A) PREDICTING LOSS-IN-WEIGHT FEEDER PERFORMANCE BASED ON A REDUCED SET OF MATERIAL PROPERTY MEASUREMENTS 13
Tianyi Li, Yi Tao, Fernando J. Muzzio, Benjamin Glasser

(96B) FLOW CHARACTERIZATION OF COMPRESSED WOODY BIOMASS 14
Ehsan Akbari Fakhrabadi, Matthew Liberatore, Jonathan J. Stickel

(96C) DEM STUDY ON THE EFFECT OF PARTICLE SIZE DISTRIBUTION ON JAMMING IN A 3D CONICAL HOPPER 15
Ya Zhao, Ray Cocco, Shiliang Yang, Jia Wei Chew

(96D) REAL-TIME MEASUREMENT OF VOLUMETRIC FILL IN A STORAGE VESSEL AND DISCHARGE RATE BY STEREOSCOPIC 3D OPTICAL SCANNING 16
Mohammad Barzegar, Gabe P. Redding, Clive E. Davies, Luke Fullard, Miles C. E. Grafton

(96E) OPTIMIZATION OF A BAG FILLING PRODUCTION LINE FOR STARCH BY DOWN SCALING TO LABORATORY SCALE 17
Gabrie Meesters, Tom Knobbe

(96F) FLOW-AIDED PNEUMATIC CONVEYING OF COHESIVE DAIRY POWDER 18
Akeem Olaleye, Gavin Walker, Harry E. A. Van Den Akker

(96G) MEASUREMENT OF SOLID MASS FLOW RATE UNDER DIFFERENT FLOW REGIMES IN PNEUMATIC CONVEYING BY NON-INTRUSIVE ACOUSTIC EMISSION DETECTION 19
Peng Zhang, Yao Yang, Zhengliang Huang, Jingyuan Sun, Jingdai Wang, Yongrong Yang

(120A) STOPPERS AND SKINS ON CLAY NANOTUBES HELP STABILIZE OIL-IN-WATER EMULSIONS AND MODULATE THE RELEASE OF ENCAPSULATED SURFACTANTS 20
Olakunle Ojo, Azeem Afarinmade, James Trout, Marzhana Omarova, Vijay T. John, Jibao He, Duy Nguyen, Arijit Bose, Donghui Zhang, Diane A. Blake, Yuri Lvov

(120B) LARGE AREA MONOLAYER COATINGS FOR JANUS PARTICLE PRODUCTION 21
James F. Gilchrist, Jinghui Gao

(120C) MAGNETIC NANOPARTICLE CHARACTERIZATION BY DARK-FIELD IMAGING AND BRIGHT-FIELD ABSORBANCE 22
Abhinav Sannidhi, Paul W. Todd, Thomas R. Hanley

(120D) YOLK-SHELL COMPOSITE NANOSTRUCTURED ELECTRODES FOR HIGH CAPACITY LITHIUM-ION BATTERIES 33
M. Silvina Tomassone, Kurt Smith

(133A) EFFECTS OF ROLLING FRICTION, COHESION AND SIZE DISTRIBUTION ON THE SHEAR FLOW OF A BINARY PARTICLE MIX 34
Jiecheng Yang, Pongpumin Bunchatheeravate, Joseph W. Bullard, Jennifer S. Curtis
(133B) SUPERQUADRIC PARTICLE METHOD FOR NON-SPHERICAL DEM SIMULATION IN OPEN SOURCE MFIX: DEVELOPMENT, IMPLEMENTATION AND VERIFICATION 35
Xi Gao, Jia Yu, Ricardo J. F. Portal, William A. Rogers
(133C) CONTINUUM SIMULATIONS OF DENSE GRANULAR FLOW AND MODEL ASSESSMENT 36
Viraj Vilas Belekar, Alberto Passalacqua, Theodore J. Heindel, Kushal Sinha, Shankar Subramaniam
(133D) COUPLED COMPUTATIONAL FLUID DYNAMICS – AEROSOL DYNAMICS MODEL FOR PREDICTION OF NANOSTRUCTURED FILMS IN AN AEROSOL CHEMICAL VAPOR DEPOSITION REACTOR 37
Clayton Kacica, Poom Sittisomwong, Anusha Kamath, Pratim Biswas
(133E) PREDICTING PRESSURE FILTRATION PERFORMANCE BY CFD-DEM COUPLING APPROACH 38
Boyang Li, Kerianne M. Dobosz, Haitao Zhang, Jessica D. Schiffman, Kostas Saranteas, Michael A. Henson
(133F) ON THE FAST FLUIDIZED BED SIMULATIONS USING RECURRENCE CFD 39
Firas Dabbagh, Stefan Pirker, Simon Schneiderbauer
(133G) SUB-GRID DECONVOLUTION APPROACH FOR FILTERED TWO-FLUID MODELS AND THE APPLICATION TO FLUIDIZED BEDS 41
Simon Schneiderbauer
(133H) SLOW GRANULAR FLOWS IN A SPLIT-BOTTOM COUETTE DEVICE 42
Peter V. Dsouza, Prabhu R Nott
(140A) STUDY OF HYDRODYNAMIC BEHAVIOR IN A SHALLOW RECTANGULAR SPOUTED BED WITH ASCENDING GAS FLOW USING BED PRESSURE SIGNALS AND IMAGE ANALYSIS 43
Jingsi Yang, Ron Breault, Justin Weber, Steven Rowan
(140B) EFFECT OF BAFFLES ON THE FLOW REGIME, TRANSITION VELOCITIES AND VOID CHARACTERISTICS IN A 6 IN. DIAMETER FLUIDIZED BED 44
Shyam Sundaram
(140C) HYDRODYNAMIC ANALYSIS OF SPOUTING CHARACTERISTICS IN A SLOT-RECTANGULAR SPOUTED BED REACTOR 45
Steven Rowan, Ronald W. Breault, Narasimhan Soundarrajan, Jingsi Yang, Samuel Bayham
(140D) DEVELOPMENT, VALIDATION AND APPLICATION OF A NEW METHODOLOGY FOR THE MEASUREMENTS OF PARTICLE STRESSES IN AN AERATED BED 46
Lyes Ait Ali Yahia, Tobias Piepke, Ross Barrett, Raffaella Ocone
(140G) ADVANCING CROSS-SCALE CHARACTERIZATION OF REACTIVE FLUID-SOLID INTERACTIONS: ACCELERATED CARBON MINERALIZATION FOR INTEGRATED CO_2 CAPTURE, UTILIZATION AND STORAGE 48
Greeshma Gadikota, Meishen Liu
(203A) DUST PARTICLE FLOW-FIELD CHARACTERIZATION USING HIGH-SPEED DIGITAL IN-LINE HOLOGRAPHY (DIH) DURING MINIMUM-IGNITION ENERGY TESTING 49
Christian Schweizer, Ankit Saini, Daniel Guildenbecher, Chad Mashuga, Waruna Kulatilaka
(203B) INFLUENCE OF PARTICLE DENSITY ON PRE-IGNITION DUST CLOUD DYNAMICS IN A MINIMUM IGNITION ENERGY TESTING APPARATUS USING DIGITAL IN-LINE HOLOGRAPHY 50
Ankit Saini, Christian Schweizer, Pranav Bagaria, Purvali Chaudhari, Waruna Kulatilaka, Chad Mashuga
(203C) EFFECT OF PARTICLE MORPHOLOGY ON DUST CLOUD DYNAMICS 51
Fnu Shrey, Pranav Bagaria, Christian Schweizer, Chad Mashuga
(203D) MINIMUM IGNITION ENERGY (MIE) PREDICTION FROM QSPR USING MACHINE LEARNING 52
Purvali Chaudhari, Nilesh Ade, Lisa M. Perez, Stanley P. Kolis, Chad Mashuga
(203E) THE CHARACTERIZATION AND USE OF LIMESTONE ROCK DUSTS IN COAL MINES TO PREVENT DUST EXPLOSIONS 53
Isaac Zlochower
(203F) MITIGATION OF IRON AND ALUMINUM DUST DEFLAGRATIONS VIA ACTIVE EXPLOSION SUPPRESSION IN 1-M^3 SPHERE VESSEL 54
Nicholas Reding, Mark B. Shiflett
(220A) ARE CROSSLINKED POLYSTYRENE BEADS AS EFFECTIVE AS ZIRCONIA BEADS TO PRODUCE DRUG NANOPARTICLES VIA WET STIRRED MEDIA MILLING? 55
Nathaniel Parker, Mahbubur Rahman, Ecevit Bilgili

(220B) UNDERSTANDING THE FLOW OF CALCIUM CARBONATE IN STIRRED MEDIA MILLS 57
Sophie Rimmer, Andrew Ingram, Richard Tamblyn, Federico Alberini
(220C) TIME AND ENERGY DEPENDENT PROPERTIES OF STIRRED WET MILLED ALUMINA-DOPED TITANIUM DIOXIDE, A DISCRETELY HETEROGENEOUS SYSTEM 60
David Austin, David Harbottle, Timothy N. Hunter, Ali Hassanpour
(220E) FROM PARTICLE TO PROCESS PREDICTING BREAKAGE OF STRUCTURED PARTICLES 63
Kerry Johanson
(220H) ANALYTICAL SOLUTIONS TO LINEAR BREAKAGE POPULATION BALANCE MODELS OF VARIOUS PROCESS CONFIGURATIONS 80
R. Bertrum Diemer Jr.
(220G) A MECHANISTIC COLLISION MODEL FOR PARTICLE BREAKAGE IN AGITATED CRYSTALLIZATION EXPERIMENTS 81
Rory Tyrrell, René R. E. Steendam, P. J. Frawley
(228A) CHEMICAL LOOPING TECHNOLOGY - A PARTICLE TECHNOLOGY PERSPECTIVE 82
L.-S Fan
(228B) MASS PRODUCTION OF WELL ALIGNED CARBON NANOTUBES AND CARBON-SILICON BASED ANODE MATERIALS BY FLUIDIZED BED 83
Fei Wei
(228C) NOVEL FLUIDIZED BED REACTORS FOR BIOMASS THERMOCHEMICAL CONVERSION 84
Xiaotao Bi, Jim C. Lim, Naoko Ellis, John R. Grace
(228D) GASIFICATION OF BIOMASS PARTICLES: BIOCHAR PRODUCTION, PARTICULATE EMISSION, AND PROCESS SIMULATION 85
Ye Shen, Zhiyi Yao, Liping Wei, Chi-Hwa Wang
(228E) DEVELOPMENT OF A TRIPLE-BED COMBINED CIRCULATING FLUIDIZED BED 86
Chihiro Fushimi, Guoqing Guan
(228F) GREEN ENERGY INNOVATION FOR SUSTAINABLE SOCIETY:EXERGY RECUPERATION TECHNOLOGY 87
Atsushi Tsutsumi
(262A) PREDICTION OF NANOPARTICLES SIZE DISTRIBUTION: THE EFFECTS OF LIGAND SURFACE COVERAGE AND NANOPARTICLE SIZE IN ALTERING THE KINETICS OF SURFACE GROWTH 88
Saeed Mozaffari, Wenhui Li, Mudit Dixit, Giannis Mpourmpakis, Ayman M. Karim
(262F) AGGLOMERATION OF FRACTAL-LIKE PARTICLES 89
Eirini Goudeli
(262C) QUANTIFICATION AND CONTROL OF AMORPHOUS CONTENTS BY RAMAN. APPLICATIONS AND CASE STUDIES IN PHARMACEUTICAL PROCESSING 90
Michelle Raikes, Fredrik Nordstrom
(262D) BREAKAGE KINETICS OF UREA CRYSTALS IN AGITATED SLURRIES 91
Priscilla J. Hill
(262E) PHOTOELASTIC PROPERTIES OF 3D PRINTED PARTICLES 92
Negin Amini, Karen P. Hapgood, David Morton, Jun Zhang
(264B) ULTRASONIC SPRAY DELIVERY OF PRECURSORS FOR LASER PYROLYSIS SYNTHESIS OF NANOPARTICLES 93
Mohammad Malekzadeh, Parham Rohani, Mark T. Swihart
(264C) TAILORING THE BIOACTIVITY OF FLAME-MADE NANOPARTICLES FOR SURGICAL APPLICATIONS 94
Martin T. Matter, Fabian Starsich, Andrea Schlegel, Sotiris E. Pratsinis, Inge K. Herrmann
(264D) CORROSION PROTECTION OF EPOXY COATING WITH CALCIUM PHOSPHATE ENCAPSULATED BY MESOPOROUS SILICA NANOPARTICLES 96
Zoi Lamprakou, Huichao Bi, Claus Erik Weinell, Kim Dam-Johansen
(264E) SELF-ASSEMBLED NANOPARTICLE ANTIREFLECTION COATINGS ON GEOMETRICALLY COMPLEX OPTICAL SURFACES 97
Peng Jiang, Calen Leverant
(264F) ELECTRONIC STRUCTURE OF CU_{2-XSE} NANOPARTICLES FOR PLASMONIC CATALYSIS: A DENSITY FUNCTIONAL THEORY STUDY 98
Daniel Harper, Jonathan Ruffley, Jonathan Klan, J. Karl Johnson
(264G) APPLICATION OF CARBON NANOTUBE & CELLULOSE NANOCRYSTAL COMPOSITE IN CONDUCTING FILMS 99
Zachariah Pittman, Christopher L. Kitchens

(264H) INFLUENCE OF DEFECT MICROSTRUCTURE ON THE STRUCTURAL COLOR OF COLLOIDAL FILMS 100
Tianyu Liu, Bryan Vansaders, Sharon C. Glotzer, Michael J. Solomon
(264I) MOS2-PT DEPOSITED ON NITROGEN-DOPED CARBON NANOTUBES AS EFFICIENT ELECTROCATALYST FOR HYDROGEN EVOLUTION REACTION 101
Alejandra De La Rosa Goméz, Juan D. Frías Hernandez, Nicté J. Perez-Viramontes, M. Luisa García-Batancourt, Sergio M. Duron-Torres, Ismailia L. Escalante-Gacia
(264J) COMBUSTION SYNTHESIZED NICKEL-COBALT CATALYST DEACTIVATION BY COKING DURING ETHANOL DECOMPOSITION REACTION 102
Anchu Ashok, Anand Kumar, Faris Tarlochan
(292A) WHAT IS HAPPENING ABOVE YOUR FLUIDIZED BEDS? 109
Reddy Karri, Ben Freireich, Ray Cocco
(292B) THE ORIGINS AND EVOLUTION OF CYCLONE DESIGN: HOW INNOVATION AND MISS-STEPS LED TO THE CURRENT STATE OF CYCLONE DESIGN 110
Jonathon Zenz
(292C) CAREER LONG WORK ON BULK SOLIDS HANDLING PROJECTS 111
Thomas F. McGowan
(292D) CYCLONES A LA ZENZ? 112
T. M. Knowlton, Reddy Karri
(292E) FRED ZENZ – A GRADUATE STUDENT PERSPECTIVE? 113
Roy Silverman
(336A) CROSS-SCALE VALIDATION AND EVALUATION OF WURSTER COATING PROCESSES 114
Thomas Forgber, Stefan Madlmeir, Martina Trogrlic, Adrian Kape, Pankaj Doshi, Alan Carmody, Avik Sarkar, Lourdes Contreras, Johannes G. Khinast, Dalibor Jajcevic
(336B) SIMULATION AND EXPERIMENTAL VALIDATION OF A LAGRANGIAN MULTIPHASE MODEL FOR ROTARY BELL CUP ATOMIZATION 115
Wanjiao Liu, Brandon Petrouskie, Kevin Ellwood, Reza Rock, Seyed Alaie, John Furar
(336C) MULTIPHASE PARTICLE-IN-CELL COUPLED POPULATION BALANCE EQUATION METHOD FOR MULTISCALE COMPUTATIONAL FLUID DYNAMICS 127
Shinhyuk Kim, Richard D. Braatz, Jay H. Lee
(336D) CYCLONES WITH AND WITHOUT AN ECCENTRICALLY POSITIONED VORTEX FINDER 129
Ray Cocco, Ulrich Muschelknautz, Ben Freireich, S. B. Reddy Karri, T. M. Knowlton
(336E) FAST CFD BASED OPTIMIZATION OF COAL MOVING BED GASIFIER 130
Jia Yu, Liqiang Lu, Mehrdad Shahnam, William A. Rogers
(336F) LARGE SCALE MULTIPHASE FLOW SIMULATION FOR TECHNO-ECONOMIC ASSESSMENT OF THE NOVEL INTERNALLY CIRCULATING REACTOR FOR ENERGY CONVERSION WITH CO_2 CAPTURE 131
Jan Hendrik Cloete, Mohammed Khan, Schalk Cloete, Shahriar Amini
(340A) ENHANCED DISSOLUTION PERFORMANCE OF AMORPHOUS SOLID DISPERSIONS BY SIZE REDUCTION 132
Kai Zheng
(340B) MODELING PARTICLE FORMATION IN SPRAY DRYING OF AMORPHOUS SOLID DISPERSIONS 133
Pedro Valente, Masahiro Nakai, Ricardo Sousa
(340C) FROM LIQUID TO DROPLET: UNDERSTANDING THE LIGAMENTS BREAK-UP IN ATOMIZATION PROCESS 134
Tiago Porfirio, Francisco J. Galindo-Rosales, Laura Campo-Deano, Joã£o Vicente, Viriato Semiã£o
(340D) ANALYSIS OF THE MICROSTRUCTURE OF PARTICLES OBTAINED BY EVAPORATING ACOUSTICALLY LEVITATED SINGLE DROPLETS USING X-RAY COMPUTED TOMOGRAPHY 137
Hassan Abdullahi, Christopher L. Burcham, Thomas Vetter
(340E) DECOUPLING TEMPERATURE AND THE KINETICS OF DYE CONVERSION VIA SCALABLE NANO-ENCAPSULATION IN A THERMAL-RESPONSIVE MEDIA 139
Douglas Scott, Jie Feng, Madeleine Armstrong, Rodney D. Priestley, Robert K. Prud'Homme
(340F) EVAPORATIVE HEAT TRANSFER MODEL FOR QBD OF AGITATED FILTER BED DRYING 140
Kushal Sinha, Prashant Kumar, Jie Chen, Raimundo Ho, Laurie Mlinar, Nandkishor Nere

(340G) NOVEL CO-PROCESSING TECHNOLOGY TO GENERATE DRUG SUBSTANCES AMENABLE TO DIRECT COMPRESSION 141
Deniz Erdemir, David Buckley, Vivek Daftary, Mark Lindrud, Gregory Lane, Adriene Malsbury, Jing Tao, Nathaniel D. Kopp, Daniel Hsieh, Whitney Nikitczuk, Joshua Engstrom

(348A) LIQUID TRANSFER BETWEEN PARTICLES IN THE IMPREGNATION OF CATALYSTS IN A ROTATING DRUM: AXIAL DISPERSION STUDIES AND THE EFFECT OF PARTICLE-PARTICLE CONTACT 142
Yangyang Shen, M. Silvina Tomassone, William G. Borghard

(348C) SIMULATION OF FLOW-DRIVEN SEGREGATION IN MIXTURES OF DISKS AND RODS 143
Ryan P. Jones, Paul B. Umbanhowar, Julio M. Ottino, Richard M. Lueptow

(348D) PREDICTING GRANULAR DENSITY SEGREGATION—A KINETIC THEORY APPROACH 144
Yifei Duan, Paul B. Umbanhowar, Julio M. Ottino, Richard M. Lueptow

(348E) MODELING SEGREGATION OF POLYDISPERSE GRANULAR MATERIALS IN HOPPER DISCHARGE FLOW 145
Zhekai Deng, Yi Fan, Jörg Theuerkauf, Karl Jacob, Paul B. Umbanhowar, Julio M. Ottino, Richard M. Lueptow

(348F) PARTICLE MOTION IN A VERTICAL BLADED MIXER USING X-RAY PARTICLE TRACKING VELOCIMETRY (XPTV) 146
Humair Nadeem, Shankar Subramaniam, Kushal Sinha, Theodore J. Heindel

(348G) EXPLORATION OF THE UNIQUE SHEAR BEHAVIOR OF BINARY BLEND 147
Pongpumin Bunchatheeravate, Jiecheng Yang, Jennifer S. Curtis, Joseph W. Bullard

(348H) SEGREGATION EFFECTS IN THREE TYPES OF MASS FLOW HOPPERS 148
Kerry Johanson

(356A) ANALYSIS OF NANOPARTICLE GROWTH AND AGGREGATION IN NON-THERMAL PLASMA AND PLASMA AFTERGLOW BY ADVANCED ION MOBILITY SPECTROMETRY AND MONTE CARLO SIMULATION 167
Xiaoshuang Chen, Takafumi Seto, Uwe R. Kortshagen, Christopher J. Hogan Jr.

(356B) MODEL IDENTIFICATION USING IN-LINE ESTIMATES OF CRYSTAL SIZE 168
Roberto Irizarry, Jochen Schoell, Lorenzo Codan

(356C) POPULATION BALANCE MODELING FOR TWINSCREW GRANULATION: INSIGHT IN THE NOTION OF DISTANCE BETWEEN DISTRIBUTIONS AND THE NEED FOR IDENTIFIABLE MODELS FOR GOOD PREDICTIVE POWER 169
Daan Van Hauwermeiren, Thomas De Beer, Ingmar Nopens

(356D) DETAILED POPULATION BALANCE MODELLING USING A HYBRID PARTICLE MODEL 171
Astrid Boje, Jethro Akroyd, Markus Kraft

(356E) SPHERICAL AGGLOMERATION PROCESS: MECHANISTIC UNDERSTANDING AND MATHEMATICAL MODELLING 173
Omid Arjmandi-Tash, Jonathan D. Tew, Siti N. Mohd Yusoff, Kate Pitt, James D. Litster, Rachel M. Smith

(356F) MECHANISM DISCRIMINATION BASED ON POPULATION BALANCE MODELS 174
R. Bertrum Diemer Jr.

(380B) MEASURING FOOD POWDERS ELECTRICAL PROPERTIES AS A RESULT OF HUMIDITY CONTENT 175
Naveen M. Tripathi, Filip Francqui, Thomas Pirenne, Geoffroy Lumay

(380C) SOLIDS METRICS FOR SUCCESSFUL 3D PRINTING OF ENERGETIC FEEDSTOCKS 176
Bryan J. Ennis, Benjamin Ennis, Brandon Ennis, Naseem Jibrin

(380D) POWDER PERMEABILITY AS A MEASUREMENT SURROGATE FOR TRIBOELECTRIC CHARGING 177
Michael Winn, Benjamin Ennis, Naseem Jibrin, Bryan J. Ennis

(380E) METAL DUST EXPLOSION RISKS: A HAZARD ANALYSIS AND REVIEW OF CONTEMPORARY INDUSTRY OBJECTIVES 178
Nicholas Reding, Mark B. Shiflett

(380G) OXIDATION OF CARBON BLACK AGGLOMERATES 179
Georgios A. Kelesidis, Sotiris E. Pratsinis

(380H) TUNGSTEN NITRIDE ALD NANOFILMS FOR REDUCING HYDROGEN DIFFUSION 180
Sarah Bull, Theodore Champ, Chanel Hill, Robert C. O'Brien, Charles B. Musgrave, Alan W. Weimer

(380I) PLANT BASED ILLUMINATION ENABLED BY FUNCTIONALIZED NANOPHOSPHOR PARTICLES 181
Pavlo Gordiichuk, Sarah Coleman, Juyao Dong, Daichi Kozawa, Tedrick Thomas Salim Lew, Michael Strano

(380J) CORRELATION OF STRUCTURE-PROPERTY RELATIONSHIP OF DRY-COATED PARTICLE FOR ALL-SOLID-STATE BATTERY 182
Risa Sakurai, Hideya Nakamura, Shuji Ohsaki, Satoru Watano
(380L) DYNAMICS OF EARLY SOOT FORMATION BY REACTIVE MOLECULAR DYNAMICS 183
Eirini Goudeli, Somesh Roy
(380N) NOVEL CLOSED SPIRAL REACTOR FOR SUPERCRITICAL HYDROTHERMAL SYNTHESIS OF NANOPARTICLES 184
Arjun Kumar Pukkella, Nagaravi Kumar Varma Nadimpalli, Sivakumar Subramanian, Venkataramana Runkana
(380O) CHEMICAL LOOPING PARTIAL OXIDATION OF METHANE: THE IMPACT OF PARTICLE SIZE OF HEMATITE ON SYNGAS SELECTIVITY 185
Yan Liu, Mengqing Guo, Josh Goetze, Lang Qin, Liang-Shih Fan
(380P) EXPLORING PARTICLE MODIFICATION STRATEGIES TO IMPROVE YIELDS FOR OXIDATIVE COUPLING OF METHANE IN A CHEMICAL LOOPING SYSTEM 186
Deven Baser, Zhuo Cheng, Sourabh Nadgouda, Lang Qin, Liang-Shih Fan
(380Q) ROBUST PEROVSKITE OXYGEN SORBENT PARTICLES FOR "LOW TEMPERATURE" THERMOCHEMICAL AIR SEPARATION 187
Emily Krzystowczyk, Jian Dou, Xijun Wang, Fanxing Li
(380R) MULTISCALE MODELING OF GROUP D IRON OXIDE PARTICLES IN CHEMICAL LOOPING SYSTEMS: FROM ATOMISITC SCALE TO CONTINUUM SCALE 188
Yu-Yen Chen, Sourabh Nadgouda, Mengqing Guo, Yitao Zhang, Andrew Tong, Liang-Shih Fan
(380S) DESIGN OF FLUIDIZED BED COMBUSTOR REACTOR IN CHEMICAL LOOPING COMBUSTION: HYDRODYNAMIC AND HEAT TRANSFER PROPERTIES 189
Yitao Zhang, Jianhua Pan, Dawei Wang, Dikai Xu, Andrew Tong, Liang-Shih Fan
(380T) IMIQUIMOD PARTICLE FORMATION USING THE SUPERCRITICAL ANTISOLVENT (SAS) PROCESS 190
Paula De Freitas Rosa Remiro, Ângela Maria Moraes, Paulo De Tarso Vieira E Rosa
(380U) SURFACE FUNCTIONALIZATION OF POLYMER LATEX PARTICLES AND ITS EFFECT ON IN-SITU DEPOSITION OF NANO-SIZED CALCIUM SILICATE HYDRATE 191
Guanzhi Cheng, Yifan Liu, Quanxiao Dong, Yongjiang Xie, Xinguo Zheng, Wei Deng
(380W) MACHINE LEARNING DETERMINATION OF RESISTANT AND TOLERANT BACTERIA GROWTH MODES INDUCED BY NANOPARTICLES 192
A-Andrew Jones, David Medina, Na Yoon Kim, Gujie Mi, Caterina Bartomeu Garcia, Lorena Baranda Pellejero, Nicole Bassous, Thomas J. Webster
(380X) OVERCOMING THE LIMITATIONS OF HYDROCHLOROTHIAZIDE MICRONIZATION USING THE PRECIPITATION WITH COMPRESSED FLUID ANTISOLVENT PROCESS 193
Hee Jeong Park, Tae Jun Yoon, Dong Eui Kwon, Kyungseok Yu, Geonhwan Park, Youn-Woo Lee
(380Y) CHARACTERIZATION OF NANOPARTICLE SIZE DISTRIBUTIONS AND ASPECT RATIO THROUGH SINGLE-PARTICLE TRACKING AND ANALYTICAL ULTRACENTRIFUGATION 194
Xun Gong, Minkyung Park, Dorsa Parviz, Kevin S. Silmore, Tedrick Thomas Salim Lew, Michael Strano
(380Z) NUMERICAL MODELING OF WET GRANULATION PROCESSES 195
Tomoya Baba, Hideya Nakamura, Shuji Ohsaki, Satoru Watano
(380AA) WALL FRICTION MEASUREMENTS ON FREE-FLOWING MATERIAL AT LOW NORMAL STRESSES 196
Alexander Cliff, Luke Fullard, Morio Fukuoka, Clive E. Davies
(380AB) CONDUCTION AND CONVECTION HEAT TRANSFER IN A ROTARY DRUM USING SIMULATIONS AND EXPERIMENTS 197
Heather N. Emady, Manogna Adepu
(380AC) IDENTIFYING THE IMPACT OF PROCESSING ON THE PHYSICAL PROPERTIES OF AM POWDERS AND THEIR SUBSEQUENT REUSE 198
Tim Freeman, Katrina Brockbank, Jeffery Kenvin, Tony Thornton
(380AE) COMPARATIVE STUDY OF MIXING IN A HIGH SHEAR AND LOW SHEAR BLENDER USING NUMERICAL SIMULATION 199
Koyel Sen, Natasha V. Rodriguez, Carl A. Anderson, Bodhisattwa Chaudhuri
(380AG) EFFECT OF VISCOSITY ON LIQUID PENETRATION IN THE CATALYST IMPREGNATION PROCESS: EXPERIMENTS AND SIMULATIONS 200
M. Silvina Tomassone, Yangyang Shen, Jiao Yang, Jing Liu, William G. Borghard
(380AH) IN-SITU SAXS STUDY OF THE NUCLEATION AND GROWTH OF PRECIPITATED SILICA 201
Yubai Liu, Yujun Wang

(380AI) A CONCEPTUAL MODEL FOR ANALYZING PARTICLE EFFECTS ON GAS-LIQUID FLOWS IN SLURRY BUBBLE COLUMNS 202
Rongtao Zhou, Ning Yang
(380AK) HYDRODYNAMICS AND MIXING IN A PSEUDO-2D RECTANGULAR SPOUTED BED 203
Ritesh Raman, Partha S. Goswami
(380AL) SOLIDS MIXING RATES OF GELDART A AND B POWDERS PREDICTED BY THE TWO FLUID MODEL 204
Mohammad-Sadegh Salehi, Maryam Askarishahi, Stefan Radl
(380AM) FLUID-PARTICLE DRAG IN LOW-REYNOLDS-NUMBER BIDISPERSE SUSPENSIONS 205
Fan Duan, Xiao Chen, Qiang Zhou
(380AN) AN INVESTIGATION OF GAS FLUIDIZED BED OF FLEXIBLE FIBERS USING THE COUPLED DEM/CFD METHOD 206
Yiyang Jiang, Zhaosheng Yu, Yu Guo, Carl Wassgren, Jennifer Sinclair Curtis
(380AO) EXPLORING AGGLOMERATE DYNAMICS THROUGH AN IMMISCIBLE TWO-FLUID MODEL 207
Julia Hartig, Alan W. Weimer
(380AP) USING A POWDER RHEOMETER TO PROBE THE DEFLUIDIZATION OF COHESIVE (GROUP C) PARTICLES 208
Ipsita Mishra, Peiyuan Liu, Christine M. Hrenya
(380AQ) FLUIDIZATION OF NANOPOWDERS: EFFECTS OF VIBRATION AND STIRRING 209
Fuweng Zhang, J. Ruud Van Ommen
(380AR) GAS-SOLID HYDRODYNAMICS AND HYSTERESIS IN A RECTANGULAR SPOUTED BED 210
Ritesh Raman, Partha S. Goswami
(380AS) EXPERIMENTAL AND SIMULATION STUDY ON BUBBLING BEHAVIOUR OF GELDART GROUP B PARTICLE PULSED FLUIDIZED BED 211
Ali Akhavan, Rosemary Clark, J. Ruud Van Ommen, Marc-Olivier Coppens, Martin Rhodes
(380BC) IMPROVING CO_2 CAPTURE EFFICIENCY IN A FLUIDIZED BED OF TIO_2 NANOPARTICLES 212
Keju An, Jean M. Andino
(380AT) COMPUTATIONAL STUDY ON BIOMASS FAST PYROLYSIS: HYDRODYNAMIC EFFECTS IN A LABORATORY-SCALE FLUIDIZED BED 213
Emilio Ramirez, Tingwen Li, Charles E. A. Finney, Mehrdad Shahnam, C. Stuart Daw
(380AU) NUMERICAL VERIFICATION AND VALIDATION OF THE BIOMASS PYROLYSIS AND GASIFICATION INSIDE A BENCH SCALE REACTOR WITH MFIX-TFM 214
Yupeng Xu, Mehrdad Shahnam, Andrea Porcu, Alberto Pettinau, Bhima Sastri, William A. Rogers
(380AV) CFD SIMULATIONS OF INTEGRATED WASTE TREATMENT UNIT 216
Yong Liu, Chris Guenther
(380AW) SINGLE BUBBLE DYNAMICS IN A DENSE PHASE FLUIDIZED BED GASIFICATION ENVIRONMENT IN PRESENCE OF BIOMASS PELLETS 217
Nicolás Torres Brauer, Hugo De Lasa
(380AZ) MFIX-EXA: ADAPTING CFD-DEM TO EXASCALE COMPUTING 220
Madhava Syamlal, Jordan Musser, Ann Almgren, John Bell, Christine M. Hrenya, Thomas Hauser, Peiyuan Liu
(380BB) A PENALIZATION METHOD FOR THE DIRECT NUMERICAL SIMULATION OF WEAKLY COMPRESSIBLE REACTING GAS-SOLID FLOWS 221
Baptiste Hardy, Juray De Wilde, Grégoire Winckelmans
(6CM) FINITE ELEMENT METHOD (FEM) MODELING OF HOPPER FLOW 223
Kunal S. Pardikar, Carl R. Wassgren, Chen Mao, Ariel Muliadi
(397A) CHEMPREN: A NEW TECHNOLOGY FOR THE CONVERSION OF MIXED MUNICIPAL WASTE TO BASIC CHEMICALS AND FUELS 224
Anne Gaffney, Gennaro J. Maffia, Amanda Peterman, Xufung Liu, Lynn M. Wendt
(397B) ANAEROBIC DIGESTION OF FOOD WASTES INTO ENERGY AND FERTILIZER 225
Yen Wah Tong, Jingxin Zhang, Kai Chee Loh
(397C) SUBCRITICAL WATER UPGRADING OF COAL REFUSE WITH FOOD WASTE 226
Shanta Mazumder, Pretom Saha, M. Toufiq Reza
(397E) POLAR SOLVENTS FACILITATE EFFICIENT MICROWAVE PRETREATMENT OF WOODY BIOMASS 227
Akshat Tanksale, Negin Amini, Victoria Haritos
(397F) COMPARATIVE LEADING PRETREATMENTS FOR IMPROVING LIGNIN WASTE VALORIZATION TO BIOBASED PRODUCTS 228
Zhi-Hua Liu, Joshua Yuan

(397G) WASTE TO RESOURCE: SYNTHESIS OF ELECTROLYTE AND ELECTRODE MATERIAL IN VANADIUM REDOX FLOW BATTERY FROM LEACHATE OF INDUSTRIAL WASTE 229
He Li, Chi-Hwa Wang

(397H) SYNTHESIS OF $LINI_{0.6}CO_{0.2}MN_{0.2}O_2$ CATHODE MATERIAL BY A CARBONATE CO-PRECIPITATION FROM SPENT LITHIUM-ION BATTERIES 230
Lei Xing, Shuying Sun, Sen Lin, Ping Li, Jianguo Yu

(397I) SUSTAINABLE PROCESS FOR THE RECYCLING OF POLYETHYLENE PHTHALATE 232
Zheng Liu

(397J) RECYCLING FE-P SLAG AS ENERGY MATERIALS VIA A GREEN TECHNICS 233
Guixin Wang, Yun Wei, Fei Xu, Jiangdong Zou, Shunxing Wang

(456A) PREDICTING CRITICAL OUTLET WIDTH OF HOPPER SYSTEMS USING FINITE ELEMENT ANALYSIS 234
Kunal S. Pardikar, Carl R. Wassgren

(456C) COMPUTATIONAL AND EXPERIMENTAL STUDIES OF MECHANICAL COMPRESSION OF FLEXIBLE FIBER PACKINGS 235
Qingzhao Liu, Yanjie Li, Dandan Xu, Yu Guo, Carl Wassgren, Jennifer Sinclair Curtis

(380AX) MODELING PARTICLE-INDUCED TURBULENCE USING SPARSE REGRESSION WITH EMBEDDED INVARIANCE 236
Sarah Beetham, Jesse Capecelatro

(456D) THE DISCRETE ELEMENT MODELING (DEM) OF ECCENTRIC WEIGHTS BASED VIBRATORY DRYER 237
Rohit Kumar, Subhash Thakur, Shrikant Swaminathan

(456E) MODELING COHESIVE POWDER FLOW IN A ROTATING DRUM USING THE FINITE ELEMENT METHOD AND A MOHR-COULOMB MATERIAL MODEL 238
Abhishek Paul, George Elias, Yu Liu, Marcial Gonzalez, Carl R. Wassgren

(456F) A CFD-DEM BASED MODEL FOR WET GRANULATION PROCESS IN A ROTARY DRUM 239
Amit Patil, John Morud, Stein Tore Johansen Sr.

(458A) REACTIVE COMPOSITES OF BORON AND BISMUTH FLUORIDE WITH VARIED FLUORIDE CONCENTRATIONS 247
Siva Kumar Valluri, Mirko Schoenitz, Edward L. Dreizin

(458B) ALUMINUM-BASED THERMITES PREPARED BY STAGED MILLING WITH DIFFERENT PROCESS CONTROL AGENTS 248
Mehnaz Mursalat, Mirko Schoenitz, Edward Dreizin

(458C) ELECTROSTATIC AND STERIC STABILIZATION OF NANOENERGETICS SUSPENSIONS 249
Mouhcine Doukkali

(458D) HETEROGENEOUS OXIDATION OF BORON POWDERS AND ITS ROLE IN IGNITION AND COMBUSTION 250
Kerri-Lee A. Chintersingh-Dinnall, Yalun Sun, Mirko Schoenitz, Edward L Dreizin

(458F) DESIGNING COMPOSITE ENERGETIC NANOMATERIALS WITH TUNABLE INTERFACIAL ACTIVITIES VIA LASER ABLATION SYNTHESIS IN SOLUTION 251
Dibyendu Mukherjee, Seyyed Ali Davari, Erick L. Ribeiro, Gerd Duscher, Jennifer Gottfried

(458G) BORON-BASED ANALOGS OF GRAPHENE EXFOLIATED FROM LAYERED METAL DIBORIDES AS CATALYTIC ADDITIVES TO IMPROVE THE PERFORMANCE OF ENERGETIC MATERIALS 252
Harini Gunda, Chinmay Ghoroi, Kabeer Jasuja

(458H) HOT-SPOT FORMATION IN COMPOSITE ENERGETIC MATERIALS DUE TO ULTRASONIC AND IMPACT EXCITATION AS A FUNCTION OF YOUNG'S MODULUS AND THE WORK OF ADHESION 253
Jason Wickham, Zane Roberts, Stephen P. Beaudoin, Steven F. Son

(458I) MICROWAVE ENHANCEMENT OF BORON COMBUSTION FOR SOLID FUEL RAMJET APPLICATIONS 255
Adam Lawrence, Travis R. Sippel

(458K) MICROWAVE CONTROLLED IGNITION AND COMBUSTION OF AN AP COMPOSITE SOLID PROPELLANT USING GRAPHENE OXIDE-WRAPPED NANOSCALE THERMITES 256
Stuart J. Barkley, Jordan Golberg, Keke Zhu, James B. Michael, Travis R. Sippel

(458L) THE MICROSTRUCTURAL EVOLUTION AND KINETICS OF THERMAL DECOMPOSITION OF SELECTED ENERGETIC COCRYSTALS 257
Joshua Dean

(458M) THERMAL BEHAVIOR OF CL-20 AT ULTRA-HIGH HEATING AND COOLING RATES USING FLASH DSC 260
Aric Denton, Rozana Bari, Zachary Fondren, Yung P. Koh, Victor Stepanov, Sindee L. Simon, Gregory B. McKenna
(476A) CERIUM OXIDE PROMOTED NICKEL NANOPARTICLES DEPOSITED BY ATOMIC LAYER DEPOSITION FOR DRY REFORMING OF METHANE 261
Baitang Jin, Zeyu Shang, Shiguang Li, Xinhua Liang
(476B) THE IMPACT OF SUPPORT IN SYNGAS CONVERSION OVER RH-MN PARTICLES ON WC 262
Florian Göltl, Yifei Liu, Lifeng Zhang, James Dumesic, Manos Mavrikakis
(476C) MODULAR SELF-ASSEMBLY OF METAL/POLYMER NANOREACTORS 263
Andrew Harrison, Tien Vuong, Michael Zeevi, Benjamin Hittel, Christina Tang
(476D) GOLD BIMETALLIC NANOPARTICLES FOR APPLICATIONS IN SELECTIVE OXIDATION CATALYSIS 264
Joseph Brindle, Michael M. Nigra
(476E) ROLE OF POLY(N-VINYLPYRROLIDONE) AND ITS INFLUENCE ON AG^+ REDUCTION KINETICS DURING FORMATION OF AG NANOSTRUCTURES 265
Suprita Jharimune, Zhifeng Chen, Justin Anderson, Ji Woong Chang, Robert M. Rioux
(476F) ENHANCEMENT OF CO_2 DESORPTION PERFORMANCES IN CO_2 -LOADED MEA SOLUTION BY USING CMK-3 BIFUNCTIONAL CATALYSTS 266
Zain Ali Saleh Bairq Sr., Zhiwu Liang, Hongxia Gao
(476G) IDENTIFICATION OF STABLE BIMETALLIC NANOCLUSTERS VIA A MATHEMATICAL OPTIMIZATION FRAMEWORK 267
Xiangyu Yin, Natalie M. Isenberg, Michael G. Taylor, Giannis Mpourmpakis, Chrysanthos E. Gounaris
(476H) OXIDATIVE DEHYDROGENATION OF ETHANE TO ETHYLENE WITH FE_2O_3@SIO_2 CORE@SHELL CATALYSTS 268
Yahui Yang, Eyram Akabua, Götz Veser
(492A) PACKING PREDICTION OF PHARMACEUTICAL POWDERS VIA BOND NUMBER ESTIMATION 269
Kuriakose Kunnath, Rajesh Davé
(492B) THEORETICAL AND EXPERIMENTAL INVESTIGATION INTO THE IMPACT OF TRIBO-CHARGING ON THE PERFORMANCE OF DRY POWDER BLENDS 270
Joana T. Pinto, Sarah Zellnitz, Michael Brunsteiner, Thomas Wutscher, Amrit Paudel
(492C) INFLUENCE OF AN EXTERNAL LUBRICANT ON THE TRIBO-CHARGING BEHAVIOUR OF CAPSULES 273
Thomas Wutscher, Sarah Zellnitz, Mirjam Kobler, Francesca Buttini, Laura Andrade Benítez, Verónica Daza Fernández, Alberto Mercandelli, Stefano Biserni, Susana Ecenarro Probst, Amrit Paudel
(492D) INVESTIGATION INTO THE EFFECTS OF ROLLER COMPACTION ON GRANULE AND API PROPERTIES 275
James Clarke, John Gamble, John Jones, Mike Tobyn, Richard Greenwood, Andrew Ingram
(492E) IN-DEPTH EXPERIMENTAL ANALYSIS OF PHARMACEUTICAL CONTINUOUS TWIN-SCREW WET GRANULATION IN VIEW OF THE DEVELOPMENT OF A GENERIC COMPARTMENTAL PBM 276
Michiel Peeters, Daan Van Hauwermeiren, Alexander Ryckaert, Christoph Portier, Tamas Vigh, Ashish Kumar, Fanny Stauffer, Jonathan Meyer, Jérôme Mantanus, Pankaj Doshi, Kai Lee, Ingmar Nopens, Thomas De Beer
(492F) THE USE OF T-PLS TO INVESTIGATE THE IMPACT OF RAW MATERIAL PROPERTIES ON GRANULE QUALITY ATTRIBUTES OBTAINED AFTER CONTINUOUS TWIN-SCREW WET GRANULATION 278
Alexander Ryckaert, Jens Dhondt, Thomas De Beer
(492G) VISUALIZATION OF GRANULE TEMPERATURE ALONG THE LENGTH OF THE BARREL USING THERMAL IMAGING TO IMPROVE PROCESS UNDERSTANDING OF PHARMACEUTICAL TWIN-SCREW MELT GRANULATION 280
Shana Van De Steene, Chris Vervaet, Thomas De Beer
(496A) 3D PRINTING: A NEW ERA OF PRECISE MANUFACTURING OF INDIVIDUALLY DEVELOPED PHARMACEUTICALS 282
Ahmed Zidan, Alaadin Alayoubi, Sarah Asfari, Nima Yazdanpanah, Thomas O'Connor, Muhammad Ashraf, Celia N. Cruz
(496B) SMART DESIGN AND VIRTUAL PRODUCT ANALYSIS FOR 3D PRINTING PHARMACEUTICAL TABLETS FOR PATIENT CENTRIC PRODUCT DEVELOPMENT 283
Nima Yazdanpanah, Ahmed Zidan, Alaadin Alayoubi, Sarah Asfari, Muhammad Ashraf, Celia N. Cruz, Thomas O'Connor

(496C) RAPID ON-SITE MANUFACTURING APPROACH WITH 3DP CAPSULES: DELAYED BURST RELEASE FOR REGIONAL ABSORPTION ... 284
Derrick Smith, Yash Kapoor, Gerard Klinzing, Andre Hermans, Rebecca Nofsinger, Kimberly Manser, Rebecca Nissley, Jessica Schlegel, Adam Procopio
(496D) PRODUCTION OF SMALL-BATCHES OF 3D PRINTED MULTI-COMPARTMENT CAPSULAR DEVICES FOR ORAL DRUG DELIVERY USING HIGH-THROUGHPUT ROBOTIC MANUFACTURING TECHNOLOGIES ... 285
Alice Melocchi, Federico Parietti
(496E) PHARMACEUTICAL APPLICATION OF INKJET BASED 3D PRINTING PROCESS ... 287
Koyel Sen, Sameera Sansare, Si W. Li, Bodhisattwa Chaudhuri, Anson Ma
(496F) 3D PRINTED TABLETS USING BINDER JETTING AND PHARMACEUTICAL EXCIPIENTS ... 289
Anthony Antic, Jun Zhang, Negin Amini, David Morton, Karen Hapgood
(496G) POTENTIAL OF 3D PRINTING IN PRECISION COMBINATION THERAPY: A CASE STUDY OF CARDIOVASCULAR POLYPILLS ... 290
Alaadin Alayoubi, Sarah Asfari, Michael Kopcha, Muhammad Ashraf, Celia N. Cruz, Ahmed Zidan
(501A) BRINGING PARTICLE SCALE PROPERTIES INTO DESCRIPTIONS OF ENERGETIC POWDER BEHAVIOR VIA THE ENHANCED CENTRIFUGE TECHNIQUE ... 291
Caralyn A. Coultas-McKenney, Contessa Norris, Andrew Roginski, Kacie Bradfish, Emma Weiglein, Stephen P. Beaudoin
(501B) NUMERICAL SIMULATION OF MUNITIONS RESPONSE FOR FRAGMENT IMPACT ... 292
Ruslan S. Mudryy
(501C) FILLED SPHERICAL COMPOSITE PARTICLES PREPARED BY MECHANICAL MILLING ... 303
Mehnaz Mursalat, Daniel Hastings, Mirko Schoenitz, Edward Dreizin
(501D) INFLUENCE OF NANOCONFINEMENT ON THE DEGRADATION OF CL-20 AND HMX ... 304
Rozana Bari, Aric Denton, Zachary Fondren, Victor Stepanov, Sindee L. Simon, Gregory B. McKenna
(501E) INFLUENCE OF CATALYST ON THE MECHANOCHEMICAL NITRATION OF TOLUENE ... 305
Ashvin Kumar Vasudevan, Edward L Dreizin, Mirko Schoenitz
(501F) IGNITION OF REACTIVE MATERIAL POWDERS IN VICINITY OF AN ELECTROSTATIC DISCHARGE ... 306
Ci Huang, Mirko Schoenitz, Edward Dreizin
(528A) PARTICLE ALD OF TUNGSTEN NITRIDE AS A HYDROGEN ENVIRONMENTAL BARRIER COATING FOR NUCLEAR SPACE PROPULSION ... 307
Sarah Bull, Theodore Champ, Chanel Hill, Robert C. O'Brien, Charles B. Musgrave, Alan W. Weimer
(528B) PREPARATION OF THE CORE-SHELL ALUMINUM NANOPARTICLE WITH A MAGNETIC NICKEL SHELL ... 308
Hongbin Shi, Tao Wang
(528C) INTERNAL STRUCTURE OF A DRYING NANOPARTICLE-POLYMER FILM ... 309
Thitiporn Kaewpetch, James F. Gilchrist
(528D) NANOCOATING OF METALS OXIDES ON NANOSCALE SUBSTRATES IN ORGANIC SOLVENTS ... 310
Ahmed Jasim, Yangchuan Xing, Xiaoqing He, Tommi White
(528E) FABRICATION AND CHARACTERIZATION OF SUPERPARAMAGNETIC HYBRID CELLULOSE NANOCRYSTALS (CNC)/IRON OXIDE (FE_3O_4) NANOPARTICLES ... 311
Mohammad J. Hasan, Frankie A. Petrie, Ashley E Johnson, Erick S. Vasquez, Esteban E. Ureña-Benavides
(528F) PARTICLE ATOMIC LAYER DEPOSITION OF ALUMINA FOR FLASH SINTERING YTTRIA-STABILIZED ZIRCONIA ... 312
Rebecca O'Toole, Bola Yoon, Christopher Gump, Rishi Raj, Alan W. Weimer
(528G) MODIFYING THE PHOTOCATALYTIC AND DISPERSION PROPERTIES OF TIO2 P25 NANOPARTICLES BY SCALABLE MOLECULAR LAYER DEPOSITION ... 313
Damiano La Zara, Maximilian R. Bailey, Dominik Benz, Mike J. Quayle, Staffan Folestad, J. Ruud Van Ommen
(540C) PSRI FLUIDIZATION AWARD LECTURE: AN APPROACH OF COUPLED KINETIC THEORY OF EULER GRANULAR PHASE WITH DISCRETE ELEMENT METHOD OF LAGRANGIAN DISCRETE PARTICLES ... 315
Huilin Lu
(567A) 3D PRINTED STATIC MIXERS FOR CONTINUOUS REACTIVE CRYSTALLIZATION ... 325
Kiran Mathew Thomas, Richard Lakerveld
(567B) ULTRA-LOW-COST CONTINUOUS MANUFACTURING PLATFORMS THROUGH 3D PRINTING AND TELESCOPED ISOLATION FREE PURIFICATION ... 327
Matthew Harding, Rafael Lopez-Rodriguez, Sarah Brady, Denis Dowling, Kevin Girard, Steven Ferguson

(567C) MODELING OF POLY(ETHER IMIDE) IN MATERIAL EXTRUSION ADDITIVE MANUFACTURING 329
Eric L. Gilmer, David Anderegg, John M. Gardner, Emilie J. Siochi, David Dillard, Michael J. Bortner
(567D) HARNESSING A HIGH ENERGY, SUPERCONDUCTING ELECTRON BEAM FOR ADDITIVE AND FAR-FROM-EQUILIBRIUM MANUFACTURING 330
Ryan D. Reeves, Adam Duzik, Justin J. Hill
(567F) CHARACTERIZATION OF SPATIAL VARIABILITY IN MATERIAL PROPERTIES IN ECHOGENICALLY-SEGMENTED METAL COMPONENTS FABRICATED BY ADDITIVE MANUFACTURING 338
Kenneth Walton, Manish Roy, Mikhail Skliar
(567G) LAB ENABLEMENT THROUGH 3D PRINTING - NITROGEN INERTION FUNNEL 339
Matthew Hagan
(568A) FORMULATION OF UV CURABLE RESINS UTILIZED IN VAT PHOTO POLYMERIZATION FOR THE ADDITIVE MANUFACTURING OF GUN PROPULSION CHARGE IN 3D PRINTERS 340
David Bird
(568B) VISCOELASTIC PROPERTIES OF ENERGETIC MATERIAL FILLED UV PHOTO POLYMER RESINS AND THERMOPLASTIC FEED STOCK FOR THE ADDITIVE MANUFACTURING OF GUN PROPULSION CHARGES 341
Robin Crownover
(568D) MGB_2/FLUOROPOLYMER FILAMENT AND PYROLANT COMBUSTION STUDIES 342
Lance Kotter, Lori Groven
(568E) SPHERICAL COMPOSITE POWDERS AS FEEDSTOCK FOR ADDITIVE MANUFACTURING 343
Daniel Hastings, Edward Dreizin, Mirko Schoenitz
(568F) SOLIDS METRICS FOR SUCCESSFUL 3D PRINTING OF ENERGETIC FEEDSTOCKS 344
Brandon Ennis, Naseem Jibrin, Benjamin Ennis, Michael Winn, Bryan J. Ennis
(572B) PERFORMANCE OF DIFFERENT TYPES OF BAFFLES USED IN GAS-SOLIDS FLUIDIZED BED STRIPPERS 345
Allan Issangya, S. B. Reddy Karri, T. M. Knowlton, Ben Freireich, Ray Cocco
(572C) DILUTE PHASE RISER DYNAMICS – A COMPARISON OF SCALE 346
Ronald W. Breault, Steven Rowan, Jingsi Yang, Justin Weber
(572D) CFD SIMULATION OF CARBON CAPTURE USING AN AMINE BASED SOLID SORBENT 347
Farnaz Esmaeili Rad, Hamid Arastoopour, Javad Abbasian
(572E) COUPLED CFD-DEM MODEL OF FLUIDIZED BED DRYING OF PHARMACEUTICAL WET GRANULES 348
Syed Ahsan, Bodhisattwa Chaudhuri, Yijie Gao
(572F) NUMERICAL AND EXPERIMENTAL INVESTIGATION OF A LAB SCALE WURSTER COATING PROCESS 350
Martina Trogrlic, Thomas Forgber, Sharareh Salar-Behzadi, Avik Sarkar, Lourdes Contreras, Dalibor Jajcevic, Johannes G. Khinast
(572G) 3D CPFD SIMULATION OF CIRCULATING FLUIDIZED BED RISER AND DOWNER: COMPARISONS OF FLOW STRUCTURE AND SOLIDS BACK-MIXING BEHAVIOR 352
Yancong Liu, Yingya Wu, Xiaogang Shi, Jinsen Gao, Xingying Lan
(598A) THERMAL DECOMPOSITION SYNTHESIS OF MONODISPERSE, PHASE-PURE IRON OXIDE NANOPARTICLES FOR MPI AND MAGNETIC HYPERTHERMIA APPLICATIONS 354
Shehaab Savliwala, Sitong Liu, Carlos Rinaldi
(598B) MAGNETIC NANOPARTICLE CLUSTERS FOR EFFICACIOUS BILIRUBIN REMOVAL FROM BODY FLUIDS 355
Alexandre Anthis, Martin T. Matter, Kerda Keevend, Lukas R. H. Gerken, Inge K. Herrmann
(598C) ANTIOXIDANT PROTECTION FROM OXIDATIVE STRESS DAMAGE IN HUMAN PULMONARY CELLS 356
Jordan A. Hoops, Timothy M. Brenza
(598D) CROSSLINKING OF OVALBUMIN NANOPARTICLES AFFECTS CELLULAR AND HUMORAL IMMUNE RESPONSES 357
Nahal Habibi, Stephanie Christau, Lukasz Ochyl, James Moon, Joerg Lahann
(598E) ENCAPSULATION OF POORLY WATER-SOLUBLE DRUGS INTO YEAST GLUCAN PARTICLES BY SPRAY DRYING: IMPROVEMENT OF DISPERSION AND DISSOLUTION PROPERTIES 358
Gabriela Ruphuy Chan, Ivan Salon, Jan Tomas, Petra Šalamúnová, Jaroslav Hanuš, František Štepánek

(598F) MULTISCALE ANALYSIS OF TOPICALLY APPLIED NANOPARTICULATE ADHESIVES: BIODISTRIBUTION AND BIOTRANSFORMATION ... 362
Martin T. Matter, Ioana Lese, Inge K. Herrmann
(598G) CONTROLLED PARTICLE SUSPENSIONS FOR DETERMINATION OF INTRINSIC DISSOLUTION RATE IN SIMULATED INTESTINAL FLUIDS AT MEDIUM THROUGHPUT ... 364
Alexandra Teleki, Christel A. S. Bergström
(600A) AVOIDING DISTRIBUTION CONFUSION: USING THE INFORMATION IN PARTICLE SIZE DISTRIBUTIONS ... 365
Jacob Albrecht, John Gamble, Ana Ferreira, Mike Tobyn, David Good
(600B) COMBINING LIQUID-LIQUID PHASE SEPARATION AND MESOPOROUS CARRIER TO DELIVER DRUG PRODUCT INTERMEDIATE ... 366
Moussa Boukerche, Sean Garner, Nandkishor Nere
(600C) OPTIMISING THE NUCLEATION OF POLYPEPTIDES USING NANOTEMPLATES: THE CASE STUDY OF INSULIN ... 367
Frederik J. Link, Jerry Y. Y. Heng
(600D) SURFACE-INDUCED NUCLEATION STRATEGIES: SEEKING SYMMETRIES BETWEEN SELF-ASSEMBLY OF HETERONUCLEANTS AND CRYSTALS ... 369
Fiora Artusio, Giacomo Ceccone, Roberto Pisano
(600E) OLANZAPINE CRYSTAL SYMMETRY ORIGINATES IN PREFORMED CENTROSYMMETRIC SOLUTE DIMERS ... 377
Peter Vekilov
(600F) DEVELOPMENT OF A HYBRID QUALITY-BY-CONTROL FRAMEWORK FOR OPTIMAL CRYSTALLIZATION PROCESS DESIGN ... 378
Ayse Eren, Botond Szilagyi, Justin L. Quon, Charles D. Papageorgiou, Zoltan K. Nagy
(603A) FLOW-INDUCED CRYSTALLIZATION IN MATERIALS EXTRUSION ADDITIVE MANUFACTURING ... 380
Anthony P. Kotula
(603B) COMPUTATIONAL FLUID DYNAMICS SIMULATION OF THE FUSED DEPOSITION MODELING PROCESS: EFFECT OF TEMPERATURE-DEPENDENT VISCOSITY ON THE SHAPE OF EXTRUDED STRAND ... 381
Behrouz Behdani, Leah Mason, Ming Leu, Joontaek Park
(603C) EFFECT OF POLYPROPYLENE MELT FLOW INDEX ON THE MATERIAL EXTRUSION BASED ADDITIVE MANUFACTURING OF POLYPROPYLENE/HYDROGENATED RESIN BLENDS ... 382
Arit Das, Emma M. Dartevelle, Eugene G. Joseph, Michael J. Bortner
(603D) ENGINEERING FILAMENT DESIGN FOR ENHANCED PRINTABILITY AND MECHANICAL PROPERTIES OF ADDITIVELY MANUFACTURED PLASTIC PARTS ... 383
Bryan D. Vogt
(603E) 3D PRINTING OF INTERDIGITATED DIELECTRIC ELASTOMER ACTUATORS ... 384
Alex Chortos, Ehsan Hajiesmaili, David R. Clarke, Jennifer A. Lewis
(603F) DIRECT PRINT ADDITIVE MANUFACTURING OF OPTICAL FIBER INTERCONNECTS FOR BOARD LEVEL COMPUTING DEVICES ... 385
Roger B. Tipton, Dianhao Hou, Thomas M. Weller, Venkat R. Bhethanabotla
(603G) CURE-THROUGH CORRECTION IN CONTINUOUS STEREOLITHOGRAPHIC 3D PRINTING ... 386
Zachary D. Pritchard, Martin De Beer, Timothy F. Scott, Mark A. Burns
(633A) CONTROLLABLE PREPARATION OF UNIFORM MAGNESIUM ETHOXIDE PARTICLES: THE EFFECTS OF N-HEXANE AND SILICONE OIL AND THEIR SYNERGIC EFFECT ... 387
Zhou Lan, Lu Yangcheng
(633B) MULTI-OBJECTIVE OPTIMIZATION OF SIMPLE CRYSTALLIZATION SYSTEMS ... 388
Hao-Jen Pan, Yu-Ti Tseng, Jeffrey D. Ward
(633C) SCALABLE FABRICATION OF MULTIELEMENT CERAMIC OXIDE PARTICLES BY AN AEROSOL PROCESS ... 389
Yujia Liang, Joseph A. Libera
(633D) CONTROLLING THE INTERNAL STRUCTURES OF POLYMERIC MICROSPHERES VIA THE INTRODUCTION OF A WATER-SOLUBLE ORGANIC SOLVENT ... 390
Yanping He, Yanlin Sun, Delong Xie, Linhua Zhu
(633G) QUANTITATIVE ANALYSIS OF HEAT TRANSFER IN A ROTARY DRUM ... 393
Manogna Adepu, Heather N. Emady

(633F) EXPERIMENTAL AND NUMERICAL STUDY OF A RADIAL MULTI-ZONE VORTEX CHAMBER SPRAY DRYER 394
Thomas Tourneur, Axel De Broqueville, Anton Sweere, Albert Poortinga, Anton Wemmers, Umair Jamil Ur Rahman, Artur Pozarlik, Juray De Wilde
(647A) SPATIALLY-AVERAGED MODELS FOR MODERATELY DENSE GAS-PARTICLE FLOWS INCLUDING A THERMAL ENERGY BALANCE 396
Stefanie Rauchenzauner, Simon Schneiderbauer
(647B) VALIDATING OPTIMAL DRAG CORRECTIONS FOR THE SIMULATION OF SAND BIOMASS CO-FLUIDIZATION 397
Liqiang Lu, Xi Gao, Mehrdad Shahnam, William A. Rogers
(647C) PARTICLE-SCALE INVESTIGATIONS OF REACTIVE FLUIDIZED BEDS BASED ON CFD-DEM COUPLING METHODS 398
Chenshu Hu, Kun Luo, Junjie Lin, Jianren Fan
(647D) A STRUCTURAL MODEL OF SUB-GRID DRIFT VELOCITY FOR FLUIDIZED GAS-PARTICLE FLOWS 399
Ming Jiang, Xiao Chen, Qiang Zhou
(647E) A CASCADING APPROACH TO DEVELOP A FILTERED DRAG FORCE MODEL FOR LARGE-SCALE GAS-PARTICLE FLOWS 400
Yundi Jiang, Jari Kolehmainen, Ali Ozel, Sankaran Sundaresan
(647F) COMPARATIVE ANALYSIS OF DRAG MODEL PREDICTIONS FOR FLUIDIZED BED APPLICATIONS 401
Casey Q. Lamarche, Ben Freireich, Ray Cocco
(647G) NUMERICAL MODELING ON GAS-SOLID TWO-PHASE FLOW INCLUDING THIN PLATES 402
Kazuya Takabatake, Mikio Sakai
(652A) COMPRESSIBLE NUTSCHE CAKE FILTRATION – RETARDING EFFECT OF RESIDUAL BASE-CAKE 403
Patrick McGough, Chris Chappelow, Madhusudhan Kodam
(652B) FREE SURFACE PROFILES IN CENTRIFUGE FILTRATION AT LOW SPEEDS – THEORETICAL AND COMPUTATIONAL SOLUTIONS 404
Madhusudhan Kodam
(652C) DEVELOPING HOLLOW FIBER AIR FILTERS FOR FILTRATION OF ULTRAFINE PARTICLES 405
Liang-Yi Wang, Liya E Yu, Neal Tai-Shung Chung
(652D) CYCLONE DESIGN AND PERFORMANCE AT RESTRICED SPACE 406
Ulrich Muschelknautz
(652E) SELF-CLEANING ELECTROSTATIC OIL MIST COLLECTOR BY USING CORONA DISCHARGE AND SURFACE DIELECTRIC BARRIER DISCHARGE 408
Myung Soo Kang, Jaeuk Shin, Jungho Hwang
(652F) OPTIMIZATION OF CLASSIFIER HEIGHT AND NOZZLE ANGLE IN A JET MILL 409
Mahesh Dhakate, Devang V. Khakhar, Jyeshtharaj B. Joshi
(652G) ENHANCEMENT OF FLUORITE FLOTATION PROCESS WITH THE COMBINATION OF IN-SITU MODIFICATION BY BALL-MILLING AND GAS-LIQUID MICRODISPERSION 410
Lu Ye
(680A) A COURSE AT PURDUE UNIVERSITY ON PARTICLE, POWDER, AND COMPACT CHARACTERIZATION 411
Kingsly Ambrose, Dhananjay A. Pai, Carl Wassgren
(680B) NON-LINEAR DEPENDENCE OF MACROSCOPIC POWDER PROPERTIES ON POWDER COMPOSITION 412
Sara Fathollahi, Eva Faulhammer, Benjamin Glasser, Johannes G. Khinast
(680C) COMPARATIVE STUDY OF SUGAR POWDER CONSISTENCY USING AN FT4 POWDER RHEOMETER AND DRAG FORCE FLOW (DFF) SENSOR 414
John Yin, Valery Sheverev, Vadim Stapaniuk, Tim Freeman
(680D) MATERIAL PROPERTY DATABASE AND ITS APPLICATIONS IN PREDICTIVE BATCH AND CONTINUOUS UNIT OPERATION MODELING 415
Pallavi Pawar, Antonio Benedetti, Jingzhe Li, Don Clancy, Fernando J. Muzzio, Robert E. Yule
(680E) WIDE RANGE PARTICLE SIZING FOR SUBMICRON BY SCATTERING LIGHT IMAGE ANALYSIS 416
Tomonari Misawa, Jun Tanaka, Yoshiki Yonamoto
(680F) POLYMER AND METALIC POWDERS FLOWABILITY & ELECTRICAL CHARGES CHARACTERISATION FOR ADDITIVE MANUFACTURING 417
Filip Francqui, Geoffroy Lumay

(680G) CHARACTERIZING THE EFFECT OF POWDER FLOW AID ADDITIVES 418
Geoffroy Lumay, Filip Francqui
(680H) ON THE FEASIBILITY OF UTILISING COMPACTION EQUATION PARAMETERS AS MATERIAL KEY ATTRIBUTE DESCRIPTORS 419
Samuel Solomon, Ahmad B. Albadarin
(688A) THE EFFECT OF AMBIENT GAS SPECIES ON TRIBOCHARGING 420
Xiaoyu Liu, Jari Kolehmainen, Ah-Hyung Alissa Park, Sankaran Sundaresan
(688C) DDPM-DEM SIMULATION OF BIOMASS GASIFICATION IN A BUBBLING FLUIDIZED BED REACTOR 422
In Sik Hwang, Jungho Hwang
(688D) KINETIC MODELING OF LOW-TEMPERATURE PM COMBUSTION WITH RELATIVE VELOCITY IN THE PRESENCE OF WATER VAPOR 423
Kento Yokoo, Tsuyoshi Yamamoto, Hideki Matsune, Masahiro Kishida, June Tatebayashi
(688E) PARTICLE ATTRITION BY CHEMICAL REACTION DURING THE REDOX PROCESS OF CHEMICAL LOOPING COMBUSTION 428
Nathan Galinsky, Samuel Bayham, Ronald W. Breault
(688F) INTENSIFICATION OF ETHYLENE PRODUCTION BY CHEMICAL LOOPING OXIDATIVE DEHYDROGENATION THROUGH EXERGY LOSS MINIMIZATION 429
Luke Neal, Vasudev Pralhad Haribal, Fanxing Li
(688G) SYNTHESIS AND CHARACTERIZATION OF AEROSOL PARTICLES AND REACTIONS FOR THE SOLAR DRIVEN HYBRID-SULFUR-AMMONIA WATER-SPLITTING CYCLE 430
Navaira Fathima, Agni E. Kalyva, Athanasios G. Konstandopoulos, Rahul Bhosale, Konstantinos E. Kakosimos
(703A) EVALUATING DIFFERENT DRY COATING PROCESSES FOR PREPARING FINE-GRADE HIGH FUNCTIONAL MICROCRYSTALLINE CELLULOSE 431
Liang Chen, Sangah Kim, Kuriakose Kunnath, Kai Zheng, Rajesh Davé
(703B) CHARACTERIZATION AND MITIGATION OF STICK-SLIP IN PHARMACEUTICAL POWDERS 432
Marissa Amichetti, Christina Pothier, Rohit Kumar, Subhash Thakur, Shrikant Swaminathan
(703C) EFFECT OF PARTICLE PROPERTIES AND FORMULATION COMPOSITION ON STICKING BEHAVIOR DURING PHARMACEUTICAL TABLETTING 433
Maxx Capece, Daniel Young
(703D) IMPACT OF PARTICLES NEEDLE SHAPE, A CRITICAL MATERIAL ATTRIBUTES (CMAS), ON MINIATURIZED PHARMACEUTICAL UNITS OPERATION 434
Mohammad Azad, Gerard Capellades, Allison Wang, David Klee, Gregory Hammersmith, Kersten Rapp, David Brancazio, Allan Myerson
(703E) COARSENING APPROACH IN MODELING GAS-PARTICLE FLOWS IN DRY POWDER INHALER 436
Xiaoyu Liu, Jari Kolehmainen, Sankaran Sundaresan
(703F) SURFACE ENGINEERING OF INHALED PHARMACEUTICAL PARTICLES BY ATOMIC AND MOLECULAR LAYER DEPOSITION 437
Damiano La Zara, Feilong Sun, Fuweng Zhang, Di Zhang, Mike J. Quayle, Gunilla Petersson, Staffan Folestad, J. Ruud Van Ommen
(703G) EVALUATING THE POTENTIAL BENEFITS OF COHESION 439
Tim Freeman, John Yin, Laura Monington, Rajeev Dattani
(703H) LINKING MOISTURE INDUCED VARIATIONS IN PHARMACEUTICAL PRIMARY POWDERS TO THEIR TABLETING PERFORMANCE 440
Rachael Shinebaum, Ian Gabbott, Gavin K. Reynolds, Hannah K. Batchelor, Andrew Ingram
(732A) COARSE-GRAINED DEM FOR LARGE-SCALE FLUIDIZED BED APPLICATIONS 442
Nico Jurtz, Oleh Baran, Ravindra Aglave, Ben Freireich, Ray Cocco, Matthias Kraume
(732B) ACQUIRING DEM VALIDATION DATA ON AN INDUSTRIAL SCALE 443
Casey Q. Lamarche, Steven R. Dahl, Ben Freireich, Ray Cocco, Christine Hrenya
(732C) EVALUATING MIXING PERFORMANCE OF A RESONANT ACOUSTIC MIXER BOTH EXPERIMENTALLY AND USING A COUPLED LBM-DEM TECHNIQUE 444
Ramon E. Lopez, Joseph J. McCarthy
(732D) NUMERICAL VERIFICATION AND VALIDATION OF THE PARTICLE TRACKING TECHNIQUE IN PREDICTING PARTICLE VELOCITY AND SOLIDS DISTRIBUTION IN A TRAVELLING FLUIDIZED BED WITH CFD-DEM 445
Yupeng Xu, Tingwen Li, Sina Tebianian, Jamal Chaouki, Thomas Leadbeater, Rouzben Jafari, David J. Parker, Jonathan P. K. Seville, Naoko Ellis, John R. Grace

(732E) CLUSTER FORMATION MECHANISM IN BOUNDLESS DOWN-FLOW GAS-PARTICLE SYSTEMS 448
Noel Gómez, Alejandro Molina Ochoa, Guy B. Marin, Kevin M. Van Geem
(732F) MODELING OF COMPLEX LIQUID-SOLID FLOW OF PARTICLE SWELLING IN SLURRY LOOP REACTORS 450
Ning Yang, Rongtao Zhou
(738A) DEVELOPMENT OF MUCOUS-PENETRATING LEVODOPA-LOADED MICROPARTICLES FOR TREATMENT OF PARKINSON'S DISEASE 451
Mahasweta Paul, Raymond Lau
(738B) ROLES OF SDS IN GRISEOFULVIN RELEASE FROM NANOCOMPOSITES AND AMORPHOUS SOLID DISPERSIONS: A COMPARATIVE STUDY 452
Mahbubur Rahman, Faustin Arevalo, Alexander Coelho, Stephanie Ahmad, James Tarabokija, Sayali Bhujbal, Ecevit Bilgili
(738C) LAYER-WISE AGGLOMERATION OF UREA GRANULES 455
Yanjie Chen, Kingsly Ambrose, Dhananjay A. Pai, Carl Wassgren
(738D) IMPACT OF DRY COATING ON DISSOLUTION PROFILE OF MICRONIZED BCS CLASS II DRUGS: INFLUENCE OF COATING MATERIAL HYDROPHOBICITY 456
Sangah Kim, Liang Chen, Kuriakose Kunnath, Ecevit Bilgili, Rajesh Davé
(738E) 3D PRINTING OF TUNEABLE COLOURED AGGLOMERATES - STRAIN DISTRIBUTION STUDY OF A SINGLE AGGLOMERATE DESIGN 458
Jun Zhang, Negin Amini, David Morton, Karen Hapgood
(738F) INTERACTION OF EUGENOL AND NOVEL LIGNIN GRAFTED SILICA NANOPARTICLES WITH SYNTHETIC LIPID BILAYERS 459
Mahsa Moradipour, Emily Chase, M. Arif Khan, Shardrack O. Asare, Bert C. Lynn, Stephen E. Rankin, Barbara L. Knutson
(738G) DESIGN OF ALL-SOLID-STATE LITHIUM-ION BATTERY BASED ON PARTICLE COATING TECHNOLOGY 460
Hideya Nakamura, Risa Sakurai, Tomoyuki Masuyama, Shuji Ohsaki, Satoru Watano
(743A) IMPACT OF DRY COATING TO IMPROVE BULK PROPERTIES OF NEEDLE-LIKE MORPHOLOGY 461
Steven H. Chan, Joshua Engstrom, Rao Mantri, Kuriakose Kunnath, Rajesh Davé
(743C) THE CALIBRATION OF DEM INPUT PARAMETERS FOR BIOMASS USING A CUBICAL TRIAXIAL TESTER 462
Nathaniel Gasteyer, Hojae Yi, Virendra Puri, Carl Wassgren, Marcial Gonzalez
(743D) UNDERSTANDING PARTICLE SCALE EFFECT OF MIXTURE SEGREGATION 463
Kerry Johanson
(743E) SCALE UP OF HEAT TRANSFER FOR DRY GRANULAR MATERIAL IN A BLADED MIXER 464
Clara Hartmanshenn, Madeline Halota, Johannes G. Khinast, Charles D. Papageorgiou, Christopher Mitchell, Justin L. Quon, Benjamin J. Glasser
(743F) INVESTIGATING THE INFLUENCE OF PROCESSING TEMPERATURE ON POWDER FLOWABILITY 465
Tim Freeman, Jamie Clayton, Laura Monington
(743G) INVESTIGATING THE MOISTURE CONTENT IMPACT ON THE FLOW BEHAVIOUR OF A WET GLASS BEADS POWDER 466
Lyes Ait Ali Yahia, Christopher Carmichael, Raffaella Ocone
Author Index

556470 Use of Bentonite Binder to Optimize Granulation Process during Fertilizer Manufacture

Bentonite is one of the abundant natural clays that have caught the interest of many industrial applications as a binder and a filler. Varying quantities of bentonite were used in the wet granulation of Compound D (NPK) fertilizer from dry powders in the granulation plant at a local plant in Zimbabwe. Samples of the product were collected from the plant, after drying (prior screening), to analyze the effect of the binding properties of bentonite on the particle size distribution of the granulated fertilizer. Higher quantities of bentonite showed a reduction in gypsum losses, as less fines were obtained upon carrying out the grist test. Results proved that bentonite can be used as an effective binder in optimizing to the right target fertilizer granule size.

562214 Compartmental Modelling of Wet Granulation in Twin Screw Using Population Balance

Wet granulation is one of the major steps in improving powder properties (flowability, friability, and size) in different industries including mining, fertilization, food production and pharmaceutical manufacturing [1,2]. Twin screw is continuous manufacturing operational unit used by pharma companies for granule production. Process parameters like liquid to solid ratio (L/S) and screw speed can influence the granule size produced from the twin screw granulator (TSG).

In this study, a compartmental population balance model (CPBM) is developed as a predictive tool of particle size distribution (PSD) produced from wet granulation in a co-rotating twin-screw granulator (TSG). The CPBM is derived in terms of liquid to solid ratio (L/S) and screw speed representing the main process parameters of the TSG. This model accounts for aggregation and breakage of the particles occurring in five compartments of the TSG with inhomogeneous screw configurations (3 conveying zones and 2 kneading zones). The experimental data is obtained for the granulation of microcrystalline cellulose in a 12 mm TSG at various L/S ratios and screw speeds. Kapur kernel is implemented for the first time as an aggregation kernel for the TSG and is compared against the commonly used sum kernel where Kapur kernel shows better fitting with 58% improvement. Also, finite volume numerical method (FVNM) that is adapted for solving the CPBM in TSG shows a dramatic improvement in solution accuracy compared to commonly used cell average numerical method (CANM). Moreover, Kriging interpolation is used to interpolate for new values of the empirical parameters at new L/S and screw speeds. The aggregation rate appeared to be higher in the conveying zones at low and medium L/S ratios while breakage appeared to be higher in the kneading zones for all L/S ratios and screw speeds. The final CPBM in terms of L/S and screw speed is validated using the experimental data.

References:

[1] A. Chaudhury, H. Wu, M. Khan, and R. Ramachandran, “A mechanistic population balance model for granulation processes: Effect of process and formulation parameters,” *Chem. Eng. Sci.*, vol. 107, pp. 76–92, 2014.

[2] A. Faure, P. York, and R. C. Rowe, “Process control and scale-up of pharmaceutical wet granulation processes: A review,” *Eur. J. Pharm. Biopharm.*, vol. 52, no. 3, pp. 269–277, 2001.

570364 Coagulation of Polydisperse Primary Particles in the Transition Regime

Coagulation of polydisperse primary particles (PPs) in the transition regime is the dominant agglomerate growth mechanism in low-temperature regions of flame reactors and high-pressure environments of combustion engines. During flame synthesis of nanoparticles, small PPs coagulate in the free molecular regime to form fractal-like agglomerates. When agglomerates exit the flame high-temperature region, they start to coagulate in the transition regime, as their mean free path becomes comparable to their diameter of gyration, d_g, or mobility diameter, d_m. The estimation of the d_m of evolving aggregates is not trivial, as it depends on their structure and the number/size of their constituent PPs, n_p. Scaling laws describing the d_m of agglomerates consisting of monodisperse PPs have been developed in free molecular, transition and continuum regimes. However, these models overestimate the d_m of mature soot aggregates up to 37% as they neglect PP polydispersity.

Here, an event-driven method [1] for agglomeration of polydisperse PPs is used to investigate the agglomerate structure and size distribution in the transition regime. The d_m, increases with increasing geometric standard deviation of the PPs, $s_{g,p}$, while the average d_g is hardly affected by $s_{g,p}$ for small compact agglomerates. The broad agglomerate size distributions formed in the free molecular regime narrow down by coagulation in the transition regime, attaining a quasi-self-preserving size distribution, regardless of $s_{g,p}$. A scaling law for the n_p is derived as function of d_m and d_g which can be coupled with method of moments or population balance models to assist aerosol reactor design.

[1] Goudeli, E., Eggersdorfer, M. L,. & Pratsinis, S. E. (2015). *Langmuir*, 31, 1320-1327.

562128 Fluidization of Graphene Nanoplatelets: From Microstructure to Hydrodynamics

Graphene is a material consisting of a single two-dimensional layer of sp2 hybridized carbon atoms arranged in honeycomb lattice. Since its discovery in 2004, graphene has been attracting considerable attention due to its exceptional physical, mechanical, thermal, electronic and optical properties. In fact, it has an extremely high surface area (theoretically 2630 m^2/g) and it is the lightest (with a mass of 0.77 mg for 1 m^2), but at the same time the strongest material known (with an ultimate tensile strength of 130 GPa and a tensile stiffness of ~1 TPa) [1]. Moreover, it is the best conductor of heat at ambient conditions (~5,000 W m^{-1} K^{-1}) and electricity (up to 200,000 cm^2 V^{-1} s^{-1}), and it is optically transparent (with a light absorption of 2.3% of white light) [2-4]. One of the most common methods to synthesize graphene is liquid-phase exfoliation, where graphite is milled into a powder whose particles are separated into tiny flakes (also called platelets) by mechanical forces in liquid. These flakes which consist of a few layers of graphene are then sorted based on their size and thickness, i.e., number of layers. Yet, controlling the number of layers in graphene manufacturing remains challenging and often the final product consists of a powder containing multilayer (i.e., $\leq$ 10 layers) flakes. Still, the properties of multilayer graphene powders are relevant to practical applications such as electric batteries and field emitters.

The number of layers of graphene determines its microstructure which affects the powder bulk behaviour (e.g., flowability, dispersibility, fluidizability and agglomerability). One of the most interesting applications of graphene is its use in composite materials and, for example, graphene powders are emerging in the field of catalysis as excellent support materials. An attractive route for the functionalization of graphene with metal and metal-oxide nanoparticle catalysts is atomic layer deposition carried out in a fluidized bed reactor. In order to achieve an uniform deposition of nanoparticles for high-performance catalysts, ensuring a good fluidization behaviour of graphene powders is crucial. Therefore, understanding how graphene powders fluidize depending on their microstructure is vital.

In this work, we study for the first time the fluidization behavior of three kinds of graphene powders: graphene nanoplatelets of 150 m^2/g (GNP150), graphene nanoplatelet aggregates of 300 m^2/g (GNPA300) and graphene nanoplatelet aggregates of 750 m^2/g (GNPA750). Graphene powders fall under group C of Geldart's classification, i.e., cohesive powders. Therefore, their fluidization is expected to be prevented by the strong interparticle forces, which exceed the drag force induced by the gas flow. However, the graphene platelets do not fluidize individually, instead they form into agglomerates, due to the aforementioned cohesive forces, as earlier observed for semi-spherical nanoparticles [5, 6]. As a result, the fluidization behavior depends on the properties of the agglomerates. We correlate the morphology of the graphene nanoplatelets at nano-, micro- and macro-scales to their hydrodynamic behaviour.

GNP150 at the nanoscale consist of agglomerates of mostly 2D flakes (6-8 nm thick and 15 μm wide). The latter, in virtue of being relatively thick, tend to be rigid, or in other words they are not prone to fold and wrinkle. As such, they truly enjoy a 2D geometry. Rigid 2D-like objects pack poorly compared to 3D-like flexible objects. As a result of that, agglomerates consisting of 2D thick nanoplatelets are loosely packed, hence GNP150 have a low bulk density. At the macroscale, GNP150 form mm-sized spherical agglomerates, whose size distribution depends on the history of the powder. Upon vibration-assisted fluidization, the bed of particles segregates into mm-sized agglomerates at the bottom, increasingly smaller agglomerates along the bed height, and fines above the splash zone. No fluidization is observed without the use of vibration at any of the tested superficial velocities (i.e., up to 16.85 cm/s). The use of vibration results into bubbling fluidization at relatively low superficial velocities (1.35 cm/s).

GNPA300 and GNPA750 at the nanoscale are not found in 2D form. In fact, higher-surface-area graphene powder translates into thinner nanoplatelets that are more prone to fold and wrinkle. As a result of that, the nanoplatelets are closely packed, and thus have an higher bulk density. At the macroscale, the powders look very fine and the size of the agglomerates is hardly discernable by naked eye. For GNPA300, fluidization occurs only when assisted via vibration. Fluidization occurs at higher gas velocities than in the previous case, i.e., 2.7 cm/s. Once fluidization occurs, the bed height remains approximately constant with increasing gas velocity. High superficial velocities result in slugging behaviour and significant elutriation. Instead, for GNPA750, fluidization does occur even without the use of assisting methods upon breaking of channels. In fact, increasingly thinner nanoplatelets translate into more cohesive agglomerates and higher agglomerate density. Even though cohesiveness deteriorates fluidization quality, higher densities promote fluidization. In light of this, the fact that GNPA750 fluidizes even without vibration is no surprise, as we expect the increased cohesiveness between nanoplatelets to result into highly packed and thus dense agglomerates that act as micron-sized Geldart-B-like powders.

This work demonstrates the large influence of the structure at the nm- to mm-scale on the large-scale fluidization behaviour. Such insight is essential when scaling up processes for gas-phase production of nanostructured materials [7].

References

1. Lee, C., X. Wei, J.W. Kysar, and J. Hone, *Measurement of the Elastic Properties and Intrinsic Strength of Monolayer Graphene.* Science, 2008. **321**(5887): p. 385-388.
2. Balandin, A.A., S. Ghosh, W. Bao, I. Calizo, D. Teweldebrhan, F. Miao, and C.N. Lau, *Superior Thermal Conductivity of Single-Layer Graphene.* Nano Letters, 2008. **8**(3): p. 902-907.
3. Bolotin, K.I., K.J. Sikes, Z. Jiang, M. Klima, G. Fudenberg, J. Hone, P. Kim, and H.L. Stormer, *Ultrahigh electron mobility in suspended graphene.* Solid State Communications, 2008. **146**(9): p. 351-355.
4. Nair, R.R., P. Blake, A.N. Grigorenko, K.S. Novoselov, T.J. Booth, T. Stauber, N.M.R. Peres, and A.K. Geim, *Fine Structure Constant Defines Visual Transparency of Graphene.* Science, 2008. **320**(5881): p. 1308-1308.
5. van Ommen, J.R., J.M. Valverde, and R. Pfeffer, *Fluidization of nanopowders: a review.* Journal of Nanoparticle Research, 2012. **14**(3): p. 1-29.
6. Yao, W., G. Guangsheng, W. Fei, and W. Jun, *Fluidization and agglomerate structure of* SiO_2 *nanoparticles.* Powder Technology, 2002. **124**(1): p. 152-159.
7. Salameh, S., J. Gómez-Hernández, A. Goulas, H. Van Bui, and J.R. van Ommen, *Advances in scalable gas-phase manufacturing and processing of nanostructured solids: A review.* Particuology, 2017. **30**: p. 15-39.

571916 Bubble Self-Arrangement in Annular Gas-Solid Fluidized Beds

Bubble Self-Arrangement in Annular Gas-Solid Fluidized Beds

Kaiqiao Wu[1], **Victor Francia[2]**, **Marc-Olivier Coppens[1]**

[1] EPSRC "Frontier Engineering" Centre for Nature Inspired Engineering & Department of Chemical Engineering, University College London, London.

[2]Institute of Mechanical, Process and Energy Engineering, Heriot-Watt University, Edinburgh

**Email of the corresponding author: m.coppens@ucl.ac.uk*

The dynamics of the bubbling gas flow are critical to the performance of gas-solid fluidized beds, but they are still hard to control, even in small, simplified systems. Gas-solid contact time and transport processes rely heavily on the bubble characteristics, such as size and velocity. Without the ability to control them properly, operation and scale-up remain largely empirical exercises. Some disadvantages associated with chaotic dynamics may be overcome by imposing more "structure" to the flow. Conventional practices entail the use of internals via baffles and heat exchanger tubes, which are typically used on an *ad hoc* basis and suffer from additional costs in energy, pressure drop and particle attrition. Instead, we have followed a more fundamentally grounded approach by using the self-organizing nature of granular systems to create ordered structures, as shown in Fig.1(a). In a pseudo-2D system, an oscillatory perturbation of the gas flow rearranges bubbles into triangular lattices, as shown in Fig.1(b) [1]. The flow structure is fully scalable and leads to much more controllable properties, such as bubble size distribution, residence time, and spatial distribution. These unique features could allow us to bypass some of the challenges in conventional units, like flow maldistribution and non-uniform contact, as well as decouple conflicting design objectives, such as good solid mixing and gas-solid contact.

Scaling up such structured flows to an industrially relevant set-up will allow engineers to design a new class of fluidized bed operation. Recently, based on the knowledge obtained from our fundamental investigations [2-4], we have been able to recreate a similarly structured flow in a cylindrical geometry, extending it from a flat-2D to an annular-3D configuration, Fig. 1(c). Nevertheless, the periodic walls and strongly curved boundaries may result in instabilities affecting the pattern formation process. In this contribution, results from experiments are shown to investigate the reproducibility of the 3D periodically structured pattern under different oscillatory flows. These results also sketch a multi-parametric operating window, which identifies a range of frequencies and amplitudes of the applied flow to produce structured flows. Comparison of flow properties in the flat-2D and the annular-3D system allows us to quantify the impact of curved and periodic boundaries on the flow behavior.

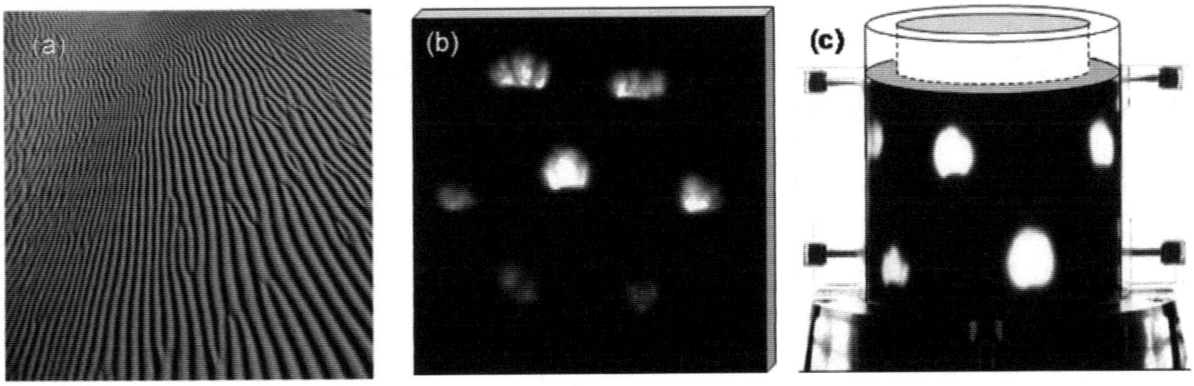

Figure 1 – Natural granular patterns (a) and formation of the structured bubble flow in a pulsed quasi-2D bed (b), and in an annular-3D bed (c).

[1] M.-O. Coppens, J.R. van Ommen, Structuring chaotic fluidized beds, Chem. Eng. J., 96 (2003) 117-124

[2] K. Wu, L. de Martín, M.-O. Coppens, Pattern formation in pulsed gas-solid fluidized beds - the role of granular solid mechanics. Chem. Eng. J., 329 (2017) 4-14.

[3] K. Wu, L. de Martín, L. Mazzei, M.-O. Coppens. Pattern formation in fluidized beds as a tool for model validation: a two-fluid model based study. Powder Technol., 295 (2016) 35-42.

[4] L. de Martín, C. Ottevanger, J.R. van Ommen, M.-O. Coppens. Phys. Rev. Fluids, 3 (2018) p.034303.

574867 Electrostatic Charging and Fluidization Behaviors of Dielectric Particles in the Presence of Noble Gases

In dynamic gas-solid particulate systems such as a fluidized bed reactor, particle-particle and particle-wall interactions can generate and accumulate significant amount of static charges and the particle surface charge can significantly influence the hydrodynamics of such particulate systems. There has been a number of interesting studies focusing on the measurement of static charges in fluidized beds and the means of reducing charge accumulation in those reactors [1,2]. One of the well-studied method of reducing static charges in multiphase reactors is the increase of humidity. In our recent study, we have found that the electrostatic charge accumulation on particles fluidized by noble gases is significantly smaller than that in air, even under dry conditions. Considering that dielectric properties of noble gases are similar to commonly used gases (e.g., air, $N_2$and CO_2), it was a surprising result. Thus, this study is designed to provide a first insight into electrostatic charge generation and dissipation phenomenon in the presence of noble gases (Ar and He) and its effect on fluidization behavior. Specifically, experiments on fluidization of glass beads (300 μm) by dry air, He and Ar are conducted in a fluidized bed equipped with an electrostatic ball probe. It is found that the particle charge dissipation rate in noble gases is significantly faster than in air, resulting in improved fluidization behaviors. Also, particle sheeting on the reactor wall completely disappears in the case of fluidization with noble gases. This interesting static reduction phenomenon by noble gases is attributed to a low breakdown potential of a noble gas. The breakdown of charges in gas-solid systems are closely linked the collision behaviors of objects (particle-particle, particle-wall and particle-probe), and a conceptual model is derived based on Matsuyama's approach [3,4] to illustrate how particles lose charges as the potential between surfaces increases to the breakdown limit of the gas phase.

[1] Guardiola J., Rojo V., Ramos G., Influence of particle size, fluidization velocity and relative humidity on fluidized bed electrostatics, Journal of Electrostatics, 37 (1996) 1-20.

[2] Yao L., Bi H. T., Park A-H., Characterization of electrostatic charges in freely bubbling fluidized beds with dielectric particles, Journal of Electrostatics, 56 (2002) 183-197.

[3] Matsuyama T., Yamamoto H., Charge relaxation process dominates contact charging of a particle in atmospheric conditions. Journal of Physics D: Applied Physics, 28 (1995) 2418-2423.

[4] Matsuyama, T. A Discussion on Maximum Charge Held by a Single Particle due to Gas Discharge Limitation. AIP Conference Proceedings 1927, (2018).

557588 An Immiscible Fluids Approach for Correctly Predicting Agglomerate Dynamics during Particle ALD

Fine particle agglomeration can occur when running particle atomic layer deposition (Particle ALD) in fluidized beds. The fine powders frequently used in Particle ALD tend to effectively fluidize as larger aggregates due to high interparticle forces, potentially blocking surface sites and inhibiting surface coating uniformity. By modeling the agglomeration process during coating, steps can be taken to facilitate agglomerate shedding/breakup and mixing, thereby enhancing surface coating uniformity. However, current models of gas-solid flows which preserve the gas-solid interface, an important component for modeling ALD, have several limitations when incorporating agglomeration. Many of these approaches fail to address agglomerate size distributions or the dynamic formation and shedding/breakup process of fluidized agglomerates, shortcomings which remain a significant challenge to properly modeling fluidized bed Particle ALD. In this work, we propose an alternative modeling approach which naturally accounts for the dynamic nature of fluidized agglomerates by treating the fluidizing gas and particles as two immiscible (non-interpenetrating) fluids. Agglomerates are modeled using dynamic "bubbles" whose interior consists of many primary particles from the solids phase. The position, shape and formation/breakup of these agglomerate "bubbles" are allowed to change with time as dictated by the corresponding transport equations. With this model, we can investigate the formation and shedding/breakup of agglomerates without prior knowledge of the agglomerate size characteristics. This study provides some preliminary agglomerate size distribution results from fluidized bed Particle ALD simulations and compares these results to experimental data from previous literature studies.

571922 Modeling Cohesive Gas-Particle Flows with a Parcel-Based Lagrangian Approach

Lagrangian approaches are often advantageous to model processes in chemical engineering applications since particle-related phenomena (e.g., the deposition and evaporation of liquids, liquid absorption in pores and asperities, or chemical reactions) can be studied in a straight forward fashion. Parcels-based approaches form a sub-group of such approaches, and track packages of particles instead of the full particle population. Consequently, parcel-based approaches can be seen as a conceptually straight forward extension of approaches that track each individual particle. However, it is clear that as parcel size (i.e., the set of particles that represents a parcel) increases, parcel-based approaches will lose accuracy, especially for cohesive powders.

Increasing parcel size can be seen as "particle coarsening", and there have been previous attempts [1,2] to probe the effect of such a coarsening strategy on the fluid-particle drag force. Still, there is no systematic investigation on how parcel interactions should be scaled with respect to parcel size to yield an identical rheological behavior of a gas-particle mixture. This is especially true in case liquid bridge-induced cohesion is considered. Our present work focusses on closing this gap by considering statistics of (i) particle velocities, (ii) stresses, and (iii) local void fractions in gas-particle flows that contain wet particles. Specifically, we present results of CFD-DEM simulations that track all involved particles, and that are then filtered (i.e., spatially averaged) considering various (particle) filter sizes (see Figure for an illustration). These statistics are then compared with results of parcel-based simulations that utilize different approaches to scale parcel interactions. These parcel-based simulations were performed with the original fluid grid resolution (i.e., that of our CFD-DEM simulations) in order to avoid effects due to under-resolved gas motion.

Our results allow us to draw conclusions on (i) the maximum permissible parcel size, the (ii) optimal particle-to-fluid mapping strategy, as well as (iii) the optimal strategy to scale cohesive interactions when performing parcel-based simulations of fluidization. We conclude our study with a comparison of our findings with that of a recent study that investigated the rheology of fluidized particles that exhibit van der Waals interactions [3].

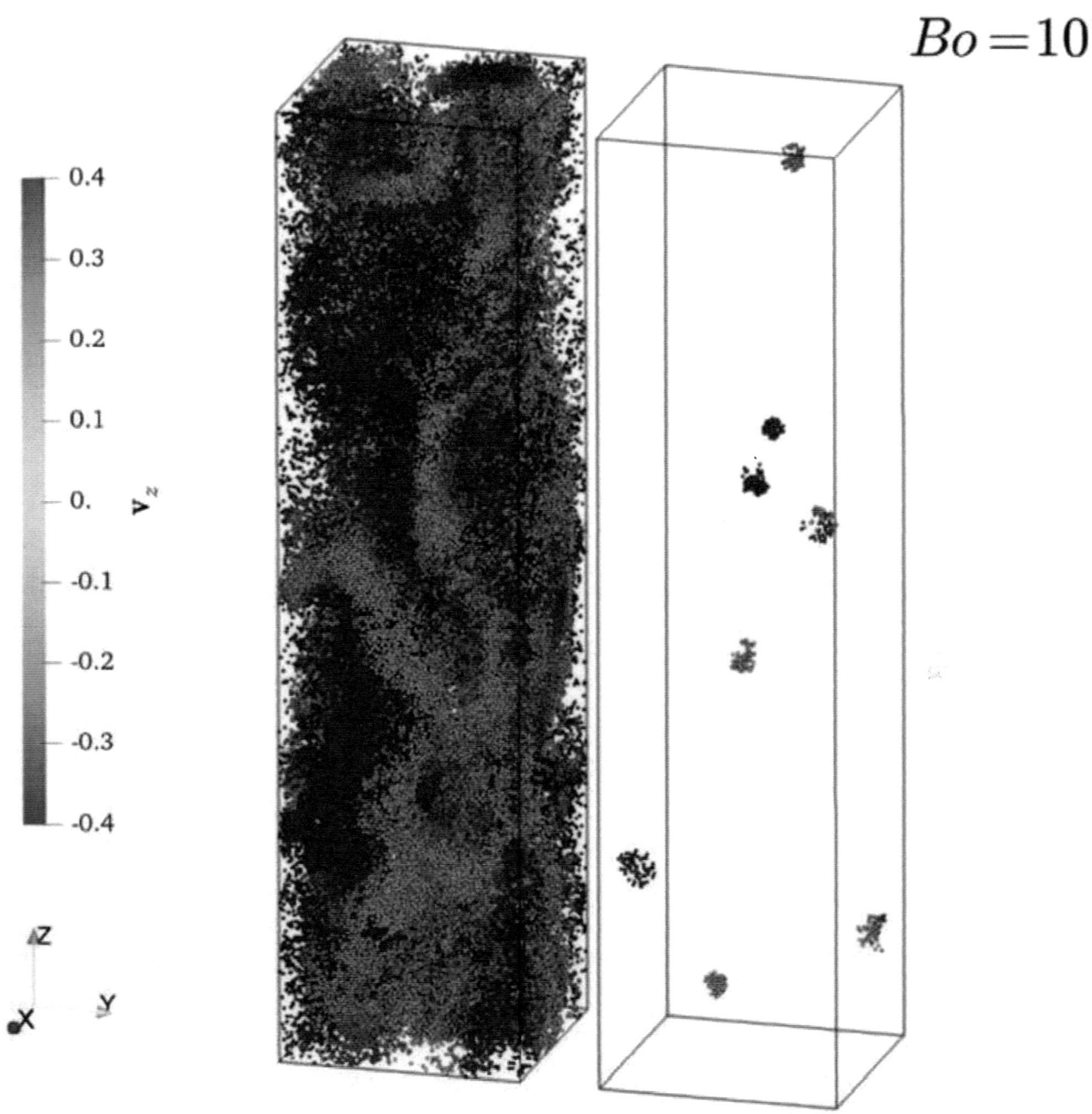

Figure: Ilustration of the simulation domain, as well as the particle-based filtering strategy to obtain parcel statistics (Bo is the Bond number that quantifies the degree of cohesion).

[1] A. Ozel, J. Kolehmainen, S. Radl, S. Sundaresan, Fluid and particle coarsening of drag force for discrete-parcel approach, Chemical engineering science 155:258-267, 2016.

[2] A. Ozel, Y. Gu, C. C. Milioli, J. Kolehmainen, S. Sundaresan, Towards filtered drag force model for non-cohesive and cohesive particle-gas flows, Physics of Fluids 29:103308, 2017.

[3] Y. Gu, A. Ozel, J. Kolehmainen, S. Sundaresan, Computationally generated constitutive models for particle phase rheology in gas-fluidized suspensions, Journal of Fluid Mechanics 860, 318-349, 2019.

560947 Rayleigh-Taylor Instability in Vibrated Gas-Fluidized Particles

A Rayleigh-Taylor-like instability[1,2] forms in a binary set of particles of different density subject to upward gas flow and vertical vibration[3]. "Fingers" and "bubbles" of light particles rise through heavy particles despite a lack of effective surface tension between the two types of particles. CFD-DEM simulations show that gas flow channels through fingers of lighter, larger particles due to a higher permeability of these particles to gas flow, locally increasing drag and causing these fingers to grow[3]. We demonstrate that the same physical mechanism causes an isolated "bubble" of light grains to rise and an isolated "droplet" of heavy grains to break up forming a binary branching pattern[3]. We explore how the combination of vibration and gas flow causes a bubble-free fluidization state in Geldart Group B and D particles and how this state can lead to other structured flow patterns.

References

(1) Rayleigh, J. W. S. B. *Scientific Papers: 1881-1887*; University Press, 1900.

(2) Taylor, G. I. The Instability of Liquid Surfaces When Accelerated in a Direction Perpendicular to Their Planes. I. *Proc. R. Soc. Lond. A* **1950**, *201* (1065), 192–196.

(3) McLaren, C.; Kovar, T. M.; Penn, A.; Müller, C. R.; Boyce, C. M. Gravitational Instabilities in Binary Granular Materials. *Proceedings of the National Academy of Sciences* **2019**, Accepted.

562642 Predicting Loss-in-Weight Feeder Performance Based on a Reduced Set of Material Property Measurements

Predicting loss-in-weight feeder performance based on a reduced set of material property measurements

Tianyi Li, Yi Tao, Fernando J. Muzzio, Benjamin J. Glasser*

Purpose: Loss-in-weight feeders plays a vital role in continuous powder processing. Material properties have a significant impact on loss-in-weight feeder's performance. Previous work has built up a correlation between loss-in-weight feeder's gravimetric feeding and refill performance utilizing a statistical model, partial least squares regression (PLSR), with 30 material flow indices. However, based on research on correlation between material properties and the weight of each property in the prediction model, a reduced set of measurement can be developed with less material characterization/properties, which can save time, powder and effort for the material characterization stage. Moreover, deeper understanding of material properties and material flow behavior can be established by the analysis in this work. **Method:** The proposed work includes techniques to characterize material flow properties, method to quantify loss-in-weight feeders feeding performance under gravimetric mode and during hopper refill, principal component analysis (PCA) to build up a material library with material flow properties, cluster analysis to classify materials investigated into groups to compare the difference between different selection of reduced set of measurements, partial least squares regression to build up a predicting model to validate the prediction made with reduced set of measurement by correlating feeding performance with material flow properties. **Results:** Experiment results shows feeder's feeding deviation is highly dependent on material properties and can be predicted with reduced set of measurement only 4 measurements of compressibility, permeability, shear cell at 6 kpa and particle size distribution. **Conclusion**: The work shows, based on the correlation between material properties and the weight of each property in the prediction of feeder performance, a good predicting model to correlate feeder performance and material flow properties, can be set up with a reduced set of measurements with 7 material properties instead of original 30. With this work, half of the time and materials can be saved from material characterization tests. Moreover, a deeper understanding between material properties and their correlation with feeding performance and be established.

571795 Flow Characterization of Compressed Woody Biomass

Global interest for using terrestrial biomass as an energy source has been increasing recently due to potentially significant socio-economic and environmental benefits relative to fossil energy. Thermochemical biomass processes converting biomass to fuels are relatively well developed, e.g. for direct combustion, gasification, and pyrolysis. However, critical problems often arise when attempting to feed biomass into commercial-scale reactors continuously and without interruption. Biomass particles, whether derived from woody or agricultural feedstock, vary greatly in size, shape, density, and moisture content. High cohesiveness, low bulk density, and poor material flow characteristics often prevent continuous feeding due to bridging, rat-hole formation, blockage, and seal failures. Among all different types of feeders utilized to feed biomass into reactors, screw feeding has received significant attention in recent years. Screw feeders are fuel-specific and require careful design to be reliable, efficient, and economical, especially for reactors operating at elevated pressures. In order to mimic the flow properties of compressed biomass inside a compression-screw feeder, rheological properties of three types of compressed lignocellulosic biomass were investigated using a co-rotating twin screw microcompounder. Experiments were conducted to find the impact of moisture content, particle size and type, and screw speed on rheological properties of forest residual particles as well as sawdust. Increasing moisture from 10 to 50% reduced the apparent viscosity of the compressed biomass. Also, higher screw speeds (10 to 90 rpm) led to a lower relative viscosity. At elevated screw speeds, significant temperature differences were measured between the initial and final states, which should be accounted for in equipment design and operation.

561854 DEM Study on the Effect of Particle Size Distribution on Jamming in a 3D Conical Hopper

Despite the advancement on the understanding on the unwelcome phenomenon of hopper jamming, the influence of the particle size distribution(PSD) width of the ubiquitous continuous PSDs remains unknown. Accordingly, this study investigated using DEM a range of PSD widths ($10\% \leq \sigma/d_p \leq 50\%$) of lognormal PSDs with a constant mean particle diameter of 10 mm in a 3D hopper. Results indicate that, although the monodisperse particles of 10 mm diameters do not jam, the wider PSDs ($\sigma/d_p \geq 30\%$) jam. The flatter arches and higher mass-averaged force that are linked to the denser packing of wider PSDs underlie the propensity to jam. To reduce the jamming probability, the friction coefficient and initial fill height can be reduced, but not changes in particle density. The results here highlight that the mean particle diameter alone is an insufficient parameter to predict jamming and the impact of PSD width is non-negligible.

570767 Real-Time Measurement of Volumetric Fill in a Storage Vessel and Discharge Rate By Stereoscopic 3D Optical Scanning

The volumetric degree of fill of a storage vessel can be important to operating and management requirements in the storage and handling of bulk solids. However, obtaining accurate and reliable measurements, particularly in real time, is challenging because the shape of the free surface is not usually known and is influenced by a variety of factors including discharge rate, filling conditions, and forces arising from motion and vibration. This study investigates the use of a stereoscopic 3D surface scanning method to monitor the volumetric degree of fill of agricultural materials in real-time; indicative flow rate out of the vessel is obtained from degree of fill measurements made in rapid succession. Point clouds generated from scans of the free surface of the material in a vessel, together with the 3D geometry of the vessel also obtained by 3D scanning, give a direct estimate of bulk solids volume, and hence the degree of fill. The work described in this paper was carried out with a cylindrical flat-bottomed vessel 300 mm in diameter and 700 mm high. Flow rates were controlled by circular orifices sized to give flow rates ranging from ~0.1 kg s^{-1} – ~3 kg s^{-1}. Mass flow rate was also determined in each trial by collecting the discharge in a pail placed on platform scales and logging to a personal computer. Mass flow and volume flow rates tracked each other closely, providing encouraging support for the accord between degree of fill estimated, in static experiments, directly by scanning and indirectly from mass and bulk density.

571413 Optimization of a Bag Filling Production Line for Starch By Down Scaling to Laboratory Scale

Bag filling of different starches was investigated on a production machine and related to the powder properties of the starches. The bag filling machine uses gravity feeding and vibration and de-aeration probes to remove excess air from the plastic bags. Bag filling time is around 10s per bag. The process was scaled down to laboratory scale and tested. The results obtained at laboratory scale were used for prediction of the bag filling on large scale and understanding of the filling and flow mechanisms involved in this production process successfully.

Measurements on production scale and lab scale were done and compared.

572184 Flow-Aided Pneumatic Conveying of Cohesive Dairy Powder

Flow-aided Pneumatic Conveying of Cohesive Dairy Powder

Akeem K Olaleye [a], Gavin M Walker [a], Harry Van den Akker [a, b]

[a] The Bernal Institute, University of Limerick, Limerick, Ireland.

[b] Department of Chemical Engineering, Delft University of Technology, Delft, The Netherlands.

Abstract

When pneumatically conveying dairy powders, a mixture of powder and air flows around pipe bends or experiences any change in flow direction, the milk particles often form a rope-like structure because of inertial, gravitational and centrifugal effects. The formation of such particle ropes can lead to stratification, re-agglomeration, and deposition of cohesive powders in the pipe. The current study focused on the development of novel, low-cost *smart flow-aids* that deliver ultra-low powder deposition and high throughput during cohesive dairy powder transport. First, several conceptual designs of such smart flow-aids were investigated in a single bends of varying orientation (i.e. H-V, V-H, and H-H) by means of a CFD-DEM. The ideas with potential were then tested in a 2-inch test rig with (i) conventional bends, and (ii) bends modified with the selected flow-aids (see Figure). In addition to measurements of pressure drop and powder deposition, an optical technique was used to measure the dynamics (probability densities) of local particle volume fractions as a function of operating conditions and conveying configurations. The results of the flow-aided and the conventional setup are compared.

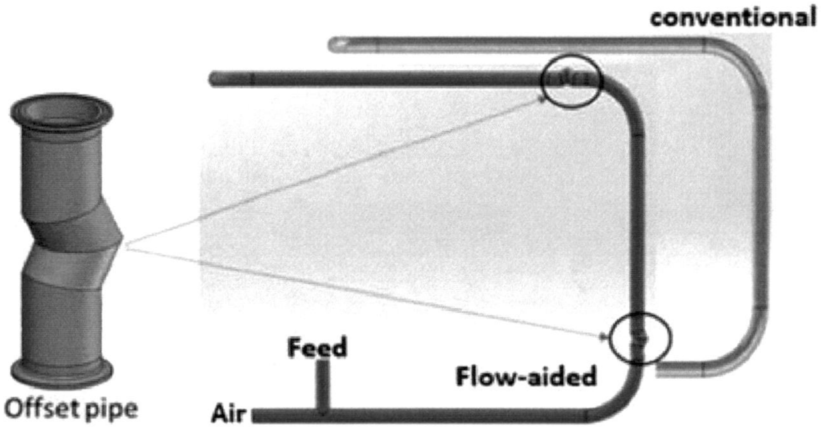

Figure: Conventional and Flow-aided pneumatic conveying setup

567389 Measurement of Solid Mass Flow Rate Under Different Flow Regimes in Pneumatic Conveying By Non-Intrusive Acoustic Emission Detection

As energy consumption is the most important in pneumatic conveying process, it is always recommended to be operated at the boundary between the dilute phase and the dense phase conveying. However, the flow regime in this operating zone is variable, which makes the real-time monitoring of the process very difficult, especially for measurement of solid mass flow rate. At present, most of the solid mass flow rate prediction models are only applicable to a specific flow regime, and when the flow regime changes, the prediction effect is poor. Therefore, we want to develop a solid mass flow rate prediction method that can be used under different flow regimes with good extrapolation by using non-intrusive acoustic emission detection. The implementation methods include the following two ways: (a) The feature quantities (including flow characterization parameters) obtained by analysis of acoustic signal are directly substituted into the prediction model for measurement of the solid mass flow rate. (b) The flow characterization parameters is used for flow regime identification, and then predict the sloid mass flow rate for each flow regime separately. The calculation results show that the accuracy of the prediction model is strongly dependent on the flow regime. For different prediction ways, the addition of flow characterization parameters can improve the prediction accuracy of the model. For prediction method (a), the predicted average relative error is 7.5%. For prediction method (b), the predicted average relative error is about 22.0%. The predicted average relative error of both methods are lower than the method which cannot be extrapolated (28.0%).

Keywords: pneumatic conveying, mass flow rate, acoustic emission detection, extrapolation

570405 Stoppers and Skins on Clay Nanotubes Help Stabilize Oil-in-Water Emulsions and Modulate the Release of Encapsulated Surfactants

This work develops the concepts of particle stabilized emulsions by using nano-tubular natural clays known as halloysites to attach to the oil-water interface and stabilize oil-in-water emulsions. Such nanotubular halloysite nanotubes (HNT) serve as reservoirs for surfactants and are able to deliver surfactants to the oil-water interface and thus lower the oil-water interfacial tension. This two-step concept of surfactant delivery and droplet stabilization by particles has significant implications to oil spill remediation. However, to deliver surfactant loaded HNTs in a water based solvent slurry it is important to stopper the nanotubes to prevent premature release of the surfactant. This work focusses on the use of an environmentally benign two-dimensional metal organic framework formed by coordinating Fe (III) with a polyphenolic as a stoppering agent. Such molecularly thin networks form a skin around the HNTs thus providing a way to effectively sequester surfactant cargo for controlled release. Cryo-scanning electron microscopy (Cryo-SEM) shows that these HNTs and HNT bundles attach to the oil-water interface with side-on orientation. Inverted drop tensiometry was used to characterize the dynamic interfacial tension resulting from the release of a model surfactant (Tween 80) from the HNTs and indicates that the stoppers are effective in sequestering the surfactant cargo for extended periods at neutral pH values. Release triggered by framework disassembly at acidic pH values can be effected just prior to delivery to oil spills. The concepts and scalability of this process have significant implications for oil spill remediation, enhanced oil recovery, biomedical and pharmaceutical applications.

564777 Large Area Monolayer Coatings for Janus Particle Production

Janus particles, having two hemispheres of different character, have widespread potential use as rheological modifiers, physical surfactants, microswimmers, and active sensors for bioapplications. They are not commonly used in industrial processes because of the lack of ability to produce them on a large scale. Specifically, the most common fabrication method is to create monolayer coatings that are functionalized through physical vapor deposition. One key limitation is the ability to coat large areas of particles. This talk will introduce our Automated Langmuir Blodgett (ALB) coating process that allows roll-to-roll coating of nanometer to micron-scale particles. This method will be discussed in light of creating large area particle arrays for functionalization of magnetic Janus particles and our initial results into making pilot plant-scale quantities of particles.

Magnetic Nanoparticle Characterization by Dark-Field Imaging and Bright-Field Absorbance

Abhinav Sannidhi[1], *Paul W. Todd*[2], *and Thomas R. Hanley*[1]
[1]*Department of Chemical Engineering, Auburn University, AL 36849*
[2]*Magnaquant LLC, 11760 Commonwealth Drive, Louisville, KY 40299, USA*

Abstract

Paramagnetic particles extensively varying in composition, size and magnetization are used in biochemical separations, cell purification, cell labeling, cell physiology, endocytosis, endosome research, ferrofluids, drug targeting, analytical biochemistry, environmental analysis, magnetic resonance imaging, hyperthermia, in vivo diagnostics and others. The particle size uniformity and colloidal stability of paramagnetic particles have been the focus of most producers, whether research or commercial, but magnetization is never specified and is typically mentioned only in terms of percent iron oxide content. Bulk measurements of magnetic susceptibility and saturation magnetization, using a SQUID magnetometer, vibrating sample magnetometer, Guoy scale, or Faraday scale provide an average value at best and do not account for the distributed nature of these variables or for particles with zero or very low susceptibility and saturation magnetization. By means of particle tracking velocimetry in darkfield illumination, we measured multiple characteristics of several thousand individual particles per sample. The Hyperflux™ velocimeter is utilized to provide quantitative video analysis of cells and particles using a high definition camera/microscope and capture of the images of the particle trajectories in an isodynamic field. Image analysis software converts the image data to the parameters of interest based on velocimetry calculations. By measuring calibrated magnetophoretic mobility and calibrated diameter of every particle we report apparent magnetic susceptibility and saturation magnetization on a particle-by-particle basis and provide distributions of susceptibility as well as saturation magnetization for numerous commercial products in a short period of time. The direct measurement of magnetophoretic mobility of the particles with size less than 400 nanometers is limited by optical resolution. In this research, to overcome the size limitation, two different approaches – the optical density method and the chain velocity method were studied to estimate the magnetophoretic mobility of magnetic nanoparticles of size < 400 nm. The optical density method is based on optical absorbance versus time measured in a bright field, using ImageJ image processing software. The chain velocity method takes advantage of chain formation of the superparamagnetic nanoparticles in an applied magnetic field, making particles visible in the dark field and extrapolates magnetophoretic mobility of nanoparticle chains. It was found that estimation of the distribution of magnetophoretic mobility and intrinsic magnetic properties of magnetic nanoparticles using dark-field particle tracking allows economical and time-efficient magnetic evaluation of a broad range of magnetic particle sizes, nano- and micro.

Keywords: magnetophoretic mobility, dark field imaging, susceptibility distribution, saturation magnetization, paramagnetic nanoparticles

Introduction

Magnetic particles of different size ranges have been used in research and production since they have wide applications in bioengineering, cell separation, drug delivery, and other biomedical applications. The particle sample quality which includes particle size and intrinsic magnetic property distributions needs to be specified for its usability in utmost applications. The particle size is usually determined by Dynamic Light Scattering (DLS) or electron microscopy techniques such as scanning electron microscopy (SEM), transmission electron microscopy (TEM), atomic force microscopy (AFM), magnetic force microscopy (MFM), and x-ray diffraction (XRD). The bulk average intrinsic magnetic properties such as magnetic volume susceptibility and saturation magnetization are measured by a superconducting quantum interference device (SQUID) (Hurt et al. 2013; Granata and Vettoliere, 2016), vibrating sample magnetometer (VSM) (Liao et al., 2018), a Faraday microbalance (Laachir et al., 1991), a Guoy's scale (Marcon and Ostanina, 2012), a torque meter (Kligfield et al., 1981), Bartington magnetic susceptibility system sensors (Dearing., 1999), magnetic magnetoresistance sensors (Little et al. 2013), and modified magnetization curves (Liu et al., 2009). These bulk average methods can be replaced by a single particle-by-particle analysis method which determines both the particle size and the magnetophoretic mobility of the magnetic particles simultaneously in the applied magnetic field (Chalmers et al., 1999; Zborowski et al., 1999; Zborowski and Chalmers, 2011). Magnetophoretic mobility depends on the particle size, magnetic volume susceptibility, and viscosity of the fluid medium. The intrinsic magnetic properties, apparent susceptibility and saturation magnetization, can be estimated with the help of their relationship to magnetophoretic mobility and particle size.

A Hyperflux™ velocimeter is used to measure the particle size, magnetophoretic mobility and other characteristics of the magnetic particles and magnetically labeled cells on a particle-by-particle basis (Zhou et al., 2016). It is also used in the study of quantitative cell analysis (Zhou et al., 2017) and in quantifying the nanoparticle endocytosis (Zhou et al., 2018). Dark field imaging is the basis of the technique to capture individual particle trajectories which are then processed by image analysis software to estimate magnetophoretic mobility and diameter of the particles.

In this present research, the estimation of intrinsic magnetic properties of micro-sized magnetic beads and their comparison between bulk method and particle-based method has been focused. Also, means of methods to estimate the magnetophoretic mobility of magnetic nanoparticles has been discussed.

Theory

Magnetophoretic mobility is defined as the ratio of the field-induced terminal velocity of the particle, v, to the gradient of the magnetic field, $\nabla\left(\frac{B_o^2}{2\mu_o}\right)$, also denoted as S_m.

$$\mathrm{U}_m = \frac{v}{\nabla\left(\frac{B_o^2}{2\mu_o}\right)} \tag{1}$$

S.I. units of magnetophoretic mobility are $\frac{m^3}{T\,A\,s}$ (cubic meter per tesla-ampere-second), also denoted as "Rm"- Rembaum, in honor of Alan Rembaum for his pioneering work on magnetophoretic mobility.

Magnetophoretic mobility derived from the force balance on a spherical particle in a viscous medium could be expressed as

$$U_m = \frac{D^2 \Delta\chi}{18\eta} \tag{2}$$

where D is the diameter of the magnetic particle, $\Delta\chi$ is the difference in magnetic volume susceptibility of particle and fluid medium, and η is the viscosity of the fluid medium.

In this study, the applied magnetic field intensity, B_o is constant with a value of 0.56 Tesla which lies within the saturation region of the particle magnetization curve (Figure 1). So, the estimated volume susceptibility is termed 'apparent susceptibility'. By rearranging the above Equation 2, apparent magnetic susceptibility and saturation magnetization of the particles can be estimated given by Equation 3 and Equation 4 respectively.

$$\Delta\chi \approx \chi_{app} = \frac{18\eta}{D^2} U_m \tag{3}$$

$$M_{sat} = \frac{18\eta B_o}{\mu_o D^2} U_m \tag{4}$$

The theoretical magnetophoretic mobility of the nanoparticle chains given by Equation 3 is derived from the study of Wise et al where the theoretical mobility (U_m^*) depends on the drag force proportionality constant [N] for 'n' number of particles of a chain, the diameter of the particle, the susceptibility of the particle (as $\Delta\chi$ is approximated to the susceptibility of the particle) and viscosity of the fluid medium (Wise et al., 2015).

$$U_m^* = \frac{nD^2\Delta\chi}{48\eta[\mathrm{N}]} \tag{5}$$

Materials and Methods

Magnetic beads

Magnetic beads designated P1 to P16 were obtained from different vendors and had varied sizes and compositions. The commercial Estapor magnetic beads (N2, N4) were obtained from Merck Chimie SAS (Estapor 2018), France; and Dynabeads (N16) were procured from Thermofisher, USA. Table 1 includes the sizes and compositions of those beads studied by both methods. FluidMAG-DXS particles of size 100 nm were used in the study of bright field imaging and chain velocity methods.

Hyperflux Velocimeter

The magnetic particle velocimeter, obtained from IKOTECH, LLC, and serviced by Magnaquant (Kentucky, USA) consists of a channel cell with magnetic assembly, video microscopy system, and image processing software. The instrument's hardware includes a channel cell, magnetic assembly, a vacuum pump used for facilitating sample movement, a high sensitivity and high-resolution Grasshopper 2.0 MP CCD camera combined with 6X Edmund optics telecentric lens used to capture images of the magnetic particles, and a Dell computer with Linux Kubuntu operating system installed with software capable of capturing and analyzing the images of particle trajectories. The image data are analyzed by the software which generates 23 different characteristics of magnetic particles based on user-selected intensity threshold and size limits.

Modifications made to the instrument for bright field technique

The instrument was modified to facilitate bright field settings with the help of a single LED bright light with cool white color and a color temperature of 6000 to 7000K to capture bright-field images with the existing video microscopy system of the velocimeter.

Results and Discussion

Velocimeter measurements: After inserting bead samples of 3 to 5 mL (with a concentration of 10^4 to 10^5 particles per mL) through the channel cell by using a syringe, the particles move from left to right in the direction of the net isodynamic magnetic force. The image acquisition tab on the screen display is selected to initiate collection of images of the particles at different times and positions of the beads. The magnetophoretic mobility, size, shape and other characteristics of the particles are estimated by analyzing 10 sets of data at a rate of one set per 2 seconds and capturing the images at a rate of 30 frames per second. The software (Cytotest™ and Magex™) tracks the particle trajectories and gives the magnetophoretic mobility, diameter, and other properties of each particle in a tracks.csv file based on the collected dark field image data. The calibration of the instrument has been reported previously (Sannidhi et al., 2018).

VSM measurements: The hysteresis loop of each dried sample (approximately 10 to 20 mg) was measured using the Vibrating Sample Magnetometer (VSM) to estimate the bulk average intrinsic magnetic properties of the beads. The required temperature (25°C) and magnetic field intensity (1 Tesla) based on the sweep rate of 0.001 Tesla per second was adjusted to allow the instrument (for about 2 hours) to generate the complete hysteresis curve by varying the applied magnetic field. Figure 1 shows the magnetization curves of six different beads obtained using VSM and apparent susceptibility estimated by velocimeter.

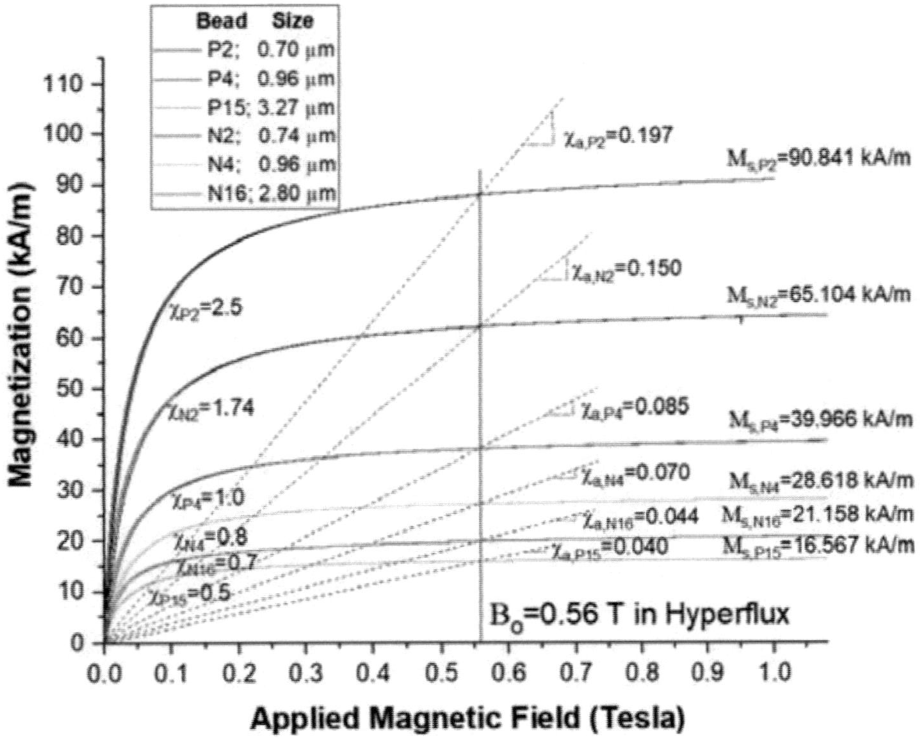

Figure 1- Magnetization curves determined by a vibrating sample magnetometer (VSM) for six different bead samples

The data in Table 1 gives the percent iron oxide composition provided by the vendors and intrinsic magnetic data comparison between VSM and Hyperflux velocimeter for six different bead samples. The intrinsic properties of the particles estimated by velocimeter are slightly higher than those obtained by VSM.

Table 1- Intrinsic saturation magnetization and susceptibility comparison between VSM and Hyperflux.

Magnetic Beads Composition			Saturation Magnetization (kA/m)		Volume Susceptibility (SI)		
Bead name	**Vendor Size (μm)**	**Fe Oxides (%)**	**by VSM**	**by Hyperflux**	**Initial χ by VSM**	**Apparent χ by VSM**	**Apparent χ by Hyperflux**
P2	0.70	41.00	90.841	104.250	2.500	0.197	0.230
N2	0.74	37.50	65.104	76.778	1.740	0.150	0.172
P4	0.96	23.70	39.966	43.356	1.000	0.085	0.097
N4	0.96	23.70	28.618	45.913	0.800	0.070	0.100
P15	3.27	14.00	16.567	23.501	0.500	0.040	0.050
N16	2.80	13.25	21.158	21.633	0.700	0.044	0.048

The saturation magnetization (M_{sat}) of the sample is proportional to the percent iron oxide of each bead as seen in Figure 2, which shows magnetization values in the range of 20 kA/m to 100 kA/m along with standard deviations.

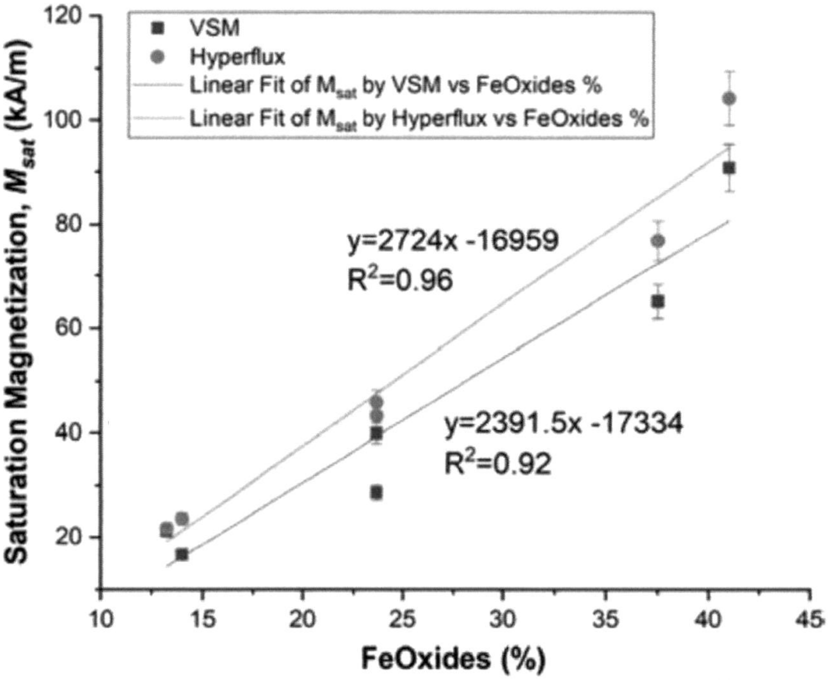

Figure 2- Saturation magnetization versus percent iron oxides for six different beads listed in Table 1

Intrinsic magnetic property distributions

The apparent magnetic susceptibility and saturation magnetization distributions of the magnetic particles as shown in Figures 3 and 4 are calculated based on their relationship to the magnetophoretic mobility (Equations 3 and 4) using the actual diameter calculated from the linear calibrated curve equation for the instrument and assuming the viscosity of the medium (distilled water) to be 0.001 Pa-s. Figure 3 shows the two-parameter calculated apparent susceptibility versus diameter distribution, where the vertical dashed line represents the vendor reported diameter. The graph confirms the presence of higher diameter particles with low susceptibility. Data points corresponding to negative or zero apparent susceptibility have been eliminated. The applied magnetic field intensity of the instrument, $B_o = 0.56$ Tesla, is well into the saturation region of the magnetization curve of the particles, which is evident from Figure 1. So, the susceptibility is named as apparent susceptibility of the sample, χ_a. Figure 4 shows the two-parameter saturation magnetization distributions where most of the beads lie in the range of 10^4 to 10^5 A/m. In this method, the intrinsic magnetic properties are directly proportional to the magnetophoretic mobility and inversely proportional to the square of the particle diameter (Sannidhi et al., 2019).

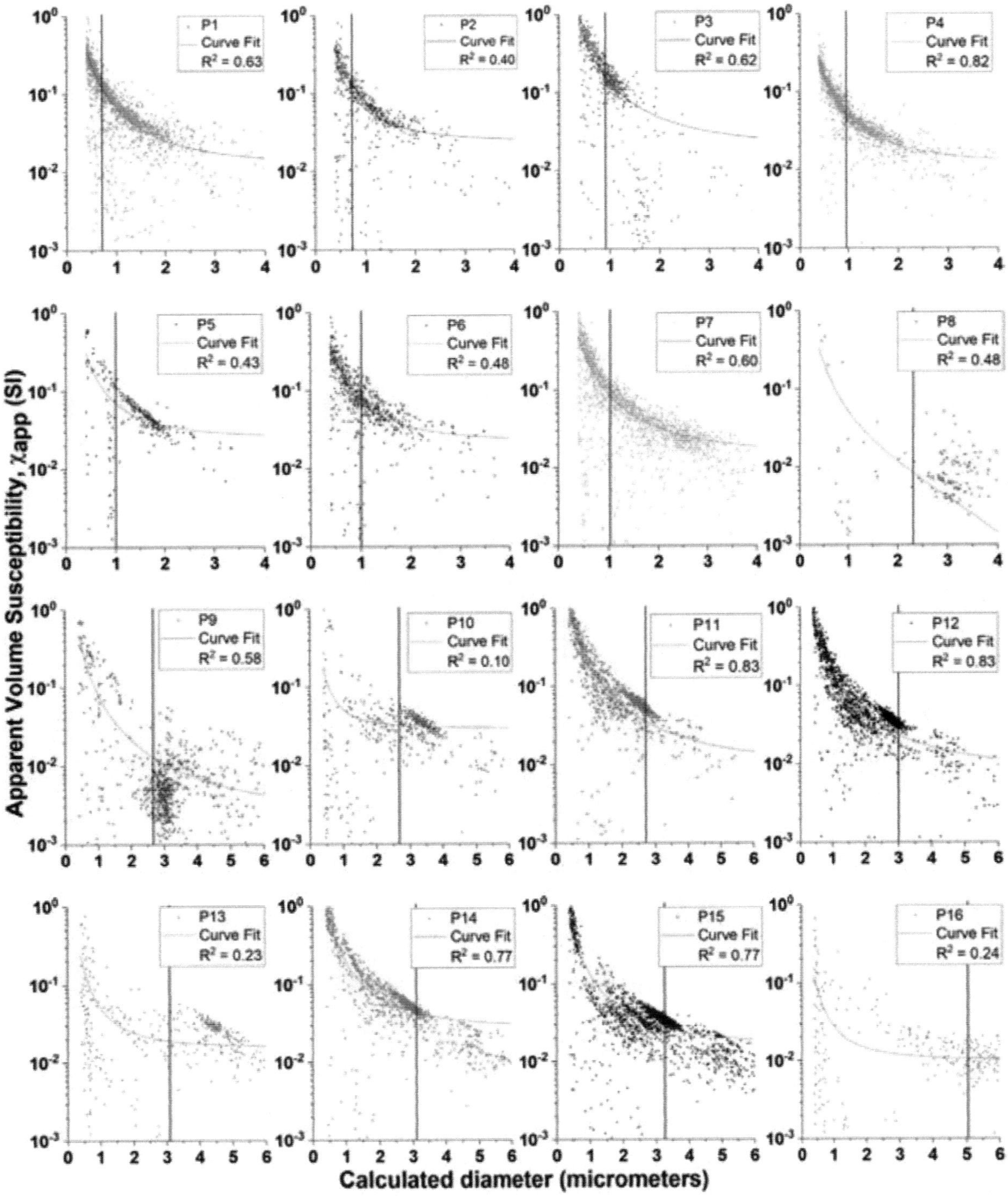

Figure 3- Apparent magnetic volume susceptibility versus diameter distribution for 16 differently sized magnetic beads. The vertical line represents vendor reported diameter.

The fitted curves follow the inverse square of the bead diameter.

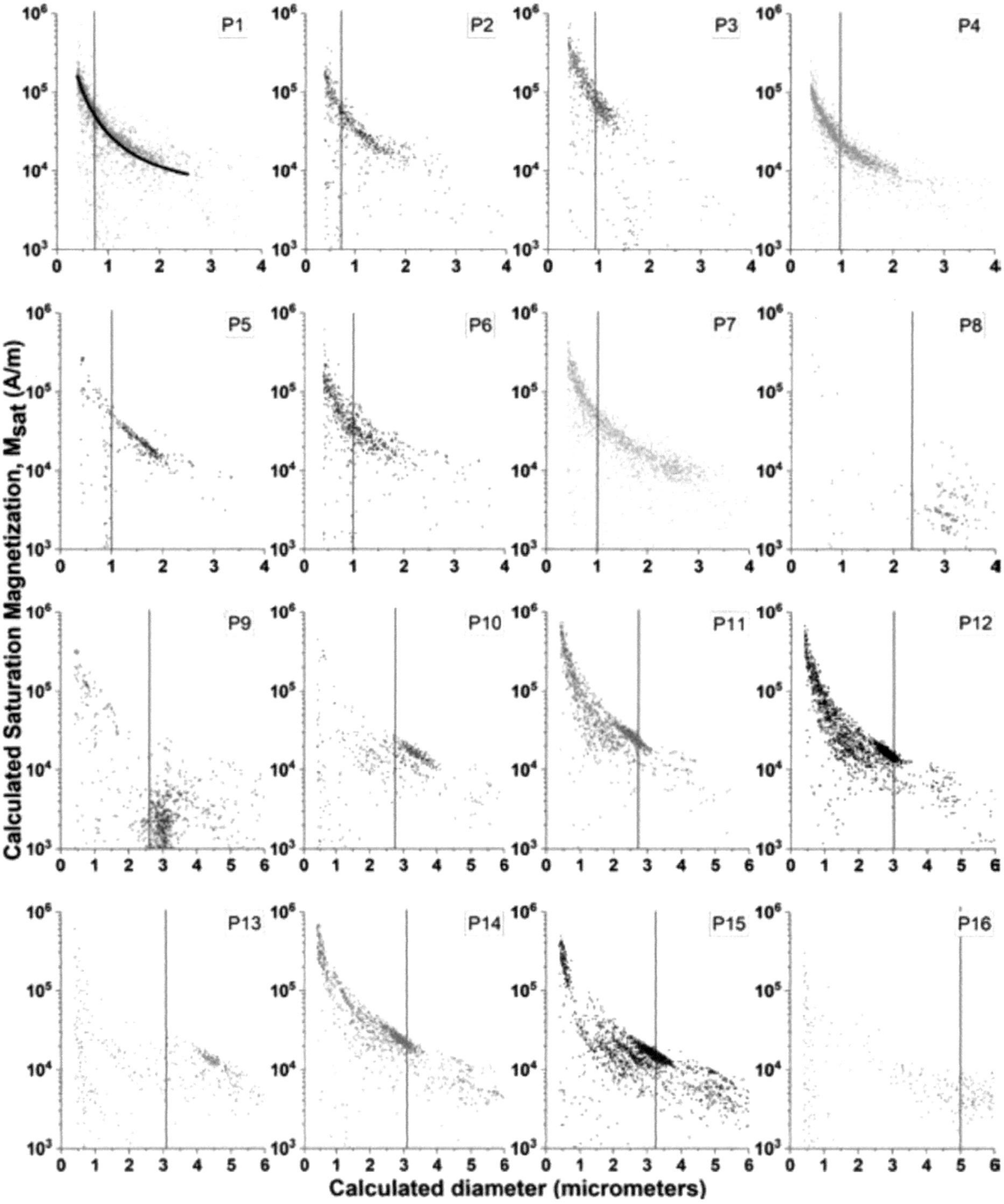

Figure 4- Calculated saturation magnetization versus calculated diameter for 16 different commercial magnetic beads. The vertical line denotes the diameter specified by the vendor for each sample. The solid curve in the P1 data fits the $1/D^2$ function with $R^2 = 0.63$.

Analysis of two different methods for the estimation of magnetophoretic mobility of nanoparticles:

Bright field absorbance method:

The objective of this study was to estimate the mobility of the 100 nm paramagnetic nanobeads from the clearing rate of the particles (Hassen 1988; Yeh 2018). In this method, we have modified the instrument to facilitate bright field imaging which was able to see the particle chain formation (Figure 5) at an applied magnetic field of 0.56 Tesla.

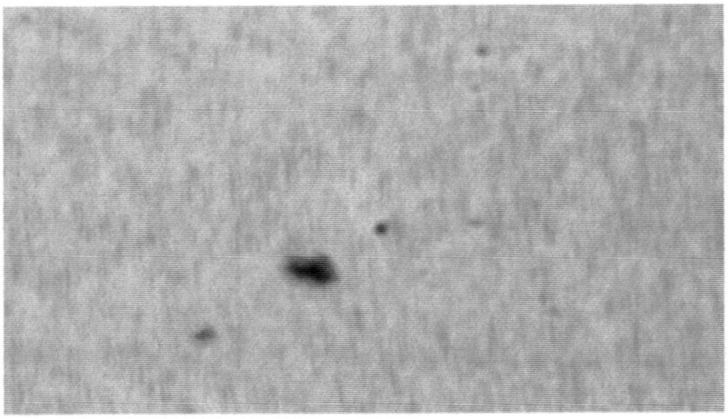

Figure 5- View of chain formation of 100 nm magnetic beads in bright field settings of modified velocimeter at an applied magnetic field of 0.56 Tesla.

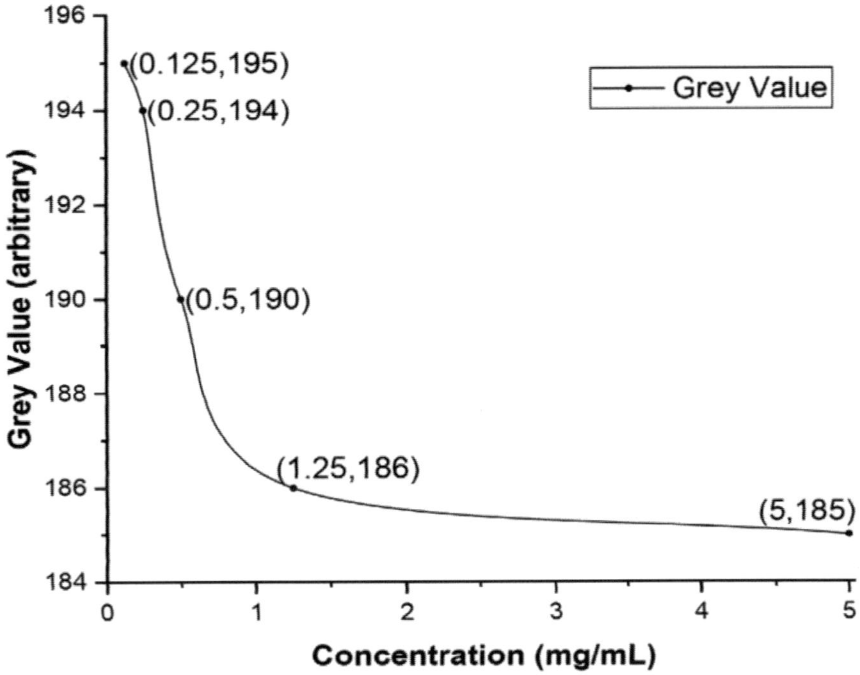

Figure 6- Grey value of 100 nm magnetic beads measured by ImageJ at different concentrations

As the greyscale bandwidth shown in Figure 6 is too small, this method could not be used to generate an optical density calibration curve. High concentrations of particles are required to sense the optical density (greyscale value) of the image with a microscope, and the clearance rate would be due to the movement of chains of particles and not the single particles.

Chain velocity method:

This method takes advantage of the SPIONs chain formation which is evident from the bright field image (Figure 5) and estimates the magnetophoretic mobility of a single nano-sized bead of the sample by curve fitting. The theoretical mobility of a single nanobead is estimated using Equation 5 and was found to be 0.0026 pRm (pico Rembaum) (Wise et al., 2015). The parameter of interest in this study is the ratio of length to width of the particle blob termed as "tAvgElongation" in the instrument software. This is estimated by the image analysis software Cytotest and tabulated in the tracks.csv spreadsheet. For a single nanobead, tAvgElongation is equal to 1. By generating the experimental mobility versus tAvgElongation curve (Figure 7), we have extrapolated the mobility of a single nanobead to be 0.0035 pRm (pico Rembaum) which is close to the theoretical value. The experimental data includes background noise resulting in poor goodness of fit. This method is still being developed and provides scope for wide applications.

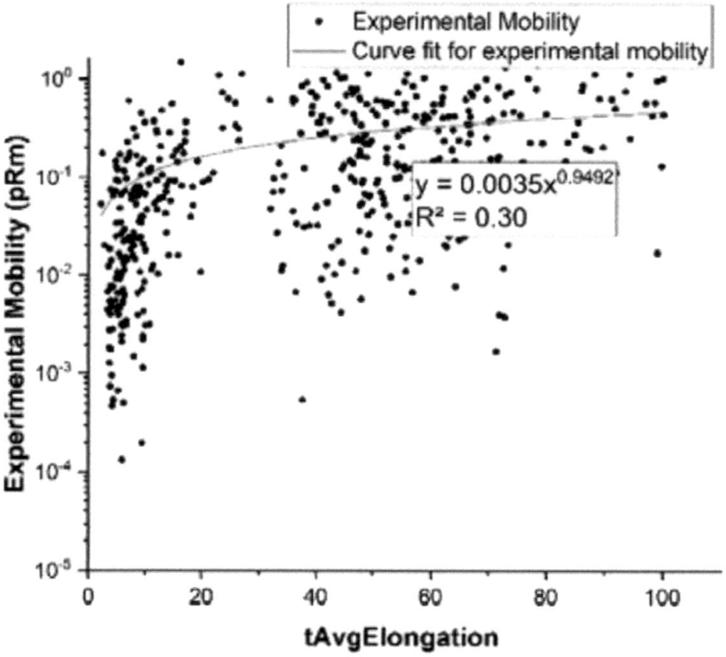

Figure 7- Experimental mobility versus tAvgElongation of the 100 nm beads using velocimeter. The red fitted curve is an empirical version of Equation 5.

Conclusions

The intrinsic magnetic property distributions of the magnetic microspheres using the Hyperflux velocimeter instrument have been estimated and their comparison to the bulk results obtained by VSM produces a reasonable agreement. The simultaneous determination of actual particle size and intrinsic magnetic properties on a particle by particle basis rapidly provides valuable information for product quality analysis and control. The nanoparticle mobility study and the agreement between theoretical and experimental mobility provide the scope to extend the method to the analysis of nanoparticles.

References

Chalmers JJ, Zhao Y, Nakamura M, Melnik K, Lasky L, Moore L, Zborowski M. 1999. Instrument to determine the magnetophoretic mobility of labeled, biological cells and paramagnetic particles. J Magn Magn Mater. 194(1):231–241. doi:10.1016/S0304-8853(98)00557-5.

Dearing JA. 1999. Environmental Magnetic Susceptibility Using the Bartington MS2 System. Interpret A J Bible Theol.:52.

Granata C, Vettoliere A. 2016. Nano superconducting quantum interference device: A powerful tool for nanoscale investigations. Phys Rep. 614:1–69.

Hassen MA. 1988. Spreading of the interface at the top of a slightly polydisperse sedimenting suspension.

Hurt D, Li S, Amann A. 2013. Versatile SQUID susceptometer with multiple measurement modes. IEEE Trans Magn. 49(7):3541–3544. doi:10.1109/TMAG.2013.2241029.

Kgaa M. SuperParamagnetic Microspheres SuperParamagnetic Microspheres from MilliporeSigma.

Kligfield R, Owens WH, Lowrie W. 1981. Magnetic susceptibility anisotropy, strain, and progressive deformation in Permian sediments from the Maritime Alps (France). Earth Planet Sci Lett. 55(1):181–189.

Laachir A, Perrichon V, Badri A, Lamotte J, Catherine E, Lavalley JC, El Fallah J, Hilaire L, Le Normand F, Quéméré E, et al. 1991. Reduction of CeO 2 by hydrogen. Magnetic susceptibility and Fourier-transform infrared, ultraviolet and X-ray photoelectron spectroscopy measurements. J Chem Soc Faraday Trans. 87(10):1601–1609.

Liao S-H, Huang H-S, Su Y-K. 2018. Scaling characteristics of the magnetization increments of functionalized nanoparticles determined using a vibrating sample magnetometer for liquid magnetic immunoassays. IEEE Sens J.

Little CAE, Pellegrino J, Russek SE. 2013. Microfluidic platform for magnetic nanoparticle trapping and detection. IEEE Trans Magn. 49(7):3402–3405. doi:10.1109/TMAG.2013.2251416.

Liu W, Zhou M, Kong L. 2009. Estimation of the size distribution of magnetic nanoparticles using modified magnetization curves. Meas Sci Technol. 20(12). doi:10.1088/0957-0233/20/12/125802.

Marcon P, Ostanina K. 2012. Overview of Methods for Magnetic Susceptibility Measurement. PIERS Proc.(5):420–424.

Sannidhi, A, P. W. Todd and T. R. Hanley: Multiparameter Paramagnetic Particle Characterization by Dark-Field Imaging, *2018 AIChE Annual Meeting Proceedings*, Pittsburgh, PA, October 2018

Sannidhi A, Todd PW, Hanley TR. 2019. Estimation of Intrinsic Magnetic Properties of Single Particles by Particle Tracking Velocimetry. IEEE Magn Lett.:1. doi:10.1109/LMAG.2019.2950298.

Wise N, Grob T, Morten K, Thompson I, Sheard S. 2015. Magnetophoretic velocities of superparamagnetic particles, agglomerates and complexes. J Magn Magn Mater. 384:328–334.

Yeh BJ. 2018. Improving the homogeneity of superparamagnetic nanoparticles and a study of their physicochemical properties and applications, Auburn University Dissertations.

Zborowski M, Chalmers J, Moore LR. 1999. Method for determining particle characteristics.

Zborowski M, Chalmers JJ. 2011. Magnetic cell separation. Elsevier.

Zhou C, Boland ED, Todd PW, Hanley TR. 2016. Magnetic particle characterization—magnetophoretic mobility and particle size. Cytom Part A. 89(6):585–593. doi:10.1002/cyto.a.22866.

Zhou C, Choi YS, David AE, Todd PW, Hanley TR. 2018. Nanomaterial endocytosis: Estimation of particles per cell by magnetic measurement. IEEE Magn Lett. 9:7–11. doi:10.1109/LMAG.2018.2830754.

Zhou C, Qian Z, Choi YS, David AE, Todd P, Hanley TR. 2017. Application of magnetic carriers to two examples of quantitative cell analysis. J Magn Magn Mater. 427(November 2016):25–28. doi:10.1016/j.jmmm.2016.11.009. http://dx.doi.org/10.1016/j.jmmm.2016.11.009.

576842 Yolk-Shell Composite Nanostructured Electrodes for High Capacity Lithium-Ion Batteries

Advances in technologies for lithium-ion batteries (LIBs) have been so spectacular in recent years that they have become one of the most popular sources for portable computing and telecommunication equipment. In the last five years there has been considerable work on the electrochemical properties of electrode Materials for Lithium Ion Batteries (LIBs). In particular, a much work has been done on the electrochemical properties of anode materials which have higher specific capacity than graphite, the most common anode material in use today. Among the top candidates for high capacity anodes is silicon, which has the highest know capacity of any material.

However, there are two well-known issues with utilizing silicon in anodes, both related to the fact that silicon undergoes a high expansion ratio upon lithiation (during the cycling of a battery). As the battery is charged (lithiated) expansion occurs in the anode, and as the battery is discharge (delithiated) contraction occurs in the anode. The volumetric expansion is only 10% percent for graphite, but is approximately 300% for silicon. The high volumetric expansion and contraction in the silicon can induce mechanical stress and leads to reformation and breakage of the solid electrolyte interface (SEI) during cycling producing loss of capacity and stability. The SEI layer is a polymeric boundary which naturally forms from the electrolyte at the surface of the anode and is known to further protect the electrolyte from further reactions at the surface of the anode. However, if the expansion ratio of the anode materials is sufficiently high, it appears that the SEI layer is broken and reformed to some extent during each cycle, thus leading to loss of efficiency and cycling stability.

Although there has been considerable research into using silicon, these works suffer from several drawbacks associated with the breakage of the Solid Electrolyte Interface (SEI) during repeated cycling. In this work we developed a new method that avoids breakage of the SEI at the silicon surface. We first created novel core shell/hollow nano-carriers made with graphene oxide and then incorporated silicon nanoparticles as the high capacity electrochemically active material which has the potential to greatly increase the capacity of these materials. By incorporating using yolk-shell particles, which have a void space between the silicon nanoparticles and the ultra-thin graphene oxide shells, the expansion of the silicon can be accommodated.

We solidify the composite to effectively trap the electrochemically active silicon nanoparticles within the graphene oxide shells, preventing segregation of the active components from the conductive matrix. These composites are thermally reduced to convert GO to a graphitic structure capable of providing high electrical conductivity as well as high lithium ion conductivity.

We present the results of both thermally reduced and chemically reduced graphene oxide based anodes. An electrode with high cycling stability was made which exhibited a storage capacity of 1450 mA h/g after 200 cycles and greater than 1300 mAh/g after 350 cycles, with a coulombic efficiency reaching 99.9%. For some anodes, capacities greater than 1200 mA-hr/g were possible at a current of 3200 mA/g. The electrode materials were characterized via laser diffraction, battery cycling, FE-SEM, TEM, Raman, and XRD.

558750 Effects of Rolling Friction, Cohesion and Size Distribution on the Shear Flow of a Binary Particle Mix

In this study, shear flows of a binary mixture of large and small spherical glass bead particles are modeled using discrete element method (DEM) in order to explore the underlying mechanism of the trend that the shear stress of a binary mixture is larger than that of mono-dispersed particles. The effects of rolling friction, particle cohesion and particle size distribution on the stress behavior are investigated. It is found that the stress for non-rotational particles is larger than that of rotational particles. By implementing the rolling friction into the model, the shear stress decreases with increasing rolling friction. Cohesion between small particles with a mean size of 40 μm is considered. The shear stress decreases with increasing cohesion, since the particles with cohesion become less compacted. The poly-dispersed particle size is represented by setting a number of bins of particles in the DEM simulation. The shear stress from poly-dispersed system matches better with the experimental result than that of mono-dispersed system. It is also found that the particle size (or particle size ratios in binary mixtures) has a significant effect on the contact structure of the whole system and consequently affects the shear stress.

568832 Superquadric Particle Method for Non-Spherical DEM Simulation in Open Source Mfix: Development, Implementation and Verification

Modeling and simulation of non-spherical granular systems is not only of academic interest but is also highly relevant for many industrial applications. Superquadric particle representation is a general and efficient way to model non-spherical particle as it can be described by beautiful mathematical expression. Also, most of the shapes (such as sphere, prolate, oblate, box, cylinder) can be easily modeled by only varying five shape parameters. This work presents the development, implementation and verification of superquadric particle method in open-source parallel CFD code MFiX, which enables new capability of the MFiX-DEM solver for non-spherical particle simulation. To reduce the computational cost of particle-particle contact detection, both bounding sphere and oriented bounding box (OBB) are employed as broad phase contact detections before entering the narrow phase contact detection. Non-linear force model with considering the equivalent radius of the particle shape curvature is used to better calculate the contact forces. Quaternions theory is implemented to model the superquadric particle orientation and the rotation between global coordinates and local coordinates. The superquadric particle DEM solver are verified and validated by extensive test cases, including particle falling, static packing, hopper discharge, etc.

570137 Continuum Simulations of Dense Granular Flow and Model Assessment

The rheology and hydrodynamics of powder flow are important in many industrial processes such as pharmaceutical production, agricultural production, and cement mixing. Key processes in the pharmaceutical industry involve the mixing of different granular materials. Heat and mass transfer considerations are also important in the drying of powders and wet granulation. Accurate and computationally feasible prediction of granular rheology, hydrodynamics, and heat and mass transfer in production-scale devices is a challenge because of the large separation of time and length scales. While Discrete Element Model (DEM) simulations allow for a reasonable representation of particle-level physicochemical processes and particle-particle interactions using a contact force model, they are computationally expensive for systems with more than 10 million particles, thereby rendering the simulation of production-scale devices infeasible with today's computers. Continuum simulations involve solving the averaged equations for conservation of mass, momentum, and energy by treating the granular medium as a continuum. In this study, continuum simulations of dense monodispersed granular flows are performed to study flows in canonical geometries using three different constitutive models (Chialvo-Sun-Sundaresan (CSS) model (Chialvo et al., 2012), Srivastava and Sundaresan (S&S) model (Srivastava et al., 2003), and the Schaeffer model (Schaeffer et al., 1987)). The results are compared with DEM simulations and experimental data. The mean particle velocity and the error in the prediction of solid stresses are also quantified, thereby permitting a comparative assessment of the different constitutive models.

574320 Coupled Computational Fluid Dynamics – Aerosol Dynamics Model for Prediction of Nanostructured Films in an Aerosol Chemical Vapor Deposition Reactor

Nanomaterials have been explored for use in countless applications over the past decade due to the enhanced properties exhibited at the nanoscale. Controlling the morphology of nanomaterials allows for optimization of these properties, such as specific surface area, conductivity, fluorescence, and chemical reactivity. Consequently, nanomaterials have been investigated in a myriad of applications, including solar energy harvesting, battery electrodes, photocatalysis, water splitting, and drug delivery. However, synthesis of high-performance nanomaterials frequently involves complex batch processes with low yields, making scale-up difficult.

Aerosol chemical vapor deposition (ACVD) is a single-step, ambient pressure technique used to deposit highly crystalline thin films consisting of oriented nanostructures. Nanostructured films synthesized using ACVD have been utilized as battery electrodes, sensors, and in CO_2 reduction. The morphology of the deposited film can be controlled by modifying easily measured reactor variables such as precursor feed rate, gas flow rate, and reaction chamber temperature. In ACVD, nucleation of particles occurs via thermal decomposition or reaction of a vapor precursor in the heated reaction chamber, followed by particle growth via coagulation and condensation reactions. Particles are deposited on a heated substrate, where the particles sinter into structured films. Film morphology may be predicted using three characteristic rates, reaction rate, particle arrival rate, and sintering rate. Due to the easily controllable morphology and atmospheric pressure operation ACVD is a promising technique for the large-scale production of nanostructured thin films.

In this work, a finite element based computational fluid dynamics (CFD) model is coupled with an aerosol dynamics moment model to predict film morphology with experimental confirmation. This model couples the large scale fluid dynamics from CFD with nano range particle nucleation, growth, and deposition to predict film morphology. Additionally, a large area deposition analysis is performed on a multi-feed geometry by calculating a spatially-resolved mass flux to the substrate surface for estimation of film height and uniformity. Reactor conditions such as reactor temperature, precursor feed rate, residence time, and feed-to-feed distance are investigated and optimized to maximize film uniformity while retaining the desired morphology.

569318 Predicting Pressure Filtration Performance By CFD-DEM Coupling Approach

Pressure filtration is an important solid-liquid separation operation in which a particle suspension is subjected to an applied pressure and the liquid is forced through a filter medium that is only permeable to the fluid phase. Common applications include separation of solid particles from a mother liquor in the pharmaceutical industry and recovery of microorganisms from fermentation broths in the biotechnology industry. The porous cakes formed by different particle suspensions can exhibit widely different filtration properties including filtration times that varying from hours to weeks under similar operating conditions. The filtrate flow rate is dependent on operating conditions (e.g. applied pressure), liquid properties (e.g. viscosity) and cake properties (e.g. resistance). Because pressure filtration often is one step in a larger batch processing plant, there is considerable motivation to improve filtration performance by reducing the filtration cycle time. Process modeling represents a powerful tool for developing improved understanding and enhanced efficiency of pressure filtration processes. Although conventional continuum models provide macroscopic predictions such as the time-dependent cake thickness and filtrate volumetric flow rate, this approach is unable to account for the effects of the particle size/shape distribution on cake structure and filtration performance.

To obtain a fundamental understanding of the various factors affecting pressure filtration performance, we developed a coupled computational fluid dynamics (CFD) and discrete element method (DEM) model for simulating the effect of the solid particle characteristics on fluid flow through the cake. The combined CFD-DEM model captured both particle-particle and particle-fluid interactions through fundamental mathematical descriptions. The model was validated using data collected by filtering spherical glass beads and deionized water mixtures using a dead-end cell over a range of applied pressures and for two mean particle sizes. Numerical experiments were then performed to study the effects of particle properties, liquid properties and operating conditions on filtration performance. The model predicted that the cake resistance and filtrate flow rate could be strongly affected by the mean size and polydispersity of the spherical particles, the presence of small particles (i.e. fines) in the particle distribution, the viscosity of the liquid, and particle deformation leading to cake compression. We extended the filtration model to account for non-spherical, polydisperse particles more representative of actual APIs (active pharmaceutical ingredients). Simulations predicted that particle shape can have a strong impact on filtration performance, especially when the shape distribution is strongly polydisperse.

568007 On the Fast Fluidized Bed Simulations Using Recurrence CFD

Fluidized beds are systems with a bed of granular particles initially resting on a perforated bottom plate. When the inlet fluid is passed upwards through the bottom plate, it suspends, or fluidizes, the particles to allow a high effective contacting process. This contributes to the chief advantage of fluidized beds in providing a rigorous mixing and favorable heat and mass transfer characteristics. With these attributes, fluidized beds have occupied a high-ranking position in the chemical processing applications, such as coating, drying, coking of fuels and polymers. The dynamics in fluidized beds includes numerous physical phenomena because of the multiscale interactions between the solid-gas phases. For example, the gas flow interacts with the solid particles by the interstitial gas drag, and the particles among themselves, induce kinetic, collisional and frictional stresses. These (spatial and temporal) small-scale (microscopic) interactions compose to larger (meso-and-macroscopic) interactions which take place between clusters of solid and gas bubbles. The temporal evolution of the (microscopic) particle's collision dynamics lasts about few milliseconds, much shorter than the bubble evolution. While the bubbles in turn, move faster than the gas slugs in the bed and phenomena as heat transfer and chemical conversions can take minutes or hours. Due to this variety in motion scales, the numerical simulations of fluidized beds are limited to a short-term duration in case of industrial-size reactors. Therein, the computational cost needed becomes extremely huge and the long-term processes as chemical species transport (like a passive scalar) is out of an appropriate reach. In order to overcome such shortcomings, an innovative approach of recurrence computational fluid dynamics (rCFD), recently introduced by Lichtenegger & Pirker [1,2], can be applied.

The idea behind of rCFD is extracting the degree of similarity between flow structures at two states, in an unsteady pseudo-periodic flow, and using a chosen global norm,

Therein, φ is an active field. By considering multiple reoccurring states, during a short-term simulation, the recurrence matrix (rMatrix) can be plotted to indicate the recurrence events of the total system in time. Afterwards, this rMatrix will be used to distinguish those recurrence statistics that can be stitched to time-advancing states, and generate a generic flow pattern upon which a passive scalar can be traced.

An infinite time-advancing recurrence path is obtained upon the rMatrix, and it's corresponding recurrent flow patterns will be used in resolving a transport equation (Eulerian model) or evaluating fluid parcel trajectories (Lagrangian model), for a passive scalar. These (postprocessing) two models were classified as flow-based versions of rCFD [2] where the recurrent velocity patterns were essentially needed therein. However, in the so-called transport-based rCFD [3], the recurrent flow patterns are stored directly to the memory as Lagrangian shift information. Namely, if we want to trace a passive scalar, we just have to shift the scalar information from the corresponding start-cell to the receiving end-cell following inertia-less tracers, which are strictly linked to the fluid's velocity. Finally, we end up to a series of cell-to-cell shifts on the computational grid that can transport the passive scalar with a temporal jump, *i.e.* recurrence time-step, bigger than the full-CFD time-step, and without resolving any equations. By this way, a huge computational demand required for the full-CFD scalar simulation can be drastically reduced to few seconds using this feasible, cheap and fast passive transportation. It is evident that this procedure of information transport is ill-conservative to the mass scalar inside the domain. In this regard, a global balance between the incoming and outgoing fluxes has to be ensured by applying a proper diffusion on the passive scalar. For further details about the methodology the reader is referred to [3].

In the present work, we aim to apply the transport-based rCFD method for the simulation of chemical species transport, which can be considered as a passive scalars, in fluidized bed systems. To do so, a lab-scale bubbling fluidized bed with a lateral injection of a two species mixture has been numerically resolved on ANSYS/Fluent using Two-Fluid model approach. The preliminary outcomes have shown a feasible perdition of species transport with a good chasing to the actual full-CFD transportation, by consuming a very low computational load using this methodology. This in turn, encourages highly the application of rCFD to be the key ingredient in fast fluidized bed simulations and exploring new frontiers in the plant-scale systems. The obtained results will be presented during the meeting.

References

[1] T. Lichtenegger & S. Pirker. Recurrence CFD – A novel approach to simulate multiphase flows with strongly separated time scales. *Chem. Eng. Sci.*, Vol. **153**, pp. 394–410 (2016).

[2] T. Lichtenegger, E. A. J. F. Peters, J. A. M. Kuipers & S. Pirker. A recurrence CFD study of heat transfer in a fluidized bed. *Chem. Eng. Sci.*, Vol. **172**, pp. 310–322 (2017).

[3] S. Pirker & T. Lichtenegger. Efficient time-extrapolation of single-and multiphase simulations by transport based recurrence CFD (rCFD). *Chem. Eng. Sci.*, Vol. **188**, pp. 65–83 (2018).

565278 Sub-Grid Deconvolution Approach for Filtered Two-Fluid Models and the Application to Fluidized Beds

Highly resolved two-fluid model (TFM) simulations of gas-solid flows in vertical periodic channels have been performed to study closures for the filtered drag force and the Reynolds-stress-like contribution stemming from the convective terms. An approximate deconvolution model (ADM, [1]–[3]) for the large-eddy simulation of turbulent gas-solid suspensions is detailed and subsequently used to reconstruct those unresolved contributions in an a priori manner. With such an approach, an approximation of the unfiltered solution is obtained by repeated filtering allowing the determination of the unclosed terms of the filtered equations directly. A priori filtering shows that predictions of the ADM model yield fairly good agreement with the fine grid TFM simulations for various filter sizes and different particle sizes. In particular, strong positive correlation ($\rho > 0.98$) is observed at intermediate filter sizes for all sub-grid terms, even though the filtered drag force has been approximated by drift velocity correction ([1], [4]).

This ADM-TFM approach is implemented in an a-posteriori manner for the coarse grid simulation of a fluidized bed of Geldart type B particles and is verified against highly-resolved TFM simulations. Furthermore, different regularization methods, which account for the sub-filter scale (SFS) contribution, are presented and discussed with respect to their predictiveness and stability. It is shown that employing an appropriate rheological model for the solid phase considerably improves stability without removing too much energy from the resolved scales. The ADM-TFM predictions are in fairly good agreement with the fine grid reference case and do not show a notable grid-dependency. Compared to TFM simulations using the same grid resolution, the ADM-TFM approach does only require marginally more computational resources [5].Finally, the application to the coarse-grid simulation of large-scale fluidized beds is discussed.

References:

[1] S. Schneiderbauer and M. Saeedipour, "Approximate deconvolution model for the simulation of turbulent gas-solid flows: An a-priori analysis," *Phys. Fluids*, vol. 30, no. 2, p. 023301, 2018.

[2] S. Vincent, M. Tavares, S. Fleau, S. Mimouni, M. Ould-rouiss, and J. Estivalezes, "A priori filtering and LES modeling of turbulent two-phase flows application to phase separation," *Comput. Fluids*, vol. 176, pp. 245–259, 2018.

[3] S. Stolz and N. A. Adams, "An approximate deconvolution procedure for large eddy simulation.," *Phys. Fluids.*, vol. 11, no. 7, pp. 1699–1701, 1999.

[4] A. Ozel, Y. Gu, C. C. Milioli, J. Kolehmainen, and S. Sundaresan, "Towards fitered drag force model for non-cohesive and cohesive particle-gas flows," *Phys. Fluids*, vol. 29, no. October, p. 103308, 2017.

[5] S. Schneiderbauer and M. Saeedipour, "Numerical simulation of turbulent gas-solid flow using an approximate deconvolution model," *Int. J. Multiph. Flow*, vol. 114, pp. 287–302, 2019.

567884 Slow Granular Flows in a Split-Bottom Couette Device

In this work, we present DEM simulations and experiments of slow shear of granular materials in a split-bottom Couette device. This device is a cylindrical cup with a split base, wherein a central disc that is flush with the bottom rotates, while all other walls of the device are stationary [1]. Shear originates from the split between this disc and the rest of the base. This device has been widely used to validate rheological models for granular flows since it exhibits wide shear bands. We show that by changing the fill height of the split-Bottom Couette device, we change the location of shear in the material. We then show the presence of large, system spanning vortices that appear in addition to the primary azimuthal flow. These vortices are like those seen in the cylindrical Couette device [2]. We show how the number and form of these vortices are highly dependent on the where the system shears. We show that the form of the vortices can be explained by accounting for shear-induced dilation in the system, validating the arguments made for the vortices seen in the Cylindrical Couette device [2]. It was shown by [2] that these vortices have a large rheological signature, wherein the components of stress increase exponentially with depth in a sheared Couette device [3]. None of the current rheological models for slow granular flow can capture these vortices because none of them allow for shear-induced dilation. We thus make the argument that dilation needs to be incorporated in any general rheological model for slow granular flows.

References

1. Dijksman, Joshua A., and Martin van Hecke. "Granular flows in split-bottom geometries." Soft Matter 6.13 (2010): 2901-2907.
2. Krishnaraj, K. P., & Nott, P. R. (2016). A dilation-driven vortex flow in sheared granular materials explains a rheometric anomaly. Nature communications, 7, 10630.
3. Gutam, K. J., Mehandia, V., & Nott, P. R. (2013). Rheometry of granular materials in cylindrical Couette cells: anomalous stress caused by gravity and shear. Physics of Fluids, 25(7), 070602.

568895 Study of Hydrodynamic Behavior in a Shallow Rectangular Spouted Bed with Ascending Gas Flow Using Bed Pressure Signals and Image Analysis

The hydrodynamic characteristics of the spouted bed play an important role in the whole bed's practical performance. The Thermal Science Team in U.S. Department of Energy's National Energy Technology Laboratory (NETL) has performed many experimental study based on the spouted bed reactors design. As a further investigation, this work combines bed pressure signals and image processing methods to study the hydrodynamic behavior in a 30.2 mm × 101.6 mm rectangular spouted bed for four different kinds of particles. The rectangular configuration is chosen due to its particular advantage in heat transfer and reaction control especially in the reactor development including multiple combustion and reaction chambers. In addition, considering the design and scaling up of the spouted bed, the ascending air flow process, which is thought to be more close to the actual industrial conditions, is applied in the experiments. First, the internal and the external spouting velocities are determined from the bed pressure drop and the bed expansion ratio curves, respectively. Then, the relationship between the internal and external spouting velocities for the four kinds of particles are obtained. At last, the correlation to predict the external spouting velocity is discussed. The present work aims to perform a fully experimental investigation on the hydrodynamic characteristics in a rectangular spouted bed so as to provide more data and guidance for the following reactor design and development.

569591 Effect of Baffles on the Flow Regime, Transition Velocities and Void Characteristics in a 6 in. Diameter Fluidized Bed

Fluidized beds are often fitted with internals that come in various shapes and arrangements. Internals such as sheds are used to improve flow behavior of solids, by distributing the bubbles across the column diameter. This study highlights the importance and advantages of fluid bed operation with baffles on the hydrodynamics of the bed. Cross hatched sheds were used as internals that were mounted inside a 6 in (15 cm) diameter fluidized bed. The unit was operated under ambient conditions with Geldart Group A FCC catalyst. An optical fiber probe was used to characterize the bubble and slug rise velocities. The identification of flow regimes was accomplished through high frequency differential pressure transmitters that were mounted across a 3ft section of the dense bed. The bed hydrodynamics were then compared and contrasted to an open bed at superficial gas velocities ranging from 1 ft/s to 6 fts/s. The sheds reduced the bed transition velocity (U_c) to the turbulent regime and bypassed the slugging regime that was dominant in an open bed. The pressure fluctuations were orders of magnitude lesser in the bed with internals suggesting their role in promoting uniform fluidization. Finally, the slug rise velocities in an open bed were widely different from the bubble rise velocities with the internals at the same bed superficial gas velocity. This is an important parameter for fast catalytic reactions that are mass transfer limited. The gas solids contacting time in various fluidization regimes can largely differ, affecting the yield of the reaction, hence care must be taken during scaleup where a pilot plant may operate in a different regime than a commercial reactor. The experimental data was then compared with previous literature studies to see if the correlations can be used to predict the bed hydrodynamics

568724 Hydrodynamic Analysis of Spouting Characteristics in a Slot-Rectangular Spouted Bed Reactor

2019 AIChE Annual Meeting

Topic/Group: Fluidization and Fluid-Particle Systems

Tentative Session: Experimental Investigation of Fluidization Processes

Title: Hydrodynamic Analysis of Spouting Characteristics in a Slot-Rectangular Spouted Bed Reactor

Authors:

Steven L. Rowan: Steven.Rowan@netl.doe.gov

Ronald W. Breault: Ronald.Breault@netl.doe.gov

Narasimhan Soundarrajan: Narasimhan.Soundarrajan@netl.doe.gov

Jingsi Yang: Jingi.Yang@netl.doe.gov

Samuel Bayham: Samuel.Bayham@netl.doe.gov

Key Words: Fluidization, Spouting, Reactors, Energy

Abstract:

An investigation of the dynamic behavior of a spouted bed reactor being designed at the U.S. Department of Energy's National Energy Technology Laboratory is currently being conducted. Previously, high speed video and time series pressure data analysis was used to identify and characterize several internal and external spouting regimes in a transparent, small scale cold flow model. This paper presents an analysis of the effects of high operating temperatures upon the hydrodynamics of these varied spouting regimes via examination of pressure data, to provide a better understanding of, and ability to recognize similar spouting conditions in a non-transparent, high temperature spouted bed reactor. The time series pressure data are analyzed via classical statistical methods, wavelet extraction, and recurrence quantification analysis techniques. The tests were conducted with 14/40 mesh (~400-1400 μm) Duralum AB hollow alumina spheres in a 1" x 4" transparent cold flow spouted bed and a 2" x 8" non-transparent high temperature spouted bed reactor. The results of this analysis show that it is possible to utilize data obtained from the cold flow unit to characterize the spouting dynamics occurring in the high temperature reactor unit.

564965 Development, Validation and Application of a New Methodology for the Measurements of Particle Stresses in an Aerated Bed

Fluidised beds are one of the most widespread methodologies in the chemical engineering industry, they can be found in various applications such as fluidised bed reactors, combustors and dryers [1]. However, despite particle stresses are of high importance to understand the hydrodynamics of fluidised bed, methodologies that allow the accurate evaluation of the particle stresses are lacking [2]. Colafigli et al., [3] developed a device that allows for the evaluation of the particle stresses and thus the viscosity in a homogeneous gas-fluidised bed using silica particles. The measurement principle of such a device is based on the Coaxial-Cylinder-Viscometer for liquids. The Coaxial-Cylinder-Viscometer (CCV) is a well-known device used to evaluate the physical characteristics of fluids. This instrument consists of a rotating inner cylinder, the so-called bob, and a static outer cylinder or cup. The fluid is placed in the annular gap between the two cylinders. The apparent torque, which can be measured during the rotation of the inner cylinder, can be converted to dynamic viscosity and shear stress [4]. The main objective of this work is to confirm whether it is possible to measure particle stresses in fluidised beds using the same measurement principle as the one used in a CCV for liquids.

In our work, the Freeman FT4 Powder Rheometer aeration test was used as a smaller version of the CCV, where the original 48 mm rotating impeller was replaced with a new cylindrical cell designed for this study. This cell, which is 70 mm in height, 36 mm in diameter and composed of 6 blades is 3D printed using plastic material. It is placed inside the FT4 cylindrical cup and is analogous to the VCC with a 7 mm gap (Marchal et al., [5]). The cell is then rotated at different rotational velocities, and the torque needed to move the aerated powder placed in the cup is evaluated. In order to validate this new FT4-VCC combination, it was firstly tested with glycerol 99% which is a Newtonian fluid with a known viscosity. The corresponding torque was registered as the shear rate ý was raised from 0.45 to 48 s^{-1}. The torque was then converted to shear stress τ using the relations given by Marchal et al., [5], this while assuming a cylindrical shear zone in the VCC and a homogeneous stress distribution along the devices height. The obtained shear stresses appeared to increase proportionally with the imposed shear rate, following the equation, with being the viscosity of the fluid, which was found to be in good agreement with the known viscosity of glycerol 99%.

The VCC was then employed using monodispersed (500 mm diameter) spherical glass beads. The corresponding torque was registered as the shear rate *ý* was raised from 1.5 to 32 s^{-1} and the air velocity was increased from 0 to 100 $mm.s^{-1}$. The first results obtained with the VCC with no aeration showed that the shear stress is independent of the increasing shear rate. The corresponding viscosity η is also evaluated as the ratio of the shear stress to the shear rate, and the obtained results are in good agreement with the one obtained by Marchal et al., [5], using the same type of particles. Furthermore, the results obtained with the air velocities used in this study showed that the glass beads shear stress increases with shear rate, showing a Bingham plastic behaviour ($\tau=\tau_0+\eta\acute{y}$ where τ_0 is the yield stress). Further measurements including different air velocities and lower shear rates will be performed in order to confirm the first measurements and observations.

References

[1] M. S. Ray, "Chemical Engineering, Volume 2: Particle Technology and Separation Processes, 4th edn, by J.M. Coulson and J.F. Richardson. Pergamon Press, Oxford, UK. 1991. 968 pp. ISBN 0-08-037957-5," *Dev. Chem. Eng. Miner. Process.*, vol. 1, no. 2–3, pp. 172–172.

[2] R. Borghi and F. Anselmet, *Turbulent Multiphase Flows with Heat and Mass Transfer*. John Wiley & Sons, 2013.

[3] A. Colafigli, L. Mazzei, P. Lettieri, and L. Gibilaro, "Apparent viscosity measurements in a

homogeneous gas-fluidized bed," *Chem. Eng. Sci.*, vol. 64, no. 1, pp. 144–152, Jan. 2009.

[4] D. S. Viswanath, T. Ghosh, D. H. L. Prasad, N. V. K. Dutt, and K. Y. Rani, *Viscosity of Liquids: Theory, Estimation, Experiment, and Data*. Springer Netherlands, 2007.

[5] P. Marchal, N. Smirani, and L. Choplin, "Rheology of dense-phase vibrated powders and molecular analogies," *J. Rheol.*, vol. 53, no. 1, pp. 1–29, Dec. 2008.

584745 Advancing Cross-Scale Characterization of Reactive Fluid-Solid Interactions: Accelerated Carbon Mineralization for Integrated CO_2 Capture, Utilization and Storage

To achieve predictive controls on reactive fluid-solid interactions, the advancement of *in-operando* cross-scale characterization methods is essential. This approach allows us to simultaneously characterize dynamic chemical and morphological changes in materials in fluidic environments and identify transient and metastable states of matter at far-from-equilibrium conditions. In this study, we harness synchrotron *in-operando* Ultra-Small, Small, and Wide Angle X-Ray Scattering (USAXS/SAXS/WAXS) and X-ray tomography measurements to elucidate the microstructural and structural features in materials from the sub-nanometer to the micrometer scales. We harness this approach to elucidate the underlying mechanisms associated with the capture, conversion and storage of CO_2 as Ca- and Mg-bearing carbonates via carbon mineralization. These characterization approaches provide insights into new approaches to design multiphase reaction environments in gas-solid, gas-liquid-solid, and liquid-solid configuration modes for accelerated carbon mineralization.

579549 Dust Particle Flow-Field Characterization Using High-Speed Digital In-Line Holography (DIH) During Minimum-Ignition Energy Testing

Improvement of dust explosion risk assessment and mitigation strategies in the process industries is a continual pursuit to reduces the frequency and severity of hazardous explosions. To improve the current understanding of dust ignitibility during laboratory-scale tests, new diagnostic techniques are required to enable the imaging and characterization of microscopic dust particles in the vicinity of the ignition spark in minimum ignition energy (MIE) testing device. In this study, high-speed digital in-line holography (DIH) is employed to obtain quantitative particle size, concentration, and velocity information from dust particles in a Kühner MIKE 3 MIE testing apparatus prior to ignition. A high-speed DIH experimental system for the volumetric, non-intrusive, and in-situ imaging and characterization of various dusts dispersed at the standard MIE test pressure are presented. Interference patterns created by suspended micron-size dust particles upon illumination of a monochromatic laser light source are imaged using a high-frame-rate CMOS camera coupled to a high-magnification lese system. Post processing of digitally refocused images provide visual characterization of the particle field, while resulting particle size and velocity distribution data quantify the central tendency and variability of each dispersion. The time evolutions of particle concentration and velocities are also presented to provide transient characterization of each dust. These results provide new quantitative dust particle data and establish a method for identifying microscopic dispersion conditions that can be related to dust cloud ignitibility and the MIE.

579761 Influence of Particle Density on Pre-Ignition Dust Cloud Dynamics in a Minimum Ignition Energy Testing Apparatus Using Digital In-Line Holography

The ASTM standard for the measurement of minimum ignition energy (MIE) of a dust cloud in air (E2019) calls for the generation of a uniform, non-turbulent (stable) dust cloud of known concentration at the time of ignition in the 1.2L Hartmann tube. The dust dispersion system employed in the MIE apparatus, generally a Kühner MIKE3 device, uses a fixed pneumatic conveyance with a pressure of 7 bar. Ignition spark delay, which is the time from the initiation of dispersion to the point of ignition, is an adjustable parameter. However, most operators usually keep this value at 120 ms. Combustible industrial dusts have a wide variety of particle density. The cloud dynamics (particle velocity, concentration, turbulence) for dusts of different particle density are expected to vary at the time of igniton when dispersed by the pneumatic system. The change in cloud dynamics can affect the MIE testing conditions.

In this study, soda lime glass and polymethyl methacrylate (PMMA), having a ratio of particle density of 2:1, are dispersed in the Kühner MIKE3 apparatus while decoupling parameters such as particle size distribution and shape. A high-speed digital in-line holography (DIH) imaging system is used to obtain measurements of the velocity field, particle size, and concentration in three dimensions around the ignition zone for each dispersed dust sample.

The quantitative results obtained showcase the pre-ignition cloud dynamics and establish dusts of different particle densities behave differently in the ignition region and at the typically selected ignition delay time. Hence, to obtain a consistent dust cloud for optimum MIE measurements in terms of turbulence and concentration around the ignition zone, use of different ignition delays is warranted for the dusts of different particle densities.

579551 Effect of Particle Morphology on Dust Cloud Dynamics

Dust explosions are a serious threat to the process industry. To tackle the problem of dust explosion, many studies have investigated the factors that influence dust explosion parameters. These studies list dust size, chemical composition, polydispersity etc. as major factors that affect dust explosions, but particle shape or morphology has often been overlooked. A recent study has shown particle shape impacts the minimum ignition energy (MIE) of a dust. This is because a higher specific surface area leads to lower conductive heat resistance, thereby enhancing ignition. While particle shape affects the heat transfer process which impacts the dust MIE, another important characteristic influenced by particle shape is cloud dynamics (cloud concentration, turbulence etc.). To understand the influence of particle shape on dust explosion parameters, understanding of both heat transfer and dust cloud dynamics as a function of particle shape is necessary. Even though attention has been given to the heat transfer aspect, work is required to understand the influence of particle morphology on the cloud aerodynamics.

In this work, a high-speed digital in-line holography (DIH) method is employed to determine the flow field of aluminum particles when dispersed in the Kühner MIKE3 MIE apparatus. This technique helps identify the impact of particle morphology (spherical and irregular) on dispersion characteristics such as velocity and volume density within the ignition zone.

Understanding cloud dynamics and identifying the role of particle morphology on dust cloud characteristics is crucial in determining the subsequent effect on explosion parameters such as MIE and the resulting risk assessment for solids processing.

579550 Minimum Ignition Energy (MIE) prediction from QSPR using machine learning

Dust and gas explosions pose a serious hazard to process industries, resulting in loss of life, property and resources. The Minimum Ignition Energy (MIE), defined as the smallest amount of energy required to ignite a flammable compound at a given temperature and pressure, is a critical safety parameter of significant importance. However, often the high value materials of concern are only available in small quantities, making destructive testing impractical during early stages of development. Because of this prediction methods should be explored and advanced.

In this study, the MIE prediction of 60 flammable hydrocarbon compounds has been conducted using a Quantitative Structure-Property Relationship (QSPR) and machine learning techniques. The prediction models were developed using Random Forests (RF), Decision Trees (DT) and Artificial Neural Networks (ANN) resulting in promising (> 0.70) R^2 values for the test sets. Decision trees were used to identify the 10 most important molecular descriptors influencing the MIE prediction model accuracy. In addition, a Genetic Function Approximation (GFA) algorithm in Materials Studio was used to develop a 10 parameter MIE prediction equation resulting in significant R^2 value. The GFA, RF and DT algorithms resulted in a more robust MIE prediction model as compared to ANN algorithm.

560052 The Characterization and Use of Limestone Rock Dusts in Coal Mines to Prevent Dust Explosions

The Characterization and Use of Limestone Rock Dusts in Coal Mines to Prevent Dust Explosions – I.A. Zlochower

Coal dust generated during the extraction of coal from underground coal seams represents a serious explosion hazard. The prevalence of high-shear equipment in Longwall mining produces finer coal dust and increases the dust explosion concern. Limestone rock dust has been used for many decades in coal mines to provide an inert and heat absorbing component to the mine dust such that the dust becomes non-explosible. Requirements on the amount and characterization of such rock dust has become more stringent in recent years. Now mine dust must be maintained at an 80% incombustible content. Recommendations to increase the fine fraction of the rock dust from 70% passing through 200 mesh sieves and to attain a minimum specific surface area have been made. This presentation will cover recent work by the Pittsburgh Mining Research & Development Branch of the National Institute of Safety and Health (NIOSH) on rock dust characterization for such use in coal mines.

553178 Mitigation of Iron and Aluminum Dust Deflagrations Via Active Explosion Suppression in 1-m^3 Sphere Vessel

Explosions induced by ignition of combustible metal powders continue to present a significant threat to metal handling and refining industries. Addition of non-combustible inert material to combustible dust mixtures, either through pre-mixing or high-rate injection as the incipient flame front begins to develop, is common practice for preventative inhibition or explosion protection via active suppression, respectively. Metal dusts demonstrate an extremely reactive explosion risk due to amplified heat of combustion, burning temperature, flame speed, explosibility parameters (K_{St} and P_{max}), and ignition sensitivity. Inhibition efficiency of suppressant agents utilized for active mitigation is shown to be reliant on fuel explosibility, discrete burning mechanism, and combustion temperature range, and thus may be increasingly variable depending on the fuel in question. For this reason, mitigation of metal powder deflagrations at moderate suppression pressures (compared to the strength of the enclosure) and at low agent concentrations remains challenging. This paper reviews recent metal dust suppression testing in Fike Corporation's 1-m^3 sphere combustion chamber, and evaluates the efficacy of multiple suppression agents (sodium bicarbonate [SBC], monoammonium phosphate [MAP], and sodium chloride [Met-L-X]) for the mitigation of iron and aluminum powder deflagrations at suspended fuel concentrations of 2,250 g/m^3 and 250 g/m^3, respectively.

564711 Are Crosslinked Polystyrene Beads As Effective As Zirconia Beads to Produce Drug Nanoparticles Via Wet Stirred Media Milling?

Nearly 40% of approved drugs and 90% in the discovery phase show poor water solubility and as a result low bioavailability [1]. A promising way of overcoming these challenges is by reducing the particle size to nanoscale, which results in an increase in drug surface area and dissolution rate. Wet stirred media milling (WSMM) has proven itself to be a robust top-down method for producing nanoparticles. However, the WSMM process is energy intensive, costly, time-consuming, and can potentially cause product contamination from bead wear. Developing a fundamental understanding of the impact of milling process parameters on particle size, breakage kinetics, and cycle time is critical to resolving some of the aforementioned issues as well as developing a robust milling process. To this end, Afolabi et al. [2] elucidated the impact of stirrer speed, concentration of zirconia beads, and drug loading; Li et al. [3] studied the impact of zirconia bead size. Both studies used a microhydrodynamic model to interpret the impact using microscale parameters such as zirconia beads oscillation velocity and frequency. Crosslinked polystyrene (CPS) beads are also widely used in pharmaceutical WSMM; but, the WSMM process with such beads has not been subjected to the same rigorous and systematic analysis as for that with the zirconia beads. Moreover, the question as to whether CPS beads are as effective as zirconia beads has not been answered in a controlled, systematic study.

We aim to gain a fundamental understanding of the impact of stirrer speed and volumetric loading of CPS beads on breakage kinetics during WSMM and compare the performance of these beads to zirconia beads. Griseofulvin (GF), a Biopharmaceutical Classification System class II drug, was chosen as the model drug for this study. Suspensions of GF were prepared using two stabilizers, hydroxypropyl cellulose (HPC) SL grade and an anionic surfactant, sodium dodecyl sulfate (SDS). Previous studies [2-4] have shown that the combination of HPC and SDS creates a synergistic effect that ensures stability of the GF suspensions. This excellent stability of the GF nanosuspensions allowed us to solely isolate the impact of process parameters on the breakage kinetics, without much confounding from particle aggregation. CPS beads with a nominal size of 400 μm was used as the milling media and compared with Zirmil Y grade yttrium-stabilized zirconia (YSZ) beads with identical nominal size. Four different bead volume fractions/loadings (c), i.e., 0.198, 0.298, 0.397, and 0.594, with respect to the total volume (80 mL) of the mill chamber were used. About 230 g pre-suspensions were prepared by dispersing 10% GF particles in an aqueous solution of 5% HPC and 0.2% SDS, where all the percentages were w/w with respect to de-ionized water (200 g). The pre-suspensions were milled for 256 min in a Netzsch Microcer recirculation mill at the stirrer speeds of 2000, 3000, and 4000 rpm with the aforementioned bead loadings. Samples were taken during milling at fixed times, particles size measurement of the samples were then determined using laser diffraction (Coulter Beckman LS 13 320). Viscometry (Brookfield R/S plus cylinder-in-cylinder rheometer) was used to characterize the apparent shear viscosity.

To quantify the breakage kinetics, time for the 50 vol.% passing size (median, d_{50}) to reach 0.5 μm (t_{d50}) and time for 90 vol.% passing size (d_{90}) to reach 1 μm (t_{d90}) were calculated via the Hermite interpolation. Moreover, an exponential decay model was fitted to the time-wise evolution of the median size and a characteristic process time constant τ_p was determined. Average power consumption, apparent shear viscosity, and density were measured for each milled suspension. Using a microhydrodynamic model developed by Eskin et al. [5], various microhydrodynamic parameters were determined such as granular temperature, average bead oscillation velocity, frequency of single bead oscillation, maximum contact pressure, and average frequency of drug particle compression. The breakage data and kinetics were then further analyzed with the insight of these parameters.

GF nanosuspensions with $d_{50} < 200$ nm were produced within 256 min using CPS beads for all stirrer speed–bead loadings (c) except for 2000 rpm and $c < 0.4$. Their sizes did not change after 7-day storage. This suggests that (i) aggregation was effectively suppressed by the stabilizers and (ii) CPS beads can be used to prepare GF nanosuspensions successfully. All three characteristic milling times t_{d50}, t_{d90}, and τ_p decreased with an increase in stirred speed and beads loading monotonically, indicating faster breakage. YSZ beads exhibited similar trends [2]; however, they appear to have caused faster breakage than CPS beads of "similar sizes" (results will not be shown for brevity). The major findings from the microhydrodynamic analysis can be summarized as follows: it was observed that as stirrer speed increased, all microhydrodynamic parameters increased monotonically. An increase in stirrer speed led to an increase in the mechanical energy that contributed to the bead–bead collisions. This resulted in more forceful bead–bead collision during drug compression and more frequent bead–bead collisions, which explains the observed faster breakage. When bead loading was varied, counteracting effects were observed. As volumetric bead loading increased, bead-bead oscillation and frequency of drug particle compressions increased, which favors faster breakage. However, less energetic/forceful collisions and compressions occurred. Apparently, the former effect was dominant, and the higher beads loading led to faster breakage overall. A systematic comparison of the performance of CPS beads and YSZ beads of the same nominal size was also performed. In this comparison, volumetric bead loading was kept constant and stirrer speed was varied. YSZ beads led to faster breakage and higher microhydrodynamic parameters than those of CPS beads under identical conditions. However, as stirrer speed increased, the process time constants associated with CPS beads and YSZ beads began to converge, and their impact on the breakage kinetics becomes similar. Overall, we conclude that while YSZ beads are more effective for breakage than CPS beads for most experimental conditions studied, at the highest stirrer speed (4000 rpm) and highly energetic conditions, both types of beads appear to perform well and similarly. From a process engineering perspective, both types of beads can be used effectively to produce drug nanoparticles; however, they may have different extent/types of product contamination, which could be a determining factor in the selection of bead type. This aspect has not been investigated in this study and warrants future investigation.

References:

1. Kalepu; V. Nekkanti, Insoluble drug delivery strategies: review of recent advances and business prospects, Acta Pharm. Sinica B, 5 (2015) 442–453.
2. Afolabi, O. Akinlabi, E. Bilgili, Impact of process parameters on the breakage kinetics of poorly water-soluble drugs during wet media milling: A microhydrodynamic view, European Journal of Pharmaceutical Sciences, 51 (2014) 75–86.
3. Li, P. Alvarez, E. Bilgili, A microhydrodynamic rationale for selection of bead size in preparation of drug nanosuspensions via wet stirred media milling, International Journal of Pharmaceutics, 524 (2017) 178–192.
4. Li, N. Yaragudi, A. Afolabi, R. Dave, E. Bilgili, Sub-100 nm drug particle suspensions prepared via wet milling with low bead contamination through novel process intensification, Chemical Engineering Sciences, 130 (2015) 207–220.
5. Eskin, O. Zhupanska, R. Hamey, B. Moudgil, B. Scarlett, 2005b. Microhydrodynamic analysis of nanogrinding in stirred media mills. AIChE Journal. 51 (2005) 1346–1358.

568092 Understanding the Flow of Calcium Carbonate in Stirred Media Mills

Stirred media mills are used extensively in industry for the fine grinding of minerals such as calcium carbonate. For use in products such as paints and papers, calcium carbonate with a d_{80} of approximately 2 micrometers is required. However, fine grinding has a low energy efficiency due to energy dissipation as heat, noise or vibrations. Hence, there is potential for a lot of improvements to be made by changing the design or operating parameters. One way of determining which parameters to change is through the investigation of flow patterns within the mill. Knowledge of the motion of grinding media and calcium carbonate particles in different zones of the mill is indicative of the grinding intensity in each zone.

Using experimental methods, flow patterns in stirred media mills have previously been visualised using positron emission particle tracking (PEPT). A grinding media bead is irradiated and added to a batch vessel [1-2]. However, usually only one grinding media bead is tracked, which means that experimental run times are long and an average occupancy plot or velocity distribution is obtained. Additionally, only the grinding media is tracked, so there is no information about the motion or shear rates of the calcium carbonate slurry.

This research looks at the potential use of particle image velocimetry (PIV) in stirred media mills to overcome the disadvantages of PEPT and provide higher resolution flow patterns. Using PIV, it is possible to look at the flow of either the grinding media or smaller seeding particles which follow the flow of the mimic fluid. The main challenge of using PIV is the requirement for a completely transparent set-up. A Perspex vessel has been produced, several mimic fluids have been tested for use in place of the calcium carbonate slurry and glass beads have been used in place of the ceramic grinding media. If the refractive index of the mimic fluid matches that of the grinding media, the grinding media are not seen in PIV experiments, enabling high quality images to be obtained.

Clove oil, cedarwood oil and refractive index adapted silicone oil are Newtonian fluids with refractive indices matching that of glass beads that could be used to represent grinding media. Additionally, their viscosities cover a wide range, enabling the effect of viscosity on flow patterns to be investigated. Differences between the transparent and actual milling set-ups may lead to PIV data being invalid for real mills. The main differences in the set-ups are the density of the grinding media and the density, viscosity and friction coefficient of the slurry. All of the transparent fluids have significantly lower densities than calcium carbonate slurries. The frictional properties of the fluids are also different, as has been confirmed using tribology techniques. The glass grinding media are of similar density to Carbolite ceramic grinding media. The effect of these parameters on the power draw of the mill have been investigated. By using glycerol-water solutions of varying concentration in place of the calcium carbonate slurries, it is possible to test fluids of vastly different viscosities and friction factors by measuring the power draw of the mill at a constant rotational speed.

To determine the extent to which the selected mimic fluids are representative of a calcium carbonate slurry, the viscosity of calcium carbonate slurries and how it changes during grinding has been investigated. Slurry viscosity is of high importance in grinding processes since it affects the amount of viscous energy dissipation occurring during grinding media collisions. Due to the reduction in particle size, viscosity changes throughout grinding, meaning that flow patterns also change.

Calcium carbonate slurries for use in the paint and paper industries have a very high solids content (typically >70 wt%) and so viscosity has been measured using a vane rheometer, since this is an effective way of eliminating wall slip [3]. Additionally, the shear rate in the mill is unknown and varies throughout the mill, so the viscosity was measured over a range of shear rates.

The viscosity of calcium carbonate slurries with 72 wt% solids has been measured using the vane rheometer, as is shown in figure 1. Since viscosity increases during milling as particle size decreases, the viscosity was measured after various amounts of energy had been consumed. The measurements were made at the temperature that the mill had reached when the sample was taken.

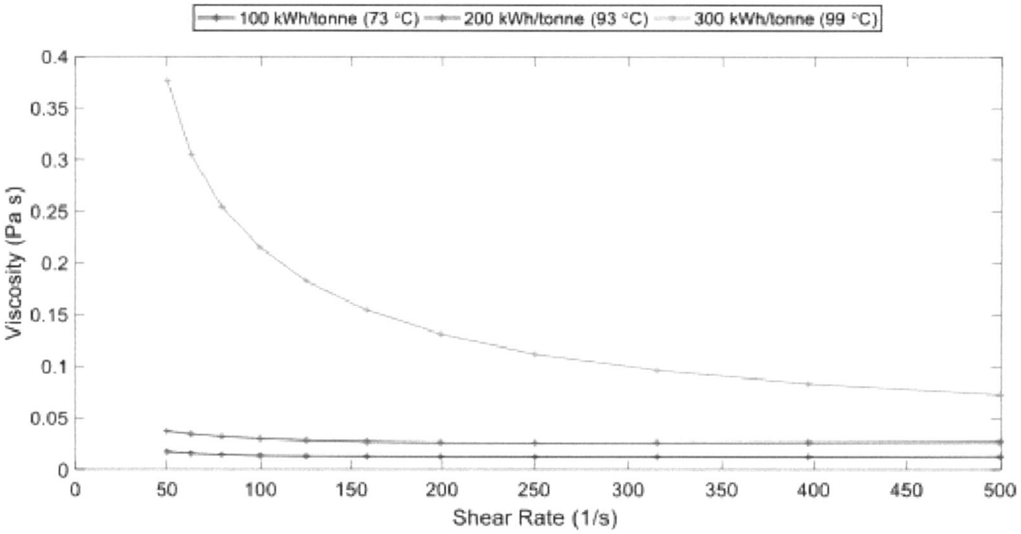

Fig. 1. Viscosity of calcium carbonate slurries after energy input by milling (energy input is per tonne of calcium carbonate being ground)

The viscosity of 72 wt% slurries during milling ranges from 0.01 Pa s to 0.4 Pa s. By using a selection of transparent mimic fluids with different viscosities within this range, the development of flow patterns over the course of a batch grind could be determined.

By using transparent mimic fluids with a refractive index matching that of glass beads, PIV can be used to experimentally determine flow patterns of grinding beads or slurry particles within stirred media mills. However, the differences between the density, viscosity and friction coefficient of the fluids and grinding media used in the transparent and actual milling set-ups mean that the results are not directly related to a specific milling set-up. Despite this, PIV can still provide valuable information about how flow patterns change with parameters such as viscosity, grinding media size and amount and impeller speed.

References

1. Barley RW, Conway-Baker J, Pascoe RD, Kostuch J, McLoughlin B, Parker DJ. (2002) Measurement of the motion of grinding media in a vertically stirred mill using positron emission particle tracking (PEPT). Minerals Engineering, 15, 53–59.

2. Van Der Westhuizen AP, Govender I, Mainza AN, Rubenstein J. (2011) Tracking the motion of media particles inside an IsaMill™ using PEPT. Minerals Engineering, 24(3–4), 195–204.

3. Pierre, A. Perrot, A. Histace, S. Gharsalli, E.H. Kadri, A. (2017) Study on the Limitations of a Vane Rheometer for Mineral Suspensions Using Image Processing, Rheologica Acta, 556, 351-367

*** Formatting Issue - Best Available Paper/Graphic ***

571740 Time and Energy Dependent Properties of Stirred Wet Milled Alumina-Doped Titanium Dioxide, a Discretely Heterogeneous System

Time and Energy Dependent Properties of Stirred Wet Milled Alumina-doped Titanium Dioxide, a Discretely Heterogeneous System

David Austin[1], Timothy Hunter[1], Ali Hassanpour[1], Stephen Sutcliffe[2], John Edwards[2], David Harbottle[1†]*

1) School of Chemical and Process Engineering, University of Leeds, United Kingdom; 2) Venator Ltd.

Titanium dioxide is widely used in coatings, inks and plastics due to the high refractive index of the two polymorphs, rutile (R.I. = 2.74) and anatase (R.I. = 2.54). Titania is often manufactured via the chloride process, during which, up to 5 mol% alumina trichloride ($AlCl_3$) is added to promote the formation of the rutile polymorph. This co-addition has led to questions around the location of alumina in the primary particles. Garbassi *et al.*[1] determined the alumina to be located in the bulk, not being observed via X-ray photoelectron spectroscopy (XPS), or through acid leaching experiments, whilst Taylor *et al.*[2] showed that with increased dopant concentrations, both the surface and bulk concentrations increased, with comparatively more alumina present at the surface.

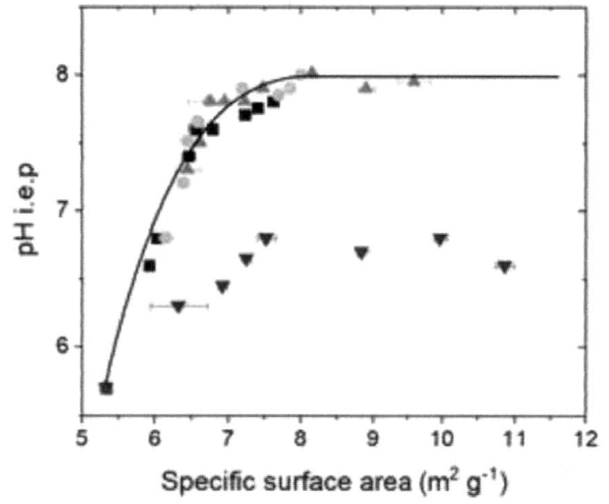

Fig. 1 pH isoelectric point of alumina doped TiO_2 milled at the stirrer speeds of 2500 to 8000 rpm, as a function of the specific surface area of the samples (determined by gaseous nitrogen adsorption). Line is drawn to guide the eye.

With the desired particle size for pigments being sub-micron, $d_{50} \sim 0.3$ mm, ultra-fine grinding is achieved by stirred wet milling. In the current study, we

***** Formatting Issue - Best Available Paper/Graphic *****

have considered the surface properties of milled alumina-doped titania. By measuring the zeta potential of the milled particles, we have shown that the particle surface properties change during milling, with the particle iso-electric point (i.e.p) shifting to higher pH values (Fig. 1). While there is good agreement for milling speeds of 2500 to 6000 rpm, the pH i.e.p. for the particles milled at 8000 rpm remains lower, albeit the specific surface area of the milled particles is consistent with lower rpm values. At 8000 rpm, reduced 1st order grinding kinetics were determined from laser diffraction analysis, suggesting a change in the grinding behavior from bead-bead impact breakage (lower rpm) to bead-bead and bead-wall shear breakage; a consequence of the relationship between turbulent dispersion and centrifugal forces with milling speed. While the high shear forces may cause mechanochemical changes, X-ray diffraction analysis showed no variance between samples (Figs. 2a and b).

Further analysis of the milled samples using XPS revealed that the changing pH i.e.p. during milling correlated linearly to the atomic % of surface alumina, ranging from 2.7 to 9.8 at. % over the pH i.e.p range of 5.7 to 8.1 (2500 – 6000 rpm). At 8000 rpm, dehydroxylation of surface sites is hypothesized, altering the zeta potential behavior, as described in the high temperature treated surface chemistry of a-alumina and rutile titania work of Yang[3] and Sedev *et al.*[4] respectively – this is an area of ongoing analysis.

This study has demonstrated the evolving surfaces properties of alumina-doped titania particles during milling, and work is ongoing to fabricate model, nanoscopically smooth composite surfaces via atomic atomic layer deposition (ALD), which can then be used to better understand the surface interactions with milling dispersants and their influence on particle lubrication/dispersion.

Fig. 2a) XRD pattern of an as received alumina doped titanium dioxide sample with ICDD database reference pattern peaks for both rutile titanium dioxide (01-076-1939) and rutile alumina doped titanium dioxide (04-22-1458) overlaid. b) XRD pattern of an as received alumina doped titanium dioxide sample (bottom) compared with 60 min milled samples at the stirrer speeds of 2500 rpm (red), 4000 rpm (blue), 6000 rpm (pink) and 8000 rpm (green).

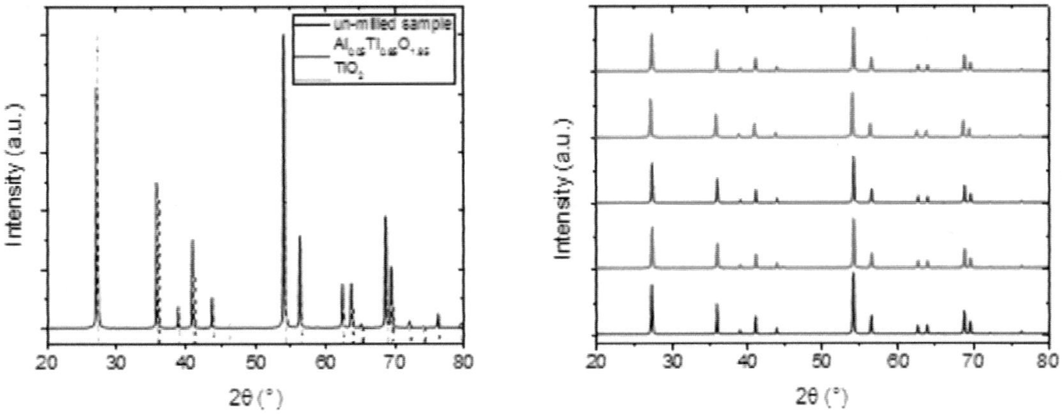

References

1 F. Garbassi, E. Melloceresa, E. Occhiello, L. Pozzi, M. Visca and D. M. Lenti, *Langmuir*, 1987, **3**, 173–179.

2 M. L. Taylor, G. E. Morris and R. S. C. Smart, *J. Colloid Interface Sci.*, 2003, **262**, 81–88.

3 D. Yang, M. Krasowska, R. Sedev and J. Ralston, *Phys. Chem. Chem. Phys.*, 2010, **12**, 13724.

*** Formatting Issue - Best Available Paper/Graphic ***

4 A. Kanta, R. Sedev and J. Ralston, *Langmuir*, 2005, **21**, 2400–2407.

From particle to process predicting breakage of structured particles

By: Kerry Johanson, PhD and Michael Brecht, MS
Material Flow Solutions

Abstract

The breakage of consumer products prior to packing or due to consumer use is a common issue, especially with structured particles. This is a complex problem spanning several size scales. On the particle scale, the structured particles can exhibit preferred breakage zones or modes due to the shape and morphology of the particles. Potato chips may break along the plane of greatest bending moment. Ice cream cones may break when the hoop stress around the perimeter of the cone exceeds a critical value. A bilayer tablet may break at the junction between two dissimilar materials. All of these breakage effects are particle structure effects that require an understanding of stresses and strains formed at the particle scale as well as solving FEM problems describing the distribution of stress in the structured particles. However, after the initial breakage of structured particles, the fragments may not have any preferred breakage modes or zones and standard breakage modeling can be useful in determining the secondary mode of breakage. Most modeling efforts in particle breakage are based on continuum descriptions of the breakage process involving many particles. However, a link must be developed between the particle scale – including structure breakage effects and when to apply them. To solve the breakage problem the effects of handling and processing on the breakage of particles must also be determined. In every process step, or during the consumer use step, the material undergoes a unique set of stress-strain events coupled with a unique set of impact events. Often these stress-strain and impact events can be computed given a description of the process. However, a link must be made between stress-strain events as well as impact events and particle breakage. If this link can be made, then the breakage in a prescribed process can be determined so that process steps to reduce breakage can be applied. Using both continuum and structured particle modeling to describe breakage, coupled with stress-strain and impact breakage events, engineers can fix breakage problems by changing the particle or by changing the process. The development of general methods and scientific frameworks to describe this multiscale effect is the discussion of the paper. The effect and use of breakage on a structured particle will be examined. The use of standard continuum-based methods for non-structured particles will be examined. The link between stress-strain and impact events on breakage will be examined, as well as the impact of processing on stress-strain and impact events.

1.0 Introduction

Understanding how particles break can aid a food engineer in designing the consumer product to mitigate this breakage. This knowledge can also help engineers design packages to minimize breakage during transport. Finally, this knowledge can help in understanding how handling process could be changed to minimize breakage that might happen. However, many food consumer products are complex structures where the breakage depends on the detailed configuration of the particles. In simple cases the particles themselves are made of a consistent material, but the shape of the consumer food product may minimize the breakage. Such is the case with potato chips. The basic material that goes into making a chip is relatively consistent and has well defined elastic and plastic properties. However, the structure of the chip itself can influence how the particle breaks. The placement of ridges or the saddle shape can influence the strength of the chip. Likewise, a pretzel consists of a relatively homogenous material, but the strength of the pretzel will depend on the number of junction points in the pretzel. The same is true with many cereal particles. The basic building block of the cereal particle is a consistent homogenous material, but the particle itself is constructed in such a way as to have a complex structure that may or may not be resistant to breakage. Examples of these types of particles are Chex® cereal where the material is uniform and relatively homogenous, but the structure is two curved mesh squares connected at the edges. Mini-Wheats® is constructed of individual fibers with relatively uniform consistency that are rolled and then pressed into a particle that is about 25 mm square. Breakage of these structured particles depends on the type of material in the primary particles as well as the general structure of the particle itself.

The goal of this project is to determine if the breakage can be explained and predicted using population balance models where the breakage selectivity coefficients are found from the structure of the particles and the breakage rate constants are found experimentally from control lab-scale breakage tests. Part of this work was to determine if these breakage tests could accurately predict the expected breakage that material may experience in a package when it is exposed to repeated handling either by a consumer or during transport.

2.0 Material and Methods

To accomplish this task the breakage characteristics of two selected structured particles subject to controlled stress-strain events were studied using a test method designed to subject the material to repeated stress-stain events. The particle size of the structured particles was measured before and after the stress-strain events. Population balance models were then used to determine the breakage rate and selectivity coefficients of structured particles. FEM analysis or simple solids mechanics models were then used on 3D models of individual particles to determine how a single particle might break. This analysis was used to determine the breakage selectivity coefficients that are part of the population balance model describing how a particle may break. Finally, breakage tests were done on these structured particles at a wide range of stress and strain events to generate the relationship between breakage and stress-strain events experienced in typical packages. The amount of stress and strain was then calculated on the material in typical packages and the amount of breakage was calculated based on the results of the breakage test. This was compared to breakage that actually occured when subjecting packages containing these structure particles to repeated stress-stain events.

When a particle is placed in a package and then transported it undergoes repeated stress-strain events. This causes a group of particles to interact, and the stress and strain imposed on the particles tend to bend or crush particles. Each stress-strain event can cause a particle somewhere in the bulk to break. The combined interaction of these bending and crushing steps on all particles is additive, creating finer particles from the well-designed structured particles. The breakage depends on both the stress applied to particles and also the total strain the particles experience while under the prescribed stress. It is possible for particles to simply break in half if the stress is great enough. However, even if the stress is not great enough to fracture the whole particle, the particles can strain or bend and break the structure of the particles causing smaller fragments of the structured particles to be liberated from the whole. One critical need is the ability to take a collection of particles and subject them to a controlled stress event while inducing a controlled amount of strain on the particles.

Particles were place in a specially prepared shear box where the bulk material can be subjected to known stress and then subjected to a known strain. Figure 1 shows a schematic of the stress-strain values used to measure the breakage of these structured particles. The material was riffled to split the sample into two identical halves. The particle size of half the material was measured to create a particle size distribution of the particles before being subjected to repeated stress-strain events. The second half of the sample was placed in the shear box and a load applied to the top material surface. The shear box is a rhombus-shaped box with hinges at the corners. The bottom of the box moves while the top of the box remains fixed in position. This action does not change the projected area of the box, but will induce a nearly perfect stain on the material in the box. The distance the bottom plate travels changes the angle of the side walls to reach some maximum angular value and can be used to compute the total strain caused by the motion of the sidewalls back and forth during one cycle. The number of times the plate goes back and forth can be used to subject the box to repeated and well-controlled stress-strain events. After subjecting the material to repeated stress-strain events, the particle size distribution was measured. Thus, the test method allows the measurement of breakage as a function of both stress and strain.

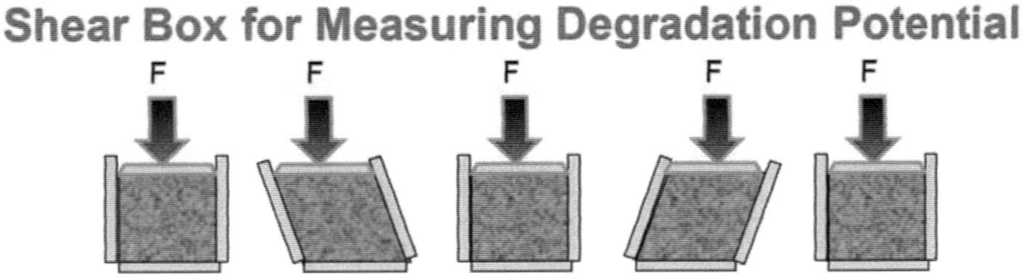

Figure 1. Schematic of stress-strain degradation box

Two materials were selected to measure the breakage as a function of stress-strain events. The first material selected was a bran fiber product. The bran fiber particle is long and cylindrical with a constant diameter of about 3.3 mm and a length of about 20 mm. The second material selected was a shredded wheat fiber material, rolled into a ball and then compressed before the packing process. The particle is a rolled mat of fibers about 26 mm by 22 mm by 13 mm, consisting of fibers about 0.8 to 0.9 mm in diameter (Figure 2).

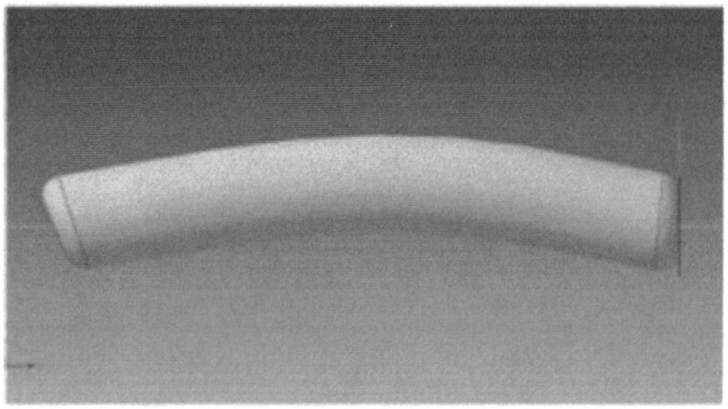

Bran Fiber

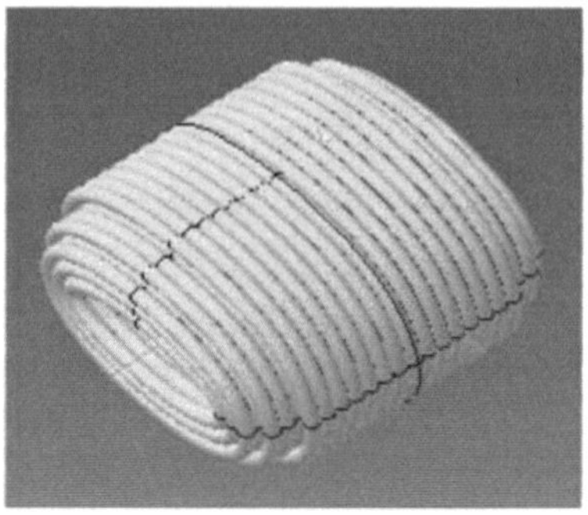

Shredded wheat

Figure 2. Schematic of typical particles studied in this paper

The common thread between these two materials is that the basic structure of the particles consists of fibers or cylinders. Part of this study is to determine if there is a relationship between the shape and structure of these two particles and a similarity concerning how they break.

3.0 Theory and Calculations

The framework of a population balance model was used to determine how these structured particles break. Consider a set of particles exposed to conditions that may cause some of them to break. This set of particles can be divided into different size bins of a given mass (m_i) and then mass balances on the broken particles and unbroken particles can be written. In this set of particles of a particular size, a certain fraction will remain unbroken. Of those that do break, they can distribute between size fractions of all particle size bins that are smaller than the original. This will give rise to a set of breakage selectivity constants ($b_{i,j}$) which determine what fraction of the broken particles become particles that would fit into each bin size smaller than the original. Each particle size bin will break at some rate or probability defined by a breakage rate constant S_j. Please note that the population balance model for breakage ignores any potential agglomeration of particles that would create particles larger than the original. This balance is summarized in Figure 3.

$$\frac{dm_i}{dt} = \sum_{j=1}^{j=i-1} \left[b(i,j) \times S_j \times m_j \right] - S_i \times m_i$$

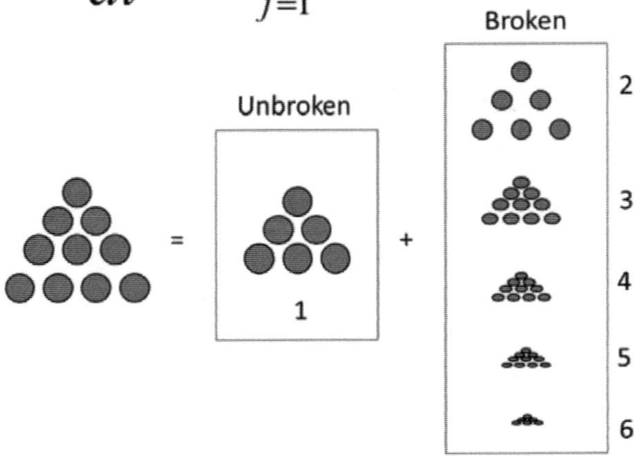

Figure 3. Schematic of the population balance model for breakage

Normally the bin sizes in a typical population balance model would be divided such that the next size down would contain particles roughly half the size of the previous bin size (i.e. the particle would break in half to enter the next bin size). However, most particle sizing methods report particles with equivalent spherical diameter. The particles studied in this paper are fibers and have a significant aspect ratio, so the bin sizes were chosen to be geometric with the next bin size having an equivalent spherical diameter of about 0.7 of the original. If the rate constants are constant over time for each particle size bin, then the formulation given by Kapur can give the approximate solution to the population balance model (Figure 4).

Size-discrete forms of the continuous PBM

$$\frac{dR_i(t)}{dt} = -S_i R_i(t) + \sum_{j=1}^{i-1} \left(S_{j+1} B_{ij+1} - S_j B_{ij}\right) R_j(t), \quad R_i(0) = R_{i0}$$

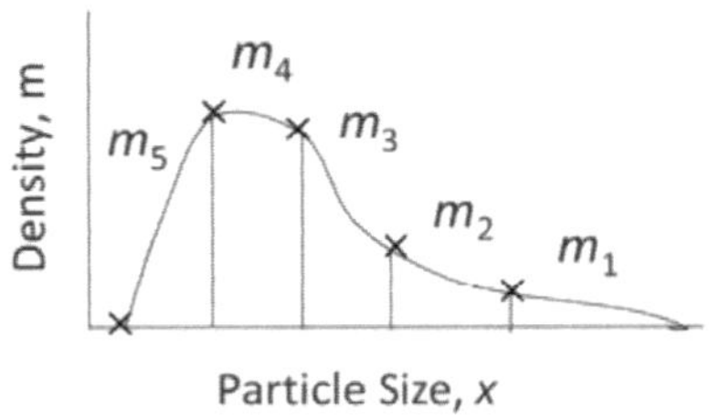

Chose the m's to be geometric with m2 being 0.7 the size of m1 etc....

Approximate solution to the cumulative formulation (Kapur, 1970)

$$R_i(t) = R_i(0)\exp\left(G_i t + H_i t^2/2\right)$$

$$G_i = -S_i + \sum_{j=1}^{i-1} \lambda_{ij} \frac{R_j(0)}{R_i(0)}, \quad H_i = \sum_{j=1}^{i-1} \lambda_{ij} \left(G_j - G_i\right) \frac{R_j(0)}{R_i(0)}, \quad \lambda_{ij} = S_{j+1} B_{ij+1} - S_j B_{ij}$$

Figure 4. Solution of typical population balance model

The basic goal of the population balance model is to find the breakage rate constants (S_i) and the breakage selectivity coefficients ($b_{i,j}$) such that the solution of cumulative breakage equation agree with measurements made for particles subjected to repeated breakage events caused by exposing the particles to a controlled stress and several increasing total strains. Typically the rate constants (S_i) have a power-law relationship with particle size. However, the contention of this paper is that the selectivity constants ($b_{i,j}$) depend on the structure of the particles and can be estimated by doing FEM analysis or structure mechanics failure of particles exposed to point loads. The selectivity constants are governed by how the primary particles within the structured particle are put together.

Consider the simple example of a single bran fiber particle in a bulk mass of particles. During a stress-strain event, the surrounding particles will contact the particle at discrete points. If the fiber particles resist rolling or have rolled into a stable configuration, then they will experience the set of forces and bending moments as shown in Figure 5. Failure of this beam/particle will occur at the point of maximum bending moment (position X2) giving rise to two particles with lengths X2 and Lp-X2. These fragments will move and find other adjacent particles that will support these fragments at different locations and break at the point of maximum bending moment.

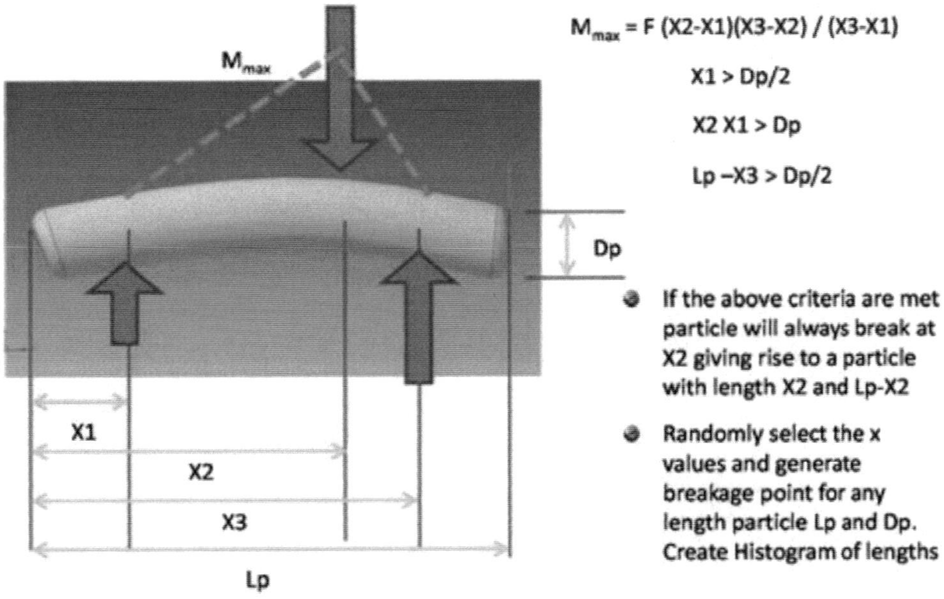

Figure 5. Simple model to describe the breakage of fiber-like particles

One can choose a give size particle and randomly select a set of allowable contact points, determine the lengths of particle breakage points, and create a histogram of the breakage fractions for each selected particle size. If particle structure is a driving force for how a particle breaks, and if the primary mode of breakage is due to a simple bending moment, then this histogram should roughly correspond to the breakage selectivity coefficients in the population balance model. It is possible that the breakage of particles occurs at the local contact points, in which case one would need to do a FEM analysis with local distribute loads to determine if smaller size fractions might break off the particle due to exceeding the local yield stress at the distributed load points. If this breakage model for the selectivity constants gives rise to a reasonable approximation to the breakage modes, then can this simple model be applied to other particles where primary particles are made of fibers?

Finally, a population balance model and breakage study are only useful if they can predict what occurs in the real world. Material is placed in containers/boxes and these containers are subjected to stress-strain events during packing, transport, and as the consumer handles the product. No one wants to receive a box of cereal where all of the cereal particles are broken. Thus, it is necessary to relate the breakage that is measured due to controlled stress-stain events in a lab-scale test to the actual breakage that might occur in the real world. Thus, FEM models describing stress-strain of material in the package can be used to estimate the range of stresses and strains that the material may experience during handling. This can be coupled with the results of the measured lab-scale breakage data measured under control stress and strain events to predict the expected breakage in the various part of the handling system subject to a range of stress and strain events. This systems approach allows the result of a lab-scale test to predict breakage in a wide range of situations.

4.0 Results

In order to accomplish the population balance model study, the bran fibers were subjected to constant stress of 14.2 KPa and then exposed to a total strain of 5 in/in, 20 in/in and 100 in/in. The cumulative volume bases particle size distribution were measured for each strain condition. There was a shift toward the finer particle sizes at increasing strain as shown in Figure 6.

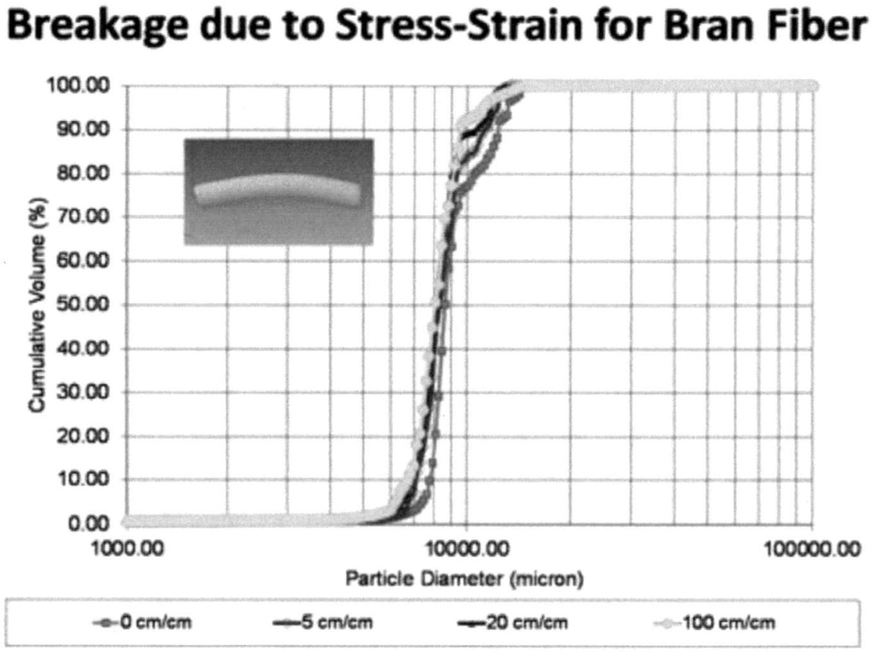

Figure 6. ***Particle size of bran fiber cereal at 14.2 KPa stress as a function of total strain***

These particle size distributions were used to compute the breakage rate constants and breakage selectivity coefficients for bran fiber cereal at 14.2 KPa (Figures 7 and 8). The rate appears to decrease significantly with time and a dual-rate model seems to fit the data well (Figure 9). There appears to be almost an order of magnitude between the breakage rates after 20 in/in strain than before. For the purpose of this paper, only the rate constant and selectivity coefficients before 20 in/in strain were important. Figure 8 shows a summary of the breakage selectivity values for each particle size bin. These selectivity constants were compared to the selectivity constants computed from the structural mechanics model described above (Figure 10). This is not too surprising since the particles are well approximated by rods and stress-strain events tend to cause the rods to be supported by adjacent particles with the movement inducing a bending moment in the rods. This suggests that the population balance model of rod-like particles can be approximated by a simple engineering analysis of the failure of structures.

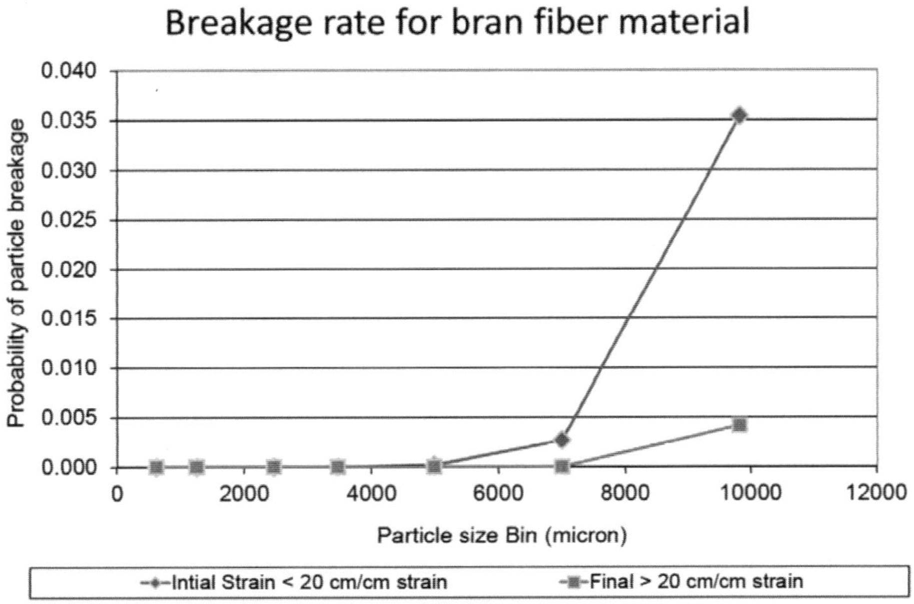

Figure 7. ***Breakage rate constants as a function of particle size for bran fiber cereal at 14.2 KPa stress***

	Bin 1	Bin 2	Bin 3	Bin 4	Bin 5	Bin 6	Bin 7
from	9818.0	6983.0	4967.0	3459.0	2460.0	1245.0	616.4
to	14710.0	9818.0	6983.0	4967.0	3459.0	2460.0	1245.0

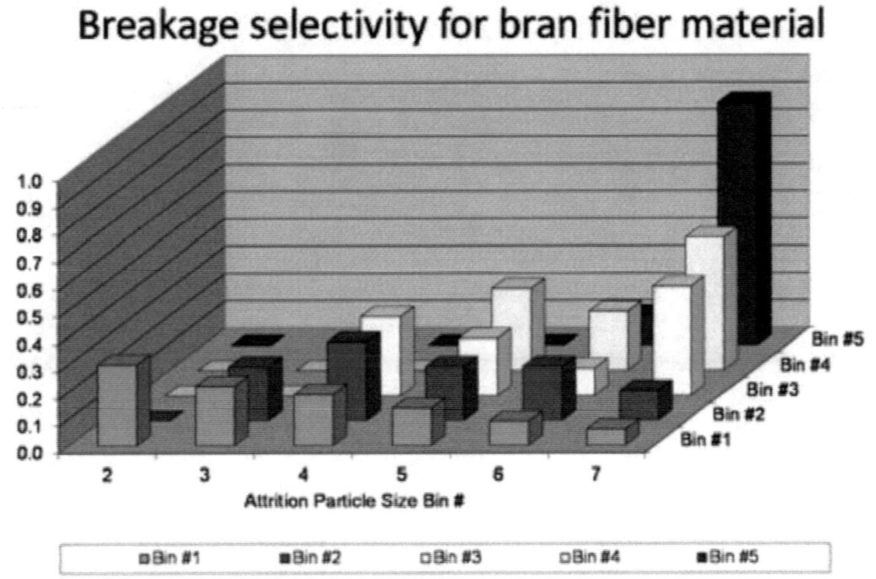

Figure 8. ***Breakage selectivity constants as a function of particle size for bran fiber cereal at 14.2 KPa stress***

	Bin 1	Bin 2	Bin 3	Bin 4	Bin 5	Bin 6	Bin 7
from	9818.0	6983.0	4967.0	3459.0	2460.0	1245.0	616.4
to	14710.0	9818.0	6983.0	4967.0	3459.0	2460.0	1245.0

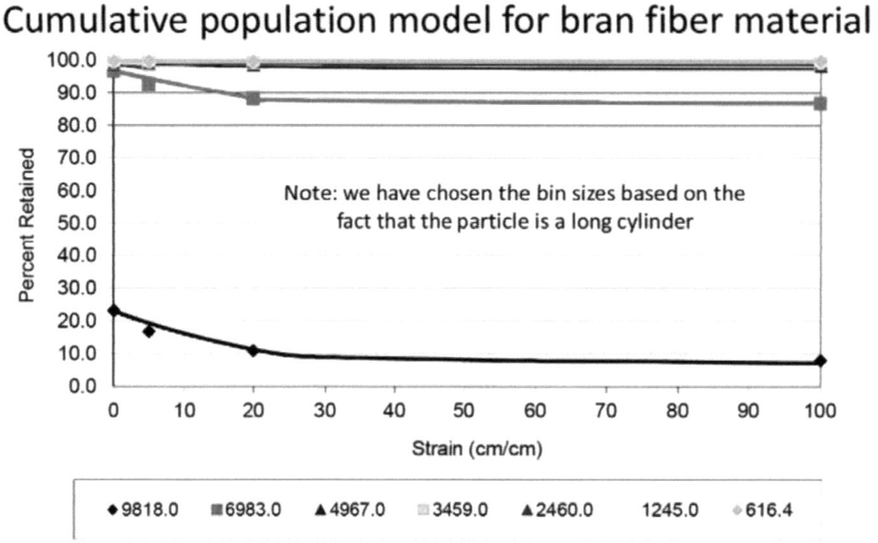

Figure 9. ***Cumulative breakage curves as a function of particle size for bran fiber cereal at 14.2 KPa stress***

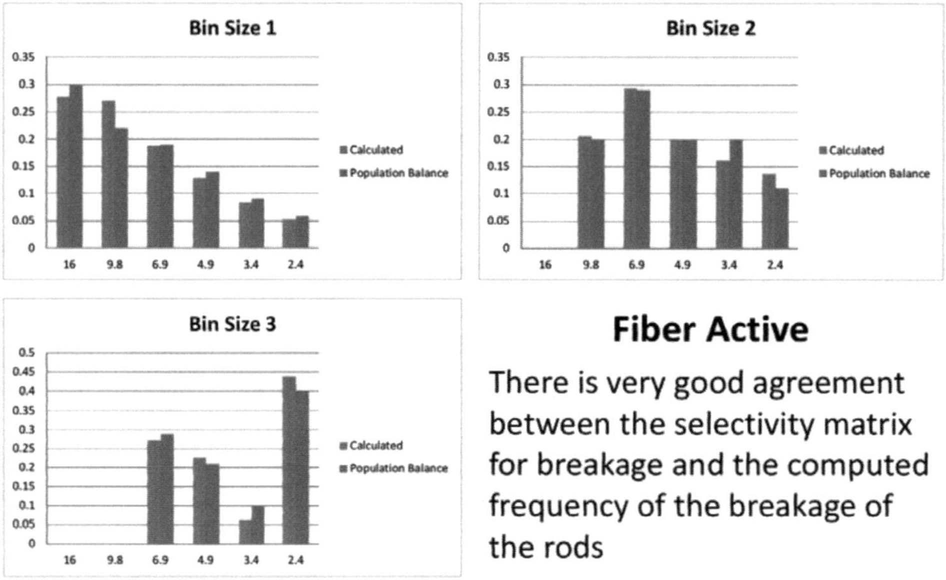

Figure 10. ***Comparision of breakage selectivity measured from population balance with those computed from a structure mechanics failure model for bran fiber cereal at 14.2 KPa stress***

The next system studied was shredded Mini Wheat cereal particles that were formed by food fibers that were rolled and then pressed before they were baked. In this case, the actual particle was not fiber shaped but the structure of the particle was made up of fibers loosely held together. Could the simple structure mechanics failure model describe the breakage selectivity of this type of particle? Shredded wheat particles were subjected to constant stress of 14.2 KPa and then exposed to various total strains. The particle size distributions were measured for each of the stress-strain conditions and are shown in Figure 11. They showed a dramatic shift toward finer material especially in the finer particle sizes. A close examination of the breakage fragments indicated that many of the fragments were fibers of various sizes, suggesting that perhaps the simple structure mechanic's failure model for simply supported rods may shed some light on the breakage selectivity of these particles.

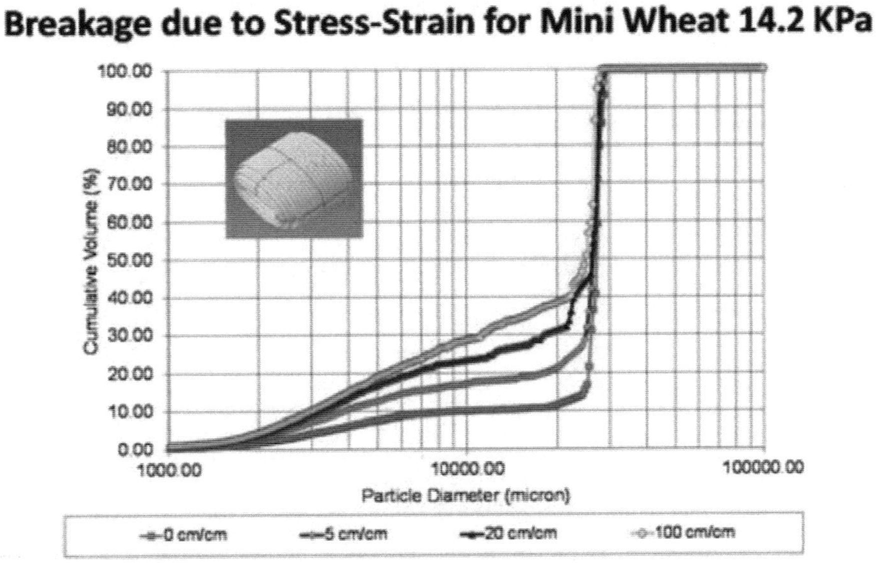

Figure 11. ***Particle size of shredded wheat cereal at 14.2 KPa stress as a function of total strain***

The breakage rate before 20 in/in strain appears to be more than an order of magnitude faster than those rates measured after 20 in/in strain (Figure 12). The breakage selectivity coefficients found in Figures 13 and 15 for the shredded wheat particles show a reasonable agreement to the histrogram computed by assuming the size of the fibers that span one side of the wrapped bundle of fibers. The dual-rate population breakage model seemed to fit the measured data well (Figure 14). This suggests that the fiber-to-fiber bond is very weak, and that the initial breakage of particles tends to break the fiber bundle at the ends of the fiber roll and then break the fiber particles in a similar manner to the simple rods for the bran fiber particles. This finding indicates the ability for the breakage selectivity to be approximated by structural models of individual particles. This suggests that coupling a population balance model with FEM analysis of individual particles may provide added benefit to determine what causes particles to break, and may give food scientists some indication about what to change in the food particle to make them more robust.

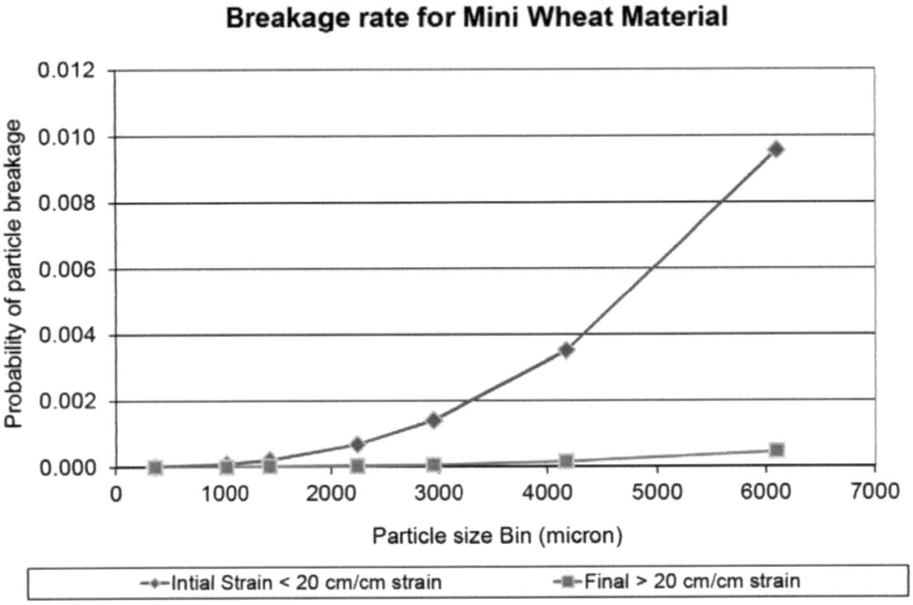

Figure 12. ***Breakage rate constants as a function of particle size for shredded wheat cereal at 14.2 KPa stress***

	Bin 1	Bin 2	Bin 3	Bin 4	Bin 5	Bin 6	Bin 7
from	6095.0	4169.0	2951.0	2239.0	1429.0	1012.0	352.8
to	30370.0	6095.0	4169.0	2951.0	2239.0	1429.0	1012.0

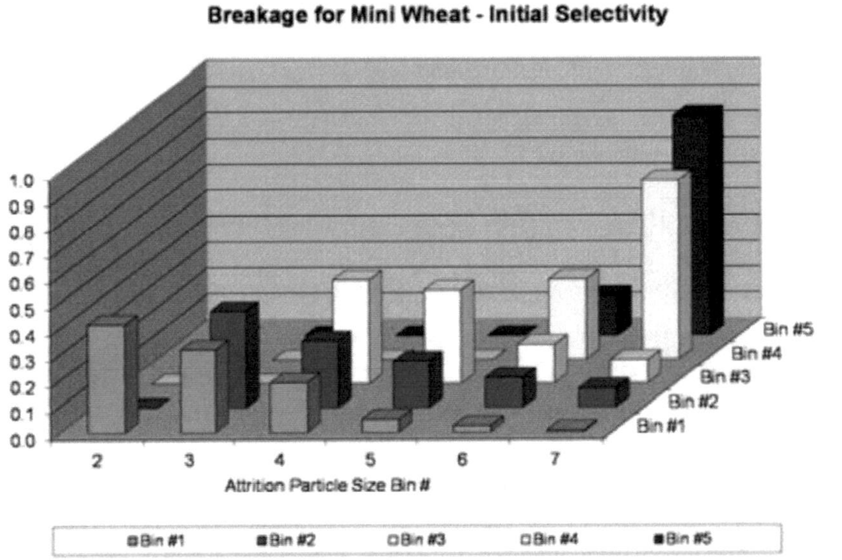

Figure 13. ***Breakage selectivity constants as a function of particle size for shredded wheat cereal at 14.2 KPa stress***

	Bin 1	Bin 2	Bin 3	Bin 4	Bin 5	Bin 6	Bin 7
from	6095.0	4169.0	2951.0	2239.0	1429.0	1012.0	352.8
to	30370.0	6095.0	4169.0	2951.0	2239.0	1429.0	1012.0

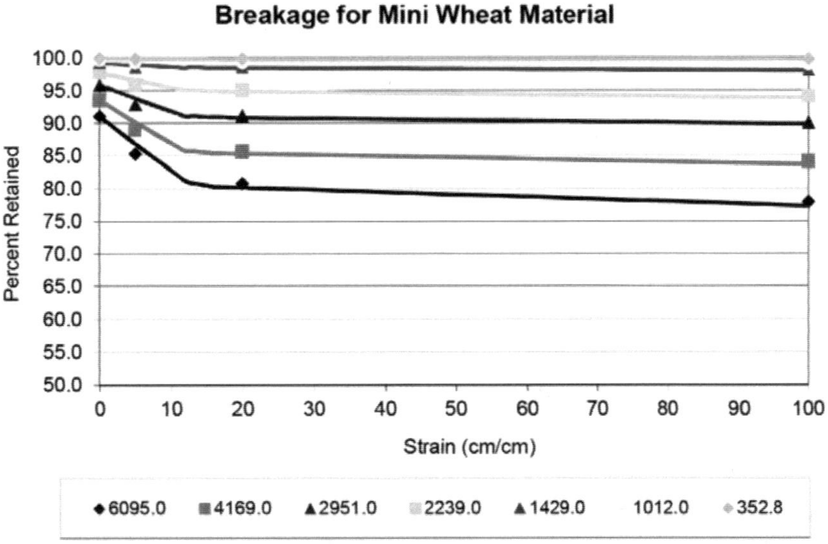

Figure 14. Cumulative breakage curves as a function of particle size for shredded wheat cereal at 14.2 KPa stress

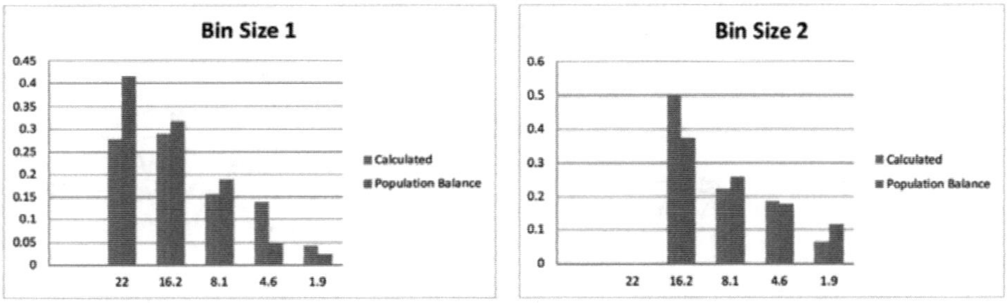

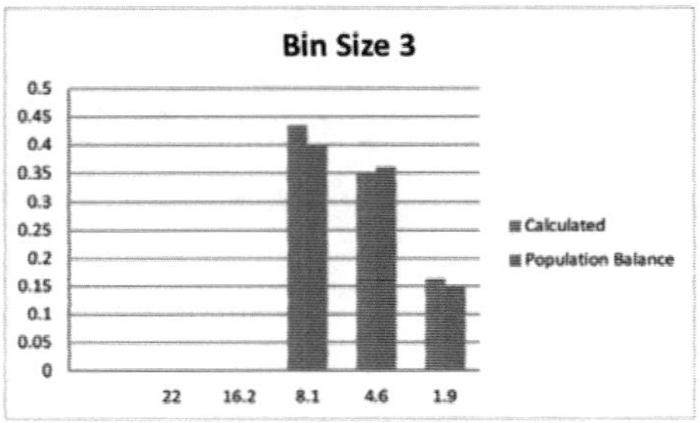

Mini Wheats

There is reasonable agreement between the selectivity matrix for breakage and the computed frequency of the breakage of the rods

Figure 15. Comparision of breakage selectivity measured from population balance with those computed from a structure mechanics failure model for shredded wheat cereal at 14.2 KPa stress

The last question to be answered is: can a lab-scale test that measures breakage due to controlled stress-strain events be used to predict the breakage that may occur during the handling of structured food particles in typical packages? To answer this question the breakage of shredded wheat particles was also measured at two other stress conditions for total strains of 5 in/in, 20 in/in, and 100 in/in. A particle size breakage parameter was defined as the difference between the original cumulative particle size distribution at a given particle size and the final cumulative particle size distribution after the stress-strain application. This generated a series of curves that describe the breakage behavior at any particle size of interest as a function of the stress and strain placed on the material. It is easier to see the relationship if the data is for a series of selected particle sizes as a function of total strain for various stress levels (Figure 17). This shows that the breakage tends to plateau to some constant value at large total strains. This data also shows a general trend to increase breakage at greater stress levels. Finally, the data suggest that the particle(s) experience breakage even without strain when a prescribed stress is applied to the material (i.e. crushing).

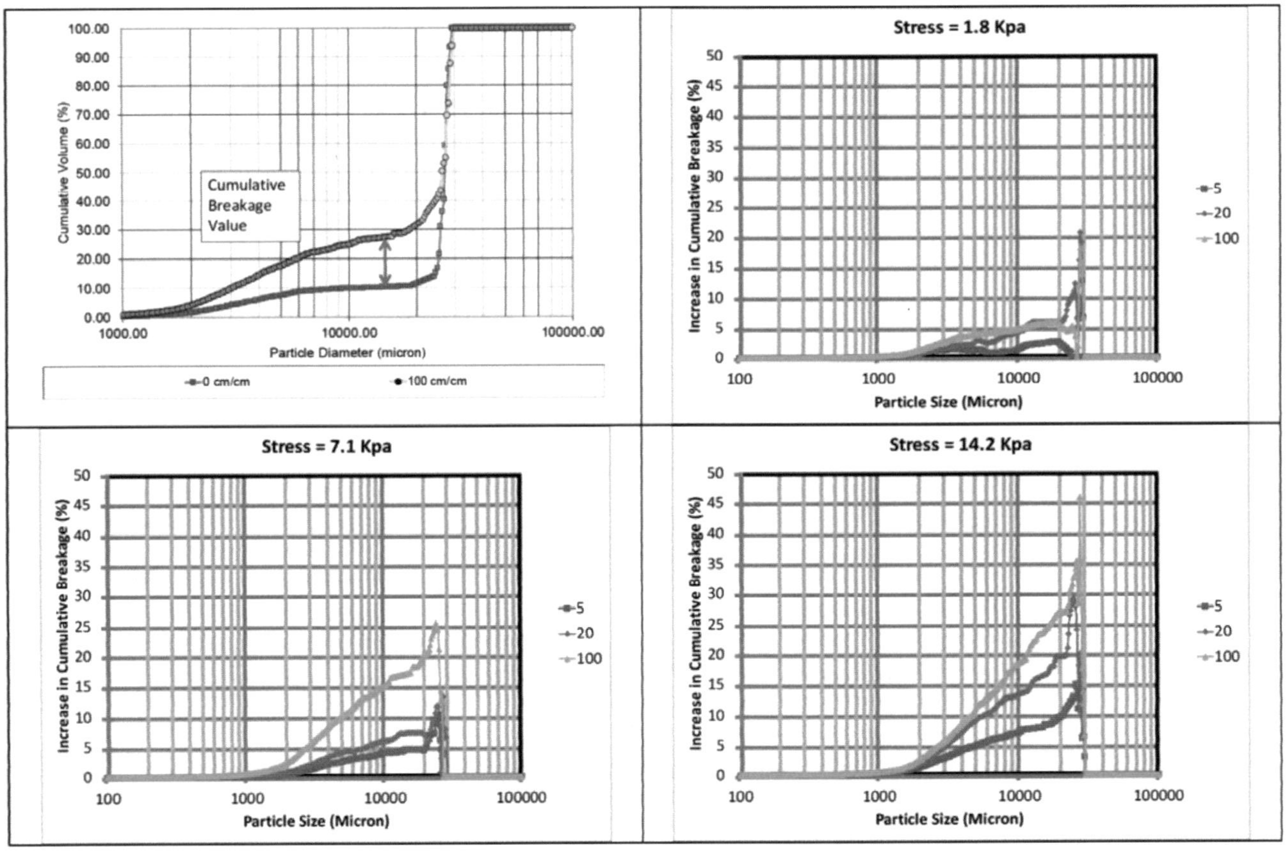

Figure 16. Summary of the increase in cumulative particle breakage as a function of particle size for various stress and total strains

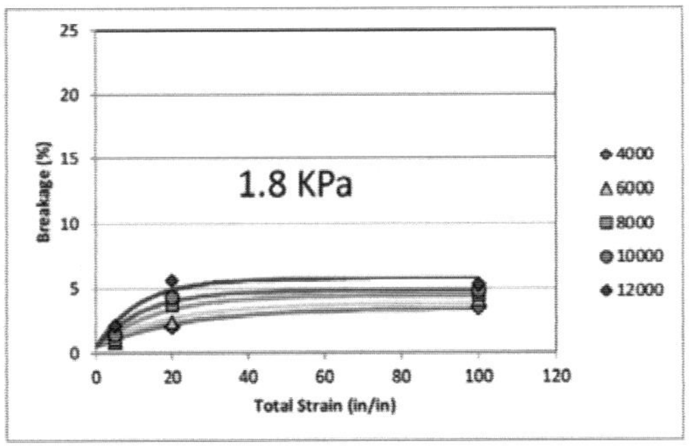

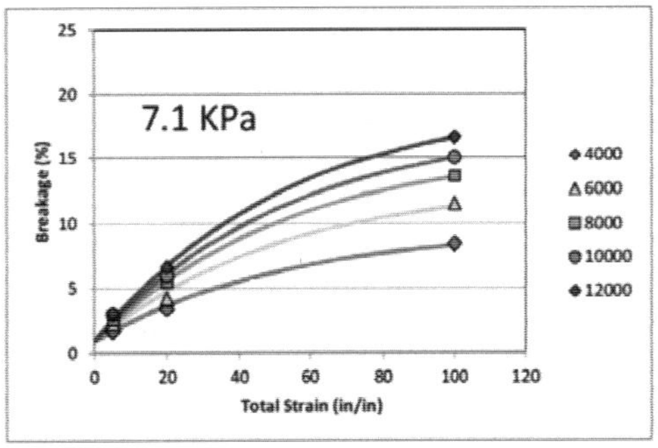

Tests can determine how the breakage varies with both stress and strain then compute the stress and strain in a process or during shipment to estimate the degradation

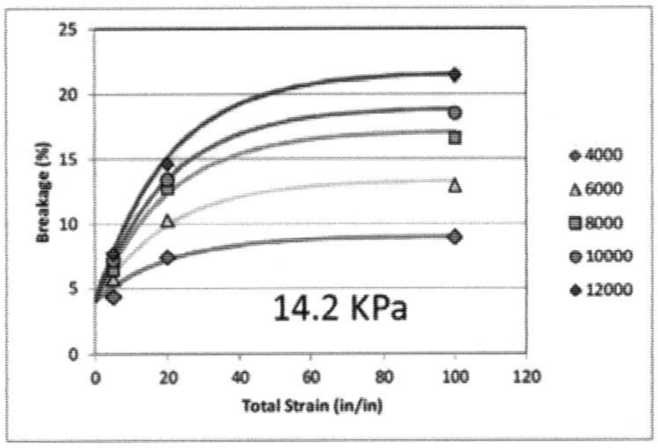

Figure 17. Summary of the increase in cumulative particle breakage for various selected particle sizes as a function of stress and total strain

Consider what occurs during transport of a cereal box from the plant to the store, or from the store to the consumer's table. The side of the box tends to flex. Thus, vibrations during transport or hands holding the cereal box tend to squish the particles in the box, inducing strain at some prescribed stress condition. The above lab-scale measurements describe the breakage that happens at a variety of stresss and strains. It is reasonable to expect that, if FEM calculations can be done describing the stresses and deformations in boxes filled with material, these could be used with the measured breakage at various stress-strain values to estimate the expected breakage of shredded wheat in boxes transported or handled by customers and, thereby, subjected to repeated flexing under a load. To test this hypothesis an experiment was constructed to subject several cereal boxes filled with shredded wheat to repeated loads where 13.3 N force was placed on the side of the box 100 times and then removed. The particle size of the shredded wheat particle was measured before and after this load application and the cumulative breakage curve as a function of particle size was measured.

A finite element model using an elastic-plastic stress model was constructed for the cereal box (Figure 18). The inside of the box was lined with wax paper and the wall friction angle of shredded wheat against wax paper was measured and found to be about 22 degrees. The bulk unconfined yield strength was measured using a large Schultz cell and found to vary between 1.5 to 1.8 KPa depending on the consolidation stress level. Uniaxial direct compression tests were done on the bulk material to find Young's modulus which varied between 130,300 Pa and 388,000 Pa as a function of compaction pressure. Poisson's ratio was assumed to be about 0.3 for the bulk material. The material was confined at the bottom of the box, but the material level at the top of the box was allowed to deform and friction on the sidewall was assumed to be activated by motion in the upward direction. It was assumed that the Tresca stress on the bulk material would be the stress level that correlates best to the breakage of material. The stress level in the entire volume of the box was averaged and gave an average Tresca stress of 1031 Pa. Likewise, the total strain of material in the box was averaged over the entire volume to give an average total strain on the material of 4.38 in/in for 100 repeated applications of a 13.3 N load to the side of the box. The measured breakage curves were used to extrapolate the expected average breakage at these average stress and strain values. This data is summarized in Figure 18. There is very good agreement in the cumulative breakage curve and the one calculated from the FEM model combined with the result of the lab-scale breakage test done on the shredded wheat material, suggesting that this systematic approach to breakage in processes and packages is useful to approximate the breakage in real systems by using lab-scale data measured at controlled stress-and strain conditions.

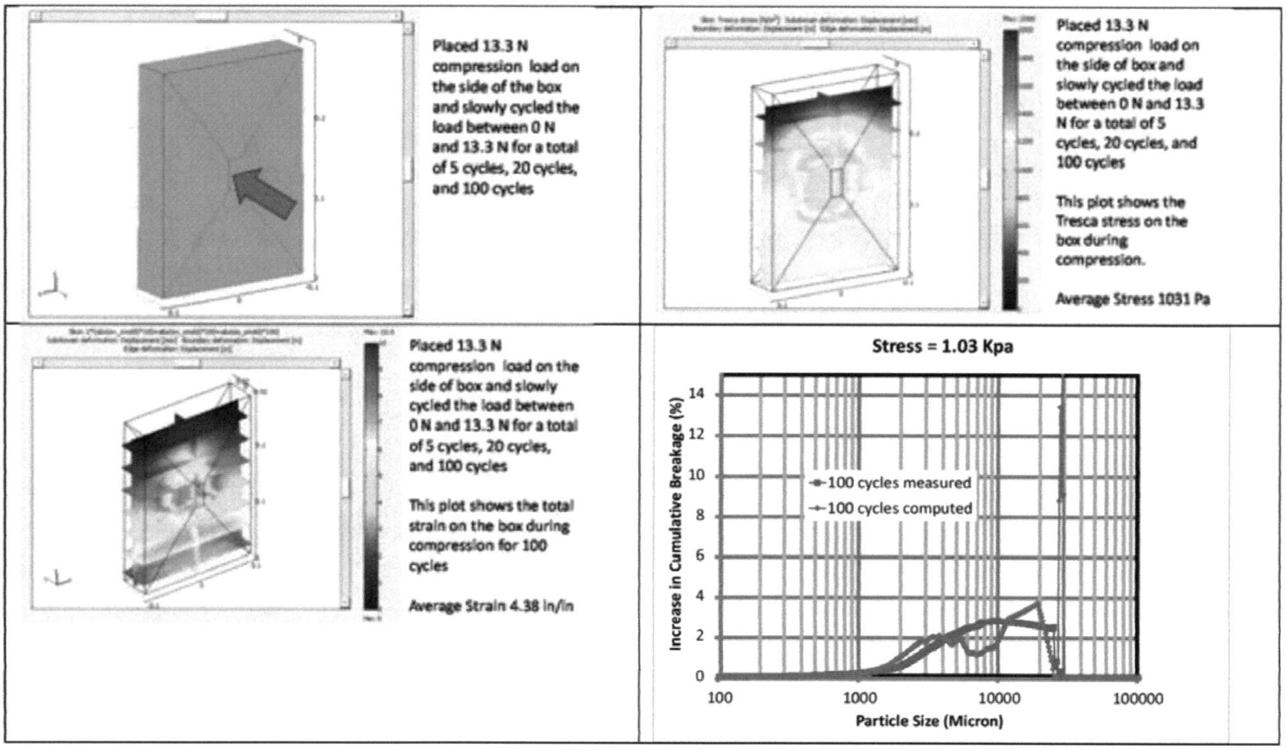

Figure 18. Summary of FEM calculation on stress-strain in cereal box during handling and compared to measured breakage of well-controlled lab-scale stress-strain events – test and analysis show a link to be able to predict real-world breakage by lab-scale measurements

5.0 Conclusions

The major conclusions of this work indicate that structural mechanics and FEM models of individual particles can be used to determine the breakage selectivity of structured particles and that the selectivity appears to be a strong function of the type or shape of primary particles that make-up the structured particles. This gives a systematic method to estimate the breakage selectivity for structured particles and can be used by food engineers to determine how a given product may break and provides direction for what might be done to the structured particles to make them more robust. Simple breakage of simply supported rods seems to approximate the breakage selectivity of structured particles that are built with fibers.

The lab-scale breakage test, when coupled with FEM analysis of real-world packaging and handling, can be used to estimate the expected breakage in cereal boxes subjected to repeated stress-strain events, thus providing a way for food engineers to design packages to minimize breakage of structured particles. When these three techniques (population balance modeling, particle scale structure failure analysis, and bulk scale stress-strain analysis of packages) are coupled together, they provide a means of understanding how particle morphology can influence the quality and robustness of a product sold to a consumer. This provides a systematic method to design consumer products and avoid poor quality products before they get to market.

588095 Analytical Solutions to Linear Breakage Population Balance Models of Various Process Configurations

For linear breakage, it is well known that one obtains an analytical solution to the resulting sectional population balance for batch milling in terms of the time trajectories of the size-class mass occupancies. The model can be expressed in vector-matrix form as **m'** = **A** ´ **m**(t) in which **m**(t) is the time-dependent vector of the size-class mass occupancies, **m'** is the vector of their time derivatives, and **A** is a triangular matrix, the diagonal of which is comprised of the negative of the selection functions for each size class, and the remainder of the non-zero triangle is comprised of products of these selection functions and fragment mass allocations showing how mass broken in one section is distributed to the sections at smaller sizes. The resulting system is a set of linear, coupled ODEs.

The system is decoupled by the classical approach of finding a similar diagonal matrix **D.** The diagonal elements of **D** are the eigenvalues of **A** and similarity is expressed via **A** = **E** ´ **D** ´ $\mathbf{E}^{-1}$. The columns of the matrix **E** are the eigenvectors of **A** while $\mathbf{E}^{-1}$ denotes the inverse of **E**. Given an initial condition of mass occupancy in terms of the vector $\mathbf{m}_0$, the analytical solution for the trajectories is simply:

m(t) = **E** ´ **J**(t) ´ $\mathbf{E}^{-1}$´ $\mathbf{m}_0$ with **J**(t) = exp(**D**t)

To be clear, **J**(t) is a diagonal matrix with elements that are the exponentials of the product of the diagonal elements of **D** and time.

It is not so well known that this decoupling through similarity to a diagonal matrix allows the solution of the sectional breakage population balance for a myriad of other process configurations besides batch. The solution for a continuous plug-flow mill is identical to the batch solution with time replaced by axial position divided by axial velocity so this will not be explored. Instead, we will illustrate employment of this approach for a perfectly mixed continuous mill (either a single mill or mills in series), and for a continuous plug-flow mill with recycle as a function of recycle ratio. These configurations are often employed commercially. For example, perfectly-mixed mills in series is analogous to the practice of "pendulum" milling.

571486 A Mechanistic Collision Model for Particle Breakage in Agitated Crystallization Experiments

Demonstrated here is a mechanistic approach to modelling particle breakage behaviour within mechanically agitated crystallization systems, gained from isolating fundamental fluid and particle behaviours. The role of fluid boundary layer effects in altering particle collision rates is coupled with a physical model for particle breakage probabilities. The proposed model combines the effects of fluid properties, operating conditions, particle impact velocities and sizes, and material properties. These factors are grouped into two governing behaviours: Hydrodynamics and Material Response. Firstly, the illustration of a material response boundary is quantified using a data classification technique.

Crystals were accelerated towards a target and impacted at various speeds and across a range of crystal sizes. This allowed construction of a failure probability heatmap, outlining the probability of damage occurring to the crystal after an impact had occurred. From this, the probability of failure for any crystal size and velocity pair was extracted and compared to theoretical forms for the expected failure probability distributions. A Multi Layer Perceptron (MLP) classifier was used to generate deterministic boundaries associated with previously observed experimental breakage data. This provides a clear proof of concept for the material response boundary, resulting in the successful mapping of size-velocity input pairs to expected breakage modes.

Limiting hydrodynamic factors from previous work are also shown alongside the new material response boundary, clarifying the role of both the fluid and solid phases at the macro scale. Through the use of shadowgraphy imaging it was shown that there exists a critical Reynolds threshold, below which collision between particles and an impeller blade is unlikely. Furthermore, those particles that do collide experience only a fraction of the nominal impeller tip-speed. Thus, this gives credit to the presence of a squeeze film boundary layer cushioning impacts at the impeller and around probes/baffles. As a result, the actual impact rate of particles with a typical crystallization system is often much lower than expected as the hydrodynamic conditions serve to protect the crystals from collision events.

From here, a Population Balance (PB) Finite Volume 1 Scheme (FVS) for pure breakage is compared against experimental breakage data for three unique pharmaceutical crystal populations in an agitated liquid-solid suspension.

The hydrodynamic and material response characteristics were combined using a physically relevant rate expression based on the characteristic circulation time of the system. It was found that by isolating important interactions in the system it is possible to construct physically based breakage modelling kinetics, without loss of generality to particulate processing as a whole. Good agreement with the mathematical model was found in all cases, even while each crystal material is chemically distinct.

This indicates that byfirst identifying key process behaviours, via such methods as the MLP classification technique used here, it is possible to construct a generalized particulate behaviour model. As such, this enables one to more accurately capture the behaviour of a range of materials without the need for more involved experimental testing. Importantly, the fundamental nature of the model is scale-independent, general across all materials well-defined by typical elastic moduli (such as Young's modulus and fracture toughness), and accounts for fluid dynamic parameters via a known Reynolds threshold for particle-plane collisions. Therefore, a key application of this methodology lay in process design, optimization, and scale-up where the need to produce reliable predictions in the smallest possible time frame is desirable.

579633 Chemical Looping Technology - a Particle Technology Perspective

The concept of chemical looping reactions has been widely applied in chemical industries. Fundamental research on chemical looping reactions has also been applied to energy systems. Fossil fuel chemical looping applications were used with the steam-iron process for coal from the 1900s to the1940s and were demonstrated at a pilot scale with the carbon dioxide acceptor process in the 1960s and 1970s. There are presently no chemical looping processes using carbonaceous fuels in commercial operation. A key factor that hampered the continued use of these earlier processes was the inadequacy of the reactivity and recyclability of the looping particles. This factor led to unsustainable operation of the process. With CO2 emission control now of great concern, interest in chemical looping technology has resurfaced due to their unique ability to generate a sequestration-ready CO2 stream.

Chemical looping technology is a manifestation of the interplay among all the key elements of particle science and technology including particle synthesis, reactivity and mechanical properties, flow stability and contact mechanics, gas-solid reaction engineering and particulates system engineering. This presentation will describe the fundamental and applied aspects of modern chemical looping technology that utilizes fossil and other carbonaceous feedstock. Specifically, it will discuss the reaction chemistry, ionic diffusion mechanisms, metal oxide synthesis and thermodynamics, reactor design, and system engineering along with energy conversion efficiency and economics of the Coal-Direct Chemical Looping Process and Syngas Chemical Looping Process being developed at Ohio State University at a pilot level. Further, CO2 emission control using the chemical looping technology will be illustrated and compared with other CO2 capture methods. Potential for selective oxidation in the production of fuels and chemicals, as well as solar based chemical looping technology, will also be discussed.

580750 Mass Production of Well Aligned Carbon Nanotubes and Carbon-Silicon Based Anode Materials By Fluidized Bed

As one of the nanomaterial representatives, one-dimensional carbon nanotubes (CNTs) possess extraordinary mechanical, electrical, thermal, optical properties, as well as controllable reactivity. This renders their broad applications in the fields of energy storage and transition, composites, heterogeneous catalysis, environmental protection, and drug delivery. This talk presents the chemical and engineering principles of CNTs mass production by fluidized bed. The detailed strategies for mass production of multi-walled CNTs, single-walled CNTs, CNT arrays as well as graphene/silicon anode material by multistage fluidized bed were presented. Meanwhile, the key engineering considerations in standardization, environment, health, and safety were analyzed, and their commercialization process was evaluated. During recent decades, CNT production capacity has reached more than thousands of tons per year, greatly decreasing the price of CNTs. The bulk applications for Li-ion battery, conductive nanocomposites, automotive, sporting goods have been achieved. Although the unique physiochemical properties of an individual CNT are stated repeatedly, manifestation of such unique properties in a macroscopic material, e.g., realization of high-strength CNT fibers, remains a great challenge. If such challenges are solved, many critical applications will be enabled. However, more efforts should be devoted on the chemical route for mass production and engineering route for commercialization of high performance CNTs, in order for versatile properties and applications. The basic research for CNTs production and application will flourish the nanotechnology industry for sustainable society.

580751 Novel Fluidized Bed Reactors for Biomass Thermochemical Conversion

Biomass residues are low-carbon renewable energy sources and have the potential to replace fossil fuels to reduce greenhouse gas emissions. Thermochemical pathways such as gasification and pyrolysis have been applied to convert biomass into drop-in biofuels, but still face both technical and economic challenges, which requires further development of novel technologies to improve the process efficiency. At the Fluidization Research Centre, we have developed and validated a dual fluidized bed reactor in pilot scale for steam gasification of biomass for the production of synthetic gas, which can be potentially converted to liquid biofuels or renewable natural gas. To produce high quality bio-oil and biochar from biomass pyrolysis, we also proposed and evaluated a gas-pulsating fluidized bed reactor in small pilot scale. The fluidization performance was optimized by adjusting the gas pulsation frequency and superficial gas velocity. The quality of bio-oil and biochar was evaluated, which showed superior performance in terms of oil water content, acidity and viscosity, and biochar surface area and water/nutrient affinity.

580752 Gasification of Biomass Particles: Biochar Production, Particulate Emission, and Process Simulation

In this work, gasification experiments have been conducted to produce biochar particles under different reaction conditions (i.e. gasifying agents, reaction temperatures). The resulting biochar particles could be further converted to soil conditioner for plant growth. Effects of biochar addition ratio on Brassica rapa chinensis (pak choi) growth was investigated from 0% to 100% to find the optimal condition. During the gasification experiments, particulate matter (PM) emissions are characterized and size-dependent particle number concentrations are measured. An empirical formula is developed to determine the PM emission rate under different operating conditions. The results could potentially serve as a basis for better understanding of possible PM formation mechanisms. he proposed model could give more precise predictions of biochar production and the deviation between experimental and simulation results is less than 10%.

580753 Development of a Triple-Bed Combined Circulating Fluidized Bed

We have developed a novel triple-bed combined circulating fluidized bed (TBCFB) gasifier which was originally proposed by Professor Atsushi Tsusumi. In this TBCFB, a downer pyrolyzer, a bubbling fluidized bed gasifier, and a riser char combustor are integrated to increase thermal efficiency of gasification at low temperatures. In this talk, we will explain 1) hydrodynamic behaviors of silica sand at very large solids circulation rate (Gs=300-500 kg/(m2 s)) in the TBCFB cold model (downer (d:0.1 m, height: 6.5 m), a bubbling fluidized bed (0.75 m×0.27 m×1.9 m), a riser (d: 0.1 m, height: 16.6 m) and a gas-sealing bed (GSB: d: 0.158 m×5 m), 2) hydrodynamics of the solids in the downer, 3) reactivity of lignite in a lab-scale downer pyrolyzer (d: 0.020 m×1 m), and 4) theoretical limit of cold gas efficiency in this TBCFB gasifier.

580754 Green Energy Innovation for Sustainable Society:Exergy Recuperation Technology

In the last several years an innovative energy-saving technology based on the exergy recuperation has been developed. For the thermal processes heat can be recirculated by providing the minimum exergy required for heat circulation to compensate for the exergy destruction in heat exchange module without any addition of heat generated by fuel combustion. This self-heat recuperation technology can be applied to various industrial thermal processes such as distillation, PSA gas separation, chemical absorption gas separation, air separation, drying, condensation, desalination, etc., leading to a considerable reduction in the energy consumption. In addition, the exergy recuperation technology has been adopted to the energy conversion processes such as hydrogen production by exergy recuperative gasification, reforming combustion and energy and material coproduction. In this lecture, a state-of-the-art innovative energy efficient utilization technology based on the concept of exergy recuperation will be discussed.

573154 Prediction of Nanoparticles Size Distribution: The Effects of Ligand Surface Coverage and Nanoparticle Size in Altering the Kinetics of Surface Growth

Controlling nanoparticle size and size distribution is vital for many applications including catalytic reactions, drug delivery and imaging, and surface coating. The process of nanoparticle synthesis often involves using ligands, where they are found to affect not only the size but also the size distribution by binding to the nanoparticle surface and/or metal precursor. In addition to the role of ligands, the nanoparticle size can affect the ligand-nanoparticle binding affinity and the rate of surface growth, and alternatively the size distribution. Despite recent advances in the field, no theoretical approach could yet accurately unravel the roles of ligand surface coverage and nanoparticle size in controlling the evolution of size distribution. This work aims towards investigating the role of size and ligand coverage in controlling the size distribution during the synthesis of colloidal metal nanoparticles. We used in-situ small angle X-ray scattering (SAXS) to measure the evolution of size distribution under different reaction conditions. We further developed a population balance kinetic model (PBM) to understand the effects of ligand coverage, nanoparticle size, and nucleation and growth pathways on the evolution of size distribution. Using palladium (Pd) as a model system, our SAXS results indicate a sharp decrease in nanoparticle polydispersity (or size focusing behavior) despite temporal overlap of nucleation and growth (i.e. increase in concentration of nanoparticles concurrent with nanoparticle growth). Our population balance modeling approach allows us to understand the main factors involved in controlling the size distribution. We show that continuous nucleation (formation of new nuclei and small population) leads to size defocusing. However, continuous nucleation results in different reaction times (or aging time) for the nanoparticle population leading to time and size-dependent ligand coverage. To capture the size focusing trend during the synthesis; we show that the smaller nanoparticles should grow at a much faster rate than the larger size nanoparticles in the population. Our modeling results along with thermodynamic calculations further suggest that the faster growth of smaller nanoparticles (relative to the larger nanoparticles in the population) is due to two factors: lower ligand surface coverage and higher surface reactivity which leads to focusing of the size distribution. Finally, we demonstrate the effects of different nucleation and growth kinetics/mechanisms (e.g. nucleation rate profile, nucleation order, etc.) on the size distribution and how these effects can noticeably impact the shape of distribution during the reaction. The presented methodology allows for the accurate prediction of size distribution and could be potentially applied to other ligated systems to obtain nanoparticles with desired sizes and polydispersity.

586726 Agglomeration of Fractal-like Particles

Agglomeration refers to the formation of physically attached primary particles by coagulation and is encountered especially at low temperature processes where particle sintering or coalescence is rather slow. Understanding agglomeration is essential for optimal industrial process design, meteorology and climate dynamics.

Here, the growth and detailed structure of particles undergoing agglomeration are investigated from the free molecular to the continuum regime by discrete element modelling (DEM). Particles coagulating in the free molecular regime follow ballistic trajectories described by an event-driven method, whereas in the near-continuum (gas-slip) and continuum regimes, Langevin dynamics describe their diffusive motion. Agglomerates containing about 10–30 primary particles, on the average, attain their asymptotic fractal dimension, Df, of 1.91 or 1.78 by ballistic or diffusion-limited cluster–cluster agglomeration, corresponding to coagulation in the free molecular or continuum regimes, respectively. A correlation is proposed for the asymptotic evolution of agglomerate Df as a function of the average number of constituent primary particles, np that can be readily used in process design for synthesis of nanomaterials or in environmental models for ambient aerosols (e.g. air pollution and climate forcing).

Agglomerates exhibit considerably broader self-preserving size distribution (SPSD) by coagulation than spherical particles: the number-based geometric standard deviations of the SPSD agglomerate radius of gyration in the free molecular and continuum regimes are 2.27 and 1.95, respectively, compared to ~1.45 for spheres. In the transition regime, agglomerates exhibit a quasi-SPSD whose geometric standard deviation passes through a minimum at Knudsen number Kn $\approx$ 0.2. In contrast, the asymptotic Df shifts linearly from 1.91 in the free molecular regime to 1.78 in the continuum regime. Population balance models using the radius of gyration as collision radius underestimate (up to about 80%) the small tail of the SPSD and slightly overpredict the overall agglomerate coagulation rate, as they do not account for cluster interpenetration during coagulation. In the continuum regime, when a recently developed agglomeration rate is used in population balance equations, the resulting SPSD is in excellent agreement with that obtained by DEM.

563860 Quantification and Control of Amorphous Contents By Raman. Applications and Case Studies in Pharmaceutical Processing

Amorphous regions in crystalline material can have significant impact on bioavailability, processability and stability. Conversely, crystalline material can act as nuclei for recrystallization and resulting in instability of amorphous systems. Raman spectroscopy is sensitive to polymorphic differences and amorphous content in pharmaceutical materials. This make Raman spectroscopy a powerful process analytical technology (PAT), when combined with multivariate modelling, to control the amorphous content in a pharmaceutical process. We will present pharmaceutical case studies showing process-induced amorphization of crystalline drugs across the DS and DP processing steps. In addition, we will compare the kinetics of crystallization of amorphous material when stored at various relative humidity and temperature conditions. The results clearly demonstrate that Raman spectroscopic techniques combined with multivariate methods is a powerful and effective tool for quantitation of amorphous content in crystalline material with applications ranging from API isolation, milling and to multiple unit operations in DP manufacturing. The information generated was critical to determine the root-causes and outline appropriate mitigation measures.

575010 Breakage Kinetics of Urea Crystals in Agitated Slurries

Crystals with high aspect ratios tend to break when stirred in slurries, and the kinetics depend on crystal properties such as the major axis length of the crystals and the aspect ratio distribution of the crystals; and on process conditions including the agitation rate and the solids concentration in the slurry. Urea is used as a model substance in these studies because urea can be grown into needles when crystallized in water. This presentation discusses experimental results, and the breakage distribution functions and parameters derived from the experimental results.

570278 Photoelastic Properties of 3D Printed Particles

3D printing can be used as a means to unlock new advances in research, particularly for particle applications. Our project investigates the photoelastic behaviour of 3D printed particle structures, to determine whether 3D printed particles can be used for stress visualisation experiments. The compression of simple spherical disc shapes to complex multiple stacked disc designs were printed using a Statasys Objet printer using VeroClear material, which is photoelastic. Compression experiments were carried out using an Instrom in a darkfield configuration to observe the stress fringes. The compression of the particles at 2mm/min were video recorded. The time lapse of the compression revealed stress fringes and force chains, similar to the previous conventional methods. It was found that complex structures fail sooner under heavier load. Furthermore, the effect of printing layer was found to have a minor effect on the photoelastic behaviour of the particle. Further work is underway to plot the exact strain distribution of the design using Digital Image Correlation (DIC) and Finite Element Analysis (FEA) to confirm the accuracy of the stress fringes. The work demonstrates that visualisation of stresses in more complex shapes can be successfully conducted using 3D printing techniques.

577928 Ultrasonic Spray Delivery of Precursors for Laser Pyrolysis Synthesis of Nanoparticles

Laser pyrolysis is a useful approach for aerosol (vapor-phase) production of nanoparticles (NPs). In vapor-phase methods, precursors must be delivered as a vapor or as small droplets or particles that can be vaporized within the process. In laser pyrolysis, a laser beam is employed to selectively heat a gas stream containing NP precursors, such that they decompose, including nucleation of NPs. Laser pyrolysis provides advantages of extremely high heating and cooling rate, which increases throughput, and flexibility in tuning particle size and composition. Gaseous, liquid or solid precursors can be used in this method. Due to issues of safety and precursor cost and limited availability of gaseous precursors, solid or liquid precursors would be preferable in many cases. The precursors can be delivered to the reaction zone as small droplets that rapidly evaporate upon laser heating. Precursors with sufficiently high vapor pressure could also be delivered as vapors from a bubbler. However, no sufficiently volatile precursor is available for some metals. Accordingly, spray-based precursor delivery is the most general approach to delivering precursors for the synthesis of NPs by laser pyrolysis. We have designed a low-cost atomizer to deliver small droplets of a precursor solution to the reaction zone in this method. Indeed, any stable precursor solution can be used to synthesize various NPs by using this atomizer to spray it to the reaction zone as small droplets. However, when the goal is production of small particles by gas to particle conversion, then using a precursor with relatively high vapor pressure is desirable. Employing this atomizer to spray a precursor solution mixture of zinc acetate and thioacetamide enabled us to synthesize zinc sulfide nanocrystalline particles with an average size below 10 nm, which makes them potentially useful for optoelectronic applications.

561158 Tailoring the Bioactivity of Flame-Made Nanoparticles for Surgical Applications

Poorly healing injuries, especially after surgeries, pose complex problems to surgeons. Current wound management solutions have several drawbacks including poor soft tissue adhesion and low bioactivity.[1] In this work, we show the potential application of flame-made bioactive nanoparticle-based tissue adhesives for a variety of wound healing applications.

In a first step, silica, iron oxide, ceria, borate glass, and bioglass nanoparticles were produced by scalable liquid-feed flame spray pyrolysis and their hemostatic properties, and adhesion to soft tissue were assessed.[2] The bioactivity of the nanoparticles was then further tailored by creating hybrid and doped versions by the same synthesis method featuring additional clinically relevant properties including antimicrobial, angiogenic, and antioxidant activities.[3]

Bioglass-based nanoparticles show exceptionally strong adhesive properties and hemostatic activity. Bioglass-ceria hybrid nanoparticles produced in a two-spray setup show unify the strong soft tissue adhesion of bioglass with the antioxidant and cytoprotective activities of ceria. The angiogenic and adhesive features of bioglass combined with the anti-inflammatory and antioxidant properties of ceria make for a promising and comprehensive wound management solution. In a first in vivo study[4], the bioglass-ceria hybrid nanoparticles significantly improved skin graft survival by improving revascularization.

In summary, we will present how the rational engineering of flame-made metal-oxide nanoparticles gives access to highly effective wound-care solutions.

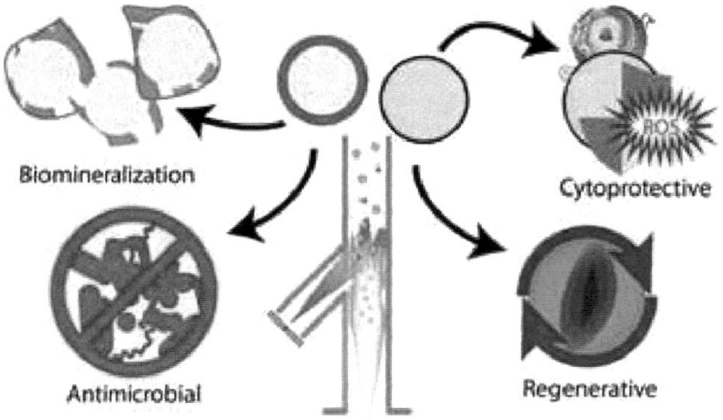

(1) Dhivya, S.; Padma, V. V.; Santhini, E. Wound Dressings – a Review. *Biomedicine (Taipei)* **2015**, *5* (4). https://doi.org/10.7603/s40681-015-0022-9.

(2) Matter, M. T.; Starsich, F.; Galli, M.; Hilber, M.; Schlegel, A. A.; Bertazzo, S.; Pratsinis, S. E.; Herrmann, I. K. Developing a Tissue Glue by Engineering the Adhesive and Hemostatic Properties of Metal Oxide Nanoparticles. *Nanoscale* **2017**, *9* (24), 8418–8426. https://doi.org/10.1039/C7NR01176H.

(3) Matter, M. T.; Furer, L. A.; Starsich, F. H. L.; Fortunato, G.; Pratsinis, S. E.; Herrmann, I. K. Engineering the Bioactivity of Flame-Made Ceria and Ceria/Bioglass Hybrid Nanoparticles. *ACS Applied Materials & Interfaces* **2018**. https://doi.org/10.1021/acsami.8b18778.

(4) Lese, I.; Graf, D. A.; Tsai, C.; Taddeo, A.; Matter, M. T.; Constantinescu, M. A.; Herrmann, I. K.; Olariu, R. Bioactive Nanoparticle-Based Formulations Increase Survival Area of Perforator Flaps in a Rat Model. *PLOS ONE* **2018**, *13* (11), e0207802. https://doi.org/10.1371/journal.pone.0207802.

571546 Corrosion Protection of Epoxy Coating with Calcium Phosphate Encapsulated By Mesoporous Silica Nanoparticles

Smart coatings are considered as a novel approach in the field of corrosion protection coatings and have drawn researchers' attention over the past years [1-3]. They are able to respond to external triggering that comes from the surrounding environment like changes in the pH, or mechanical stress, which leads to the release of healing or inhibitive species [4].

In the present work, mesoporous silica nanoparticles were successfully synthesized. A smart corrosion inhibition system was produced by formulating an epoxy coating with calcium phosphate encapsulated by the mesoporous silica nanoparticles as nanocontainers. The anti-corrosive performance of the epoxy coatings with calcium phosphate and calcium phosphate encapsulated in the nanocontainers after exposure to the salt spray chamber were compared. Electrochemical Impedance Spectroscopy (EIS) was used for the characterization of the coating properties and Scanning Acoustic Microscopy (SAM) was applied for the examination of the coating failure from the artificial defect. Results revealed the effect of the application of mesoporous silica nanoparticles as nanocontainers on the anti-corrosive performance of calcium phosphate on epoxy coatings.

References

[1] M. Yeganeh, M. Omidi, and T. Rabizadeh, "Anti-corrosion behavior of epoxy composite coatings containing molybdate-loaded mesoporous silica," *Prog. Org. Coatings*, vol. 126 (2019), pp. 18–27.

[2] T. Matsuda, K. B. Kashi, K. Fushimi, and V. J. Gelling, "Corrosion protection of epoxy coating with pH sensitive microcapsules encapsulating cerium nitrate," *Corros. Sci.*, vol. 148 (2019), pp. 188–197.

[3] M. F. Montemor, "Functional and smart coatings for corrosion protection : A review of recent advances," *Surf. Coat. Technol.*, vol. 258 (2014), pp. 17–37.

[4] J. Simancas, M. Morcillo, D. De Fuente, C. Zea, and J. Alc, "pH-dependent release of environmentally friendly corrosion inhibitor from mesoporous silica nanoreservoirs," *Microporous and Mesoporous Materials,* vol. 255 (2018), pp. 166–173.

574567 Self-Assembled Nanoparticle Antireflection Coatings on Geometrically Complex Optical Surfaces

Antireflection (AR) coatings are of great technological importance in a wide spectrum of applications ranging from improving conversion efficiencies of solar cells to reducing unwanted light reflection at optical surfaces. However, traditional vacuum deposition technologies suffer from high cost, limited material selection, and low throughput. Critically, it is formidably challenging in coating nonplanar optics with complex geometries, especially those with enclosed concave surfaces. Here we report a simple and scalable electrostatics-assisted colloidal self-assembly technology for fabricating nanoparticle AR coatings on geometrically complex optical surfaces. Surface-functionalized silica nanoparticles with designed surface charge densities can be electrostatically adsorbed on various material surfaces with opposite charges, forming quarter-wavelength AR coatings. Compared with other bottom-up approaches (e.g., layer-by-layer deposition), this single-step, solution-based process enables much simpler and faster assembly of uniform monolayer nanoparticle AR coatings on curved and enclosed optical surfaces. Complementing systematic microstructural and optical characterization, we have also performed numerical simulations to gain fundamental insights into the effects of nanoparticle surface coverage ratio on antireflection performance of self-assembled nanoparticle AR coatings. Importantly, we found that optimal nanoparticle surface coverage predicted by the theoretical model can be readily achieved by this simple bottom-up technology.

575760 Electronic Structure of $Cu_{2-X}se$ Nanoparticles for Plasmonic Catalysis: A Density Functional Theory Study

Copper selenide ($Cu_{2-X}Se$) is a non-precious metal plasmonic material with potential for many applications in sensing and catalysis. A detailed understanding of the electronic properties of $Cu_{2-X}Se$ is important for tailoring and optimization of this material for specific applications. However, calculation of the electronic structure is very challenging because of the high mobility of the Cu atoms. We have used density functional theory to compute the band structure of ideal and defective structures of $Cu_{2-x}Se$ for x=0, 0.25, and 0.5. The HSE06 hybrid functional is used in conjunction with the +U approach. It is known experimentally that the Cu atoms are disordered in $Cu_{2-x}Se$ at room temperature. We use ab-initio molecular dynamics simulations explore the impact of Cu disorder on the electronic structure, particularly the band gap. We compare our calculations with experimental data for the band gap as a function of the composition

576153 Application of Carbon Nanotube & Cellulose Nanocrystal Composite in Conducting Films

The fabrication of cellulose nanocrystals (CNCs) into high performance films has gained increasing attention as a potential material for the manufacturing of micro-scale structures and devices due to their high mechanical strength, thermal stability and anisotropic properties. However, CNCs alone are an insulating material, limiting their use in sensor applications but introduction of carbon nanotubes (CNTs) allows for conductive composite films. CNTs are also rod-shaped nanomaterials with high mechanical strength and electrical conductivity. The hydrophobic regions of surface functionalized CNCs interact with the hydrophobic CNTs creating a favorable association between CNTs and CNCs. Additionally, CNC's long-range electrostatic repulsion contributes to stable CNT/CNC water dispersions. CNC assisted single-wall and multi-wall CNT dispersion capacities in aqueous solution have been measured by UV-vis spectroscopy and atomic force microscopy. Conductive and anisotropic composite films are casted on glass microscope slides and sheared by blade coating with greater than 12 wt% CNT/CNC aqueous dispersions. The films were then examined for orientation for both CNTs and CNCs by polarized light and Wide-Angle X-ray Scattering. It was observed that solutions with different CNC concentrations yield different dispersion capacities for multi-walled CNTs. Additionally, sheared films exhibited an anisotropic conductive property compared to non-sheared films. CNCs with different charge densities via surface functionalization have been introduced as alternatives for improving composite film dispersity and conductivity.

578263 Influence of Defect Microstructure on the Structural Color of Colloidal Films

Self-assembled films of colloidal scale particles can exhibit structural color, intense wavelength-specific reflection derived from geometric diffraction. Due to their lack of dye agents, structural color materials have potential to serve as pigments or sensing materials in chemically degrading environments. While the relationship between colloidal particle length scale and structural color reflection peak wavelength is well studied for self-assembled films, the connection between assembly quality and intensity of color is not. Thickness, crystal defects, and other imperfections impact the perceived color purity of an assembled film. Here, we study this connection by investigating the structural color reflection peak shape as a function of film microstructure *via* simulation and experiment.

By combining Molecular Dynamics simulations, Transfer Matrix Method calculations, and Finite Difference Time Domain calculations we explore and explain experimental results obtained for evaporative assembly of polystyrene spheres. We find that different commonly observed assembly defects (stacking faults, vacancies, film cracking, etc.) have different magnitudes of effect on colloidal film structural color. Furthermore, we find that films with small quantities of defects may be approximated as multilayer materials. These findings can guide the design of optical materials to tune structural color intensity.

578658 MoS_2-Pt Deposited on Nitrogen-Doped Carbon Nanotubes As Efficient Electrocatalyst for Hydrogen Evolution Reaction

Renewable energy sources, such as solar and wind, can supply a large amount of electrical energy in a sustainable mode. However, their massive use has been limited by their variability and intermittent nature. Energy storage systems can be an effective solution to ease these disadvantages. Redox flow battery (RFB) systems have been pointed out as an attractive option for energy storage applications that, also, can enable the integration of renewable energy sources with the current electricity grid. A RFB is an electrochemical device that stores electrical energy as soluble redox electro-active species in a liquid electrolyte for longer cycle life, and quick response times. Recently, the hybrid H2-Br2RFB system (E_{cell}=1.09 V) has been subject of different investigations since the redox reactions of Br2and H_2at the positive and negative electrode, respectively, own fast kinetics and great reversibility. However, the kinetics of the H_2reactions at a Pt/C electrocatalyst is affected by crossover through the membrane of bromide species, reducing the overall performance of the device.

In this research, MoS_2 and MoS_2-Pt nanoparticles were synthesized on nitrogen-doped carbon nanotubes (N-CNT) by a single-step surfactant assisted via. Scanning electron microscopy (SEM) was carried out to study the distribution, particle size and morphology of the nanocomposites. The kinetics and electrocatalytic activity of the nanocomposites were studied by cyclic voltammetry (CV), rotating disk electrode (EDR) and electrochemical impedance spectroscopy (EIS) towards the hydrogen reactions in acidic media with additions of HBr. SEM micrographs showed that MoS_2 or MoS_2-Pt were deposited at the surface of the N-CNT as spherical-particles with 20 nm diameter approximately. Element mapping by SEM confirmed the formation of MoS_2active materials. A decrease in the overpotential for hydrogen evolution reaction was observed when using a one pot synthesis compared to mechanical blend. Finally, MoS_2-Pt @ N-MWCNT exhibited high stability in the presence of HBr as compared with Pt/C electrodes for further use as hydrogen electrode in a H_2-Br_2 RFB.

19AIChE
Annual Meeting, Orlando, FL

Combustion synthesized nickel-cobalt catalyst deactivation by coking during ethanol decomposition reaction

Anand Kumar, Anchu Ashok
Department of Chemical Engineering
College of Engineering
Qatar University, Doha, Qatar
akumar@qu.edu.qa

Faris Tarlochan
Department of Mechanical and Industrial Engineering, Qatar University, Doha, Qatar,

Abstract

In this work, we report the catalytic dehydrogenation of ethanol over porous cobalt oxide and nickel-cobalt bimetal oxide ($NiCoO_2$). Catalyst deactivation and the nature of carbon formation on these two catalysts are studied in detail to correlate the active phase with carbon formation. In-situ FTIR analysis was conducted between 50°C to 400°C on Co_3O_4 and $NiCoO_2$ catalysts to understand the reaction mechanism and product selectivity. Addition of Ni improves the activity for ethanol decomposition by achieving complete ethanol conversion at 350°C as compared to 420°C for cobalt alone. The crystallinity, morphology and particle analysis of the spent catalyst after reaction was identified using XRD, SEM and TEM respectively. The XRD shows a phase change of porous $NiCoO_2$ to NiCo alloy, whereas SEM indicates the presence of fibrous structure on the surface with 91.7 % of carbon while keeping 1:1 ratio of Ni and Co after the reaction. The detailed analysis of carbon structure using HRTEM-STEM shows simultaneous growth of carbon nanofibers (CNFs) and multiwalled carbon nanotubes (MWCNTs) that were favored by larger and smaller crystallites respectively.

Keywords: Solution combustion synthesis, Nickel cobalt catalysts, Hydrogen production, Ethanol decomposition

Introduction

In past decade nanoparticles gained high attention in the field of industry, medicine and research due to its specific property and characterization [1-12]. The application of nanoparticles in the field of industry and science varied as catalyst, energy storage, biomedicines and optics. Nickel and cobalt are one of the promising catalysts for research studies owing to its tunable physical and chemical properties when mixed in appropriate ratio to form bimetallic compounds or mixed oxides. It is used in the major applications of gas sensing, battery storage, heterogeneous catalyst, pigments and fuel cells [2,13-19]. Several techniques were employed to synthesize nicke, cobalt nanoparticles with various sizes and shape includes pyrolysis, wet impregnation, sono-chemical process, microwave irradiation, precipitation and sol-gel process etc. In our work we follow simple, economical and fast mode of synthesis, known as combustion method, in which highly crystalline particles were prepared in single step without any post preparation such as calcination[4,5,8,17,20-27]. In SCS homogeneous mixture of solution is prepared by the molecular level mixing of the metal nitrates and fuel and heated to initiate the combustion process. The redox reaction between the reducing agent (fuel e.g.: glycine, cellulose, citric acid, glucose, urea, hydrazine) and oxidizing agent (metal nitrate) initiate combustion process and produce a self-propagating flame over the mixture of solution. The stoichiometric reaction for the synthesis of metal nanoparticles using SCS technique can be found in many published articles[5,8,28-31], involving a key parameter φ, known as fuel/oxidizer ratio. The parameter φ is non to determine the oxidation state of the metal in the output. $\varphi > 1$ is fuel rich condition, $\varphi < 1$ is fuel lean condition and $\varphi = 1$ is the stoichiometric condition with no need of atmospheric oxygen for fuel oxidation.

Herein, we utilized solution combustion synthesis to prepare cobalt oxide catalysts, and nickel-cobalt mixed oxide catalyst. These catalysts were evaluated for their performance towards ethanol dehydrogenation reaction. A detailed account of the reaction conditions, product distribution and fresh/spent catalyst characterization can be obtained from the published article[27]. Here we briefly summarize the outcomes of those experiments.

Experimental

Cobalt nanoparticles were prepared with metal nitrate and glycine as the initial precursor with a metal-fuel ratio $\phi = 0.5$, which is suitable for synthesizing oxides[26,32,33], to obtain a 3g of product as output. These mixtures was dissolved in 25 ml of deionized water and kept it for 1 h to obtain a homogenous solution. Thus prepared solution was heated over a hot plate heater at 300ºC in air at atmospheric pressure until all the water get evaporated and the mixture reaches to its ignition temperature. The nickel-cobalt mixed oxide was prepared following the same steps, expect that nickel precursor (nitrate) was also present along with the cobalt.

Crystalline phase of the particles were identifies with XRD analysis using Rigaku MiniFlexII Desktop X-ray powder diffractometer (Cu-Kα radiation) at a scan range of 10 – 90 degrees with a wavelength of 0.154056 nm. Morphology study was done with Scanning Electron Microscope (Nova Nano 450, FEI, 200 kx magnification) coupled with EDS for the elemental analysis. A high resolution transmission electron microscopy (HRTEM, TECNAI G^2 F20, FEI) has been used for the particle size identification of the derived nanopowders. The ethanol decomposition property of prepared catalyst was conducted with in situ Diffused Reflectance Infrared Fourier Transform Spectroscopy (DRIFT) technique. Thermo Nicolet 6700 FTIR with a reaction chamber (Harrick, HVC) along with an optic assembly of praying mantis was used to carry out the DRIFTS study. The sample was reduced first by placing in the chamber and heated the temperature at 300ºC in H_2 (100 sccm) atmosphere. Once the reduction is over, approx. after one hour, N_2 was purged over the chamber to collect the background spectrum. Ethanol was passed over the catalyst with N_2 as carrier gas through an ethanol bubbler at two different temperatures of 50ºC and 400ºC, and the spectra was collected with 256 scan between 700-4000 cm^{-1} at a resolution of 4cm^{-1} .

The ethanol conversion studies were conducted using a customized tubular flow reactor consisting of a quartz tube (internal diameter of 10.5 mm, and 35 cm length) and a K-type thermocouple to measure the catalyst bed temperature. The freshly prepared catalyst powder were agglomerated to form pellets using a hydraulic press by applying a load of 25 MPa for 10 min. These agglomerates were broken up and sieved to obtain particle size between 600μm – 1mm, out of which 0.4 g was loaded in the tubular reactor. The catalyst was reduced at 300°C in presence of H_2 for 1 h at a flow rate of 30 sccm. Once reduced, the gas flow was changed to N_2 to remove the excess H_2 residuals from the catalyst surface while the reactor was cooled down to room temperature. Thereafter ethanol was introduced to the reactor by passing 100 sccm of N_2 through an ethanol bubbler. The gaseous product from the outlet of the reactor was connected to a gas chromatography (GC) to analyze the product gases. Perkin Elmer Clarus 580 GC containing Porapak Q column was used to measure the conversion of ethanol from room temperature to 500°C. Carboxen 1010 column was used to separate hydrogen, methane, carbon dioxide and carbon monoxide in product mixture. The ethanol conversion was calculated using following equation (1):

$$EtOH_{conv}\,(\%) = \left(\frac{EtOH_{in} - EtOH_{out}}{EtOH_{in}}\right) * 100 \qquad (1)$$

$EtOH_{in}$ and $EtOH_{out}$ are the molar flow rates of ethanol at the inlet and outlet of the reactor respectively. The hydrogen selectivity was calculated using the following equation-

$$H_2 \text{ selectivity } (\%) = \frac{H_{2-produced}}{3 \times EtOH_{converted}} \times 100 \qquad (2)$$

Summary of Results

A summary of the results is presented here and the full details of analysis can be accessed in the published journal article[27]. A detailed XRD analysis indicates the formation of Co3O4 and NiCoO2 after the combustion reaction. The mixed oxide was further confirmed using the EDS analysis that indicated an approximate ratio of 1:1:2 for Ni:Co:O elements. Temperature programmed reduction (TPR) experiments were conducted to test the reducibility of each catalyst, which indicated low temperature reduction of NiCoO2 as compared to the Co3O4 catalyst. TEM along with detailed phase mapping indicates a nanoscale homogeneity in phase distribution of NiCoO2 as uniform distribution of all the three elements were observed without any phase segregation of Ni, NiO, or oxides of cobalt.

XPS spectrum of O1s showed three Gaussian peaks associated with O_L, O_v and O_c centered at 529.2 eV, 531.25 eV and 532.5 eV respectively[27]. The O_L peak indicates the existence of metal-oxygen bonding on the lattice of Co_3O_4. The O_v peak is associated with defects or oxygen vacancies and the O_c peak is related to chemisorbed oxygen or dissociated species (O^-, O^{2-}, O_2^-) on the surface. The O_v/O_c ratio was measured as 1.287 and 0.92 for $NiCoO_2$ and Co_3O_4 representing the presence of more chemisorbed oxygen on the monometallic cobalt species. Moreover, oxygen vacancies increase and higher surface defects are expected with alloying. The mesoporous structures with more oxygen defects could improve the adsorption capacity of ethanol and allow ethanol activation at lower temperature.

A detailed DRIFTS analysis of ethanol decomposition reaction was conducted on both, the cobalt and nickel-cobalt catalysts[27]. Cobalt catalyst indicates the adsorption of ethanol and formation of ethoxy intermediate that subsequently generates aldehyde and acetate before completely decomposing to H_2, CH_4 and CO_2 gases at high temperature. Nickel-cobalt catalyst also displays the formation of acetaldehyde at intermediate tempeatures, however it decomposes quickly to produce CO, CH_4 and H_2 at $T > 400°C$.

SEM analysis of the used samples indicates the formation of fibrous structures on the surface of nickel cobalt catalyst with 91.7 % of carbon while maintaining the 1:1 ratio of active metals Ni and Co respectively, after the reaction. The detailed analysis of carbon structure using HRTEM-STEM shows the simultaneous growth of carbon nano fibers (CNFs) and multiwalled carbon nanotubes (MWCNTs) that were favored by larger and smaller crystallites respectively. A detailed mechanism of carbon deposition is also proposed and the readers are referred the published journal article[27].

Acknowledgement.

This publication was made possible by NPRP grant (NPRP8-145-2-066) from the Qatar national research fund (a member of Qatar foundation).The statements made herein are solely the responsibility of the author(s). The authors also wish to acknowledge the technical services granted by Central Laboratory Unit (CLU) and Gas Processing Centre (GPC) at Qatar University.

References

1. Varma A, Mukasyan AS, Rogachev AS, Manukyan KV. Solution combustion synthesis of nanoscale materials. *Chem Rev*. 2016;116(23):14493-14586.

2. Ashok A, Kumar A, Bhosale R, et al. Influence of fuel ratio on the performance of combustion synthesized bifunctional cobalt oxide catalysts for fuel cell application. *Int J Hydrogen Energy*. 2019;44(1):436-445.

3. Matin MA, Kumar A, Saad, Mohammed Ali H Saleh, Al-Marri MJ, Suslov S. Zn-enriched PtZn nanoparticle electrocatalysts synthesized by solution combustion for ethanol oxidation reaction in an alkaline medium. *MRS Communications*. 2018;8(2):411-419.

4. Ashok A, Kumar A, Tarlochan F. Preparation of nanoparticles via cellulose-assisted combustion synthesis. *International Journal of Self-Propagating High-Temperature Synthesis*. 2018;27(3):141-153.

5. Thoda O, Xanthopoulou G, Vekinis G, Chroneos A. Review of recent studies on solution combustion synthesis of nanostructured catalysts. *Adv Eng Mater*. 2018:1800047.

6. Matin M, Kumar A, Bhosale R, Saad MS, Almomani F, Al-Marri M. PdZn nanoparticle electrocatalysts synthesized by solution combustion for methanol oxidation reaction in an alkaline medium. *RSC Advances*. 2017;7(68):42709-42717.

7. Wang L. Tailored design of architecturally controlled pt nanoparticles with huge surface areas toward superior unsupported pt electrocatalysts. *ACS applied materials & interfaces*. 2012;4(6):2865; 2865-2869; 2869.

8. Kumar A. Current trends in cellulose assisted combustion synthesis of catalytically active nanoparticles. *Ind Eng Chem Res*. 2019;58(19):7681-7689.

9. Qi Z, Xiao C, Liu C, et al. Sub-4 nm PtZn intermetallic nanoparticles for enhanced mass and specific activities in catalytic electrooxidation reaction. *J Am Chem Soc*. 2017;139(13):4762-4768.

10. MacKenzie KJ, Dunens OM, Harris AT. An updated review of synthesis parameters and growth mechanisms for carbon nanotubes in fluidized beds. *Industrial engineering chemistry research*. 2010;49(11):5323.

11. Xia Y, Xiong Y, Lim B, Skrabalak SE. Shape- Controlled synthesis of metal nanocrystals: Simple chemistry meets complex physics? *Angewandte Chemie International Edition*. 2009;48(1):60-103.

12. Aruna ST, Mukasyan AS. Combustion synthesis and nanomaterials. *Current opinion in solid state materials science*. 2008;12(3-4):44-50.

13. Schmitz,. Highly active methanol dissociation catalysts derived from supported molten salts. *Energy fuels*. 1994;8(3):729.

14. A.D. Schmitz, D.P. Eyman, K.B. Gloer. Highly active methanol dissociation catalysts derived from supported molten salts. *Energy fuels*. 1994;8(3):729.

15. M.A. Pena, J.P. Gomez, J.L.G. Fierro. New catalytic routes for syngas and hydrogen production. *Applied catalysis*. 1996;144(1-2):7.

16. Ashok A, Kumar A, Matin MA, Tarlochan F. Probing the effect of combustion controlled surface alloying in silver and copper towards ORR and OER in alkaline medium. *J Electroanal Chem*. 2019;844:66-77.

17. Ashok A, Kumar A, Ponraj J, Mansour SA, Tarlochan F. Highly active and stable bi-functional NiCoO2 catalyst for oxygen reduction and oxygen evolution reactions in alkaline medium. *Int J Hydrogen Energy*. 2019;44(31):16603-16614.

18. Matin MA, Saad MS, Kumar A, Al-Marri M, Mansour SA. Effect of fuel content on the electrocatalytic methanol oxidation performance of pt/ZnO nanoparticles synthesized by solution combustion. *Appl Surf Sci*. 2019;492:73-81.

19. Nazir R, Kumar A, Ali S, Saad MAS, Al-Marri MJ. Galvanic exchange as a novel method for carbon nitride supported CoAg catalyst synthesis for oxygen reduction and carbon dioxide conversion. *Catalysts*. 2019;9(10):860.

20. J.B. Wang, C.H. Li, T.J. Huang. Study of partial oxidative steam reforming of methanol over Cu–ZnO/samaria-doped ceria catalyst. *Catalysis Letters*. 2005;103(3):239.

21. N. Iwasa, T. Mayanagi, W. Nomura, M. Arai, N. Takezawa. Effect of zn addition to supported pd catalysts in the steam reforming of methanol. *Applied catalysis A, General.* 2003;248(1-2):153.

22. M. Scott, M. Goeffroy, W. Chiu, M. A. Blackford and H. Idriss. Hydrogen production from ethanol over Rh–Pd/CeO_2 catalysts. *Topics in Catalysis*. 2008;51:13-21.

23. Wang Z, Wang W, Lu G. Studies on the active species and on dispersion of cu in cu/SiO2 and cu/zn/SiO2 for hydrogen production via methanol partial oxidation. *International journal of hydrogen energy*. 2003;28(2):151-158.

24. R.M. Navarro, M.A. Pena, J.L.G. Fierro. Production of hydrogen by partial oxidation of methanol over a cu/ZnO/Al2O3 catalyst: Influence of the initial state of the catalyst on the start-up behaviour of the reformer. *Journal of catalysis*. 2002;212(1):112.

25. Ashok A, Kumar A, Bhosale RR, Saleh MAH, van den Broeke, Leo JP. Cellulose assisted combustion synthesis of porous Cu–Ni nanopowders. *RSC Advances*. 2015;5(36):28703-28712.

26. Ashok A, Kumar A, Bhosale RR, et al. Cobalt oxide nanopowder synthesis using cellulose assisted combustion technique. *Ceramics International*. 2016;42(11):12771-12777.

27. Ashok A, Kumar A, Ponraj J, Mansour SA, Tarlochan F. Effect of ni incorporation in cobalt oxide lattice on carbon formation during ethanol decomposition reaction. *Applied Catalysis B: Environmental*. 2019;254:300-311.

28. Wolf E, Kumar A, Mukasyan A. Combustion synthesis: A novel method of catalyst preparation. In: Spivey J, Han Y, Shekhawat D, eds. *Catalysis.* Vol 31. Royal Society of Chemistry; 2019:297-346.

29. Patil KC, Hegde MS, Tanu R, Aruna ST. *Chemistry of nanocrystalline oxide materials : Combustion synthesis, properties and applications.* Singapore: World Scientific Publishing Co. Pte. Ltd; 2008:345.

30. Patil K., Aruna S., Mimani T. Combustion synthesis: An update. *Current opinion in solid state materials science*. 2002;6:507-512.

31. S. T. Aruna, S. Ghosh, K. C. Patil. Combustion synthesis and properties of Ce1− xPrxO2− δ red ceramic pigments. *International journal of inorganic materials*. 2001;3(4-5):387.

32. Kumar A, Wolf EE, Mukasyan AS. Solution combustion synthesis of metal nanopowders: Copper and copper/nickel alloys. *AIChE Journal*. 2011;57:3473-3479.

33. Kumar A, Wolf EE, Mukasyan AS. Solution combustion synthesis of metal nanopowders: Nickel- reaction pathways. *AIChE Journal*. 2011;57(8):2207-2214.

581636 What Is Happening Above Your Fluidized Beds?

Fred is one of pioneers in envisioning analogies in matters of science. He developed several mechanistic based correlations relied on the analogies between matters of particles and matters of fluids or molecules. He foresaw the freeboard of fluidized beds contain “clusters of galaxies”.

Predicting catalyst entrainment in fluidized bed regenerators is essential to the design and operation of these units. Entrainment rates dictate cyclone and dipleg design. However, entrainment rates are difficult to measure at operating conditions and difficult to predict. Most entrainment rate correlations assume that as the particle size in the bed becomes smaller, the probability for elutriation becomes higher. This may not be the case. High speed video imagining of particle clusters in and above a fluidized bed suggests that clustering is significant for fine FCC catalyst particles. Fluidized bed experiments at varying fines concentration, bed heights and bed internal locations, indicate that these clusters form in the bed. Some of these clusters are then ejected into the freeboard region. Hydrodynamics do not appear to be solely responsible for cluster formation. Cohesive forces such as electrostatics, capillary, and van der Waals forces, may play a significant role in particle cluster formation and ultimately the entrainment rate.

581638 The Origins and Evolution Of Cyclone Design: How Innovation and Miss-steps Led To The Current State Of Cyclone Design

As Fluid-bed technology grew in popularity after being introduced nearly 100 years ago, the need for condensing and return of entrained particulates in a cost effective and reliable method increased. Inertial separators known as cyclones gained popularity due to their simplicity, reasonable pressure drop and good efficiency. As demands for ever higher efficiency at ever higher loadings and smaller particle sizes challenged the "simple" design criteria for cyclone sizing, innovations and new sizing criteria had to be developed. Testing performed at PSRI and PemmCorp, geared toward both trouble shooting and research-oriented intentions exposed flaws in previous thinking and led the way to new design criteria.

Underlying accepted design practices is a theory based equation for efficiency calculation that is clearly not polished since data contradict it. As highly loaded cyclones became more widespread, the accepted notion that tangential entry cyclones were more efficient than volute entry cyclones eroded away, and the reason was eventually determined, starting with an "accidental" observation. The new demands placed on cyclones required sizing criteria based on a combination of science and observation and allowances made for the duty under which the cyclone would be operated.

Understanding how a cyclone actually functions and the miss-conceptions associated with cyclone design are the subject of the presentation, which includes video of the actual tests performed, examining such features as inlet designs, vortex length etc.

581639 Career Long Work On Bulk Solids Handling Projects

I began work for Fred Zenz and PSRI in 1970 to 1976, including completing my master's thesis on feed point and elbow losses in pneumatic conveying.

My presentation will include a few stories about my time with Fred and my PSRI work, and then move into a PowerPoint presentation on career-long work on bulk solids handling projects, including sizing, conveying, drying, and associated air pollution control.

581640 Cyclones a la Zenz?

Dr. Fred Zenz was a one of the first people to chronicle flows in fluidized beds, risers, solids transport systems and cyclones in his famous pioneering Zenz/Othmer book "Fluidization and Fluid-Particle Systems:. Cyclones (which separate solids from an exit gas stream) are a necessary piece of equipment in nearly all fluidized-bed processes, and Fred was one of the first to develop a technique to predict cyclone efficiencies and cyclone pressure drops. The method that he developed is still in use today by many organizations and was widely publicized in the "Manual on Disposal of Refinery Wastes" in 1975.

The Zenz method of characterizing cyclone operation consists of taking the particle size distribution (PSD) in the carryover (entrained) solids from a fluidized bed and subdividing the PSD into several "cuts". Each "cut" is then analyzed to determine if it can be collected by the cyclone relative to a collection model. The model is then used to calculate cyclone efficiency at extremely low loadings, and the effect of loading on cyclone efficiency is predicted.

Cyclone pressure drop is determined by summing several individual components of the pressure drop. A little-known aspect of cyclone pressure drop is that as solids loading increases from near zero loading, cyclone pressure drop initially decreases as solids loading is increased. The Zenz method cyclone analysis takes this unique characteristic of cyclones into account.

In this presentation, the Zenz method of cyclone analysis will be described in detail.

581823 Fred Zenz – A Graduate Student Perspective?

Dr. Fred Zenz was many things to many people: a researcher, teacher, engineer, innovator, consultant, businessman, father, husband. I knew him as a graduate advisor when I was a graduate student at Manhattan College, and it is my honor to share some experiences from that perspective. At the time, in 1980-81, Fred was balancing multiple roles as a consultant, Technical Director of PSRI, VP of Ducon, and professor of chemical engineering at Manhattan College. So he was busy, very busy, and anyone needing his attention needed to work to get it. But once you had it, he was all in, 100% focused on you and your particular issue. He would help, guide, advise, inform, and do it all with seemingly effortless grace, and good humor. We were all were better for having known him.

567826 Cross-Scale Validation and Evaluation of Wurster Coating Processes

The Wurster process is of major interest throughout the pharmaceutical industry and is commonly used to modify the release rate or the taste of granulated beads. The coating process takes place in the Wurster tube, in which the particles are transported centrally upwards followed by a fountain region and subsequent downwards movement outside the tube. The beads are gathered in the downbed region before transported through the Wurster tube again. Thus, the particle flow resembles a torus-shaped recirculation pattern.

Due to various equipment scales and complex thermal effects, a process transfer, e.g., from lab to production scale, might be challenging. Furthermore, several combinations of operating conditions can lead to similar outcome in terms of thermodynamic behavior. Basic scale-up routines rely on the scaled mass flow through the bottom plate with correction factors for batch mass, distribution plate and process time. Nevertheless, they neglect differences with respect to the specific coating solvent, thermal operating conditions and particle trajectories (i.e., cycle and residence times).

Due to the complexity of the process, recent numerical investigations were limited to simplifications regarding geometry, batch size or spray modeling (see e.g. Askarishahi et al. [1] and Pietsch et al. [2]). To our knowledge, also the direct comparison between different real scale application devices has not been demonstrated before.

In our work a CFD-DEM framework is presented where a commercial code XPS (eXtended Particle System) is used for the DEM simulation and is coupled to AVL-FIRE™ for the CFD simulation. Both implement state-of-the-art models for fully coupled momentum, heat and mass exchange. For the lab scale, a numerical model of the Glatt GPCG-2 is built, whereas for the production scale, the Glatt GPCG-30/60 Wurster coater is used. Most importantly, we aim for a novel view on the scale up between the two devices. This is based on (i) particle trajectories, (ii) particle temperature distribution and (iii) drying times. All numerical analysis are validated with large scale experiments and performed at both scales which enables a direct comparison.

[1] M. Askarishahi, S. Mohammadsadegh, S. Radl. Full-Physics Simulations of Spray-Particle Interaction in a Bubbling Fluidized Bed. AIChE Journal, 63(7), 2569-2587, 2017.

[2] S. Pietsch, S. Heinrich, K. Karpinski, M. Müller, M. Schönherr, F. Kleine Jäger. CFD-DEM modeling of a three-dimensional prismatic spouted bed. Powder Technology, 316, 245-255, 2017.

Simulation and Experimental Validation of a Lagrangian Multiphase Model for Rotary Bell Cup Atomization

Wanjiao Liu[1], Brandon Petrouskie[2], Kevin Ellwood[1], Reza Rock[2], Seyed Alaie[1] and John Furar[2],
(1)Ford Motor Company, Dearborn, MI, (2)PPG Industries, Allison Park, PA

Abstract

Rotary bell cup atomization is the primary method within the automotive industry to apply coating onto car bodies. The spray formed from rotary bell cups is electrically charged and directed towards the target surface by shaping air as well as an oriented electric field. The droplet transport process will have great impact on transfer efficiency and the final appearance of the coating. Simulation of the droplet transport process allows one to form a better understanding of the complex flow field in rotary atomization, as well as providing a virtual manufacturing capability resulting in reduced design cost and cycle. In this work, we developed multiphase Computational Fluid Dynamics(CFD) approach to simulate the droplet transfer process in rotary atomization. Lagrangian multiphase modelling is used to simulate the discrete droplets and is two way coupled to the continuous phase. Paint layer thickness and droplet characteristics are discussed, and the results are compared against experimental data obtained by high speed imaging. Furthermore, secondary particle interaction models and their impact on the simulation results are also investigated.

Introduction

In order to apply the coatings required for aesthetics and surface protection, the automotive industry primarily uses rotary bell cup atomization. The final paint layers, including primer, basecoat and clearcoat are all applied through this type of applicator and thus the final appearance of the surface coating is greatly affected by the atomization process. To achieve atomization of the paint, the bell cup rotates at high speeds (up to 60krpm), while the paint is supplied through the center of the bell cup. This results in the formation of a thin paint film on the surface of the bell cup under centrifugal force. When the thin film leaves the edge of the bell cup, it typically forms ligaments and eventually break up into droplets. Once the droplets are formed, they are traveling within the same plane as the rotating surface and are not directed towards the spraying target. Thus, additional shaping forces are needed to re-direct them to the target. Shaping air tubes are placed behind the bell cup and high-speed air is directed towards the edge of the bell cup where the droplets are forming. In addition, the droplets are directly charged by contacting the bell cup surface that is held at high electric potential, and are pushed towards the substrate by the electric field formed between the bell cup and the grounded target. Fig. 1 shows the configuration of a typical rotary bell cup.

Both droplet formation and droplet transport processes are complicated multiphase, multi-physics problems. There have been several past studies with the goal of developing simulation methods [1,2,3,4,5]. Droplet transport and spatial distributions are important characteristics of spray and will affect the final appearance of the coatings. However, most studies have been focused on the film thickness validation, and there lacks a thorough study and validation on the spatial droplet transport and distribution. In this work, we will be focusing on the role of shaping air in the droplet transport process. Shaping electric field will be the subject of future studies, due to the additional complexities and the lack of knowledge on droplet charge distributions. Our goal is to develop a numerical tool based on Lagrangian Multiphase approach to simulate the droplet transfer process when shaping air is used to shape the spray. Two main characteristics of the spray, including paint thickness distribution and droplet spatial distribution are investigated and compared with experimental results. The paint thickness distribution is a key indicator for spray pattern uniformity and finishing quality. Additionally, the droplet spatial distribution greatly influences the final appearance of the coating, since droplets of different sizes might have different characteristics and behavior. We will investigate the secondary physics model, including turbulence model, wall interaction model, drag force model, collision model and discuss their impact on the results.

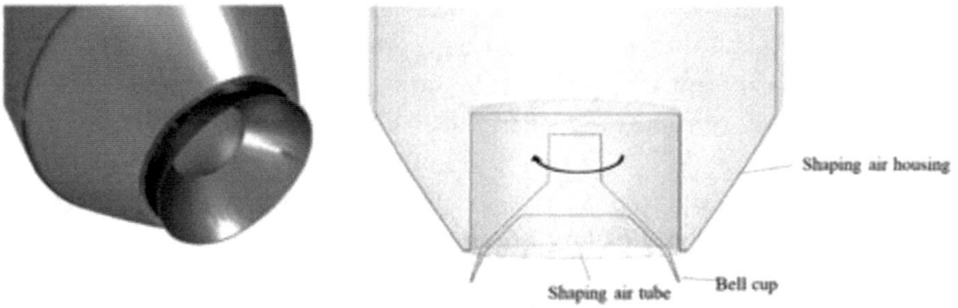

Fig. 1 Rotary bell cup and shaping air tube configuration

Simulation approach

Computational fluid dynamics(CFD) software package STAR-CCM+ is used for the simulation. Lagrangian multiphase flow approach is used to track the particle motion, and the continuous air phase is solved by Detached Eddy Simulation(DES). The particle motion and the air flow field are two way coupled. Descriptions of solvers are given below but the details and formulations are omitted here and can be found elsewhere [6].

Particle Solver

The basic concept for the implementation of a Lagrangian multiphase approach is to, first, understand that tracking all individual particles is unrealistic, given the large number of particles present in the field. Rather, a more practical implementation is to bin particles into parcels, where each parcel is viewed as representing a collection of particles with the same properties. The parcel trajectories are solved by the same equation governing individual particle trajectories. The only force particles are subject to is the drag force, where the drag coefficient is described by the Schiller-Naumann model. Further discussion of the drag force model can be found in the results section. To reduce the computational cost and prevent particles from accumulating over time, particles are removed from the computational domain once they reach either the substrate or a large distance from the center of the spray.

Continuous phase solver

The continuous air phase is solved by compressible Navier-Stokes equation because the shaping air operates with compressed gas leading to an injection velocity close to the speed of sound. The particle and air phase are two way coupled and the force from the particle to the air flow is the negative of the particle drag force. In the momentum equation this force is reflected as a body force that is averaged in space.

The air flow field of the bell cup atomization process is highly turbulent. Turbulent modeling is therefore required and is critical for correctly capturing the particle trajectories. Strong interaction of turbulence structures and particles are also present, thus requiring additional care when modeling the interaction between the turbulence and particles. Detached Eddy Simulation(DES) is used in this work as the flow is generally unsteady. DES can be briefly described by the following: areas with high resolution are generally solved with Large Eddy Simulation(LES), while close to walls and in areas with low resolution RANS is used. This strategy ensures that the inherent unsteadiness is resolved while the simulation will still be computationally efficient. Interaction between parcels and flow are resolved instead of being modeled. The current implementation in the solver, however, does not include a subgrid scale turbulent-particle interaction model, which will be discussed further in the result section.

Secondary physics model

The paint emitted from bell cup will form a dense particle field, leading to a strong coupling between the air phase and the particle phase. Furthermore, the particle field can be dense enough that droplet-droplet interactions are not negligible. A droplet collision model is, therefore, used to capture the coalescence of droplets in the spray. The effects of other secondary physics, including turbulence modeling, droplet wall interaction, and drag force model are investigated and the details will be discussed in the results section.

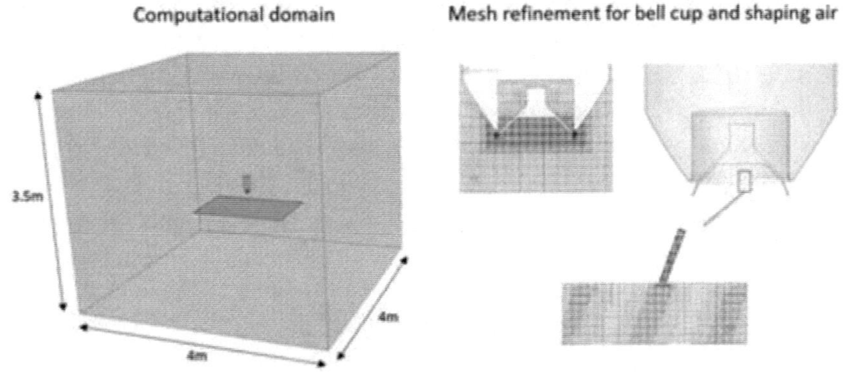

Figure 2 Computational domain and mesh

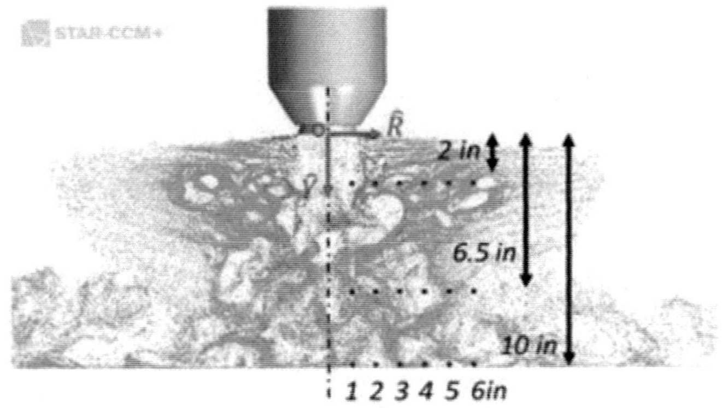

Figure 3 Coordinate system and sampling locations. Particles shown are colored by diameter.

Table 1. Simulation conditions

Paint density(kg/m^3)	1200
Paint flow rate(ccm)	400
Shaping air flow rate(nlpm)	300
Bell cup diameter(mm)	65
Bell cup rotation speed(rpm)	40000
Bell cup edge to target distance(in)	10
Total simulation time(s)	1.5

Simulation setup and conditions

The input parameters to the model can be found in Table *1*. Figure 2 illustrates the simulation domain with boundaries created so that they are far enough from the spray to not interfere with the flow near the bell cup. The top boundary is set as a velocity inlet, which corresponds to the draft air in the spray booth, with a value of 1m/s. The rest of the boundaries are defined as pressure outlet boundary conditions. Particles are emitted from 30 point injectors 1mm from the edge of the bell cup. The particle size distribution is obtained from experiment and then used as a source for each of the point injectors. To produce a more uniform injection process near the cup edge, the injector locations are randomized at each time step. In order to have a quasi-steady initial flow, the Lagrangian phase injection into the domain is delayed by 0.2s. Film thickness growth rates from the simulation are calculated from the time averaged mass flux of Lagrangian phase on the target, which are subsequently converted to "dry film" thickness from the mass loss during baking for comparison against experimental data. In addition to film thickness, the droplet size distributions are collected at 18 different locations, at R=1, 2, 3, 4, 5, 6*in* and Y=2, 6.5, 10*in*, as shown in Figure 3. The droplet size distribution is collected from the instantaneous result at 1.5s, but is spatially averaged around the annular volume so that it is still a statistically averaged result.

Experimental Setups

The overall droplet size distribution used as simulation input was measured using a laser diffraction spray analyzer for which the setup is shown in Figure 4. In this system, an incident light beam passes through a sample volume where it can be either absorbed or scattered. A Fourier lens redirects the scattered light to the detector. The droplet size distribution is then interpreted from the signal based on Mie light scattering theory[7]. High speed imaging is also carried out close to the bell cup to validate the measured size distribution.

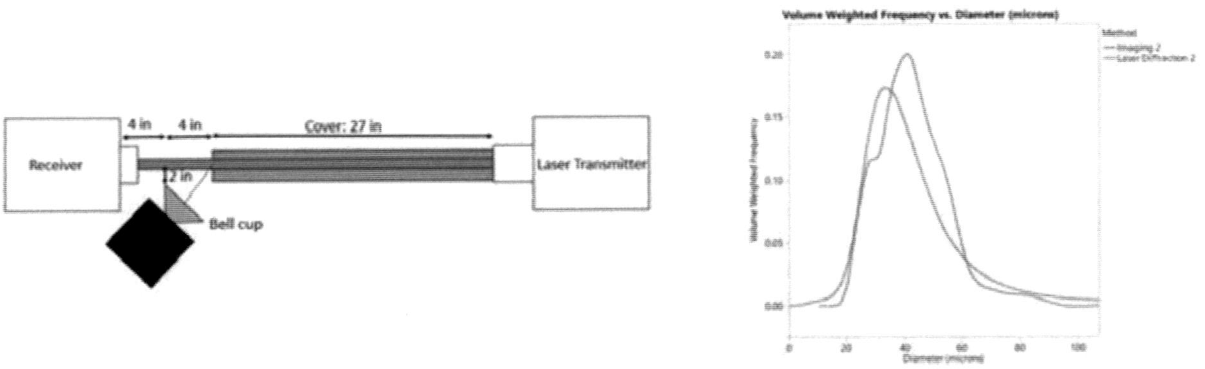

Figure 4 Droplet distribution measurement setup, and the measured droplet size distribution

The droplet spatial distribution was measured using highspeed imaging. This technique uses a nanosecond pulse laser to illuminate the background as a camera captures images of the spray. Image analysis techniques extract droplet size distributions from the resulting images. Measurements were taken at the same locations as in Figure 3 to compare the experimental and simulation droplet spatial statistics.

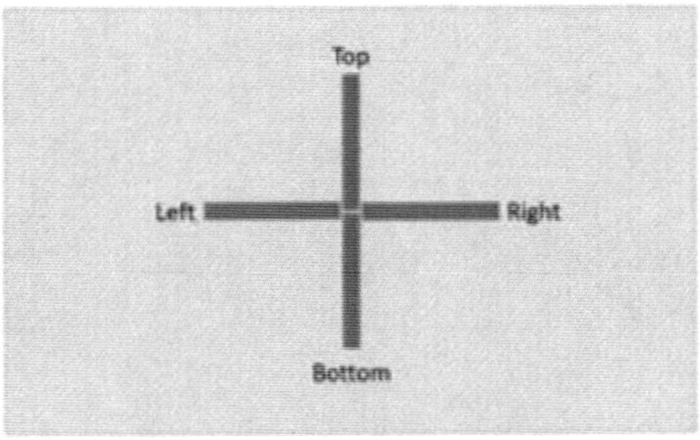

Figure 5 Spraying target setup

The rotary bell cup was positioned 10*in* from the spraying target. Four 2*in* by 12*in* panels were placed on the target directly in front of the atomizer in a grid pattern with 4 distinct panel positions (top, bottom, left, right), as shown in Figure 5. The atomizer was sprayed for a specified amount of time and the panels were placed in a 140°C oven for 20 minutes to cure the coating and evaporate any solvent. Dry film thickness measurements were taken in increments of an inch on each of the panels. The film thickness growth rates in micron per second were reported for each radial position averaged across each panel position.

Results and discussion

Air flow pattern and spray plume shape

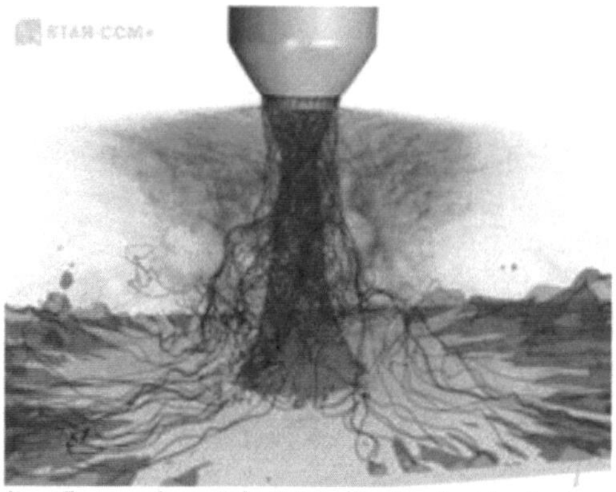

Figure 6 Recirculation zone, swirling flow and particle field. Region inside the isosurface has negative axial velocity,

The general air flow structures of bell cup atomization have been studied previously[8,9]. The high rotation rate of the bell cup creates a strong swirling motion leading to low pressure recirculation zone in the center and negative axial velocity towards the applicator. Outside the recirculation zone the shaping air creates a strong positive axial momentum towards the target. The length of the recirculation zone can vary depending on the ratio of the swirling flow to the shaping air flow [10]. For the condition in this study, the recirculation zone extends all the way to the target due to the strong shaping air flow rate. This can be seen in Figure 6, where the areas inside the iso-surface indicate negative mean axial velocity region. The diameter of the recirculation column experiences a contraction upstream and then an expansion closer to the target. Outside the recirculation zone, the flow has a high positive axial velocity and is highly turbulent and unsteady as seen in Figure 7. Such unsteady motion will have a great influence on particle deposition since most particles are concentrated in that area.

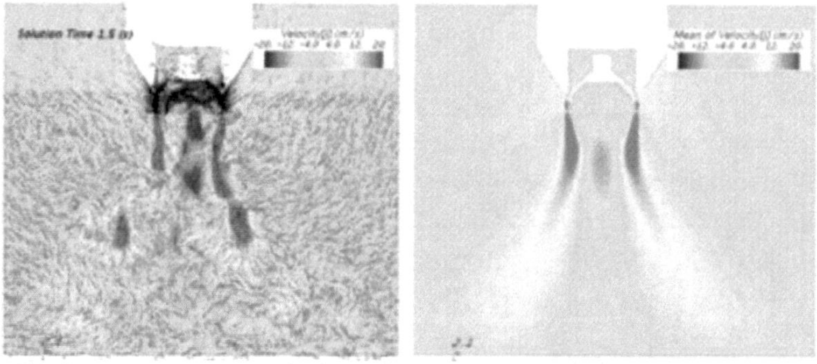

Figure 7 Left: Instantaneous axial velocity field and vector. Right: mean axial velocity field

The plume shape of the spray is compared against experimental data and the result is shown in Figure 8. The plume shape is obtained from the simulation by plotting the mean volume fraction of Lagrangian phase. In general, the shape agrees very well with experiment with a slight overprediction of the widest portion of the plume width.

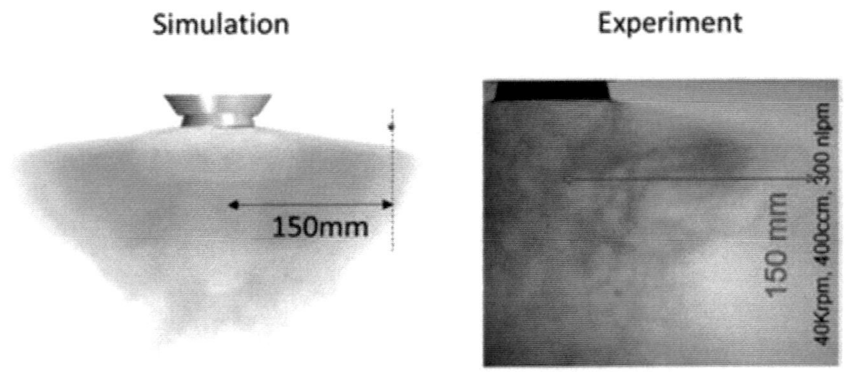

Figure 8 Shape of the plume from simulation and experiment

Film thickness growth rate profile

Figure 9 shows the mean axial velocity field together with the film growth rate distribution on the target. It can be seen that peak film growth location occurs at the same location as the peak axial velocity. Inside the recirculation zone, the axial velocity is negative which leads to droplets traveling backwards and away from the target, resulting in low deposition rates at the center. The film growth rate distributions results are shown in Figure 9 as a function of radial distance from the center. As a comparison, the results from steady Reynolds Averaged Navier-Stokes(RANS) and Unsteady Reynolds Averaged Navier-Stokes(URANS) are shown. The k-omega model with turbulent dispersion is utilized within both RANS and URANS simulations. Both model formulations correctly capture the general "ring" shape of the distribution seen in the experiments. However, the peak location of the film growth rate profile is different, and DES results show better agreement than RANS and URANS. In addition, steady state RANS failed to reach a converged steady state solution due to the inherent unsteady nature of the flow as seen in Figure 7.

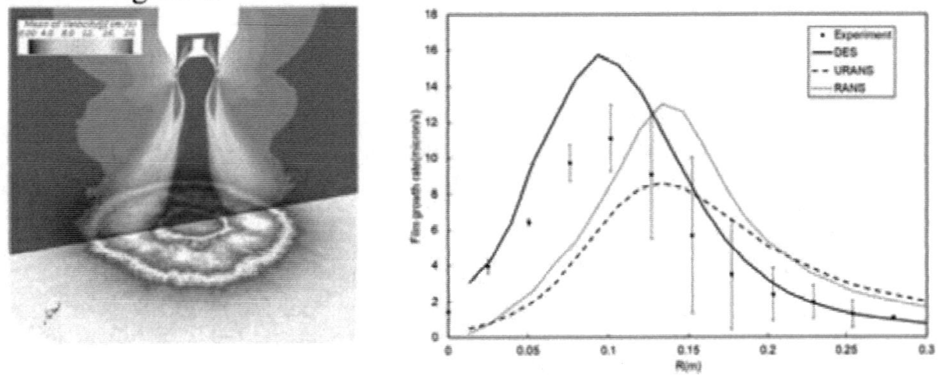

Figure 9 Left: Mean axial velocity field and spray pattern on the target. Right: Film growth rate with different turbulent models vs experiment results.

Despite better agreement in the shape of the film growth rate profile, the DES results were approximately 20% larger than the results from experiments. This indicates that there is additional loss of particles in space that is not captured in the simulation. The additional loss must occur near the target or from the transport in space.

In the current model, it is assumed that a droplet will "stick" when comes into contact with the target. However, for a liquid droplet, depending on the impact condition and whether the wall is dry or wetted, a droplet can have different behaviors. If a droplet rebounds, it will be likely to be entrained again in the vortices near the wall and could be carried away from the target. A standard model for wall interaction is Bai-Gosman wall interaction model, in which the droplet can adhere, rebound, spread or splash depending on the droplet Weber number[11]. The critical droplet Weber numbers used are the default settings within the STAR-CCM+ solver, and rebound will happen when particle Weber number is between 2 and 20. Figure 10 shows the film growth rate when the Bai-Gosman wall interaction model is utilized, and Figure 11 shows the distribution of rebound droplet mass flux. Adding the wall interaction model has significantly reduced the peak film growth rate from 16 micron/s to 6 micron/s. The reduction in film thickness with Bai-Gosman wall interaction suggests that a significant portion of particles rebound and are carried away by the flow. From the results, it seems that the wall interaction does have a significant impact on the droplets behavior near the wall and the film thickness distribution magnitude. Further study is needed to investigate the specific behavior of paint droplets when interacting with walls.

Another particle loss mechanism that is not currently considered is the subgrid scale turbulent dispersion. DES resolves all large eddies and their interaction with particles. However, subgrid scale models are lacking and could lead to underprediction of the particle dispersive motion[12,13]. Small particles are particularly susceptible to the subgrid scale dispersive forces, which could also explain some discrepancies later seen in the particle spatial distribution.

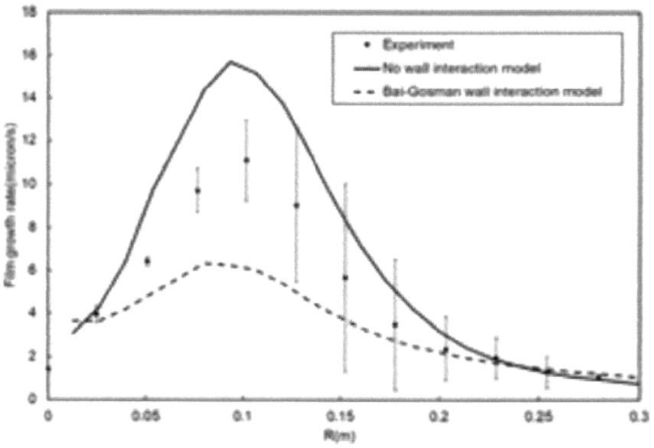

Figure 10 Film growth rate for simulations with and without particle wall interaction model, vs experiment

Figure 11 Mean mass flux of rebound droplet on the target

Particle size spatial distribution

Droplet spatial distribution is of interest because droplets of different sizes might behave differently during transport in the air. They could also behave differently after they hit the target where secondary flow in the paint film is induced. It is therefore important to ensure that the simulation models predict the trajectories and spatial distributions of the droplets correctly.

The distribution of droplets with small, medium and large sizes are shown in Figure 12. Droplets of different sizes clearly have a spatial separation with the smaller droplets distributed towards the center and the larger droplets distributed in the outer areas. This is due to the fact that larger particles have higher inertia and, thus, more difficult to be redirected by aerodynamic forces from shaping air flow. Results showing the volume weighted particle probability density function(PDF) are shown in Figure 13. The initial droplet separation can be seen at Y=2*in*. Further downstream at Y=6.5*in*, the PDF from radial location of 1*in* to 4*in* shows that the droplets are well mixed and thus have very similar PDFs, while at radial location of 5*in* and 6*in*, a bimodal distribution is seen where a second peak occurs close to 60microns. Near the target where Y=10*in*, all particles are well mixed and PDF are the same among all radial sampling locations. Comparing the trend qualitatively with experimental measurements, the separation at Y=2*in* and the final mixing at Y=10*in* are well captured. However, the experimental data does not show the second peak in PDF in the outer sampling regions. This means the larger droplets in the experiments turn back towards the center sooner leading to earlier mixing of different size particles.

Figure 12 Distribution of droplets with small, medium and large sizes

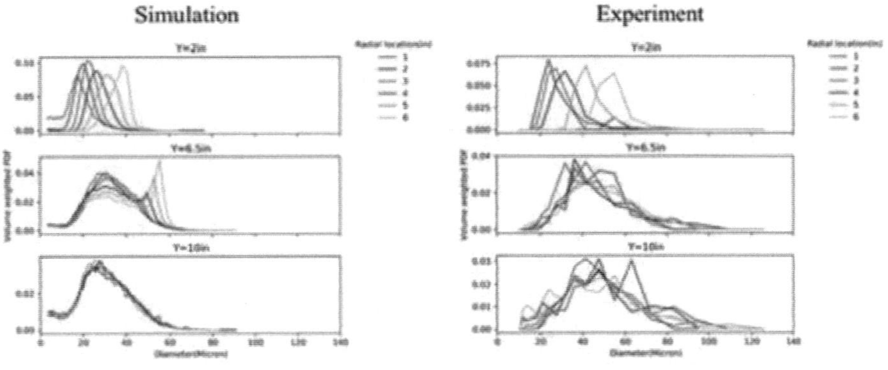

Figure 13 Spatial particle distribution at 18 sample locations, simulation vs experiment

Past research[14,15] has shown that liquid droplets can undergo deformation, or even oscillation when traveling at high speed, leading to a deviation in the drag coefficient relative to a solid particle, which is described by the Schiller-Naumann relationship. An empirical relationship for this deviation is[15]:

$$C_D^{liquid} = \left(\frac{Re}{Re_{crit}}\right)^{1.12} \cdot C_D^{crit}$$

where C_D^{crit} is the Schiller-Nauman drag coefficient:

$$C_D^{crit} = \begin{cases} \frac{24}{Re}\left(1 + 0.15Re^{0.687}\right) & Re \leq 1000 \\ 0.44 & Re > 1000 \end{cases}$$

and Re_{crit} is defined by

$$We_{crit} \cdot Re_{crit}^{0.65} = 165.$$

This relationship suggests that larger deviations from the solid particle drag occur for larger liquid droplet Reynolds numbers. Furthermore, larger droplet Weber numbers lead to a deviation from the solid drag coefficient model. Implementing such drag force correction, however, can be non-trivial because the original study was conducted for droplets reaching a steady or quasi-steady state. The droplet transport in rotary bell cup spray however is highly dynamic and unsteady, and one cannot assume the drag force of a liquid droplet released from the edge of the bell cup will immediately reach the corrected liquid droplet drag coefficient. The relaxation time for the droplet to reach the deformed state is also unclear and could be determined by other factors, such as droplet diameter. A simpler assumption is made so that the particles all experience the same relaxation time, after which the particle drag will change from solid particle drag to liquid particle drag. A hyperbolic tangent relationship is assumed:

$$C_D = C_D^{solid} + 0.5\left[tanh\left(\frac{t-t^*}{A}\right)+1\right]\cdot\left(C_D^{liquid} - C_D^{solid}\right)$$

Where t^* is the relaxation time and A controls how quickly the transition occurs. Three cases are carried out with three different relaxation times, chosen based on the travel time of the larger particles before reaching the widest point of the spray plum. The results are shown in Figure 14. Changing the relaxation time significantly changes the predicted particle spatial distribution at Y= $2in$ and Y = $6.5in$, where earlier mixing of larger particles is achieved in the cases with 0s and 0.002s relaxation times. Beyond that the bimodal distribution starts to show up in the PDF at Y=$6.2in$. The particle distribution at the target and the film thickness distribution are not much affected by the modified drag model, however, because by the time the particles reach the target they are all well mixed in all cases.

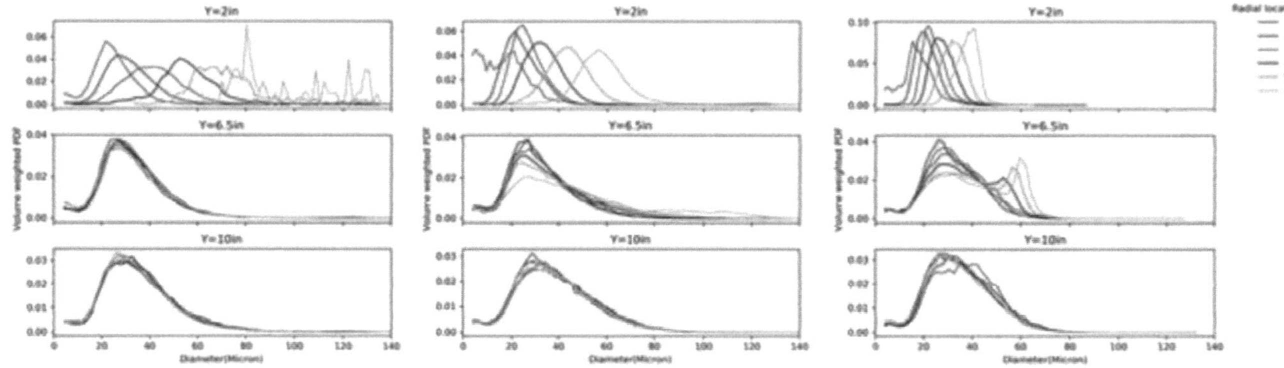

(a) (b) (c)

Figure 14 Liquid droplet drag correction with different relaxation times. (a) $t^* = 0s$, *(b)* $t^* = 0.002s$, *(c)* $t^* = 0.004s$

To look at the droplet size distribution as a function of target distance, PDF results are shown for three Y locations by adding the radial PDFs together, which are shown in Figure 15. The results show that there is a shift of droplet size distribution towards larger diameter closer to the target. When comparing the PDF results at Y=6.5in and Y=10in with the experimental data, it shows that the predicted PDF has a slight shift of profile towards smaller size. The subgrid scale dispersion mentioned earlier can be a contributing factor to this shift, which can result in overprediction of the landed small particles and a shift of PDF towards larger sizes. Simulation without the collision model is also carried out and the results are plotted in the same figure. It is shown that the standard collision model creates more larger particles and has improved the predictions. Further analysis on the collision model parameter as well as the effect of subgrid scale dispersion is needed in the future to further improve the accuracy of the model.

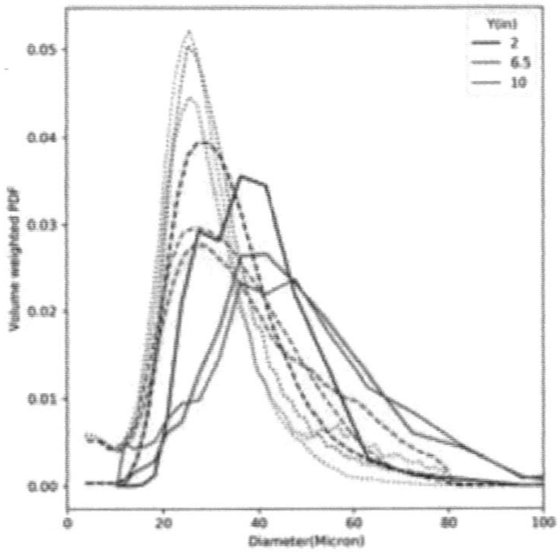

Figure 15 Volume weighted PDF are three Y locations. Solid line: experiment. Dashed line: simulation. Dotted line: simulation without the collision model.

Summary

In this work, we carried out simulations for the droplet transport process in rotary bell cup atomization. The results are compared against experimental data for plume shape, film thickness growth rate, and spatial particle size distribution. We investigated the influences of various secondary physics models, including turbulence modeling, wall interaction model, drag force model, and collision models. The film thickness growth rates show a good overall agreement in capturing the ring shape of the profile with both DES and RANS/URANS, while RANS/URANS show an offset in the peak film growth rate location compared to the experimental data. With DES, the peak location is captured correctly while there is approximately a 20% overprediction in the film growth rate. The wall interaction model was shown to significantly reduce the film thickness growth rate since droplet rebound effects allowed particles to be carried away by vortices near the wall. Corrections to particle drag force model due to liquid droplet deformation were investigated. The drag force model was shown to impact the particle spatial distribution, and using the liquid droplet drag correction led to early mixing of larger particles and better agreement with experiments. Subgrid scale turbulent dispersion is currently missing in the model and will be the subject of future studies, since it will likely impact both the film thickness and the droplet spatial distribution.

References

1. Ellwood, K. R., & Braslaw, J. (1998). A finite-element model for an electrostatic bell sprayer. Journal of Electrostatics, 45(1), 1-23.
2. Andersson, B. (2013), Modeling and simulation of rotary bell spray atomizers in automotive paint shops, Ph.D. Thesis, Chalmers University of Technology.
3. Domnick, J., Scheibe, A., & Ye, Q. (2003, July). The electrostatic spray painting process with high-speed rotary bell atomizers: influences of operating conditions and target geometries. In Conference on Liquid Atomization and Spray Systems.
4. Osman, H., Adamiak, K., Castle, G. P., Fan, H. T. C., & Simmer, J. (2015, November). Comparison between the numerical and experimental deposition patterns for an electrostatic rotary

bell sprayer. In ASME 2015 International Mechanical Engineering Congress and Exposition (pp. V02BT02A003-V02BT02A003). American Society of Mechanical Engineers.

5. Mark, A., Andersson, B., Tafuri, S., Engstrom, K., Sorod, H., Edelvik, F., & Carlson, J. S. (2013). Simulation of electrostatic rotary bell spray painting in automotive paint shops. Atomization and sprays, 23(1).
6. https://www.plm.automation.siemens.com/global/en/products/simcenter/STAR-CCM.html
7. https://www.oindpnews.com/wp-content/uploads/2012/10/Malvern-Laser-Diffraction-and-Automated-Imaging-Complementary-Techniques-for-Nasal-Spray-Development.pdf
8. Toda, K., Salazar, A., & Saito, K. (Eds.). (2012). Automotive Painting Technology: A Monozukuri-Hitozukuri Perspective. Springer Science & Business Media
9. Im, K. S., Lai, M. C., Liu, Y., Sankagiri, N., Loch, T., & Nivi, H. (2001). Visualization and measurement of automotive electrostatic rotary-bell paint spray transfer processes. Journal of Fluids Engineering, 123(2), 237-245.
10. Stevenin, C., Béreaux, Y., Charmeau, J. Y., & Balcaen, J. (2015). Shaping air flow characteristics of a high-speed rotary-bell sprayer for automotive painting processes. Journal of Fluids Engineering, 137(11), 111304.
11. Cossali, G. E., Marengo, M., & Santini, M. (2005). Single-drop empirical models for spray impact on solid walls: a review. Atomization and Sprays, 15(6).
12. Vinkovic, I., Aguirre, C., Simoëns, S., & Gorokhovski, M. (2006). Large eddy simulation of droplet dispersion for inhomogeneous turbulent wall flow. International journal of multiphase flow, 32(3), 344-364.
13. Weil, J. C., Sullivan, P. P., & Moeng, C. H. (2004). The use of large-eddy simulations in Lagrangian particle dispersion models. Journal of the atmospheric sciences, 61(23), 2877-2887.
14. Hu, S., & Kinter, R. C. (1955). The fall of single liquid drops through water. AIChE Journal, 1(1), 42-48.
15. Greene, G. A., Irvine, T. F., Gyves, T., & Smith, T. (1993). Drag relationships for liquid droplets settling in a continuous liquid. AIChE journal, 39(1), 37-41.

571706 Multiphase Particle-in-Cell Coupled Population Balance Equation Method for Multiscale Computational Fluid Dynamics

The Multiphase Particle-in-Cell coupled Population Balance Equation (MPPIC) Method for CFD of Processes with Micro-Sized Particulates

Shin Hyuk Kim[¢Ó], Richard D. Braatz[¢Ô], and Jay H. Lee[¢Ó*]

[¢Ó]Department of Chemical and Biomolecular Engineering, Korea Advanced Institute of Science and Technology, 291 Daehak-Ro, Yuseong-Gu, Daejeon, 34141, Republic of Korea

[¢Ô] Massachusetts Institute of Technology, 77 Massachusetts Avenue, Cambridge, MA 02139

(jayhlee@kaist.ac.kr[¢Ó*], braatz@mit.edu)

Many chemical processes, e.g., crystallizers, fluidized bed reactors and polymerization reactors, involve micro-sized particulates. The multiphase particulate flow exhibit complex patterns that are not seen from gas and liquid fluids. In order to facilitate the simulation of the multiphase particulate flow in the computational fluid dynamics (CFD) setting, there exist several CFD solvers, which can be categorized into the Euler-Eulerian and Euler-Lagrangian methods. The Euler-Eulerian expresses the flows of all phases using the continuum governing equations, but modeling the flow of particles of different types and sizes greatly complicates the continuum formulation because it requires a separate model for each type and size. The population balance equation (PBE) is one way to reflect the particle size distribution in the Euler-Eulerian setting. However, solving both the PBE and fluid dynamics simultaneously is costly because it is expressed as a partial differential equation involving a total of five variables: time, 3-dimensional space, and particle distribution. The Euler-Lagrangian uses the Lagrangian description for the particulate phase and the Euler description for the continuum phase. Using this approach, the particles can have different sizes, shapes, densities, and velocities. However, when the volume fraction of particles in the system is higher than 5%, the frequency of particle collisions is unrealistically increased and accuracy is drastically lowered[1]. Also, expressing chemical reactions or mass transfer for all the particles represented by the Lagrangian method and simulating particle size changes at the micro-scale can require unrealistically high computational cost.

To overcome the limitations of the conventional CFD solvers for such chemical engineering problems, this work proposes the multiphase particle-in-cell (MPPIC) model[2] coupled the population balance equation, which we refer to as MPPIC-PBE. Specifically, the current state-of-the art involves the consideration of PBE within the Lagrangian frame in an effort to combine with the particle distribution function used in MPPIC in a practical manner. This allows for the preservation of mass and energy between Eulerian and Lagrangian frames. The PBE in this procedure is directly linked to each of the discrete parcels and retains all the original information while maintaining the homogeneity. This characteristic makes the calculation of the particle size distribution more efficient than in the PBE approach in the Euler-Eulerian framework. For validation, a crystallizer[3] is simulated by both the newly developed method and the conventional method.

The simulation result demonstrates that the developed approach is more efficient and potentially more accurate than the conventional method.

1. O'Rourke PJ. *Collective drop effects on vaporizing liquid sparys*, Princeton University; 1981.
2. Andrews MJ, O'Rourke PJ. The multiphase particle-in-cell (MP-PIC) method for dense particulate flows. *Int J Multiphas Flow.* Apr 1996;22(2):379-402.
3. da Rosa CA, Braatz RD. Multiscale Modeling and Simulation of Macromixing, Micromixing, and Crystal Size Distribution in Radial Mixers/Crystallizers. *Ind Eng Chem Res.* Apr 18 2018;57(15):5433-5441.

558130 Cyclones with and without an Eccentrically Positioned Vortex Finder

With fluidized bed, circulating fluidized bed (CFB) and even pneumatic conveying operations, reverse flow cyclones are typically the critical component in a successful process. Although the designing principle for cyclones is straightforward, the hydrodynamics in these units are not. This poor understanding of the hydrodynamics has limited cyclone capabilities which why cyclone designs has not changed significantly in 150 years.

This is especially true with the symmetrical design of the cyclone. With the exception of the inlet, everything else on a cyclone is symmetric. For example, the gas outlet tube immersed into the cyclone, also called the vortex finder, is located concentrically, i.e. its axis coincides with the axis of the cyclone. However, it has been shown in numerous recirculating cyclones of CFB power plants that eccentric positioning of the vortex finder axis improves the cyclone efficiency without affecting the pressure drop [1,2]. By applying this measure the circulating material became finer and its quantity increased resulting in a considerable improvement of the boiler performance: The heat transfer in the combustion chamber and, if installed in the fluid bed cooler, improved the optimum combustion temperature to ~900°C and reduced the amount of unburnt filter ash. In addition, the optimization led to cost savings of operational costs.

Trefz [3] gave a qualitative explanation of the positive effect of eccentrically shifting the vortex finder axis on the cyclone efficiency based on flow measurements in cyclones. At the top of the cyclone barrel region, there is a secondary flow through the boundary layer of about 10% of the total flow carrying particles towards the center. The vortex finder forces this secondary flow around the tube down to the opening allowing to remove particles from that flow by centrifugal force. In cyclones with a tangential inlet and a concentric vortex finder, however, this separation is not optima. The axis of the vortex flow is pushed away from the geometrical center of the cyclone by the incoming gas flow and as a consequence the tangential velocity of the flow around the vortex finder is not uniform. In the section of the vortex finder surface which is closer to the cyclone axis the tangential velocity stagnates and the secondary flow streams vertically downwards with practically no circumferential velocity. In this section, the particle separation is poor. By shifting the vortex finder axis into the "eye" of the vortex a uniform flow around the vortex finder is achieved and its separation efficiency improves considerably.

Recently, numerical granular-fluid simulations of high loaded cyclones with a concentric and with an eccentric vortex finder have been performed by applying the CFD code Barracuda Virtual Reactor® of CPFD, LLC. The CFD results confirm that eccentrically shifting the vortex finder results in a better collection efficiency without compromising the pressure drop. The calculations are compared with measurement data of commercial high loaded cyclones in CFB boilers and the origin of the improvement due to the eccentric shifting the vortex finder is studied.

References

[1] Muschelknautz U, Muschelknautz E (1999). Separation efficiency of recirculating cyclones in CFB combustions. VGB PowerTech 4/1999, 48-53.

[2] Ipsen C, Roschek D, Muschelknautz U (2014) Optimierung eines Zyklonabscheiders einer zirkulierenden Wirbelschichtfeuerung. VGB PowerTech 4/2014:75-79

[3] Trefz M (1992). Die verschiedenen Abscheidevorgänge im höher und hoch beladenen Gaszyklon unter besonderer Berücksichtigung der Sekundärströmung. VDI-Fortschr.-Ber., VDI-Verlag, Düsseldorf, Reihe 3, No. 295.

574635 Fast CFD Based Optimization of Coal Moving Bed Gasifier

A coal gasification model has been developed, based on Coarse Grained Discrete Element Model in MFiX. The gasification model has been validated with data from a 200 KWh moving bed gasifier. The validated CFD model is used to optimize the syngas composition of the gasifier for use in a Combined Heat and Power (CHP) process and for use in a Fischer-Tropsch (F-T) process. To accelerate the time to solution of the simulations, a 1-dimensional gasification model was developed to model the transient start up process of the moving bed gasifier. The 1-dimensional model accounts for the drying, pyrolysis, gasification and char combustion during the startup period. Once the gasifier has reached its operating temperature, the results from the 1-dimensional model is used as initial conditions to the 3-dimensional CFD model. This workflow led to achieving the steady state operating condition within 10 minutes of simulation time, using 3-dimensional MFiX-CGDEM.

573885 Large Scale Multiphase Flow Simulation for Techno-Economic Assessment of the Novel Internally Circulating Reactor for Energy Conversion with CO_2 Capture

This work presents application of filtered Two Fluid Model (fTFM) for modelling of large scale fluidised beds in cost estimation and techno-economic assessment of a novel energy system in commercial scale. This energy system; internally circulating reactor, is based on chemical looping combustion (CLC) for energy conversion with integrated CO2 capture. The internally circulating reactor (ICR) has been developed at SINTEF and NTNU with high potential for scale-up. ICR simplifies the conventional chemical looping reactor configuration by combining the functionality of two reactors, cyclones, loop seals and solids transport lines into a single reactor unit. This is achieved by dividing the reactor into two sections, connected with specially designed ports for circulating the oxygen carrier particles between the two reactor sections. This reactor concept which has been successfully demonstrated in laboratory scale, promises to simplify the scale-up process, especially under pressurized conditions. To demonstrate the benefits of this concept in large scale, this study will conduct large scale reactor simulations for ICR using a recently published anisotropic filtered Two Fluid Model (fTFM) formulation. The fTFM is capable of providing reasonably accurate predictions of fluidized bed dynamics using the coarse computational grids required for industrial scale simulations within reasonable computational times. The costs estimation will be used in the techno-economic assessment of ICR concept in large scale.

571017 Enhanced Dissolution Performance of Amorphous Solid Dispersions By Size Reduction

Amorphous solid dispersions (ASDs) are widely used in pharmaceutical industries in order to enhance the drug absorbance, solubility, dissolution rate. Although the ASDs have higher solubility, the corresponding dissolution rate naturally depends on the nature of the polymeric matrix and particle size. Unfortunately, particle size reduction promotes recrystallization rate, because it increases the available surface areas of the ASDs to environment, both during storage and during exposure to dissolution medium. The effect of particle size on the dissolution behavior of amorphous griseofulvin (GF) particles was investigated using 0-50% Kollidon® VA 64 as a polymeric crystallization inhibitor. Results show that both the final dissolved GF concentration, and the dissolution rate of GF amorphous solid dispersions (ASDs) are inversely proportional to the particle size. The solution concentrations for the smallest (45-75 μm) size group with different polymer loading at 2 hours were 1.5 to 7 and 1.3 to 4 times higher than those for the largest (250-355 μm) group under non-supersaturating and supersaturating conditions, respectively. The dissolution rate of GF ASDs with 50% polymer loading for the finest group was 2.7 times higher than for the largest group under supersaturating conditions. The individual rates of dissolution and recrystallization were assessed through surface concentration and Avrami recrystallization rate kinetics. That indicated that particle size reduction enhanced ASD drug loading by reducing the amount of polymer necessary as finest size ASDs initially dissolve faster negating their higher recrystallization rate. Kollidon® VA 64 at 30% loading was sufficient to inhibit the GF recrystallization. Thus the combination of particle size reduction and recrystallization inhibition is effective for improving the dissolution behavior of GF ASDs.

574108 Modeling Particle Formation in Spray Drying of Amorphous Solid Dispersions

In this presentation we present a mathematical model of the various concurrent physical phenomena that a polymer solution with an Active Pharmaceutical Ingredient (API) undergo prior to forming an amorphous solid dispersion particle, such as atomization, co-solvent evaporative dynamics, propensity for phase separation and shell formation.

The numerical model is benchmarked against single-droplet drying data obtained with a suspended droplet apparatus indicating that the model captures, to a first approximation, the dominant physical phenomena (see Figure 1).

The motivation of the work is to support the development of Amorphous Solid Dispersion powders of poorly water soluble APIs in hydrophilic polymeric matrices which is a commonly used strategy to increase the solubility and dissolution rate of oral-dosage forms. However, successful development of amorphous solid dispersions for optimal performance and stability depends on both the formulation (e.g. drug load, polymer system and surfactants) and on the manufacturing technology/process conditions. Optimizing the formulation and the process requires extensive laboratory work, which is often compromised by time and cost constraints. Therefore, the ability to perform virtual experiments is highly advantageous and cost-effective.

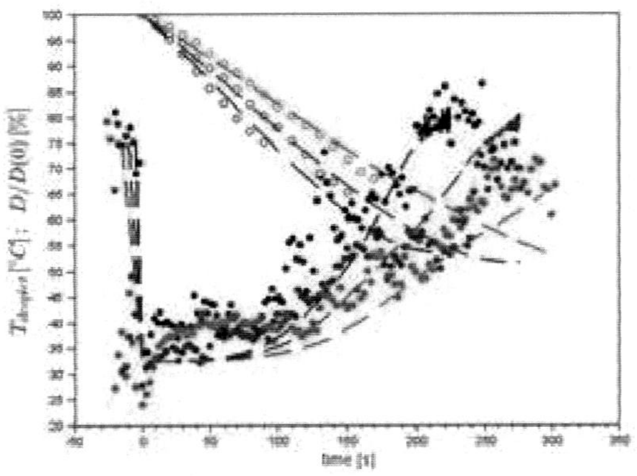

Figure 1: Model prediction (dashed-line) benchmarked against experimental data (filled circles represent Temperature measurements, whereas hollow circles represent droplet size).

*Vulcanus in Japan-Europe program for supporting Mr. Masahiro Nakai's internship

574286 From Liquid to Droplet: Understanding the Ligaments Break-up in Atomization Process

From liquid to droplet: understanding the ligaments break-up in atomization process

T. Porfirio[1,2*], F. J. Galindo-Rosales[3,4], L. Campo-Deaño[4], J. Vicente[1], V. Semião[2]

[1] Hovione Farmaciência SA, Estrada do Paço do Lumiar, 1649-038 Lisbon, Portugal; *tporfirio@hovione.com

[2]LAETA, IDMEC, Mechanical Engineering Department, Instituto Superior Técnico, Universidade de Lisboa, Av. Rovisco Pais 1, 1049-001 Lisbon, Portugal

[3]Instituto de Ciência e Inovação em Engenharia Mecânica e Industrial, Rua Dr. Roberto Frias 400, 4200-465 Porto, Portugal

[4]CEFT, Departamento de Engenharia Mecânica, Faculdade de Engenharia da Universidade do Porto, Rua Dr. Roberto Frias, 4200-465 Porto, Portugal

The droplet formation is a key phenomenon in spray drying process in which the particle size is mainly dependent of atomization performance. The spray formation is described as the breakup of a liquid jet into droplets in a gaseous atmosphere. Atomization has been studied by many scientists from more than a century in which different mechanism were proposed for the several applications. Currently, the atomization process is viewed as the instability and collapse of simple liquid-jet systems [1]. The whole process can be divided in three steps: sheet formation, sheet destabilization into ligaments and droplet formation. Firstly, the ejection of the liquid through the nozzle orifice forms a fine sheet of liquid in conic shape that contacts with the surrounding gas. The cohesive and disruptive forces in the formed sheets rise and result in perturbations and instability of the liquid resulting in breakup into ligaments. The third phase is the further breaking of such ligaments into droplets due to capillary force. This process is referred as primary atomization. If the produced droplets exceed the critical size, further disintegration can occur to form smaller droplet – secondary atomization [2].

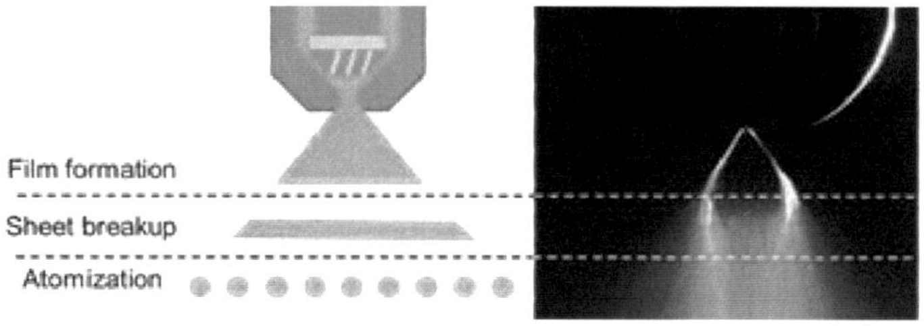

Figure 1 – Droplet formation mechanism [3].

The droplet formation process is strongly influenced liquid properties mainly viscosity, surface tension and density [4]. Two types of viscosities must be distinguished in droplet formation process: shear viscosity η_s

and extensional viscosity η_e. The extensional viscosity is related with the viscoelasticity and has special relevance in the ligaments break-up and in secondary atomization. The effect is not visible in the early stages of sheet break-up close to the nozzle discharge. The difference is usually observed far from the discharge in which the ligaments are stretched and elongated [5]. Depending of fluid viscoelasticity, the droplet formation mechanism is affected, and it may lead to poor atomization. The viscosity and rheological behavior are highly dependent of the polymer (important for the formulation of amorphous solid dispersions), mixture of solvents and concentrations. The proposed work explores different case studies explored to link the spray drying feed solution variables (polymer, solvents and concentration) with the rheological behavior and atomization performance. The test fluids comprised the most used polymer (HPMCAS, HPMC, copovidone and Eudragit) to formulate amorphous solid dispersions and dissolved in organic solvents and water.

The rheology of the fluids was also measured in shear and extensional flow. The extensional experiments were conducted in a Capillary Breakup Extensional Rheometer (CaBER™ 1 Thermo Haake GmbH, Karlsruhe, Germany) in which the sample is placed between the two plates at an initial position and is elongated by a plate separation following the Slow Retraction Method [6]. The sample forms a filament / ligament bridging the two plates that thins due to surface tension until it breaks. The steady shear viscosity curve was determined by means of a rotational rheometer with a plate-plate geometry.

Two sample fluids are here compared. The same polymer with a solid concentration of 6% was dissolved in two systems: 1) acetone and water, 2) acetone. Both fluids showed a shear thinning behavior in which the viscosity of acetone solution was lower than the acetone/water solution. However, the extensional experiments show a contrary result for the viscoelasticity of the fluids: the acetone-based fluid presented a high viscoelasticity and relaxation time (Figure 2). In fact, this fluid presents a bead-on-string effect that can lead atomization problems. Higher deviations in the prediction of particle size distribution (mainly Dv10 and Dv90) were observed for this case.

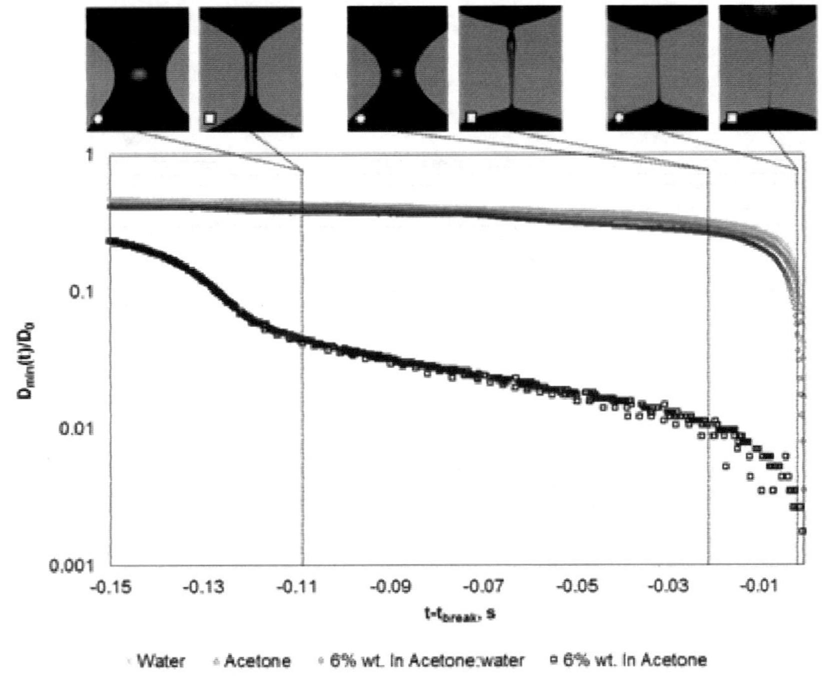

A set of case studies will be explored to show the impact of feed solution variables, including polymer type, on the atomization and spray drying process. The incorporation of these data in atomization models to increase the accuracy of spray drying scale-up will be discussed.

[1] K. Masters, *Spray drying handbook*, Third edit. New York: John Wiley & Sons, 1979.

[2] V. Lefebvre, A., McDonell, *Atomization and Sprays*, Second edi. CRC Press, 2017.

[3] R. J. Schick, "Understanding Drop Size," Wheaton, 2008.

[4] A. H. Lefebvre and X. F. Wang, "Mean drop sizes from pressure-swirl nozzles," *J. Propuls. Power*, vol. 3, no. 1, pp. 11–18, 1987.

[5] B. Keshavarz *et al.*, "Studying the effects of elongational properties on atomization of weakly viscoelastic solutions using Rayleigh Ohnesorge Jetting Extensional Rheometry (ROJER)," *J. Nonnewton. Fluid Mech.*, vol. 222, pp. 171–189, 2015.

[6] L. Campo Deaño and C. Clasen, "The Slow Retraction Method (SRM) for the determination of ultra-short relaxation times in capillary breakup extensional rheometry experiments", *J. Nonnewton. Fluid Mech.*, vol 165, pp 1688-1699, 2010.

573159 Analysis of the Microstructure of Particles Obtained By Evaporating Acoustically Levitated Single Droplets Using X-Ray Computed Tomography

Analysis of the microstructure of particles obtained by evaporating acoustically levitated single droplets using X-ray computed tomography

Hassan Abdullahi [1], Christopher L. Burcham [2] Thomas Vetter [1*]

[1] *School of Chemical Engineering and Analytical Science, University of Manchester, Manchester, UK;*

[2] *Eli Lilly & Company, Indianapolis, IN, USA*

thomas.vetter@Manchester.ac.uk

The flowability and tabletability of powders is an important consideration in the manufacturing of pharmaceutical products. These properties can be successfully tuned by crystallising particles of a desired morphology, or, more flexibly, by engineering the microstructure of the individual particles in a spray drying process it can be tuned by altering the formulation and process conditions [1][2]. Therefore, spray drying offers a unique opportunity to design tailor made particles in a single step process involving simultaneous droplet drying and particle formation. However, carrying out an in-depth investigation of particle morphology and microstructure directly on the scale of a typical spray dryer (that contains millions of droplets at any time) is infeasible. Instead, single droplet studies are often used to understand how drying conditions affect the microstructure of the resulting dried particles [3][4]. In the present work, an acoustic levitation approach is used to suspend single droplets in an atmosphere of controlled relative saturation and temperature (Figure 1). Through a camera and automated image analysis, we study the droplet drying dynamics, as well as the macroscopic morphology of the resulting particle. We have further applied Raman spectroscopy and scanning electron microscopy to study the solid state form, as well as the surface characteristics. By applying X-ray computed tomography, we have elucidated the internal microstructure of the formed particles and have uncovered a fascinating variety of structures, depending on the mixture dried (we have investigated combinations of different solutes, solvents, and polymers acting as excipients), as well as the processing conditions (temperature and relative saturation of nitrogen used as drying gas). The results can be correlated with formulation and process conditions to produce drying maps for different systems.

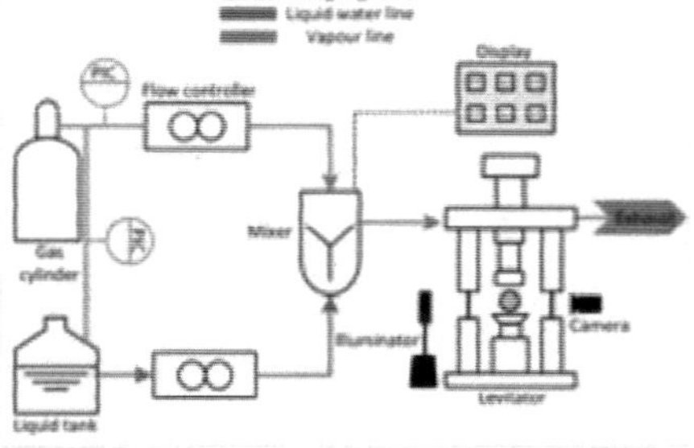

Figure 1: Suspension of a droplet in a controlled atmosphere using an acoustic levitator.

The droplet drying history during evaporation is shown in Figure 2a for an aqueous formulation of mannitol (4.5 wt%) and PVP (0.9 wt%). In the first drying stage, the diameter of the droplet decreases as solvent is evaporated from the droplet surface. In the second drying stage, a shell forms and evaporation occurs through interstices in the shell. The evaporation rate hence reduces significantly due to the added resistance of the crust. When presented as a plot of normalised squared diameter D^2/D_0^2 against time, the two drying regimes can be clearly identified. We have performed such characterisation for a variety of process conditions and the particles obtained can be subsequently analysed to provide useful information on particle properties such as size, shape, porosity, roughness and crystal structure. Figure 2b and 2c shows exemplary images of particles obtained using SEM and micro X-ray computed tomography. By investigating the droplet drying dynamics and resulting particle microstructure, for a variety of mixtures and operating conditions enables us to gain deep insight into the process behaviour and the properties of the resulting particles.

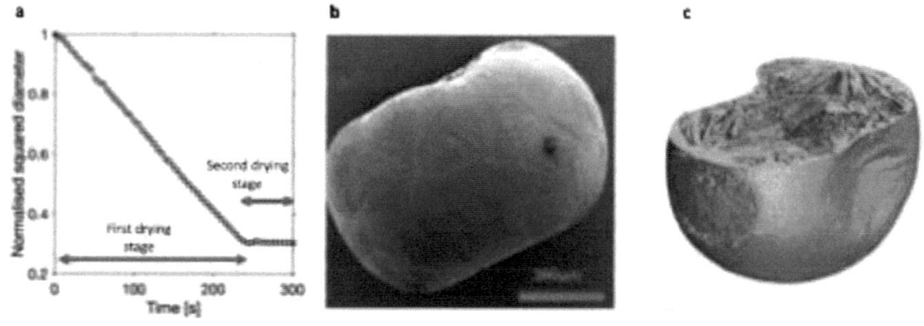

Figure 2: **a)** Droplet drying history and particle obtained from drying an aqueous formulation containing 0.9%w/w polymer, 4.5%w/w mannitol at 60C and 0% relative humidity **b)** SEM image **c)** cut-away surface from micro X-ray tomography.

References

[1] Carver, K. M., & Snyder, R. C. (2012). Ind. Eng. Chem. Res, *51*(48), 15720.

[2] Mšnckedieck, M., Kamplade, J., Fakner, P., Urbanetz, N. A. (2017). *Drying Technology*, *35*(15), 1843.

[3] Osman, A., Goehring, L., Patti, A., Stitt, H., & Shokri, N. (2017). Ind. Eng. Chem. Res, *56*(37), 10506.

[4] Sugiyama, Y., Larsen, R. J., Kim, J., & Weitz, D. A. (2006). Polymer Suspensions, (8), 6024.

572842 Decoupling Temperature and the Kinetics of Dye Conversion Via Scalable Nano-Encapsulation in a Thermal-Responsive Media

Chemical treatment processes in which substrates are exposed and interact with chemicals require quality control to ensure process specifications such as temperature, dosage, and duration are satisfied. One approach for quality control is the use of colorimetric dyes which undergo chemical reactions in response to an applied treatment, thereby shifting their absorbance spectrum. When incorporated in a process run alongside samples, operators can observe these post-treatment color changes as qualitative success metrics. While this approach is simple to implement, the reliance on the kinetics of dye conversion presents challenges including dye aging during storage which can lead to false positive results as well as the inability to decouple the effects of dosage and temperature on the final result.

We seek to overcome this disadvantage by decoupling the kinetics of dye conversion from temperature through the implementation of a wax matrix surrounding dye nanocapsules. The wax matrix serves as a barrier which melts and exposes dye-encapsulated nanoparticles at a specific melting temperature, commensurate with that of the treatment process. To form dye nanocapsules directly in wax, the versatility of flash nanoprecipitation (FNP), a scalable nanoparticle fabrication technique, is expanded. In this work, an organic solution of dye and polymeric stabilizer is impinged against a molten wax to create turbulent mixing, yielding the super-saturation, precipitation, and subsequent two-level encapsulation of the dye. Characterization of the resultant nanoparticle-imbued waxes is presented as well as the response of encapsulated dye to treatments at temperatures above and below the melting point of various wax matrices. As a result, the ability of these novel materials to thermally trigger exposure and initiate the kinetic response of the dye is demonstrated. Nanoparticle-imbued waxes are shown to be amenable to further post-processing including emulsification, spray prilling, and printing.

569768 Evaporative Heat Transfer Model for QbD of Agitated Filter Bed Drying

Drying of an active pharmaceutical ingredient (API) is an important unit operation in the pharmaceutical industry as it is used to control residual solvent content, in addition to controlling appropriate physical properties. Drying operation is often time consuming and may turn out to be a bottleneck in the overall process cycle time. Many APIs are potent compounds that require filtration and drying to be done in a single contained vessel, known as an agitated filter dryer (AFD), to minimize operator exposure.

Scale-up/tech-transfer of drying process is quite challenging as different AFDs have different surfaces that could transfer heat that ends up impacting effective heat transfer coefficients across the bed as we scale the process. Understanding of bed temperature and residual solvent across the filter cake is critical for reliable and robust scale-up of AFD drying protocol.

In this work, we present results from a novel 3D heat-transfer model that allows us to understand the temperature evolution across the filter cake considering the equipment variability at scale. We also present experimental techniques that can be employed to extract relevant model parameters. In the end, we would provide the decision tree to guide efficient design and reliable scale-up of agitated filter drying.

Disclosures:

All authors are AbbVie employees and may own AbbVie stock. AbbVie sponsored and funded the study; contributed to the design; participated in the collection, analysis, and interpretation of data, and in writing, reviewing, and approval of the final abstract.

560684 Novel Co-Processing Technology to Generate Drug Substances Amenable to Direct Compression

Many active pharmaceutical ingredients (APIs) display poor powder properties and cannot be directly compressed into intact tablets with sufficient strength. The desired powder properties are often difficult to achieve through conventional particle engineering approaches, such as particle size and habit modification during crystallization. Co-processing of API with excipients can significantly improve the functional properties to overcome these challenges. In this talk, we will present a novel co-processing technology that we developed at Bristol-Myers Squibb to improve powder properties in which polymer is precipitated and coats the crystalline particles resulting in discrete, nearly spherical agglomerates. We identified critical process parameters and demonstrated scalability up to 50 kg scale in the pilot plant using theophylline. Furthermore the co-processed theophylline generated under various process conditions were formulated into a blend at 50-90% loading and successfully processed by direct compression.

576956 Liquid Transfer between Particles in the Impregnation of Catalysts in a Rotating Drum: Axial Dispersion Studies and the Effect of Particle-Particle Contact

Impregnation of active metals onto a porous catalyst support is an important step in the preparation of heterogeneous catalysts. In a typical impregnation process, metal solutions are sprayed over a particulate bed in a mixing vessel until the pore volume is reached. The inter-particle variability of the impregnated liquids inside the particles and metal content may significantly affect the activity and selectivity of the resulting catalyst. The dispersion of material in the axial direction is the most critical factor determining the final uniformity of metal content in the particle bed. The aim of this work is to characterize water/metal transfer in an alumina particle system, which is essential for a desired content uniformity in the catalyst impregnation process.

A series of mixing experiments and matching simulations were designed to characterize the axial dispersion in a horizontally rotating cylinder. The experiments are initially set up with a 3-inch band of the impregnated particles located at the center of the cylinder. At each side of the center band, a band of dry particles is placed in a symmetric way. The cylinder is set to rotate for 16 minutes at a constant rotational speed, with samples being taken at 5 evenly spaced locations along the rotational axis, every 2 minutes. The concentration profiles of water or metal content start with a maximum value at the center, and then as time progresses the profiles approach a flat line. Based on a "wide band" model [Sherritt et al, 2003], axial dispersion coefficients are calculated for the cases of both metal transfer and water transfer in the rotating drum. Results are compared between the same experiment made with water or metal solution and show that in the rolling regime the dispersion coefficient increases with the rotational speed.

In addition, a particle-particle contact experiment was designed to further understand the impregnation of liquid with different concentrations and to determine simulation parameters to be incorporated in the liquid transfer algorithm. A fully impregnated particle (saturated with water or nickel nitrate solution of different concentrations) is placed next to a dry particle on the bottom of a small vial. Transfer of liquid is observed during different lengths of contacting time (range from 1min to 20min). Results show that the rate of transfer for a concentration of 3M is 2 times larger than that of 0.1M. Impregnation simulations incorporating different rates of transfer are considered and different metal concentrations are compared.

564456 Simulation of Flow-Driven Segregation in Mixtures of Disks and Rods

While much effort has been devoted to understanding granular segregation in mixtures of spherical particles, e.g. size- and density-driven segregation, non-spherical particle mixtures are much more common in real-world applications. Here, we characterize the orientation and segregation velocity in bidisperse mixtures of equal density disk-like and rod-like particles. We use discrete element method simulations of a quasi-2D bounded heap geometry, which includes a wide range of concentrations and shear rates, with particles modeled as super-ellipsoids to evaluate the effect of varying the length ratio and diameter ratio between particle species on segregation. The simulations reveal that particles orient to minimize the cross-sectional area perpendicular to the flow direction and that the segregation velocity in mixtures of disk- and rod-like particles can be predicted by the volume ratio. Funded by The Dow Chemical Company.

569340 Predicting Granular Density Segregation—a Kinetic Theory Approach

In dense granular flows, particles of different densities segregate as they flow. Here we use momentum conservation equations for individual constituents to derive an expression for the segregation velocities of density bidisperse segregating species in a dense granular flow. These segregation velocities are then used in a proposed relation based on the kinetic theory of granular flow that was originally derived for dilute granular flows. The relation for dilute flows is modified to include terms to account for the effects of the local inertial number, the relative species concentrations, the particle surface friction, and the coefficient of restitution that are typical of dense granular flows. The segregation velocities derived from this approach agree well with DEM simulations in both uniform and nonuniform shear flows of density bidisperse particles, accurately predicting the difference between upward species velocity and downward species velocity for different density ratios and relative constituent concentrations. The ongoing challenge is to extend this model to size bidisperse particles, a much more difficult problem.

560718 Modeling Segregation of Polydisperse Granular Materials in Hopper Discharge Flow

Predicting mixing and segregation of polydisperse granular materials in hopper discharge flow is of potentially wide application in bulk solids handling across many industries. However, because of the limited understanding of polydisperse segregation and the complexity of flow kinematics during hopper discharge, only marginal progress has been made in modelling the segregation of polydisperse granular materials in hopper flow. In this research, we used a modified transport equation that captures the effects of advection, diffusion, and segregation to model material transport and polydisperse segregation in quasi-two dimensional hopper flow. Coupled with an accurate hopper filling model which provides the initial condition for the hopper discharge process, the model accurately predicts polydisperse segregation during hopper discharge. Quantitative agreement of continuum model predictions with DEM simulation and experiment over a range of hopper geometries and particle size distributions validates our modeling approach. The research is funded by The Dow Chemical Company and NSF Grant #CBET-1511450.

569269 Particle Motion in a Vertical Bladed Mixer Using X-Ray Particle Tracking Velocimetry (XPTV)

Particle motion in a vertical bladed mixer is dependent on blade-particle interactions, particle-particle interactions, and wall-particle interactions. These interactions are in turn a function of different operating parameters and system characteristics such as blade velocity, bed depth, particle size, and particle density. Despite the heavy deployment of bladed mixers in industry, the effects of these operating and system conditions on individual particle trajectories and velocities is still not completely understood. In order to study the mixing process in bladed mixers, it is imperative to study the motion of particles under different mixing mechanisms arising from various operating conditions.

In this study, X-ray particle tracking velocimetry (XPTV) is performed by utilizing two simultaneous X-ray projections to acquire images 90° apart. Using the two sets of images, the 3D location of tagged particles as a function of time can be accurately obtained. Tagged particles for X-ray imaging are manufactured by inserting small, but high density, metal inserts into a host particle. Studies are performed in a transparent 15.2 cm ID acrylic cylinder fitted with two 45° blades. The effects of particle density, initial location, bed height, and impeller speed on the tagged particle trajectory will be reported.

558722 Exploration of the Unique Shear Behavior of Binary Blend

Previous efforts to employ simple mixing rules to predict the flow properties of multi-component systems revealed that in some instances, flow function (FFc) values determined for binary mixtures were lower than those of either of their components. The behavior of these systems could not be described using conventional mixing rules, which necessitate that the properties of a mixture be bound by the values of those same properties exhibited by its constitutive materials. Here we present flow measurements from a series of such systems, and generalize characteristic features of the shift in flow behavior across multiple examples. In annular shear cell measurements of these systems, the shear force necessary to fail the bulk material is greater than that of the individual materials. The interaction of the two components appears to hinder the flow behavior of the bulk material. This effect is observed to be heavily influenced by the relative size and shape of the two components in a given system. The results indicate an approach to understanding the basis of certain cases of the flow behavior of mixtures, and offer an experimental framework for further *in silico* studes through methods such as DEM.

Segregation effects in three types of mass flow hoppers

By Kerry Johanson
Material Flow Solutions

Abstract

Mass flow is often suggested as a means of solving all segregation issues [1], [2], [3]. Although traditional mass flow may be a necessary condition to solve several segregation issues, it is not a sufficient condition. The solution to segregation in process equipment requires matching the segregation pattern and magnitude to the velocity profile in the specific mass flow bin or hopper. Many mass flow hoppers produce flow along the walls. However, not all of these designs are sufficient to mitigate segregation. This paper addresses segregation prevention in typical conical hoppers, cone-in-cone hoppers, and Diamondback® or plane flow hoppers. All of these bins can be designed as mass flow devices. Nonetheless, because of differences in velocity profiles in each of these bins, they offer differences in the ability to mitigate segregation. The focus of this paper is to compare the segregation prevention aspects of using these three types of mass flow bins with radial segregation patterns as well as axial segregation patterns. The radial stress theory was used to compute velocities in the bin. These velocities were then used to estimate the expected segregation mitigation for radial as well as axial segregation patterns. Six conditions were compared. These include conical hoppers, cone-in-cone hoppers, and Diamondback® hoppers. In each case, the effect of both radial and axial segregation patterns was examined. In each case, segregation mitigation plots were generated and compared to determine which hopper configurations yielded the best segregation prevention. It was determined that the cone-in-cone geometry offers the best segregation prevention, provided it is designed to have an overall velocity profile that is slower in the center than in the annular region.

1.0 Introduction

Two modes of segregation that are important to understand in processing situations. The first mode of segregation is the segregation that occurs when a piece of process equipment is filled through a batch filling mode and then emptied completely, or at least mostly, during the course of the processing. In this case, the goal is to assure that all of the material leaving the process equipment from a fill and then empty mode is within allowable content uniformity targets.

The second mode of segregation pccurs in a process that is mostly continuous in nature. In this case, we are concerned about the segregation that happens at a particular point in the process over some characteristic time frame. The goal is to minimize the variation of key components with time as material passes through the system in a continuous fashion. Ideally, all the material passing through the continuous process should be within the allowable content uniformity range.

In both modes of segregation knowing the velocity profile, process geometry, and segregation pattern in the process equipment can help determine the variation in concentration of key components in a segregating mixture. The velocity profile is dependent on the flow properties of the bulk material and the process geometry. The segregation pattern in the process is dependent on the measured radial segregation due to pile formation and the process geometry (i.e. where the piles form). There is a secondary relationship between the flow properties and segregation of the material that may be important in some cases. Local segregation could change the key flow properties that induce the velocity profiles in the first place. However, for the purposes of this paper, we will consider this effect a minor influence.

There is a tendency to believe that the best way to handle a segregation problem is to approach it from a brute force computational approach using DEM [4], FEM, CFD, and other calculation techniques to model the particle scale segregation, bulk scale flow, and process scale effects caused by operation parameters and geometry changes. This is actually a very complex problem and many numerical simulations would need to be done to correlate the effect of key variables in the process segregation. But at the end of the day, if we know the segregation pattern and the velocity profile in the bin and the process geometry we can estimate the segregation in the process and thereby determine the cause of the majority of the segregation of material leaving a process. We do not need to know the causes of particle scale segregation, just the pattern and magnitude to estimate what occurs in the process. Furthermore, this paper is limited to consider what process geometries would reduce the expected segregation due to fill-then-empty and continuous operations. This paper will consider three process geometries (simple cone, cone-in-cone, and plane flow hoppers) and rank their benefit as segregation reduction devices.

2.0 Methods

We will consider the velocity profile in a piece of process equipment where bulk solids flow through it. It is a well-established fact that the flow pattern in a piece of process equipment depends on the wall friction angle, the effective angle of internal friction, and the shape of the process geometry. This gave rise to the concept of a mass flow limit in the process geometry where the material ceases to flow along the walls in some geometry, with some solids possessing, and a set of adverse flow properties. It is well established that there is a relationship between the friction angle at the hopper wall and the hopper angle that suggests that, as the friction angle increases, the flow in a conical piece of process equipment will eventually stop flowing at the walls (Figure 1). This effect describes the transition from mass flow to funnel flow where the flow at the walls ceases, resulting in a funnel flow active flow channel that expands at some critical angle up from the outlet.

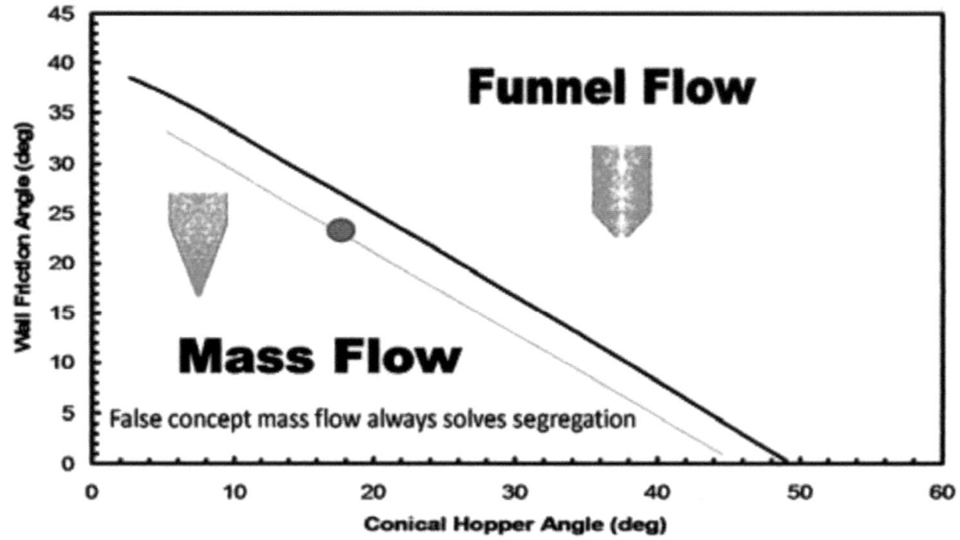

Figure 1. Typical mass-flow-limiting line for bulk material

This mass flow limit was the direct result of applying the radial stress theory to the flow of bulk material in conical hopper geometries [5], [6]. The radial stress theory is well established and a very brief summary of the method is given here for those less familiar with its use. The theory starts with the equations of motion with the acceleration terms neglected equations 1 and 2.

$$\frac{1}{r^2}\times\frac{\partial\left(r^2\times\sigma_r\right)}{\partial r}+\frac{1}{r\times\sin(\theta)}\times\frac{\partial(\tau_{r\theta}\times\sin(\theta)}{\partial\theta}+\frac{1}{r\times\sin(\theta)}\times\frac{\partial\tau_{r\phi}}{\partial\phi}-\frac{\sigma_\theta+\sigma_\phi}{r}=\gamma\times g_r-\frac{\partial P}{\partial r} \qquad (1)$$

$$\frac{1}{r^2}\times\frac{\partial(r^2\times\tau_{r\theta})}{\partial r}+\frac{1}{r\times\sin(\theta)}\times\frac{\partial(\sigma_\theta\times\sin(\theta))}{\partial\theta}+\frac{1}{r\times\sin(\theta)}\times\frac{\partial(\tau_{\theta\phi})}{\partial\phi}+\frac{\tau_{r\theta}}{r}-\frac{\cot(\theta)}{r}\times\sigma_\phi=\gamma\times g_\theta-\frac{1}{r}\times\frac{\partial P}{\partial\theta} \qquad (2)$$

The Mohr-Coulomb criteria, or constitutive relation, is then used to relate the stresses within the bulk material to each other as shown in Figure 2. The stress is assumed to be a linear function of the distance from the apex of the converging conical geometry with some unknown function s(θ) that describes how the stress changes with the theta position in the hopper. The analysis also includes the angle omega (ω(θ)) which is the direction of the major principal stress relative to the spherical coordinate system in a conical geometry.

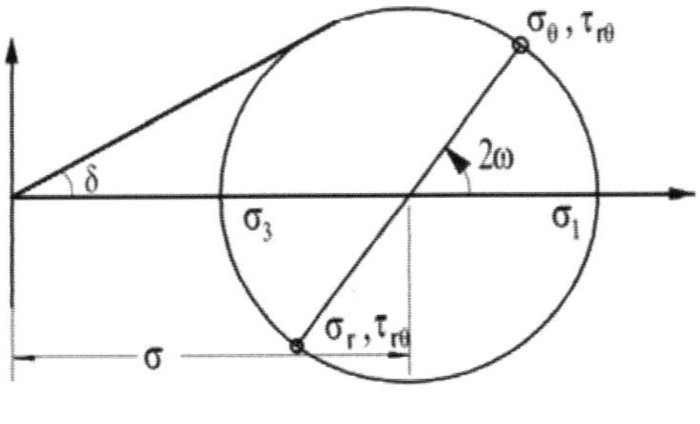

$$\sigma_r = \sigma \times (1 - \sin(\delta) \times \cos(2 \times \omega))$$

$$\sigma_\theta = \sigma \times (1 + \sin(\delta) \times \cos(2 \times \omega))$$

$$\tau_{r\theta} = \sigma \times \sin(\delta) \times \sin(2 \times \omega)$$

$$\sigma_1 = \sigma \times (1 + \sin(\delta))$$

$$\sigma_3 = \sigma \times (1 - \sin(\delta))$$

$$\sigma_\phi = \sigma \times (1 + \sin(\delta))$$

$$\sigma = \gamma \times g \times r \times s(\theta)$$ Radial Stress Assumption

Figure 2. Mohr Columb Constitutive equations for radial stress theory

Application of the Mohr Columb constitutive relation and the radial stress assumption gives rise to two ordinary differential equations (3) and (4) that when solved with boundary conditions equation (5) and (6) determines the stress profile in the conical hopper during steady flow.

$$\frac{d\omega}{d\theta} = \frac{1}{2} \times \frac{\begin{array}{c}\left[\cos(\delta)^2 - (\cot(\theta) \times \sin(2 \times \omega) + 1 + \cos(2 \times \omega)) \times \sin(\delta) \times (\sin(\delta) + 1)\right] \times s + \\ \cos(2 \times \omega) \times \cos(\theta) - \sin(2 \times \omega) \times \sin(\theta) \times \sin(\delta) + \cos(\theta)\end{array}}{(\cos(2 \times \omega) + \sin(\delta)) \times s \times \sin(\delta)} - 1 \quad (3)$$

$$\frac{ds}{d\theta} = \frac{\begin{bmatrix}\sin(2 \cdot \omega) + \sin(\delta) \cdot \cot(\theta) \cdot \cos(2 \cdot \omega) - \\ \sin(\delta) \cdot \cot(\theta) - \sin(\delta) \cdot \sin(2 \cdot \omega)\end{bmatrix} \cdot s + \cos(2 \cdot \omega) \cdot \sin(\theta) + \sin(2 \cdot \omega) \cdot \cos(\theta)}{\cos(2 \cdot \omega) + \sin(\delta)} \quad (4)$$

$$\omega = 0 \quad at \quad \theta = 0 \quad (5)$$

$$\omega = \frac{1}{2} \times \left(\phi_{we} + a\sin\left(\frac{\sin(\phi_{we})}{\sin(\delta)} \right) \right) \quad at \quad \theta = \theta_w \quad (6)$$

The application of the radial stress theory can also be used to estimate the velocity profile in the hopper but requires the use of a flow rule which states that the direction of major principal stress coincides with the direction of major principal strain rate and gives rise to equations (7) and (8).

$$\tan(2\times\omega)=\frac{\frac{1}{r}\times\frac{\partial u_r}{\partial\theta}+\frac{\partial u_\theta}{\partial r}-\frac{u_\theta}{r}}{\frac{\partial u_r}{\partial r}-\frac{u_r}{r}-\frac{1}{r}\times\frac{\partial u_\theta}{\partial\theta}} \qquad (7)$$

$$u_r=\frac{-V(\theta)}{r^2} \qquad u_\theta=0 \qquad (8)$$

This flow rule can be solved to yield the velocity profile in a conical hopper during a steady flow, equation (9). Please note that this velocity is a function of the direction of principal stress relative to the coordinate direction.

$$V=V_0\times\left(\frac{r_0}{r}\right)^2\times\exp\left(-3\times\int_{\theta_0}^{\theta_w}\tan(2\times\omega)\times d\theta\right) \qquad (9)$$

The application of the radial stress and velocity theory to hoppers leads to the computation of velocity profiles as a function of the wall friction angle and effective angle of internal friction as shown in Figure 3. Please note that all of these velocity profiles are mass flow profiles, but that the velocity ratio between the center and the side depends very strongly on the wall friction angle. If the wall friction angle increases, the velocity profile tends to become much steeper across the hopper. This difference in mass flow velocities is one key reason why just using a mass flow hopper is not sufficient to prevent the segregation of powder and granular materials.

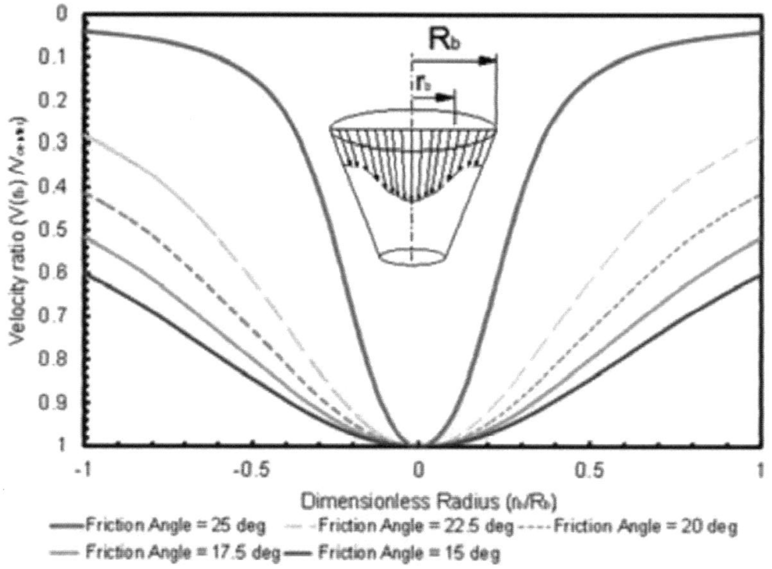

Figure 3. Typical mass flow velocities in conical bins given by radial stress theory

The radial stress theory can also be applied to the cone-in-cone geometry to compute the mass flow limit for that geometry [7]. Figure 4 shows a typical result for the case where the inner cone is at 10 degrees and the outer cone slopes at 20 degrees measured from the vertical. The solid line is the traditional mass flow line for the inner cone. The dotted line is the solution of the radial stress theory for the case where the inner cone is 10 degrees. Note that the dotted line approaches an asymptotic solution which is the boundary between mass flow and funnel flow behavior in the annular region. Combining the two radial stress solutions gives rise to the new mass flow limit for flow in a cone-in-cone device. It is clear that this device can extend the mass flow behavior to flatter hoppers than a simple cone can obtain.

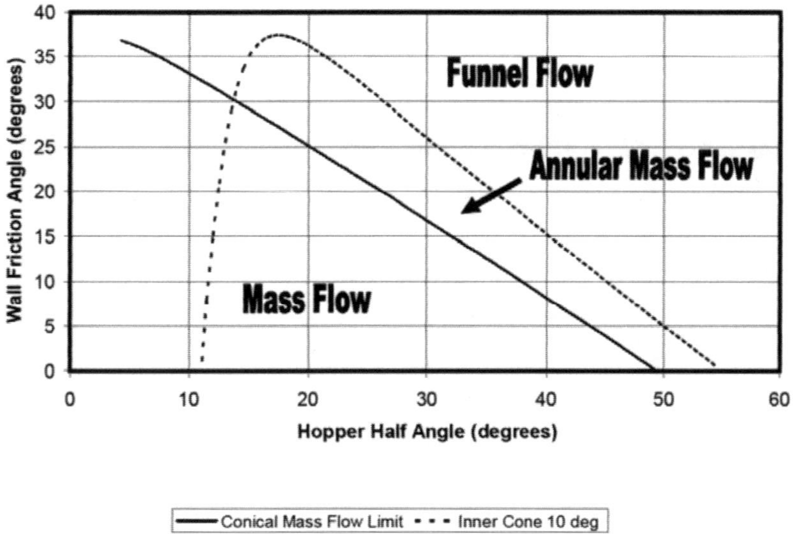

Figure 4. Typical radial stress solution for cone in cone geometry

The radial stress and velocity theory can also be used to compute the velocity profiles in cone-in-cone hoppers. However, the dimensions of the inner and outer cones can be adjusted to create overall velocity profiles where the flow in the inner cone equals the flow in the annular region or where the flow in the inner cone is some multiple of the flow in the annular region as shown in Figure 5.

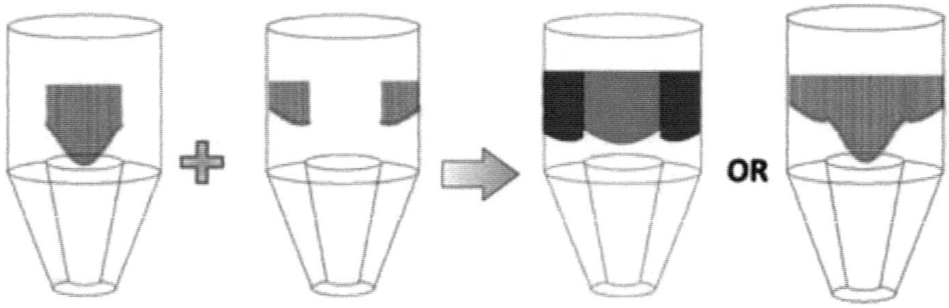

Figure 5. Potential flow profiles in cone-in-cone geometries

The velocity profiles can be computed from the application of the radial stress and velocity theory for plane flow devices such as the Diamondback® hopper. This hopper is a combination of two plane flow hoppers placed on top of each other. The first plane flow hopper is chisel-shaped with a circular cross-section and transitions down to an oval cross-section with convergence in only one direction. The second hopper is an oval hopper that transitions down to a circular cross-section with convergence in only one direction. In this case, the velocity in each hopper can be computed and then combined to give a velocity in the composite geometry. It is important to note the overall velocity above a Diamondback® hopper can be computed by first normalizing the velocity in the chisel-shaped and oval hopper section such that the average area under the velocity curve equals 1.0. The velocity profile in the oval hopper propagates up from the hopper outlet in accordance with the dimensionless velocity profile, the average flow rate, and the cross-sectional area, yielding fast flow in the center of the oval bin and a slower flow at the edge. However, this induces a nonuniform velocity along the length of the oval outlet at the bottom of the chisel-shaped outlet which propagates up through the chisel-shaped hopper. The velocity in the chisel-shaped hopper is normalized and applied to the overall hopper geometry by multiplication of two normalized velocity profiles by first rotating the chisel-shaped hopper velocity by 90 degrees and multiplying it by the velocity of the oval hopper. This yields a velocity profile that has the combined effect of both hopper sections (Figure 6).

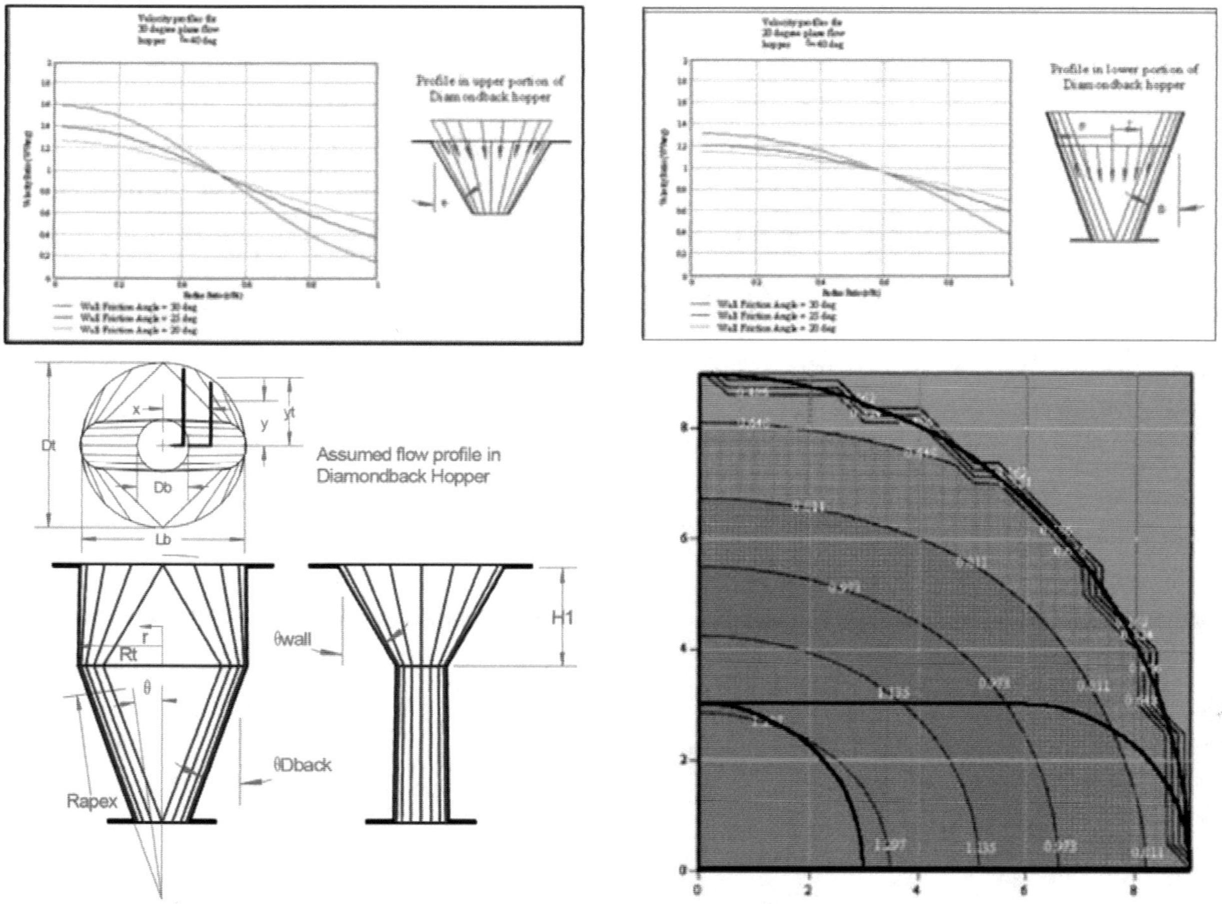

Figure 6. Typical velocity profile in Diamondback® hopper

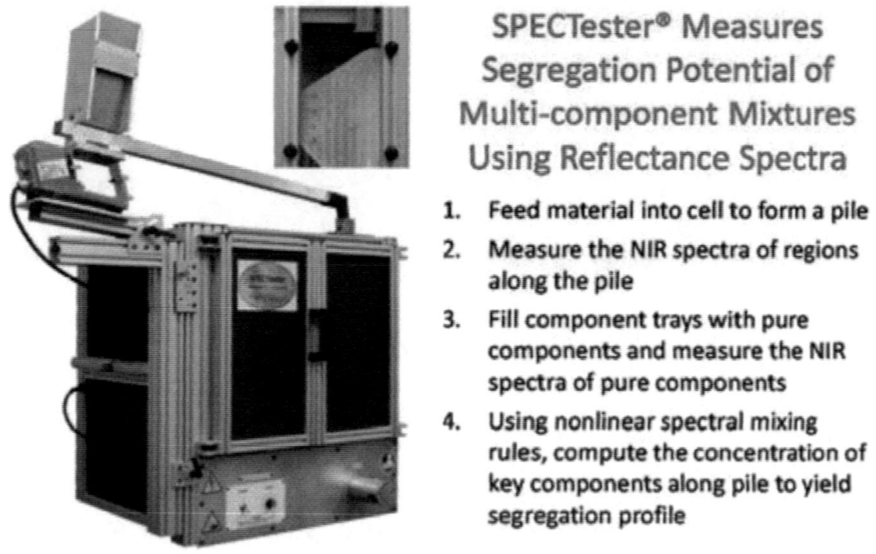

Figure 7. Segregation tester used to measure the segregation pattern of the material

The next piece of information required is to understand how a particular material might segregate when charged onto a pile. We measured that profile using the SPECTester® available commercially from Micromeritics [8]. This tester forms a pile and then uses differences in reflectance spectra [9] to compute the concentration profile of key components along the length of the pile and expresses the segregation pattern as a function of dimension radius where 0.0 is the top of the pile a 1.0 is the bottom of the pile. Please note we have not said anything about the cause of segregation. Fines can sift down through the material of coarse particles causing segregation. Differences in frictional characteristics between particles can cause particles to slide down the pile at different relative velocities, resulting in segregation. Air currents induced in free fall streams can be liberated during pile impact and carry fine particles down the pile in the freeboard space, resulting in segregation. Differences in coefficients of restitution can cause particles to bounce to different relative locations down the pile. All of these effects combine to give the segregation pattern for a particular material. Some effects happen within the thin shearing layer on the top of the pile, some effects happen in the freeboard space above the pile, and some effects happen right on the pile surface. Regardless of the reason, each material will have a characteristic segregation pattern that can be measured. A typical API mixture consisting of MCC, lactose, API, glycolate and magnesium stearate was used to evaluate the segregation in conical, cone-in-cone, and Diamondback® hoppers as shown in Figure 8.

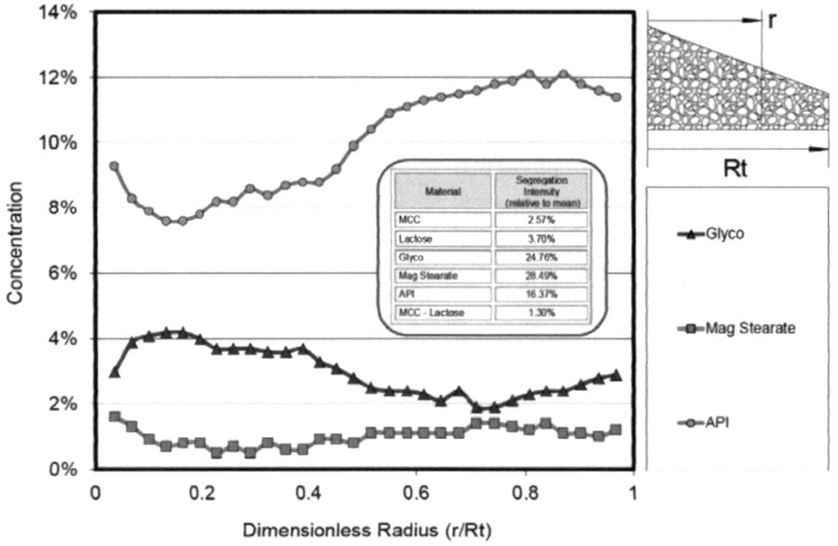

Figure 8. Segregation profile measured for a typical pharmaceutical mixture

3.0 Results

The measured segregation profile was used, along with the calculated hopper velocities given by radial stress and velocity theory, to estimate the segregation leaving hoppers when the hopper was first filled and then emptied. This was done numerically by deviding the hopper into zones of equal volume and placing a marker randomly within that small differential volume at some x, y, and z location (Figure 9). The dimensionless position from the top of the charge point to the location of interest was computed and the concentration of key materials from the segregation test was assigned to this spatial coordinate. The radial stress and velocity theory was then used to compute the local velocity in the hopper and the markers were moved using particle tracking methods to generate the flow of segregated material through the hoppers. However, in some cases during discharge the velocity profile would generate a top surface that sloped at an angle steeper than the angle of repose. In this case the material should cascade down the pile and not follow the typical mass flow velocity profile. Thus, whenever this situation was detected a cascade velocity in the direction of the repose angle was employed to move the marker position to a new location [10]. The position of the markers leaving the bin was monitored and the average concentration in a preselected time exit range was computed.

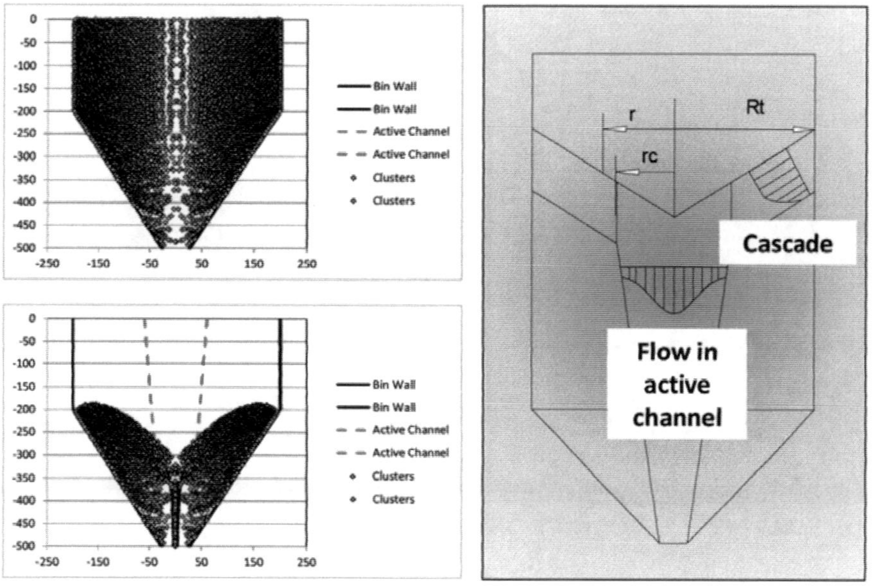

Figure 9. Position of markers in bin during flow based on radial stress theory and cascade velocity

The result of this analysis is generation of concentration profiles as a function of discharge time resulting from an operation scenario where a powder mixture was first charged into a bin, segregated during filling, and then completely emptied from the bin. This senerio is common practice in batch processes where a drug formulation is created, placed in a surge bin, and then transported to a tablet press and used completely. All the material flowing from this process must be within allowable content uniformity ranges for the process to be validated. Typically acceptable content uniformity ranges are between 95% and 105% of the mean concentration for European standards and between 90% and 110% of the mean concentration for American standards. Thus, a content uniformity standard can be selected and then the analysis performed to determine when and how often the material leaving the feed system is outside the bounds of the content uniformity range (Figure 10). It is important to note that the selection of allowable content uniformity range is an itergral part of determining if a particular process is acceptable to prevent segregation of a mixture. This method provides a systems approach to estimate the benefit of using a mass flow design to mitigate segregation. Prior art and design among particle science and technology practitioners would rely on the general comments stating that segregation can be best helped by using mass flow designs but did not offer any guidance as to what type of mass flow is best. This approach will answer that question.

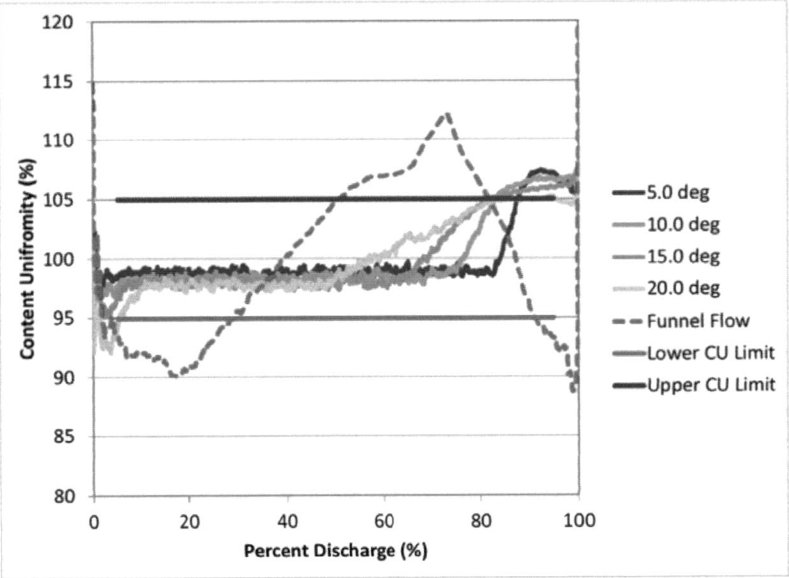

Figure 10. Typical concentration profiles resulting from radial stess velocity and cascade velocity profiles in conical hoppers

The concentration profiles were computed for a condition where the wall friction was a particular value (25.4 degrees) and the angle of the hopper was changed to give different velocity profiles. The concentration of API was used as the segregation indicator for this example. Note that the funnel flow profile causes the concentration of API to initially drop below allowable bounds, then increase above allowable bounds, and finally decrease again to below allowable bounds. The amount of material that was within allowable bounds increases as the hopper becomes steeper. However, even if the conical hopper was 5 degrees measured from the vertical, there is still some material at the end of the discharge cycle that falls outside the allowable content uniformity bounds. If we take the amount of material that is within the allowable content uniformity range as the segregation indicator, then 100% would be a perfect operation. However, in the case of placing this drug mixture in a conical hopper with an allowable content uniformity range of 95% to 105% and operating in first fill and then empty mode, the best we can expect is that 87.2% of material will be within acceptable limits.

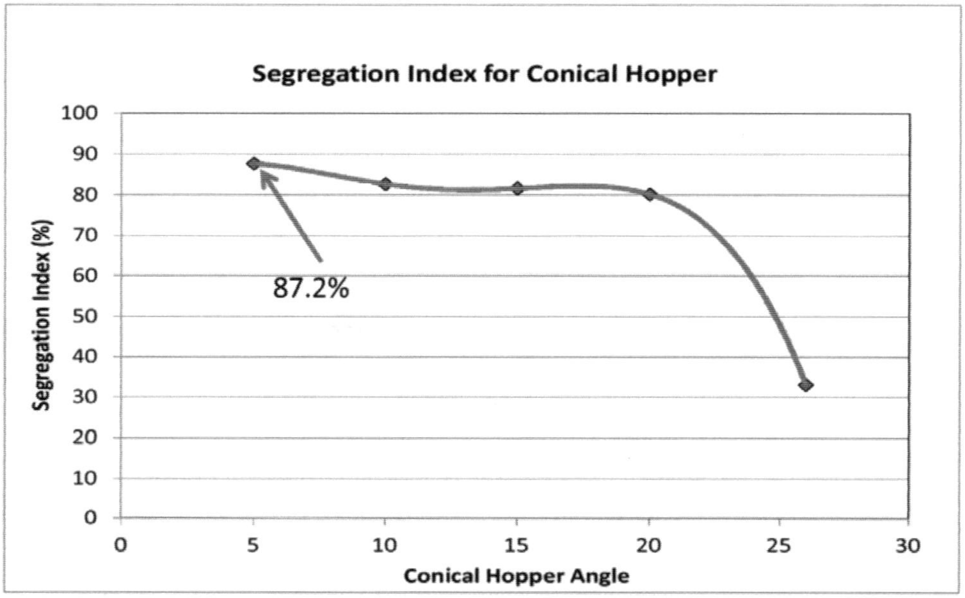

Figure 11. Segregation limits for drug mixture place in a conical hopper subject to first fill and then empty mode of operation

This type of behavior is also found with other types of materials that have a nearly linear segregation profile of key components. Consider the segregation index as described above as the fraction of material within acceptable range, 1.0 being perfect operation and 0 being fully segregated. In this case, the segregation profile is a linear function of position on a pile where the fraction of good material was plotted as a function of the wall friction angle and the angle of the conical hopper as shown in Figure 12. The point where the contour lines bunch up is the funnel flow limit. Obviously, funnel flow designs are very poor in segregation prevention. However, there is still significant segregation in conical hoppers that are first filled and then emptied, even in very steep conical hoppers.

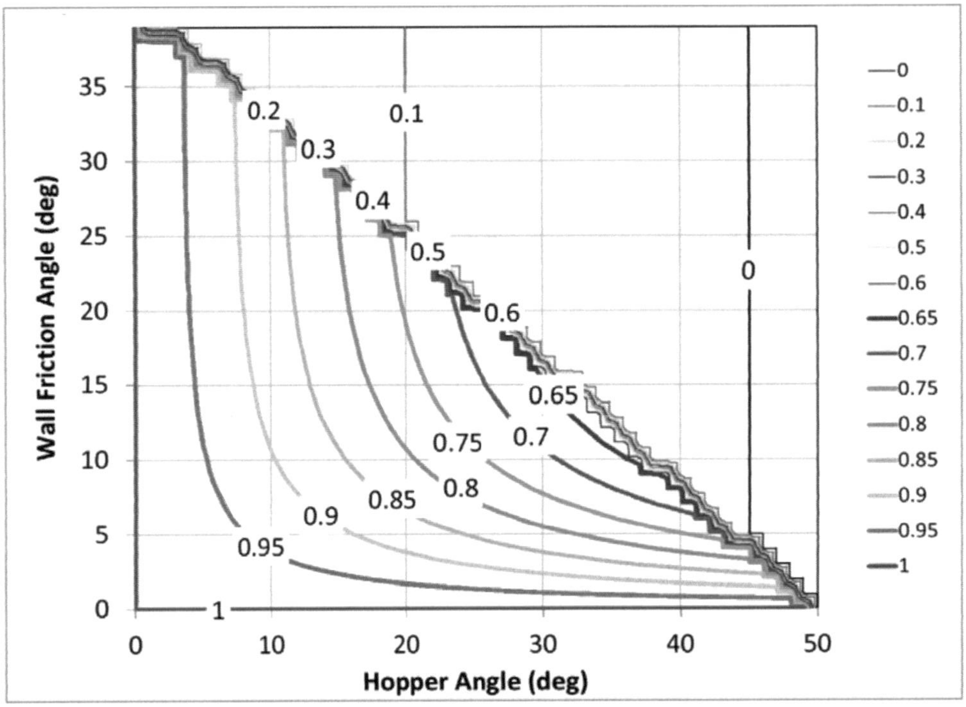

Figure 12. ***Typical segregation in conical hoppers that operate in a fill then empty mode: 1.0 is a perfect operation and 0.0 is fully segregated material***

This paper also considers the flow in a cone-in-cone geometry where the outer cone is 20 degrees measured from the vertical and the inner cone is 10 degrees measured from the vertical. In this case, the analysis was carried out to investigate the effect of the ratio of the average velocity in the inner cone and the annular region. Velocity ratios between 0.5 to 1.5 were investigated using the radial stress and velocity theory to determine the best configuration. The results of this analysis are summarized in Figure 13. It is clear that operation in a funnel flow regime is not a good idea to prevent segregation. If the velocity profile between the inner and outer cones is 1.5, then the degree of segregated material leaving the system is also very bad and is actually slightly worse than operation in funnel flow. However, as the ratio of velocities between the inner and outer cone approachs 1.0, the amount of material leaving the bin that is outside the allowable content uniformity range is very low. The best segregation profile occurs when the ratio of the velocity is 0.9 and this results in 98.3% of the material leaving the bin to be within the allowable content uniformity limits (Figure 14). Operation at velocity ratios much less than 0.9 also causes bad segregation during a fill then empty mode of operation.

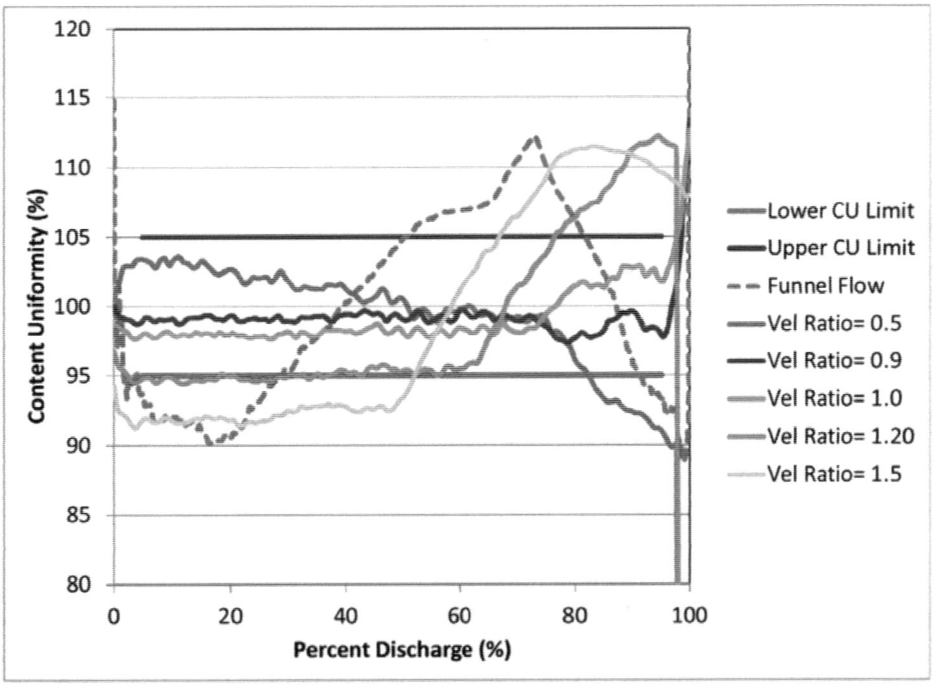

Figure 13. Segregation of drug mixture leaving cone-in-cone hopper as a function of the velocity ration between the outer and inner cones for bins that operate in a fill then empty mode: 1.0 is perfect operation, 0.0 is fully segregated material

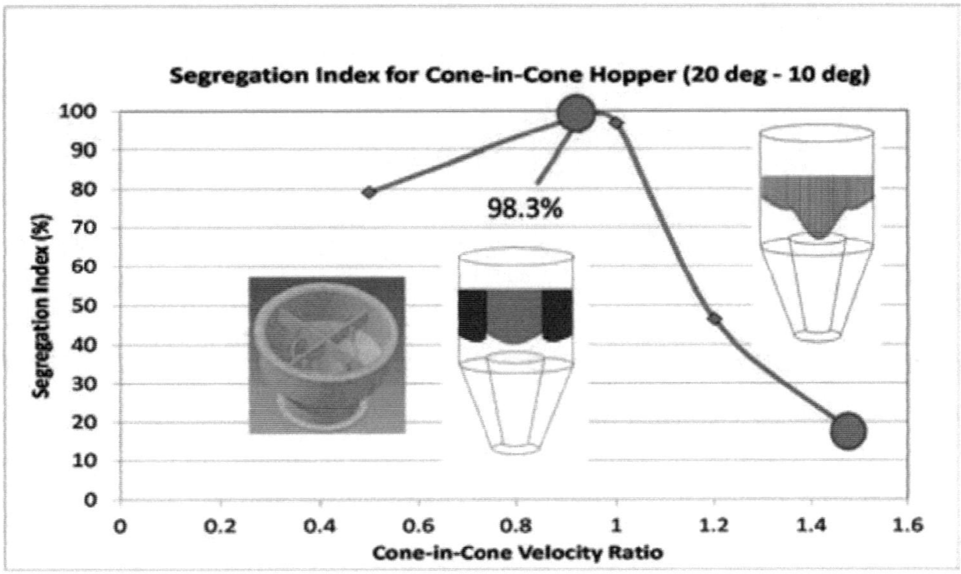

Figure 14. Segregation index of drug mixture leaving cone-in-cone hopper as a function of the velocity ratio between the outer and inner cones for bins that operate in a fill then empty mode: 100% is a perfect operation, 0% is fully segregated material

The method used above is a reasonable method based on a systems approach using the radial stress and velocity theory to characterize segregation of material leaving the bin during a fill then empty mode of operation. However, in some cases, the plant operates the process in a mostly continuous mode where material is made in a batch mode then dumped into a surge bin where the level in the surge bin is maintained at a more or less consistent height. The material leaving the batch system can segregate as indicated in the analysis above but the main surge bin rarely operates in a complete discharge mode. This mode of operation requires a different analysis to estimate the segregation leaving the system. The goal, in this case, is to maintain the content uniformity to some allowable value during all operation times where the input concentration may vary from batch dump to batch dump. In this case the computed radial stress velocity profiles can be used to estimate the residence time distribution for continuous operation through the bin (Figure 15). This residence time distribution function can be used to compute the expected concentration profile leaving/passing through the bin and then examined to determine when and if the concentration profile is outside allowable content uniformity ranges. The continuous flow mode analysis requires that the concentration during a fill and then empty batch mode be used as an input concentration to the continuous surge bin. The analysis above was done for the drug mixture segregation profiles for a conical hopper where the feed location was offset from the centerline of the hopper, but the bin was filled and then completely emptied. It was assumed that as soon as the feed bin was emptied another identical feed bin of the same size and geometry was placed above the surge bin in such a way as to maintain the level in the surge bin to be relatively constant. This gave a cyclic input concentration profile that had a frequency equal to the size of the batch feed hopper. One of the important parameters was found to be the relative size of the batch bin to the surge bin. The concentration of API leaving the surge bin was plotted as a function of the number of surge bin volumes that were passing through the process. The size of the surge bin was divided by the size of the batch feed bin to give a parameter that described the number batch cycles passing through the surge bin. If this ratio was 1.0, then the batch bin was the same size as the surge bin. If this ratio was 2.0, the batch bin was ½ the capacity of the batch bin. The output concentration was then calculated given the cyclic input batch discharge concentration and residence time distribution. The ratio of the maximum concentration fluctuation (C_{max} –C_{min}) between the input and output concentrations was used to determine the effectiveness of the continuous operation on reducing the concentration. Based on the assumed input concentration variation, the maximum concentration span (C_{max} –C_{min}) would need to reduce to 31% of the current value to assure that the output concentration profile was always within the 95% to 105% content uniformity range (Figure 16).

The residence time distribution function E(θ) can be used with the input concentration profile to compute the concentration profile leaving the bin as a function of time

$$C_{out}(t) = \int_{\theta=0}^{\infty} C_{in}(t-\theta)\cdot E(\theta)\, d\theta$$

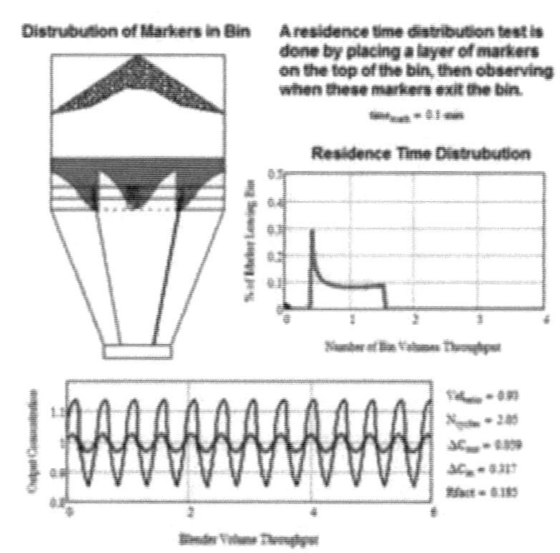

Figure 15. Typical residence time distribution functions and output concentration profiles computed from the radial stress velocity profiles

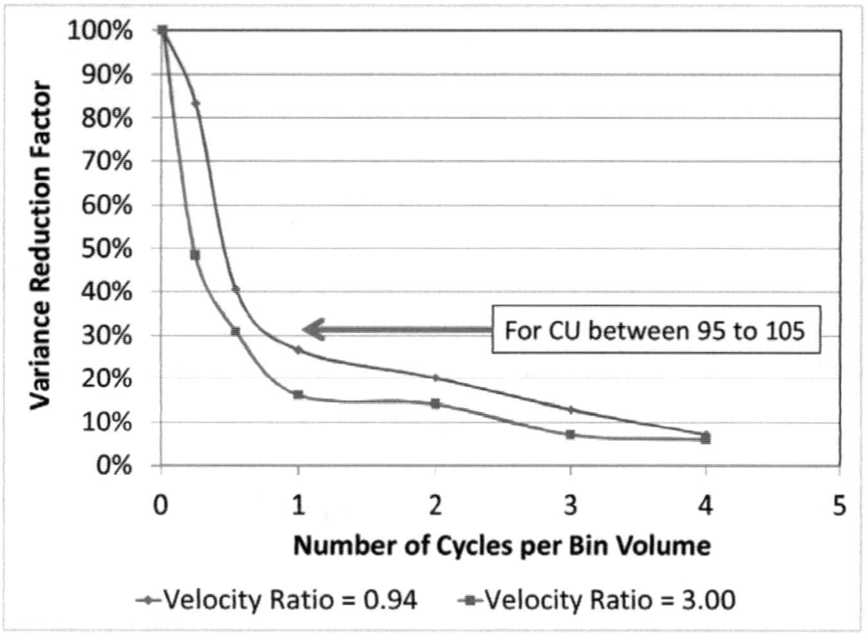

Figure 16. Expected reduction in the concentration variation for flow of assumed input concentration through a cone-in-cone bin as a function of the number of input cycles per bin volume

In the case of the cone-in-cone, changing the velocity profile to create a more uniform velocity actually reduces the ability of the bin to mitigate the segregation passing through the surge bin. Thus, the segregation velocity profile that helps prevent segregation of a fill and then empty mode of operation is much worse when applied to the case of trying to mitigate segregation during continuous flow modes of operation. Segregation mitigation in this mode of flow requires significant differences in velocity within the bin.

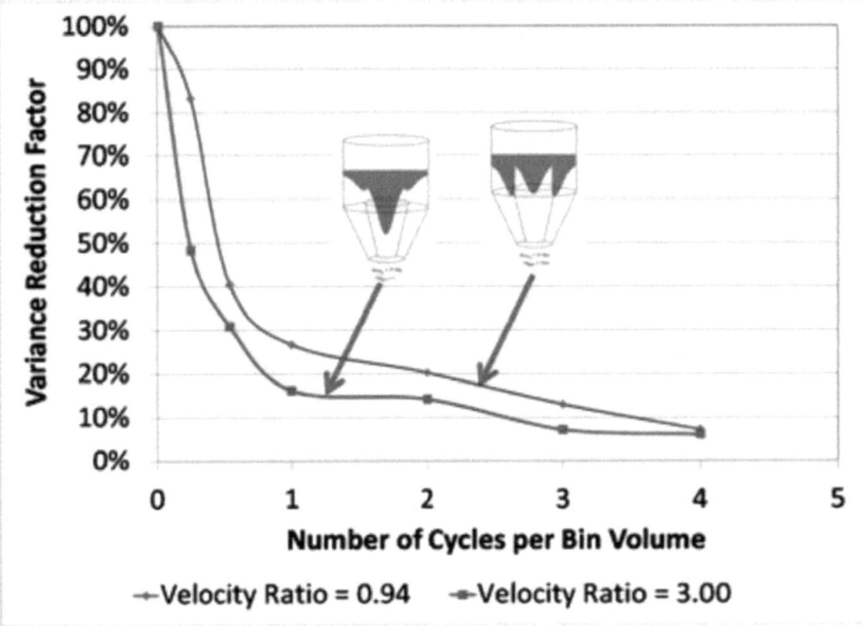

Figure 17. Expected reduction in the concentration variation for flow of assumed input concentration through a cone-in-cone bin as a function of the number of input cycles per bin volume showing the effect of velocity profile in cone-in-cone bin

The same is true when the segregation model is applied to flow in a conical hopper. The situation where the velocity profile across the bin is relatively steep (high friction angle) gives better segregation mitigation than the case were the velocity profile is more uniform (low friction angle). This is summarized in Figure 18. It is important to note that, depending on the number of input concentration cycles per bin volume, the flow through the conical hopper with a large friction angle may result in better segregation mitigation in continuous mode than a cone-in-cone hopper with a large velocity ratio between the inner and outer cones.

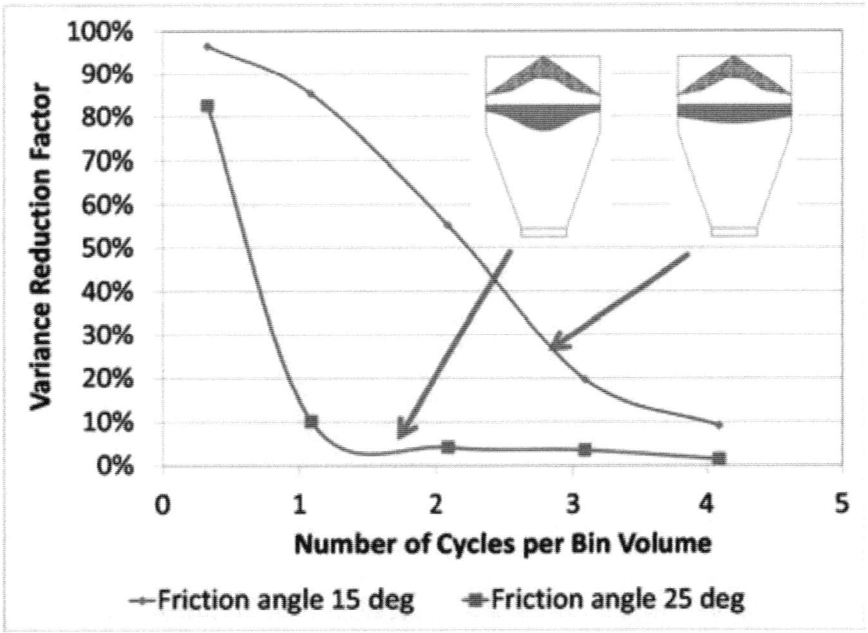

Figure 18. Expected reduction in the concentration variation for flow of assumed input concentration through a conical bin as a function of the number of input cycles per bin volume showing the difference in the wall friction angle

4.0 Conclusions

The radial stress and velocity calculations computed from standard flow property measurements, when combined with measured segregation patterns, can provide a reasonable system approach to determine if a particular bin can successfully mitigate segregation during processing. Two modes of segregation mitigation must be considered. First is the segregation that occrs when a piece of process equipment is filled and then completely emptied. In this mode of operation, the cone-in-cone bin shows a significant ability to mitigate segregation where conical or plane flow hoppers would still posses significant segregation. Second is the case where segregation occurs during a repeated batch operation passing through a surge bin in a continuous manner. In this mode of operation, the best velocity profile is one that shows significant velocity difference in the velocity across the bin. This will occur in conical or plane flow bins with a large friction angle or in cone-in-cone bins with a large difference between the average flow between the inner and outer cones. The best choice bin seems to depend on the number of concentration variations in the surge bin relative to the volume of the surge bin. If the batch size variations are large relative to the surge bin, the cone-in-cone geometry appears to give better segregation mitigation. However, if the batch size variation is small relative to the surge bin (i.e. many variations in a surge size volume), then the conical bin with a large friction angle may be a better choice to mitigate segregation in a continuous mode of operation.

5.0 References

1. Johanson, J.R., *Solids segregation: case histories and solutions*. Bulk Solids Handling, April 1987.

2. Engblom, N., H. Saxén, R. Zevenhoven, H. Nylander, G.G. Enstad, *Segregation of powder mixtures at filling and complete discharge of silos*. Powder Technology 215–216 104–116. 2012.

3. Bertulo, D., S. Volpato, P. Canu, A.C. Santomaso, *Prediction of segregation in funnel and mass flow discharge*. Chemical Engineering Science Volume 150 pp. 16–25. 2016.

4. Ketterhagen, W.R., J.S. Curtis, C.R. Wassgren, A. Kong, P.J. Narayan, B.C. Hancock, *Granular segregation in discharging cylindrical hoppers: a discrete element and experimental study*. Chemical Engineering Science Volume 62, pp. 6423–6439. 2007.

5. Johanson, J.R., *Stress and velocity fields in the gravity flow of bulk solids*. ASMA Journal of Applied Mechanics, September 1964 .

6. Nedderman, R.M., *Statics and kinematics of granular materials*. Cambridge University Press. 1992.

7. Johanson, K.D., *Predicting cone-in-cone blender efficiencies from key material properties*, Powder Technology, Volume 170, Issue 3, pp. 109-124. December 2006

8. Asachi M., A. Hassanpour, M. Ghadiri, A. Bayly, *Assessment of near-infrared (NIR) spectroscopy for segregation measurement of low content level ingredients*. Powder Technology, Volume 320, pp. 143-154. October 2017.

9. Johanson, K.D., *Review of new segregation tester method by Dr. Kerry Johanson*. Powder Technology, Volume 257, pp. 1-10. 2014.

10. Johanson, K.D., *New Mass Flow Limiting Lines Based on Segregation Pattern and Magnitude*, Powder Technology, August 28, 2019. https://doi.org/10.1016/j.powtec.2019.08.079

564550 Analysis of Nanoparticle Growth and Aggregation in Non-Thermal Plasma and Plasma Afterglow By Advanced Ion Mobility Spectrometry and Monte Carlo Simulation

Non-thermal plasma synthesis method is an established method for producing nanoparticles in a controllable manner with potential semiconductor, solar cell, and bioimaging applications. Non-thermal plasmas are non-equilibrium systems; the high electron energy in a non-thermal plasma can dissociate vapor phase precursors, initiating the nucleation and growth of nanoparticles. Numerous studies have shown that nanoparticles in non-thermal plasmas are negatively charged due to the large gap of mass and mobility between electrons and ions. The unipolar charging of the nanoparticles also prevents them from coagulation and leads to spherical, monodispersed nanoparticle production.

On the other side, we recently characterized size distribution of Si nanocrystals synthesized in a low-pressure radiofrequency Ar non-thermal plasma by a uniquely developed low pressure differential mobility analyzer (LPDMA). We detected both negatively and positively charged nanocrystals at the plasma reactor outlet, and modest aggregation of nanocrystals was also observed. To further study this phenomenon, we have developed and applied a constant-number Monte Carlo population balance model to simulate nanoparticle charging and growth in the plasma afterglow region. The simulation incorporates the effects of electron desorption from nanoparticles, particle-particle collision, ion-particle collision, electron-particle collision, particle wall deposition and decay of ion and electron concentrations and electron temperature in the afterglow. The simulation results were compared with LPDMA experimental data where a wire mesh at the plasma exit was used to control the extent of nanoparticle aggregation. The result shows that in the plasma afterglow region, since the concentration and energy of electrons and ions decay rapidly, the nanoparticles are neutralized, and can become bipolarly charged through collisions with remaining ions. The change in nanoparticle charging also leads to post plasma aggregation, as we have observed experimentally.

567901 Model Identification Using in-Line Estimates of Crystal Size

Application of population balance models (PBM) to crystallization depends on the ability to identify the correct crystal formation mechanism and the corresponding parameters. This is accomplished by solving an optimization problem that reduces the error between a model's prediction and the experimental values derived from crystallization experiments. Another approach for estimating the correct mechanism and parameters is by maximizing the likelihood function, or probability of observing these phenomena, using Bayesian theory. Industrial applications of this workstream are limited by the data available, which in most cases consists of a solution concentration as a function of time, and initial and final crystal size distributions (CSD). In many cases, this may lead to several mechanisms (and/or parameter sets) capable of explaining the same set of data. In this work we use numerical experiments to address the issue of the amount of data needed to correctly identify the correct mechanism and parameters. In particular, we study the use of in-line estimators of crystal size like the ones introduced in [1,2], which allows for an estimation of the inline CSD as a function of time. We study the applicability of this methodology with actual experimental data.

[1] Irizarry R, Chen A, Crawford R, Codan L, Schoell J. Data-driven model and model paradigm to predict 1D and 2D particle size distribution from measured Chord-length distribution. Chem Eng Sci. 2017; 164: 202-218.

[2] Schoell J, Irizarry R , Sirota E, Mengel C, Codan L, Cote A. Determining particle size distributions from chord length measurements for different particle morphologies. AIChE J. Vol 65.

568769 Population Balance Modeling for Twinscrew Granulation: Insight in the Notion of Distance between Distributions and the Need for Identifiable Models for Good Predictive Power

Traditional pharmaceutical solid oral dosage processes comprise of a series of batch unit operations. More recently, a transition is developing from batch unit processes to continuous manufacturing application, to cope with the inefficiencies and high cost involved in process development & scale up. Twin-screw wet granulation is a developing pharmaceutical continuous process that is being assessed for its performance in solid dosage manufacturing. In this research, the twin-screw wet granulation unit is a unit operation of the ConsiGmaTM-25 continuous powder-to-tablet process line from GEA Pharma Systems. However, since these continuous processes are still under development in the pharmaceutical industry, detailed process knowledge and understanding is still evolving. Application of mechanistic models can help in filling knowledge gaps by thoroughly investigating gathered detailed experimental data and unravelling the underlying mechanisms.

Previous work of the authors focussed on population balance modelling of the continuous pharmaceutical twin-screw wet granulation unit [1]. In that work, a novel two compartmental population balance model was calibrated and validated using measurements of particle size after the wetting zone and at the end of the granulator.

To further develop the practical applicability of this model, some more insight is needed. From an application perspective, the model parameters need to be linked with the process conditions. To ensure that this linkage is sound, the obtained parameter set needs to be unique. In other words, every single simulated distribution can only be simulated using one parameter set. In mathematical terms, it is said that the function is injective. This is in fact the basis of the identifiability concept, being viewed as a parameter estimation problem. First, the parameter set is estimated using simulated data (structural identifiability). Next, noise is added to find the maximum allowed experimental error under which the model can be reliably calibrated (practical identifiability). The questions to be answered are: are the current kernel formulations sufficiently identifiable? Is the intermediate data after the wetting zone needed to perform a calibration?

In the parameter estimated problem, an objective function is used to quantify a "distance" between the measurement (in the calibration, this is measured particle size data; in the identifiability, this is synthetic data) and the model simulation. In this application field (including our previous work), the standard approach is to use a L_2 norm or similar variant (sum of squared errors or root mean squared error). The question is if this distance is an ideal objective function for the comparison of two distributions? For comparing distributions, a plethora of choice is available: comparing the d10, d50 and d90 values, comparing the moments of the distribution, etc.

A number of different scenarios were worked out to assess the applicability of objective functions: Gaussians, Gaussian mixtures and distributions from simulated population balance models. These cover all the distributions encountered in this application field: unimodal as well as bimodal distributions. The parameter space of these parametric distributions was sampled to generate a visualisation of the objective functions. The assessment of the quality of an objective function was based on the ability to find the original parameter set, the smoothness around the optimum, the occurrence of local minima and the appearance of valleys or oscillations in the objective function.

Two key metrics were identified to perform optimal for distributions which are encountered in this application field: Kullback-Leibler divergence from information theory, also called the relative entropy and the Earth Movers distance from optimal transport theory. These metrics can now be applied in the parameter estimation problem of the identifiability analysis.

The results of the identifiability analysis state that the model from previous work ([1]) is usable, however, under certain constraints. Some parameters need to be constrained, others should be left out. The data in the wetting zone is crucial for understanding the dynamics in the system as well as for the calibration of the population balance model. This is an important result, as in recent years, more and more papers are being published on compartmental population balance modelling of the twin-screw wet granulator without using intermediate data, e.g. data inside the barrel: after the wetting zone or after the first kneading zone. This implies that the predictive power of such models is low and thus not usable in practical applications.

A high predictive power is the most important goal of these model efforts. As such, the identifiability procedure described in this work should be performed first before attempting a calibration. Without the confirmation of a sound identifiable model, there is no guarantee that the calibrated results can be used for further applications such as model predictive control or optimal experimental design.

[1] Van Hauwermeiren, D., Verstraeten, M., Doshi, P., am Ende, M. T., Turnbull, N., Lee, K., ... Nopens, I. (2018). On the modelling of granule size distributions in twin-screw wet granulation: Calibration of a novel compartmental population balance model. *Powder Technology*, *341*, 116–125. https://doi.org/10.1016/j.powtec.2018.05.025

569160 Detailed Population Balance Modelling Using a Hybrid Particle Model

Population balance modelling provides a powerful tool to study formation and growth of particulate species in a wide range of systems from crystallisation to combustion and atmospheric processes. In many applications, the particle structure is an important determinant of its properties; thus, it is necessary to study the relationship between process conditions and the evolution of particle morphology. Detailed particle models have been developed[1] to track the structure of complex particles formed in combustion synthesis, for example of soot, silica and titania. These can be used to study materials synthesis, supplementing experimental understanding. A Monte Carlo approach is required to resolve particle structure in more than a few dimensions. This treats the formation and growth processes as events that occur in proportion to their respective rates, evolving a discrete ensemble of computational particles to resolve the particle size distribution (PSD). Operator splitting can be used to couple gas-phase kinetics and the particle dynamics. For industrially relevant process conditions, the particle dynamics are incredibly rapid – which can lead to very high number densities and broad particle size distributions[2]. This heterogenous mix of complicated aggregates and many newly-incepted, simple primary particles is challenging to treat efficiently with fixed computational ensemble sizes.

We recently developed a hybrid particle-number and particle (PN/P) model[3] to facilitate use of a stochastic algorithm and a detailed particle model under aggressive particle process rates. This model aims to resolve the particle size distribution more efficiently under high rate conditions, using different levels of details for different regions of the particle type space. Small particles are tracked using a particle-number model which requires only a single internal coordinate and complex particles are tracked using a discrete ensemble of computational particles in which particle structure is described in up to hundreds or even thousands of dimensions (i.e. primary particle connectivity and sizes). In the current work, we present the new algorithm and demonstrate its use to study a wide range of process conditions and track detailed variation in particle product morphology with a significant improvement in computational efficiency over the standard, single-ensemble approach. We also show that the algorithm is exact; thus, the improved performance does not compromise prediction accuracy.

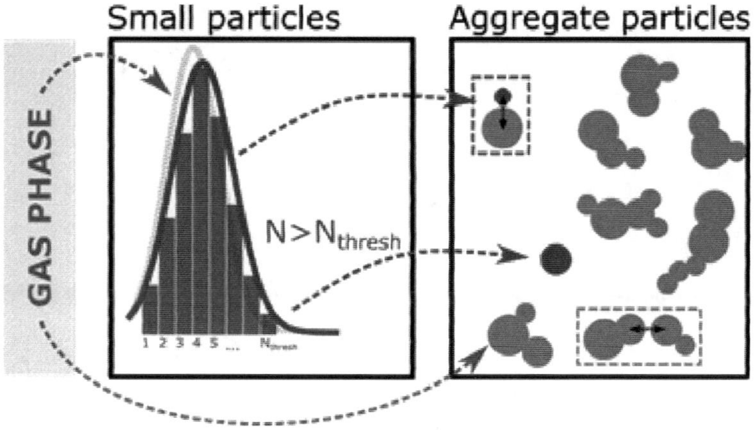

Figure 1: Mass transfer from the gas-phase to the particle phase and between the particle-number and particle models in the hybrid PN/P algorithm[3].

---

[1]Shekar, S., Menz, W. J., Smith, A. J., Kraft, M., Wagner, W. On a multivariate population balance model to describe the structure and composition of silica nanoparticles, Computers and Chemical Engineering 43 (2012) 130–147. doi: 10.1016/j.compchemeng.2012.04.010.

[2]Boje, A., Akroyd, J., Sutcliffe, S., Edwards, J., Kraft, M, Detailed Population Balance Modelling of TiO_2 Synthesis in an Industrial Reactor, *Chemical Engineering Science* 164 (2017) 219–231. doi: 10.1016/j.ces.2017.02.019.

[3]Boje, A., Akroyd, J., Kraft, M, A hybrid particle-number and particle model for efficient solution of population balance equations, *Journal of Computational Physics*, in press (2019). doi: 10.1016/j.jcp.2019.03.033.

571966 Spherical Agglomeration Process: Mechanistic Understanding and Mathematical Modelling

Spherical agglomeration is a specialised process which has recently been successfully applied in pharmaceutical manufacturing, specifically to address problematic properties of irregular platelet- or needle-like crystals such as poor flowability, compressibility and packability. The process involves the precipitation and agglomeration of crystals to produce compact and dense spherical agglomerates. The precipitation is typically achieved by an anti-solvent crystallisation in which an anti-solvent is added to a solvent-drug solution. The agglomeration is achieved by addition of an immiscible bridging liquid into the solution in which the crystals are suspended.

The focus of the current work is on the agglomeration in suspension rate processes as there is no comprehensive mechanistic understanding of this process. Based on the similarities between wet granulation and spherical agglomeration, the following rate processes may be proposed: wetting/nucleation of primary crystals by the bridging liquid, consolidation/growth of agglomerate nuclei, and attrition/breakage of the agglomerates. The aim of the current study is to identify the influence of key material properties and process parameters on the final properties of agglomerates. This was achieved by a) integration of different targeted, systematic small-scale experiments designed to observe and quantify agglomerate evolution, and b) development of predictive and robust mathematical models to inform future process design tools through a bottom-up approach.

Spherical agglomeration experiments were conducted in a stirred vessel and an oscillatory baffled reactor. A novel microfluidic system and a contracting nozzle device were proposed and devised to respectively investigate wetting/nucleation and attrition/breakage of the agglomerates in isolation. A completely novel mathematical model is developed to investigate the kinetics of wetting and nucleation (via an immersion mechanism) and to identify different regimes of the nucleation process. gPROMS FormulatedProducts® was used to develop a population balance model for the agglomeration in suspension process, including a customised agglomeration kernel. The effect of material properties and process conditions on different rate processes, as well as size distribution of the produced agglomerates, was determined through the experiments and the developed models.

557450 Mechanism Discrimination Based on Population Balance Models

When observing the evolution of particle populations, it may not be obvious which of the possible mechanisms are influencing the system trajectory. For 1-D populations in which particle size is the internal dimension, the mechanisms most likely at work are some combination of nucleation, accretional growth, collisional growth and breakage. If we do not consider nucleation, there are eight possible combinations of these mechanisms, some of which exhibit stationary states and others similarity solutions. In this paper, we class these eight combinations based on the direction of particle concentration change, the direction of mean particle size change, and the value of the scaled 2nd moment of the population at either the stationary state or similarity solution. We find that when all three mechanisms reach similarity solutions when occurring in isolation, consideration of these three measures allows identification of the particular combination of mechanisms causing the observed behavior. For certain collisional growth rate kernels, a similarity solution is not reached. In these cases, there are two mechanism combinations that cannot be distinguished from each other this way, but one can still discriminate this pair from the other six possible combinations and those remaining combinations from each other.

581109 Measuring Food Powders Electrical Properties as a Result of Humidity Content

Granular materials and fine powders are widely used in many industrial applications. In particular, excipients are necessary in dry powder pharmaceutical formulations, such as dry powder inhalers (DPIs), tablets and capsules. Thus, any recommendations regarding their behaviour and the effects of using of additives could have significant consequences for the optimisation of industrial processes, including avoiding technical issues like granulation, noncompliance and non-conformity.

In this paper, we show how the influence of anti-static agent additions affect excipients behaviour. For this purpose, two lactose products provided by the Meggle company (Germany) were used during this study: Inhalac 400 and Flowlac 90. The anti-static agent studied is a commercial talc powder (brand "Care"). The influence of Talc addition in the lactose products will be studied with five different talc mass fractions 0%, 5, 10, 15 and 20%.

Electrical properties are finding increasing application in agriculture and food processing industries. Knowledge of dielectric properties of foods as a function of moisture content and temperature is essential in the design and control of drying systems.

580486 *Solids Metrics for Successful 3D Printing of Energetic Feedstocks*

Traditional manufacturing methods for producing warheads, propellants, and pyrotechnic systems are timely and costly. Additive manufacturing (AM) approaches offer one path of alternative manufacturing technologies. Among various technologies, powder bed fusion (PBF) is one promising approach. A PBF printer is essentially a miniature manufacturing facility in a box, with significant powder handling challenges that must be overcome for robust production performance and part quality [1, 2]. Feedstock handling in a powder bed fusion printer is no different than and faces the same challenges as full-scale solids handling processes.

For a feedstock to function properly in a powder bed fusion 3D printer, it must be designed to meet several material characteristics. Some of the most important material properties include: powder flowability & cohesion, powder and roll friction, permeability, particle size & shape distribution (PSSD), triboelectric charging propensity, wetting characteristics, compaction characteristics and thermal softening point.

This paper presents select studies of developing feedstocks for PBF printing of energetic and other materials. Key attributes benchmarked against included powder flowability, aeration attributes, wetting behavior, spreadability, and fusion, as well as final part performance (bulk density, detonation, sensitivity, and part tolerances.)

[1] Ennis, B.J. *et al.* (2008) Section 21: Solids-Solids Operations & Processing. *Perry's Chemical Engineers' Handbook*. Perry, R. H., & Green, D. W. (eds.), New York: McGraw-Hill.

[2] Ennis, B.J. et al. (2017), HPC4Mfg Modeling of Powder Dynamics in Metal Additive Manufacturing, Austin, TX. Link for presentation:

https://hpc4mfg.llnl.gov/pdfs/events/powder2017/Powder_Dynamics_EG_17oct2017_FINAL.pdf

[3] Ennis, B.J. et al. (2017). Navy SBIR Phase I and Phase II Awards. Development of Explosive Feedstock for Commercial-off-the-Shelf (COTS) 3D Printers. Link for award:

https://www.navysbirprogram.com/NavySearch/Award/award.aspx?pk=869CC83B-5945-4E75-A598-AA2527D51045

[4] "*Chattanooga firm to help U.S. Navy figure out how to 3-D print explosives*" (2017), T. Omarzu,

Chattanooga Times Free Press. Story link:

http://www.timesfreepress.com/news/business/aroundregion/story/2017/dec/10/chattanoogfirm-help-us-navy-figure-out-how-3-/458777/

580487 Powder Permeability As a Measurement Surrogate for Triboelectric Charging

Triboelectric charging is common for many powders in solids handling and other processing applications, sometimes intentional as with laser printing but more often than not unintentional and a processing nuisance. Such particle charging can lead both to wall deposition and increased interparticle agglomeration with substantial increases in minimum bin opening sizes to prevent stable arching and ratholing.

Furthermore, the exact nature of triboelectric charging remains poorly understood, with severe limitations in readily made characterization to provide reliable engineering and process guidance. For example, powders may charge by surface interactions (traditional triboelectric charging) or by particle-particle interactions (self-charge by frictional rubbing). Generation of frictional charge can also lead to (i) a net bulk powder charge, (ii) distributions in particle net charge, or even (iii) surface distributions in charge across single particles. Net bulk charge is readily measured through transfer or Faraday cups; however, the latter two mechanisms involving charge distribution are more likely relevant to powder agglomeration, and far more difficult to assess.

Recent additive manufacturing research has shown that bed permeability is readily impacted by the extent of powder handling and related charging. Bulk powder permeability is the relationship between interstitial gas velocity and the resulting bed pressure drop as typically governed by Darcy's law [1,2]. This paper presents a novel approach to using powder bed permeability as a rapid surrogate measurement to the extent of powder triboelectric charging. Permeability is shown to track with the extent of drying, humidity conditioning, and time of handling. Limited confirming measurements are also made by Faraday cup for varying processing time and transfer material, as well as by shear cell measurements of minimum opening size and friction.

References:

[1] Darcy, H. (1856). *Les fontaines publiques de la ville de Dijon*. Paris: Dalmont.

[2] Ennis, B.J. *et al.* (2008) Section 21: Solids-Solids Operations & Processing. *Perry's Chemical Engineers' Handbook*. Perry, R. H., & Green, D. W. (eds.), New York: McGraw-Hill.

561513 Metal Dust Explosion Risks: A Hazard Analysis and Review of Contemporary Industry Objectives

Throughout recent years, metal dust deflagrations continue to pose a critical threat toward safety in the metal processing and refining industry. Following ignition of a metal dust cloud within a contained volume, deflagration can quickly transition to detonation if the process enclosure operates unprotected. Such extreme pressure generation is capable of producing devastating loss of life and property. Metal dusts demonstrate an exceedingly reactive explosion risk due to amplified heats of combustion, burning temperatures, flame speeds, explosibility parameters (K_{St} and P_{max}), and ignition sensitivities. In comparison to fuel types of organic composition, the spontaneous nature of metal powder oxidation mechanisms exhibits capacity for characteristically intensified explosion severity. These conceptions will be exemplified through both examination of several case studies involving metal dust explosion incidents and single-particle fuel combustion modeling. Addition of non-combustible inert material to combustible dust mixtures, either through pre-mixing or high-rate injection as the incipient flame front begins to develop, is common practice for preventative inhibition or explosion protection via active suppression, respectively. However, proper mitigation techniques for enclosures conveying metal powders have posed challenging; commonly, design calls for saturation of the protected volume with exceedingly high concentrations of suppressant agent to achieve moderate maximum observable pressures. Thus, this review additionally outlines the application of new materials toward a more efficient metal dust explosion suppression design, with computational fluid dynamic (CFD) simulation as an effective supplement.

575064 Oxidation of Carbon Black Agglomerates

Oxidation of soot and carbon blacks determines their emitted mass concentration and production rate as well as mobility diameter, d_m. Soot and carbon black oxidation takes place mostly by O_2 and OH radicals at high temperatures, T (> 1100 K) and competes with agglomeration in determining its particle sizes [1]. In diesel engine exhaust and regenerative traps of particulate filters, single or packed beds of soot agglomerates are oxidized at lower temperatures [1]. Even though accurate oxidation kinetics over a wide T range are essential in both industrial uses and environmental impact of carbon black, they are often derived neglecting internal particle oxidation and the structure of such carbon black agglomerates.

Here, the detailed evolution of the fractal-like agglomerate carbon black mass, m, and mobility diameter, d_m, during both internal and surface oxidation is determined by a moving sectional model. The oxidation mode index, a, given by the ratio of the characteristic O_2 reaction and diffusion times is used to quantify the contributions of internal and surface oxidation of carbon black [2]. At low T (e.g., < 1100 K), O_2 diffuses into the primary particles and reacts with bulk carbon black, hardly altering the d_m and yielding a > 3. As T increases, surface oxidation becomes dominant, decreasing both d_m and a. The surface oxidation model for spheres underestimates the measured d_m [3] by 50 % on average over the whole T range. Accounting for the fractal-like agglomerate morphology during surface oxidation improves substantially the agreement with data for T > 1100 K, but underestimates carbon black d_m up to 24 % for lower T, as oxidation takes place mostly within carbon black primary particles. Accounting for both internal and surface carbon black oxidation of agglomerates results in good agreement with mature carbon black data [3] over the whole T range. Coupling this detailed moving sectional model with mobility size distributions of nascent carbon black [4] yields a specific oxidation rate that is on average more than 50 % smaller than that obtained when neglecting internal oxidation and the fractal-like carbon black morphology at $[O_2]$ = 0.2 - 0.78 vol %.

References:

[1] Stanmore, B. R., Brilhac, J. F., & Gilot, P. (2001). Carbon, 39, 2247-2268.

[2] Essenhigh, R. H. (1988). Symp. (Int.) Combust., 22, 89-96.

[3] Ma, X. F., Zangmeister, C. D., & Zachariah, M. R. (2013). J. Phys. Chem. C, 117, 10723-10729.

[4] Camacho, J., Tao, Y. J., & Wang, H. (2015). Proc. Combust. Inst., 35, 1887-1894.

557617 Tungsten Nitride ALD Nanofilms for Reducing Hydrogen Diffusion

Nuclear thermal propulsion is an attractive in-space propulsion technology due to its large specific impulse about twice that of a chemical combustion engine. This specific impulse is achieved by using energy released from uranium fission to heat hydrogen gas to over 2400ºC before expelling it out a supersonic nozzle. However, hydrogen diffuses into the nuclear fuel elements and embrittles them, hindering the use of nuclear thermal engines. Here, we investigate the use of nanoscale tungsten nitride (WN) films as hydrogen environmental barrier coatings to address this issue.

Ultrathin WN films were deposited by atomic layer deposition (ALD) on zirconia nanoparticles and yttria stabilized zirconia (YSZ) micropowders using a fluidized bed reactor. The as-deposited film contained tungsten (W) in a variety of oxidation states, but was shown to dissociate nitrogen and form a crystallized W film upon annealing. The film's intended use as a hydrogen barrier coating motivated the investigation of hydrogen diffusion in W using density functional theory (DFT) and differential thermal analysis (DTA). The diffusion pathway was determined from the lowest energy adatom site and the hydrogen charge state at various sites along the pathway were calculated. DTA was used to investigate the efficacy of the W/WN films in preventing hydrogen attack experimentally. The temperature at which hydrogen reacted with the sample increased with film thickness, thereby indicating that the film inhibited this reaction. However, the existence of a hydrogen reaction peak in the thicker film indicated a material with greater than 2.0 eV energy barrier is necessary for a substantial decrease in hydrogen interaction with the substrate. This is the first study for ALD of WN on particles from bis(t-butylimido)bis(dimethylamino)tungsten(VI) and the use of WN ALD films as environmental barrier coatings (EBCs) for high-temperature H_2 above 1000ºC.

574437 Plant Based Illumination Enabled By Functionalized Nanophosphor Particles

The concept of a light-emitting plant, or plant exhibiting chemiluminescence powered from its own stored chemical energy, offers promise to advance off grid illumination and other autonomous photonic applications. In this work, introduce and investigate an additional nanoparticle designed to augment plant light emission in the form of silica coated strontium aluminate nanoparticles as nanophosphore elements. These nanoparticles can adsorb and re-emit generated light at longer times, increasing the duration of light emission. Moreover, such nanophosphores can also scavenge additional energy from solar fluence, increasing and augmenting total light emission from the plant. Infiltrated strontium aluminate particles showed homogeneous distribution inside plants leaves in spongy mesophyll region without penetration inside plants cell, preserving their intact structure, as well as efficient particles infiltration deep into the plant's stem.

560787 Correlation of Structure-Property Relationship of Dry-Coated Particle for All-Solid-State Battery

Recently all-solid state lithium ion secondary batteries have attracted much attention because of high-energy density and safety. All-solid-state batteries are composed of cathode, anode and inorganic solid electrolytes instead of organic liquid electrolytes. For commercial application, it is necessary to construct a solid-solid contact interface between solid electrolytes and active materials in order to produce a pass of lithium ion transport during charge and discharge. To construct a solid-solid contact interface, we have proposed to use a dry coating process. By using this process, a coated particle in which electrode are coated with solid electrolytes is able to be produced. In this study, the effect of thickness of the solid electrolyte layer on the performance of all-solid-state battery cells was investigated. In this study, $LiNi_{1/3}Co_{1/3}Mn_{1/3}O_2$(NCM) and Li_3PS_4(LPS) were used as an active material and solid electrolyte, respectively. SEM images of the coated particle showed that NCM particles were uniformly coated with LPS and its surface was smooth. Additionally, width of the particle size distribution of the coated particles was unchanged form the original NCM. These results suggest that all LPS particles uniformly covered the NCM particles. Result of particle size distribution showed that coated particles with difference thickness of coating layer were successfully prepared. The thickness of LPS layer was 0.03 μm to 0.61 μm. From the result of charge/discharge test, the cells exhibited significant different performance depending on thickness of LPS layer. Cross-sectional structure of the electrode was analyzed to investigate an essential reason of the influence of the thickness of LPS layer. As a result, we revealed that the balance of both of lithium ion and electron conducting paths is a key for enhancing the high–rate cell performances of the all-solid-state lithium ion secondary batteries.

570378 Dynamics of Early Soot Formation By Reactive Molecular Dynamics

Soot formation is of fundamental importance in combustion systems as the presence of soot influences the efficiency of combustion. Soot emission necessitates post-combustion control technologies, as soot can have deleterious effects on human health and the environment. Developing an accurate representation of early soot formation stages has proven challenging. The initial steps in soot nucleation are thought to be due to clustering reactions of different gaseous precursors species. However, there are still large uncertainties in the reaction rates involved in soot nucleation.

Here, reactive molecular dynamics (MD) simulations are used to capture accurately the dynamics of soot nuclei at the early stages of their formation, to determine its structure and composition of incipient soot nuclei and to obtain precise soot inception rates. Bond breakage and new bond formation upon collisions are accounted for by employing a reactive ReaxFF interatomic potential. A newly-developed MD framework is applied to elucidate the collisions between nascent soot nuclei of various sizes and reactant precursor species. The above atomistic method allows for accounting for soot cluster morphology and composition during its collisional growth. Furthermore, the effect of process conditions on soot dynamics is quantified. The above method predicts accurately the sticking probability that can be used to correct the collision frequency of soot aggregates and can be readily employed into larger scale models, including mesoscale methods, such as discrete element models for nascent soot (Kelesidis et al., 2017) and computational fluid dynamics.

Kelesidis, G. A., Goudeli, E., & Pratsinis, S. E. (2017). *Proc Combust Inst*, 36, 29-50.

579029 Novel Closed Spiral Reactor for Supercritical Hydrothermal Synthesis of Nanoparticles

A novel spiral device that combines the mixing and phase separation for liquid-liquid applications is presented conceived [1]. Its application to the supercritical hydrothermal synthesis of nanoparticles involving mixing of supercritical water and metal salt solution producing nano-particles is studied. There is a growing interest for scale-up of the hydrothermal synthesis of nanoparticles. It is a challenging problem in this field. Therefore, there is a need for novel reactors for hydrothermal synthesis of nanoparticles. The mixing of reactants plays a crucial role in controlling the hydrothermal reactions and particle growth rates.. The novel reactor is explored computationally for the synthesis of nanoparticles.

A multiphase coupled Computational Fluid Dynamics(CFD)-Population Balance Model(PBM) model based on the Eulerian-Eulerian approach is employed to simulate the flow, heat transport, reactions and particle population inside the novel reactor. The model is built using, ANSYS Fluent 16.0 and tested with published experimental data of ceria nanoparticle synthesis.

The reactor configuration, wherein metallic precursor solution is introduced through the inner tube of the mixer whereas supercritical fluid is introduced through the outer tube is studied for nanoparticle synthesis. The mixing inside the reactor is driven by centrifugal force due to the curvature of a spiral turn, which makes the heavier fluid (metal salt solution) penetrating into the lighter fluid phase (supercritical water) resulting in the efficient mixing followed by chemical reaction. The subsequent counter spiral causes the mixing to repeat enabling greater mixing of the phases.

Based on our simulation studies, we observed that flow ratio, ratio of metallic precursor flow rate to supercritical fluid flow rate (FR) provides better insights into the mixing and chemical reactions inside the reactor. Figure 1 shows the exit concentration of ceria at different FR. It can be observed that ceria concentration is unaffected for different flow rates for FR<1. However, it is more sensitive to flow rates when FR>1. The maximum ceria concentration is obtained at FR=1 for different flow rates of the reactants. Hence, the production rate of the ceria nanoparticles is independent of flow rates (up to 3400 ml/min) for FR=1, which facilitates the design and scale-up of the reactor for nanoparticle production. Then we have evaluated the product PSD at the exit of the reactor for the best operating conditions (FR=1) to control and optimize the properties of ceria nanoparticles. Thus, the current study is helpful in the design and scale-up of a novel reactor for scaling-up the production of nanoparticles using the hydrothermal method.

References:

1. "INTEGRATED APPARATUS FOR MIXING AND SEPARATING FLUID PHASES AND METHOD THEREFOR" A Patent Application Filed at Indian Patent Office on 14th March 2018. Application Number 201822007888 Inventors: Arjun Kumar Pukkella, Sivakumar Subramanian and Raviraju Vysyaraju

575231 Chemical Looping Partial Oxidation of Methane: The Impact of Particle Size of Hematite on Syngas Selectivity

Chemical looping partially oxidation (CLPO) of methane is one the most promising techniques to generate syngas. A major challenge of this process is the development of oxygen carriers that have high reactivity, selectivity, and recyclability.

In this research, the size effect of Fe2O3@SBA-15 was carefully studies. Samples with different Fe2O3 loading were synthesized by wet impregnation method and applied to generate syngas from methane. Samples with different nanoparticle distribution were also synthesized by adjusting the amount of surfactant. Techniques, including X-ray diffraction (XRD), N2 physisorption, transmission electronic microscope (TEM), scanning electronic microscope (SEM), and thermogravimetric analysis (TGA), were used to characterized the material. Experiment results showed that nanoparticles embedded inside of the SBA-15 silica tubes have high reactivity and selectivity. Samples with Fe2O3 nanoparticles on the surface of silica support showed lower selectivity as well as worse recyclablility. The mechanism of Fe2O3@SBA-15 as a good performance oxygen carrier was also discussed. The diffusion limitation of mesoporous support can be attributed to the high selectivity of the Fe2O3@SBA-15.

575291 Exploring Particle Modification Strategies to Improve Yields for Oxidative Coupling of Methane in a Chemical Looping System

Direct valorization of methane to value added products is a challenging and highly sought-after technology. Oxidative coupling of methane (OCM) is one such attractive technology, where methane molecules under the presence of an oxygen species, couple to form ethylene and water. Chemical looping oxidative coupling of methane (CL-OCM) provides this oxygen via lattice oxygen from a catalytic oxygen carrier (COC). CL-OCM has several advantages over the traditional OCM and thus is the focus of this study. Detailed energy analysis has been performed previously on the CL-OCM process over a proprietary Mg_6MnO_8 based COC, which gave a hydrocarbon yield of 23.2%. The objective of this study is to intuitively develop COCs which can achieve high yields in OCM by modifying the base COC.

The particle modifications are carried out using density functional theory (DFT) calculations, which would help in the screening of different metal oxide dopants that can increase the OCM hydrocarbon yield. These results have been coupled with experimental verification to develop an optimal COC configuration. Thermogravimetric analysis (TGA) has been used to investigate the reactivity of these COC's under methane reduction. These results have been combined with fixed bed tests of the COCs, which provide gas-composition data as the reduction proceeds. A Fourier transform infrared (FTIR) gas analyzer along with a gas chromatography analyzer has been used for this purpose. Parametric studies have been carried out to find the optimal reaction temperature and space velocity to obtain the highest hydrocarbon yield. Additionally, the recyclability of the modified COC has also been tested for 50 redox cycles. Solid characterization techniques such as X-ray diffraction (XRD), X-ray photoelectron spectroscopy (XPS) and Scanning electron microscope (SEM) were used to understand the changes in the solid phase. Results from this study provide key insights into the development of such COCs for CL-OCM technology.

573771 Robust Perovskite Oxygen Sorbent Particles for "Low Temperature" Thermochemical Air Separation

Perovskite Oxygen Sorbents for "Low Temperature" Thermochemical Air Separation: Correlating Compositions with Redox Performance

As a potentially more efficient alternative to cryogenic air separation, thermochemical air separation via cyclic redox reactions of an oxide-based oxygen sorbent has been extensively investigated recently. Although a number of promising oxygen sorbents have been proposed and investigated, further improvements in sorbent performance through fundamental understanding of the relationships between oxygen sorbents' structural/compositional properties and redox performance are highly desirable.

In this study, we systematically investigated the effects of A and B site dopants on the oxygen uptake/release properties of $SrFeO_3$ family of perovskites as oxygen sorbents. Addition of dopants to the A site, namely Ba and La, or the B site, Cu or Mn, increased the oxygen capacity for temperature and pressure swings when compared to undoped $SrFeO_3$ with the best sample showing an increase of oxygen capacity of 0.16%. Based on these results, the Mn sample underwent 1000 redox cycles where the sorbent was subjected to oxygen partial pressure swings between 20% O_2 and inert at 600 °C. In this scheme, the material was oxidized for 4 minutes, and reduced for 6 minutes. The sample proved to be stable in terms of oxygen storage capacity, with less than a 3% decrease in capacity. The redox kinetics were similar for the last few cycles when compared to the first few cycles. DFT calculations, in conjunction with the experimental results, indicated that oxygen vacancy formation energies are correlated to initial decomposition temperature for the samples investigated. However, the oxygen storage capacities of the doped samples did not show good correlation with the vacancy formation energies. It was further determined that the dynamic nature of the redox process makes it difficult to use a single vacancy formation energy as the descriptor for the oxygen storage capacities of the sorbents. Instead, a systematic approach was developed to correlate the oxygen storage capacities with the sorbents' compositional properties and vacancy formation energies. The findings from this study can potentially be used to develop improved sorbents for thermochemical air separation.

577380 Multiscale Modeling of Group D Iron Oxide Particles in Chemical Looping Systems: From Atomisitc Scale to Continuum Scale

Chemical looping (CL) technology, which utilizes solid oxygen carriers to enable full or partial oxidation of fossil fuel, is one of the most promising CO_2 capture platform in the field of clean energy due to its capability in simplified inherent CO_2 capture and value-added chemical production. Process modeling with Aspen Plus has been conducted on a specific coal-based CL power production process. The results show a better energy and exergy efficiency of CL process compared to conventional power production process with CO_2 capture. Reactivity of oxygen carriers is crucial for CL processes. In the CL technology developed at The Ohio State University (OSU), iron oxide is employed as the active oxygen carriers due to their abundance and low cost. In addition to process simulations that are based on thermodynamics, kinetics modeling for the gas-solid reaction of iron oxide with hydrogen (H_2), carbon monoxide (CO), methane (CH_4) and oxygen (O_2) in both atomistic and continuum scales were developed. In the atomistic scale, *Ab initio* density functional theory (DFT) simulations were employed to unravel the reaction pathways of CO and CH_4 oxidation with Fe_2O_3. The first-principles simulations provide a detailed insight of the reaction mechanisms and the rate-limiting steps (RLS), and hence the rational for material modifications. Results from DFT simulations were further validated by temperature programmed reduction experiments (TPR). Copper was identified as the most efficient dopant among 1st-row transition metals as it loweres the reduction initiation temperature with CO and CH_4 by 50 and 150 °C, respectively. In continuum scale, the development of kinetics model is essential for the optimum design of operating conditions and reactor configurations. In this work, reduction/oxidation kinetics models that can tackle multi-step reactions (i.e., Fe^{3+}->Fe^{2+}->Fe) without presuming any rate-controlled mechanism for the gas-solid reactions of group D iron-titanium composite particles were established. The obtained kinetics models were verified in a thermogravimetric analyzer (TGA) under common operating temperatures and gas compositions in CL processes. The proposed models and insights gained from the simulations under different scales could be valuable in the development of future CL technology.

577663 Design of Fluidized Bed Combustor Reactor in Chemical Looping Combustion: Hydrodynamic and Heat Transfer Properties

The chemical looping combustion (CLC) technology is one of the most promising CO_2 capture technologies for fossil fuel power generation applications. CLC utilizes a metal oxide as an oxygen carrier (OC) to supply oxygen from air to the fuel through a reduction and oxidation reaction pathways of the OC, thus avoiding contact between fuel and air to allow effective separation of O_2/N_2, CO_2, and H_2O. In the CLC process developed by The Ohio State University (OSU), an iron-based OC is circulated between a counter-current moving bed reducer reactor for full combustion of fossil fuel and a fluidized bed combustor reactor for OC regeneration by air. High quality heat can be extracted from reducer flue gas, combustor flue gas and combustor for power generation. While heat recovery from flue gas has been extensively studied and employed in existing power plants, the properties of the granular flow in the fluidized combustor is the key aspect remain unresolved in the design of CLC power plant. Two-fluid model (TFM) was adopted in the present work for its high efficiency in large-scale simulation when compared with CFD-DEM method. Experiments in the fluidized bed were conducted as a baseline to calibrate the hydrodynamic and heat transfer coefficients of the TFM. Specifically, the validated TFM model was applied to the fluidized combustor of a coal-based CLC power plant. We can then configure the design of combustor itself and heat transfer surface in combustor based on the hydrodynamic and heat transfer properties and information of energy and material balance from process modeling by Aspen Plus.

570717 Imiquimod Particle Formation Using the Supercritical Antisolvent (SAS) Process

One of the main problems emerging from the pharmaceutical industry is the low solubility of hydrophobic drugs, which can cause insufficient bioavailability. Thus, approaches able to improve the solubility of these drugs are of high interest. An alternative to this problem is the use of supercritical fluid micronization, a particle formation method, which has large potential to be used in the pharmaceutical field because of its low toxicity, low cost and relatively mild processing conditions, since carbon dioxide is one of the most common antisolvents used for this purpose. In the present study, imiquimod, a topical chemotherapeutic used to treat skin tumors such as basal cell carcinoma, anogenital warts, and keratosis, was processed using the supercritical anti-solvent (SAS) technique aiming at improving its solubility in aqueous media. This drug has poor solubility in apolar solvents and is soluble only in acidic aqueous solutions. Thus, glacial acetic acid was chosen as solvent. The processing conditions used were 120 bar, 60°C, CO_2 flow rate of 15 g/min and a solution flow rate of 0.5 mL/min. The characterization of the material obtained was performed by X-Ray diffraction (XRD), Fourier-transform infrared spectroscopy (FTIR), thermal gravimetric analysis (TGA), differential scanning calorimetry (DSC) and scanning electron microscopy (SEM). It was possible to produce particles with a thin, porous, sheet-shape d structure that was quite different from the untreated drug, as observed by SEM analysis. The FTIR analysis showed that the functional groups were maintained, ensuring that the drug has not been degraded. TGA and DSC analysis showed that the drug degrades only at 300°C. Thus, the supercritical antisolvent process was adequate to produce particles with high surface area and with preserved drug structure.

The authors are grateful for the financial support of this work by the São Paulo Research Foundation (FAPESP) and the National Council for Scientific and Technological Development (CNPq), both from Brazil

570926 Surface Functionalization of Polymer Latex Particles and Its Effect on in-Situ Deposition of Nano-Sized Calcium Silicate Hydrate

Polymer latexes have been widely used in cementitious materials to improve their workability and mechanical properties. However, the reduced hydration activity is often found after the polymer latexes are added, especially during the first hours and days. Nucleation seeding with synthetic nanoscale calcium silicate hydrate (C-S-H) is a possible method to compensate for the retarded effect, as it has enormous potential to accelerate the cement hydration. But how to play the role of them effectively and conveniently are still the challenges. In this work, the polymer latex particles with hydrophilic carboxylic/sulfonic chains grafted from the surface were first fabricated via soap-free emulsion polymerization. And then, some of them were modified by amino/phosphate groups respectively. The nanoscale C-S-H was anchored on the particle surface by in-situ deposition reaction of $Ca(NO_3)_2$ and Na_2SiO_3 afterwards. The effects of different functional groups on the deposition process of nanoscale C-S-H were investigated, and its application in cement paste was attempted.

The surface hydrophilic groups content of the latex particles was measured by conductometric titration. The interaction between particles and Ca^{2+} was studied by treatment with $Ca(NO_3)_2$ individually. The deposition process was analyzed by centrifugal titration and inductively coupled plasma optical emission spectrometry (ICP-OES). The morphologies of the latex particles and composite microspheres were characterized by transmission electron microscopy (TEM) and scanning electron microscopy (SEM). Thermos gravimetric analyzer (TGA) and dynamic light scattering (DLS) were employed to characterize the composition and the dispersibility of the composite microspheres, respectively. After adding the composite microspheres into paste, the hydration process of the fresh mixture was monitored by cement hydration heat analyzer. Its workability and mechanical properties were tested via conventional methods. The porosity of the hardened specimens was characterized by mercury injection method.

The results show that the surface hydrophilic groups content of P(St-*co*-MAA) latex was around 2×10^{-4} *mol*/g. The enrichment of Ca^{2+} on the particle surface can be observed, and that with carboxylic groups has the highest content. The C-S-H gel always exhibits an amorphous phase, and the maximum mass ratio of C-S-H is about 8.0 *wt*% which also appears in the carboxylated particles. The accelerated hydration can be observed for all types of composite microspheres, in which those with exposed sulfonic chains have the most significant effect. No obvious effects on the paste fluidity and the water absorption of the hardened specimen can be observed. The modified mechanism of the composite microspheres for paste was proposed.

569646 Machine Learning Determination of Resistant and Tolerant Bacteria Growth Modes Induced By Nanoparticles

Nanoparticles are promising alternatives to antibiotics since nanoparticles are easy to manufacture, non-toxic, and do not promote resistance. Nanoparticles act via physical disruption of the bacterial membrane and/or the generation of high concentrations of reactive-oxygen species locally. Potential for physical disruption of the bacterial membrane may be quantified by free energy methods, such as the extended Derjuan-Landau-Verwey-Overbeek theory, which predicts the initial surface-material interactions. The generation of reactive-oxygen species may be quantified using enthalpies of formation to predict minimum inhibitory concentrations. Neither of these two quantitative structure- activity values describes the dynamic, in situ behavioral changes in the bacteria's struggle to survive. In this paper, borrowing parameters from logistic, oscillatory, and diauxic growth models, we use principal component analysis and agglomerative hierarchical clustering to classify survival modes across nanoparticle types and concentrations. We compare the growth parameters of 170 experimental interactions between nanoparticles and bacteria. The bacteria studied include Escherichia coli, Staphylococcus aureus, Methicillin-Resistant Staphylococcus aureus, Staphylococcus epidermidis, Pseudomonas aeruginosa, and Helicobacter pylori, and were tested across multiple concentrations of liposomal drug delivery systems, amphiphilic peptide, and silver and selenium nanoparticles. Clustering reveals specific pairs of bacteria and nanoparticles where the nanoparticle induced growth dynamics could potentially spread the infection through the development of resistance and tolerance. This rapid screening also shows that bacteria generated nanoparticles do not induce growth modes indicative of the development of resistance. This methodology can be used to rapidly screen for novel therapeutics that do not induce resistance before using more robust intracellular content screening. This methodology can also be used as a quality check on batch manufactured nanoparticles.

554084 Overcoming the Limitations of Hydrochlorothiazide Micronization Using the Precipitation with Compressed Fluid Antisolvent Process

When a patient is prescribed with medication, the dosage considers how much medication is needed to exhibit its therapeutic effect while not inducing toxic or lethal effects. However, many essential drugs are poorly soluble. These drugs may not be available to system when needed due to the slow dissolution rates or may even be excreted from the body. Therefore, many times, the drugs are prescribed at higher dosages than the needed amount. To lower the dosage of these drugs, chemical and physical modifications can be applied. The latter option is preferred as chemical modifications, which alter the chemical structures to enhance solubility, require time consuming repetitions of the drugs' clinical trials. Therefore, many researches are focused on physical modifications to improve the dissolution rate of the poorly soluble drugs, which is most often the rate determining step for drug absorption. Conventionally, milling and spray drying have been used to decrease the particle size of the drugs. However, limitations of these processes include drug degradation, residual solvent, and high drug yields (when excipients are utilized).

To overcome these restrictions, this research utilized the precipitation with compressed fluid antisolvent process under sub-and supercritical conditions. The model drug, Hydrochlorothiazide(HCT), is classified as poorly soluble by the Biopharmaceutics Classification System. Miconization of neat HCT yielded agglomerated, irregular microparticles. While the particle size decreased, not much difference was observed in the dissolution rate. Therefore, to further decrease the particle size, HCT was coprecipitated with a water soluble polymer, polyvinylpyrrolidone (PVP). As a result, uniform, spherical nanoparticles that significantly enhanced the dissolution rate of HCT were obtained. FT-IR results show no sign of residual solvent, which is crucial for the drug's stability and toxicity. A combination of HR-TEM and the dissolution tests revealed that the particles are composed of PVP particles that contain HCT. Even though an excipient was used, the products showed high normalized drug loadings and precipitation yields. These results suggest that HCT particles precipitated on the PVP particles by heterogeneous nucleation.

577547 Characterization of Nanoparticle Size Distributions and Aspect Ratio through Single-Particle Tracking and Analytical Ultracentrifugation

Colloidal dispersions of nanoparticles have been a focus of recent study due to their unique properties and potential applications. However, a varying degree of polydispersity is inevitable in most non-ideal systems where the nanoparticles are derived from natural sources or top-down methods. Combined with the fact that colloidal dispersions are inherently unstable and are also sensitive to environmental influences, the characterization these materials with respect to the evolution of their size distributions should be routine for their biological applications.

The most commonly employed methods in the literature, such as dynamic light scattering, only estimate the first or second moment of narrow distributions. In this contribution, we leverage single-particle tracking (SPT) to evaluate the sample preparation and purification of single walled carbon nanotubes (SWCNT) and graphene oxide (GOx). Noncovalent conjugations of SWCNT with DNA and sodium dodecyl sulfate (SDS) were studied due to their differences in surface coverage and dispersion protocol. Changes in particle hydrodynamic size distribution with varying degrees of centrifugation were monitored. The SPT data collection and processing were optimized for adequate sampling and noise cutoff. Comparison of samples of increasing centrifugation duration show overall left shift of distribution toward smaller sizes. Analytical ultracentrifugation (AUC) on a similar set of samples show a corresponding decrease in particles of large sedimentation coefficients.

The two analytical techniques provide independent sets of information including the aspect ratio of the particles *via* the frictional coefficient, which for SWCNT and GOx can be modeled as simple geometric shapes with two dimensional parameters. Thus, for the highly centrifuged relatively monodisperse samples, we were able extract additional information through the combination of the AUC and SPT to provide reasonable estimates of particle hydrodynamic dimensions.

560326 Numerical Modeling of Wet Granulation Processes

Granulation is an important powder handling process to produce granules from fine powder. Due to many operating parameters, a number of trial-and-error tests are required for a granulation process, and this is a bottleneck for design, optimization and scale-up. Modeling of granulation is expected for design of a granulation process. Discrete element method (DEM) and population balance model (PBM) are widely used for the numerical modeling. DEM is a simulation method to calculate the powder flow in the granulator. However, DEM does not account for particle aggregation. On the other hand, PBM is a process modeling framework for tracking changes in particle properties over time. However, PBM does not consider the powder flow in the granulator. In recent years, a coupling approach between DEM and PBM (DEM-PBM model) has been proposed. DEM-PBM model is able to calculate both of the powder flow and the particle growth. However, more efforts are required to develop an accurate coalescence kernel for DEM-PBM model. In this study, DEM-PBM model including a novel coalescence kernel for granulation process was proposed. As a result, the model was able to predict the experimental result.

570921 Wall Friction Measurements on Free-Flowing Material at Low Normal Stresses

Recent experiments in gravity-driven hopper flow have necessitated the quantification of wall friction angles between the fill material, mustard seeds between 2.00 and 2.36 mm in diameter, and three different wall materials. The wall material samples consisted of flat pieces of 3D-printed ABS plastic, one 'as printed', one smoothed with acetone vapour, and the other roughened by adhesion of a monolayer of mustard seeds. Angles of wall friction were estimated using two methods: firstly, a translational shear cell was constructed, in which the wall material was moved against a stationary sample of fill material, to avoid issues which could arise from moving cell containing the test sample across a very rough surface, as is usual practice. Forces associated with the movement were related to varying normal forces to establish the wall friction angle for each condition. These values were compared to those obtained by slowly inclining a wall friction sample until seeds on its surface overcame friction and began to move. The latter method resulted in markedly higher angles, though little difference was shown between the 'as printed' and smoothed surfaces; also, the angle of friction was significantly higher for the roughened surface, but could not be measured in the shear cell, due to jamming of seeds falling into crevices in the rough surface.

570454 Conduction and Convection Heat Transfer in a Rotary Drum Using Simulations and Experiments

Granular material subjected to agitation is encountered in many practical applications of material processing. Furthermore, many of these applications also involve heat transfer whereby solids face cold or hot surfaces and heat is exchanged not only between individual particles but also between the particles and external surfaces via the particle–particle, particle–wall, particle-fluid, and particle-environment contacts. This work focuses on understanding the conduction and convection heat transfer in the granular bed inside a rotary drum, one of the most commonly used process equipment. Experiments are performed using silica gel beads inside a 3-inch radius and 3-inch long stainless-steel rotary drum. One side of the drum is closed with a transparent quartz window, which can handle high temperatures. Another side of the drum is closed with a sapphire window, compatible with an IR camera, specifically chosen to give a high transmittance to the infrared light. The setup is designed to handle up to 1000 °C, making it possible to study all modes of heat transfer at a wide range of temperatures. The wall of the drum is maintained at a desired high temperature using heat guns placed around the drum. Heat transfer is quantified under varying operation conditions to establish a strong understanding of the heat transport. For this, both experiments and CFD-DEM simulation techniques are used to analyze the thermal behavior. Experiments are conducted to validate the CFD-DEM code in the first phase. In the second phase, the validated code is used to study the effect of rotation rate, fill level, particle size, and particle size distribution on the wall-particle bed heat transfer.

572291 Identifying the Impact of Processing on the Physical Properties of AM Powders and Their Subsequent Reuse

Whilst the term 3D printing is used as a synonym for all additive manufacturing process, additive manufacturing specifically covers a multitude of processes which vary the method of layer manufacturing. Powder Bed Fusion requires layers of powder to be built up, which are then selectively fused together, using either a laser or electron beam, depending on the process and substrate been utilised

Typically, the stock powder is contained in a hopper or reservoir and a blade or roller is used to spread a discrete amount of powder over the build platform, delivering a 0.1mm / 1-2 particles layer of powder which is then fused to the previous layer. The process is then repeated until the entire model is completed. Once complete any loose, unfused powder is removed. Whilst this process is suitable for a wide range of powders, it is typically used with common metals, for example, Stainless Steel, Titanium, Aluminium and thermoplastic polymers including nylon and polypropylene.

The volume of the actual component is typically considerably less than the volume of powder deposited during the build process, resulting in a large volume of loose powder at the end of the printing process. In order to make the process economically viable this material is usually recovered and reused. The recycled material is mixed with virgin material in order to achieve acceptable process performance and final product, with the required ratio of virgin to recycled material varying across different instruments.

The effect of continued reuse can affect both the chemical and physical properties of the powder, for example, powder exiting an AM machine may contain splatter from the melt pool in the form of larger particles, or the powder surface may be contaminated. Importantly it's not just material close to fabrication that's affected by going through process, with the process itself influencing the powder properties. How the physical and chemical changes impact the powder-bed fusion process is less well understood although previous studies have indicated changes in the powder bulk behaviour – including the flowability which may explain why the recycled material exhibits unacceptable process behaviour.

In order to better understand both the changes to the powder physical properties and their subsequently influence on the process performance, samples of virgin and recycled materials alongside mixtures with known process performance where characterised using a wide range of techniques including particle size analysis and surface area measurements. The powder bulk properties including flowability where also characterised using powder rheology. By comparing the particle and powder properties as well as their relative performance it was possible to identify both the key changes to the powder physical properties resulting from the initial processing of the virgin material and the resultant changes in the bulk properties, thereby providing an understanding of why the recycled powders exhibit poor process performance.

566516 Comparative Study of Mixing in a High Shear and Low Shear Blender Using Numerical Simulation

Comparative study of mixing in a high shear and low shear blender using numerical simulation

Koyel Sen [1], Natasha V Rodriguez [4], Bodhi Chaudhuri [1,2,3], Carl Anderson[4]

[1]Department of Pharmaceutical Sciences, University of Connecticut [2]Department of Chemical and Biomolecular Engineering, University of Connecticut [3]Institute of Material Sciences, University of Connecticut [4]Duquesne University, Graduate School of Pharmaceutical Sciences

Correspondence: koyel.sen@uconn.edu

Background:

Mixing is one of the crucial unit processing widely used in Pharmaceutical manufacturing to produce homogenized mixture which further goes for other treatments e.g. tablet press, wet granulation for tableting, capsule filling machine for capsules etc. The purpose of this comparative study is to understand and evaluate the effect of shear on mixing in a high shear and low shear blender using numerical simulation.

Methods:

Mixing was simulated in a high shear and a bin blender using Discrete element modeling or DEM. The computational geometries were generated according to lab-scale blenders to minimize the experimental variability. The impeller of high shear blenders was generated by gluing cohesive particles together. A multicomponent mixture was used as the blend formulation. The characterization of mixing was carried out using 'box counting method', measuring relative standard deviation or RSD as a function of mixing time. The effect of operational parameters such as fill percentage, loading pattern and rotational speed were observed to determine the optimized blending process.

Results:

Simulated result shows higher degree of mixing can be achieved in high shear compare to low shear blender. It also shows that fill percentage, rotational speed and loading pattern has a direct effect on the degree of mixing. The result shows the impeller in high shear blender promotes diffusive transport of particles which leads to quicker mixing.

Conclusions:

A comparative study was carried out to explore the effect of shear on mixing using Discrete Element Modelling. The blend used in the simulations consists a multicomponent mixture. Simulation shows that high shear blender provides quicker mixing comparatively. In both cases the operational parameters such as loading pattern, rotational speed and fill percentage have shown significant effect on the degree of mixing which complies with the previous single component mixing literature.

575151 Effect of Viscosity on Liquid Penetration in the Catalyst Impregnation Process: Experiments and Simulations

Dry catalyst impregnation of active metals onto a porous catalyst support is an important step in the preparation of heterogeneous catalyst. In a typical dry impregnation process, metal solutions are sprayed over a particulate bed in a mixing vessel until the pore volume is reached. The inter-particle variability of the impregnated liquids inside the particles and metal content may significantly affect the activity and selectivity of the resulting catalyst. In this process, metal salts or complexes are dissolved in a typically aqueous solution and added to porous catalyst support particles. The viscosity of the metal solution can affect the penetration and diffusion of the liquid into the pores. The aim of this work is to understand how viscosity of the metal solution affects the impregnation process and how the viscosity affects the uniformity of the liquid content in the particle bed.

We studied impregnation of alumina particles with viscous solutions consisting of different level of viscosities and concentrations of Nickel Nitrate and Polyethylene Glycol (PEG, molecular weight of 400 g/mol). Polyethylene Glycol was considered as the viscous agent in our experiments. The viscosities of the PEG solution with different concentrations were measured by a Brookfield DV3T viscometer. Impregnation of viscous solutions on alumina particles were performed and compared with one to one DEM simulations.

We developed a new viscous liquid transfer algorithm that is based on a Langmuir isotherm. We performed a series of adsorption experiments with the goal of determining the adsorption parameters used in the Langmuir isotherm with two adsorption parameters: K_L is the equilibrium constant and q_{max} is the maximum adsorption capacity.

We developed a predictive model for the impregnation and transfer of a viscous solution in a rotating vessel. We performed experiments to obtain the Langmuir parameters (K_L and q_{max}) for different values of viscosities, and we observed that for low values of the viscosities qmax does not show a lot of variation until 2cp, but for viscosities > 2cp, it decreases with increase in viscosity. In general, experimental results show that the viscosity delays the impregnation process. The higher the viscosity the longer it takes to reach the equilibration phase. Metal absorbed in the pores depends strongly on the viscosity of the solution. In general, higher the viscosity, smaller the amount of metal absorbed. We also found the viscosity of the solution affects the Langmuir equilibrium parameters K_L and q_{max}, which decrease with increasing viscosity. DEM simulations incorporating these parameters for a viscous solution were considered to reveal the amount of metal absorbed and the uniformity of the particle bed. Using the Langmuir model coupled with the mass balance, the total amount of metal absorbed on solid and the amount metal in the liquid phase can be calculated. Our simulation results show good agreement with the experiments. Both DEM simulations and experiments confirm that metal absorbed in the pores depends strongly on the viscosity of the solution.

In addition, we studied the penetration time of viscous PEG solution droplet into alumina particles. The drop penetration time is defined as the time taken for the drop to penetrate completely into the porous particle with no liquid remaining on the surface. Results show that that the penetration time increases with solution viscosity.

564771 In-Situ SAXS Study of the Nucleation and Growth of Precipitated Silica

The nucleation and growth of precipitated silica has been drawing attention. Compared with the hydrolysis reaction, the precipitation of silica is much quicker, so we focused on a millisecond reaction level. In this work, the in-situ measurement was made to follow the initial stage of the formation of silica particles by using Small Angle X-ray Scattering (SAXS). The particle size was obtained as a function of time. The experiment was carried out at different temperatures and mixing conditions, based on which, we combined the kinetic process of nucleation-growth as well as micromixing effects and established a nucleation-growth model of precipitated silica particles.

568077 A Conceptual Model for Analyzing Particle Effects on Gas-Liquid Flows in Slurry Bubble Columns

Particles play significant roles in the flow structure of gas-liquid flows in slurry bubble columns. The Dual-Bubble-size (DBS) model based on the Energy-Minimization Multi-Scale (EMMS) approach has been developed to characterize gas-liquid flows by incorporating a stability condition in our previous work. A Particle-dependent Dual-Bubble-Size (PDBS) model is thereby developed in this work to analyze the different particle effects on gas-liquid flows. The particle effects on apparent properties of liquid phase and drag coefficients of bubbles are considered for systems of wettable particles. The model calculation indicates that the regime transition is delayed due to the increase of apparent density and viscosity. On the other hand, the particle effect on bubble drag coefficients decrease the gas holdup. The PDBS model taking account of both the effects can reproduce the decrease of gas holdup and energy consumption of bubble breakage arising from the increase of solid concentration. A new drag model is then derived from the PDBS model for CFD modeling. The CFD predictions of gas distribution under high solid loadings are in conformity with experimental data. The particle effect on stabilizing bubble surface is also analyzed for systems of non-wettable particles. The model reflects the increase of gas holdup with addition of non-wettable particles, and reveals also that small bubbles may dominate the system in such cases.

573748 Hydrodynamics and Mixing in a Pseudo-2D Rectangular Spouted Bed

Hydrodynamics and mixing behavior of polydispersed particles in a pseudo-2D rectangular spouted bed has been investigated using coupled CFD-DEM technique. Eulerian method (computational fluid dynamics, CFD) has been used to solve the Navier-Stokes equation for the gaseous phase and Lagrangian method (discrete element method, DEM) for particle phase. k-e two-equation turbulence model has been used to incorporate the gas phase turbulence effect. In the present work, the minimum spouting velocity is predicted numerically for a fixed bed height of polydispersed glass beads. Hysteresis in time-averaged mean pressure drop and bed expansion have been observed. Solid velocity and mixing pattern have been studied for different superficial gas velocities in three different regions of the bed i.e. at annulus, spout, and fountain zones. The effect of particle phase restitution coefficient and gas velocity on the temporal evolution of particle mixing are analyzed. We observe a double recirculation pattern in the fountain region, where particle axial velocity is insignificantly influenced by the particle restitution coefficient. But in the spout zone, the particle velocity increases with a decrease in the restitution coefficient. Solid mixing improves with increasing superficial gas velocity due to the combined effect of convective and shear-induced mixing.

568496 Solids Mixing Rates of Geldart a and B Powders Predicted By the Two Fluid Model

The influence of particle density and fluidization velocity on mixing behavior of Geldart A and B particles was numerically investigated in bubbling fluidized beds for various operating conditions. An Eulerian-Eulerian approach based on the kinetic theory of granular flow (i.e., the Two Fluid Model, TFM) was applied to perform the simulations. Boundary conditions, grid resolution, as well as the rheological model were carefully chosen to match experimental observations. A variety of systems differing in particle density, particle size, and fluidization speed were investigated to collect data on flow and dispersion rates of particles in the bed. This was done in order to expand the parameter space investigated in previous studies [1,2], as well as to investigate the anisotropy of solids dispersion rates.

To examine the particles' mixing behavior, characteristic mixing times were computed based on the temporal evolution of a solids-based scalar variance. In addition, scalar dispersion rates were extracted from the simulation data, both in the horizontal and vertical direction. Finally, it was demonstrated that the characteristic mixing time can be correlated with the particle Froude number, the expanded bed height and the solid dispersion coefficient with high accuracy. This new correlation allows practicioners to estimate solids mixing rates for a wide range of fluidization parameters of Geldart A and B powders, without necessitating the knowledge of bubble parameters as is the case in other expressions for the solids mixing time [2].

[1] M. Banaei, N.G. Deen, M. van Sint Annaland, J.A.M. Kuipers, Particle mixing rates using the two-fluid model, Particuology. 36 (2018) 13–26.

[2] A. Bakshi, A.F. Ghoniem, C. Altantzis, Mixing dynamics in bubbling fluidized beds, AIChE J. 63 (2017) 4316–4328.

572137 Fluid-Particle Drag in Low-Reynolds-Number Bidisperse Suspensions

A second-order immersed boundary-lattice Boltzmann method (IB-LBM) is used to study the drag force of freely evolving bidisperse suspensions at low Reynolds numbers. The total solid volume fraction of the suspension is varied between 0.1 and 0.3, and the particle size ratio is from 1.2:1 to 2:1. The main characteristics of freely evolving suspensions, compared to fixed beds that were reported in the literature, is that the two dispersed phases have different slip velocities relative to the fluid phase. Hence, in freely evolving bidisperse systems, the drag force on each particle phase is normalized by the Stokes drag force based on their corresponding slip velocities. It is found that the drag force in freely evolving bidisperse suspensions is significantly different from previous results based on computations of systems with fixed particles. The drag force on small particles on average decreases much slower than that in fixed beds as the particle size ratio increases. The difference of the drag force on the small and large particles is profoundly smaller than that predicted using previous relations obtained from fixed-bed simulations. Based on the present simulation results, new relations for the drag forces on both small and large particles are proposed for bidisperse suspensions. The effects of the slip velocities of the small and large particles are explicitly accounted for in the new relations.

572694 An Investigation of Gas Fluidized Bed of Flexible Fibers Using the Coupled DEM/CFD Method

Gas fluidized bed of flexible fibers is investigated using the coupled approach of discrete element method (DEM) and computational fluid dynamics (CFD). In the numerical simulation, a flexible fiber is formed by connecting a number of spheres using elastic bonds. The fiber can deform as the bonds undergo stretching/compressive, bending, and twisting deformations. The resultant drag on a fiber by a gas is a collection of drag forces on each sphere component. The fibers of the same volume but various aspect ratios are examined for the effect of fiber elongation. It is observed that the minimum fluidization velocity increases with increasing fiber aspect ratio (AR) for AR =2, due to the increase in porosity. However, the minimum fluidization velocity for the spheres (AR = 1) with the same volume as the elongated fibers is between those of AR = 3 and AR =4 fibers, because the sizes of the voids are larger for the spheres compared to the AR=2 and 3 fibers and also the effect of gas drag on the fiber motion is reduced as the sphere component size increases. At last, the effects of fiber flexibility and cohesive liquid bridge force on the fluidization behavior are reported.

559444 Exploring Agglomerate Dynamics through an Immiscible Two-Fluid Model

Fine particle agglomeration can occur when running particle atomic layer deposition (Particle ALD) in fluidized beds. The fine powders frequently used in Particle ALD tend to effectively fluidize as larger aggregates due to high interparticle forces, potentially blocking surface sites and inhibiting surface coating uniformity. By modeling the agglomeration process during coating, steps can be taken to facilitate agglomerate shedding/breakup and mixing, thereby enhancing surface coating uniformity. However, current models of gas-solid flows which preserve the gas-solid interface, an important component for modeling ALD, have several limitations when incorporating agglomeration. Many of these approaches fail to address agglomerate size distributions or the dynamic formation and shedding/breakup process of fluidized agglomerates, shortcomings which remain a significant challenge to properly modeling fluidized bed Particle ALD. In this work, we propose an alternative modeling approach which naturally accounts for the dynamic nature of fluidized agglomerates by treating the fluidizing gas and particles as two immiscible (non-interpenetrating) fluids. Agglomerates are modeled using dynamic "bubbles" whose interior consists of many primary particles from the solids phase. The position, shape and formation/breakup of these agglomerate "bubbles" are allowed to change with time as dictated by the corresponding transport equations. With this model, we can investigate the formation and shedding/breakup of agglomerates without prior knowledge of the agglomerate size characteristics. This study provides some preliminary agglomerate size distribution results from fluidized bed Particle ALD simulations and compares these results to experimental data from previous literature studies.

571100 Using a Powder Rheometer to Probe the Defluidization of Cohesive (Group C) Particles

Given the abundance of cohesive particles in various industrial applications, accurate prediction of inter-particle cohesion is crucial for the efficient design and optimization of many operations. However, predicting particle-level cohesion is non-trivial at best. For example, isolation of normal (cohesive) force in direct atomic force microscopy (AFM) measurement is difficult even for spherical particles. The indirect estimation of particle-level cohesion via force models relies upon the AFM surface maps of particles. However, carrying out AFM scans on individual particles is not conducive to an industrial setting where feed stocks change frequently. An attractive alternative is bulk measurements from which cohesion can be easily extracted. In recent work carried out by Liu et al. [1], defluidization was identified as one example of a bulk experiment that can provide particle-level cohesion. In that work, a simplified square-force cohesion model is proposed, and with the help of discrete element method (DEM) simulations, the parameters for the square-force model are determined from a defluidization (pressure drop vs. gas velocity) curve. However, as the level of particle cohesion increases (i.e., Group C particles), standard defluidization curves cannot be obtained. In the present work, a powder rheometer is used to carry out a careful exploration of defluidization behavior of particles with a range of cohesion levels. By adding energy through a rotating impeller, the rheometer is able to fluidize cohesive (Group C) particles without channeling, which is not possible in conventional fluidized beds. Both pressure drop and torque are measured during defluidization. The torque profile indicates a clear change demarking the transition from a fluidized to packed bed state, unlike the pressure drop profile where the curvature of the transition increases with cohesion level. Hence, determination of the characteristic velocity (U_{cf}), a key parameter for the square-force model [1], is more straightforward using the torque profile compared to the pressure profile. Another advantage of the methods described is that the rheometer can be used by industry as quick gauge for the relative level of cohesion as particle feedstocks and/or ambient conditions change.

Reference

[1] P. Liu, C.Q. LaMarche, K.M. Kellogg, C.M. Hrenya, A square-force cohesion model and its extraction from bulk measurements, AIChE Journal. 82 (2018) 1156–11. doi:10.1002/aic.16089.

571777 Fluidization of Nanopowders: Effects of Vibration and Stirring

Due to strong inter-particle interactions (such as Van der Waals, electrostatic and moisture-induced surface tension forces), fluidization of most nanoparticles is difficult to achieve, which may require the introduction of external forces. In this study, we investigate both mechanical vibration and stirring to enhance the fluidization of nanopowders. Hydrophilic nanosilica is used as the testing nanopowders, which shows strong agglomeration behavior leading to poor fluidization hydrodynamics. Compared to using a stirring blade in previous studies, we use a fishbone structure as the stirring bar. This modification can greatly reduce the possibility that induces flow turbulence in the fluidized bed and particles' attaching to the blade surface as a result of using the blade which has a relatively big surface area. Effects of the vibration direction (horizontal or vertical), vibration conditions (amplitude range from 0.5 to 5 mm and frequency from 20 to 100 Hz) and rotational speed of the stirrer (range from 30 to 200 r/min) on the fluidization quality of nanoparticles, the agglomerate formation (structure), pressure drop curve, bed expansion, and the minimum fluidization velocity, have been experimentally investigated.

573551 Gas-Solid Hydrodynamics and Hysteresis in a Rectangular Spouted Bed

A spouted bed is a special case of the fluidized bed in which the gas is injected into the granular bed through a single centered orifice. It is used as a gas-solid contactor in various industrial applications (such as combustion, drying, chemical vapor deposition) due to better heat and mass transfer. In the present work, particle dynamics in a pseudo-2-D rectangular tapered spouted bed has been studied numerically by coupling computational fluid dynamics (CFD) for fluid phase and discrete element method (DEM) for particle phase. In the gas phase momentum equation, k-e two-equation turbulence model has been used to capture the effect of turbulence. Fluid-particle interaction force has been incorporated using Gidaspow drag model. The minimum spouting velocity is obtained numerically for a fixed bed height. Experimental measurements are also carried out in a similar spouted bed for different particle size and static bed height. The simulated pressure drop has been compared with the experimental results. Hysteresis in the pressure drop in numerical simulations has been compared with experimental findings when the superficial gas velocity is cycled up and down successively. We have also investigated numerically the effect of superficial gas velocity on bed expansion, the longitudinal profile of particle maximum axial velocity, and voidage profile. The maximum solid velocity and its fluctuation are observed to occur in spout zone. The effect of fluctuating particle-particle contact force is found to be significant in the dilute spout regime.

568490 Experimental and Simulation Study on Bubbling Behaviour of Geldart Group B Particle Pulsed Fluidized Bed

Pulsed fluidization is a special mode of fluidization in which the flowrate of the fluidizing medium is intermittently changing. Pulsed fluidization has been proven to offer some unique advantages compared to steady-flow fluidization. These advantages include better mixing for cohesive particles [1], improved drying of wet particles in fluidized beds [2] and a more controlled bubble pattern in some bubbling fluidized beds [3].

Bubbling behaviour of group B particles in a two-dimensional column was studied experimentally. Solenoid valves were used to generate pulsation in the gas flow. Unlike the conventional concept of pulsed fluidization, total gas flow was not subjected to pulsation; a pulsing gas flow was imposed on a constant steady gas flow to the bed. In this paper, we are presenting the CFD simulation results of the system alongside the experimental findings. Barracuda VR was used for the CFD simulation. Glass beads with density of 2520 kg/m3 were used for particles. Air was used as the fluidizing medium and Wen-Yu was the drag model applied in most simulation cases. Effects of flow pulsation in general, and pulsation conditions such as frequency in particular, on bubbling characteristics were studied.

We found that pulsation in general leads to an increase in average bubble size, especially at lower pulsation frequencies; Up to 100% increase in average bubble size could be observed. The bubble size decreases with increasing pulsation frequency. This demonstrates that pulsation gives additional degrees of freedom to control fluidized-bed hydrodynamics.

[1] Hadi, B., van Ommen, J. R., & Coppens, M.-O. (2011). Enhanced particle mixing in pulsed fluidized beds and the effect of internals. Industrial & Engineering Chemistry Research, 51(4), 1713-1720.

[2] Akhavan, A., van Ommen, J. R., Nijenhuis, J., Wang, X. S., Coppens, M.-O., & Rhodes, M. J. (2008). Improved drying in a pulsation-assisted fluidized bed. Industrial & Engineering Chemistry Research, 48(1), 302-309.

[3] Wu, K., de Martín, L., & Coppens, M.-O. (2017). Pattern formation in pulsed gas-solid fluidized beds - the role of granular solid mechanics. Chemical Engineering Journal 329, 4-14.

583902 Improving CO_2 Capture Efficiency in a Fluidized Bed of TiO_2 Nanoparticles

Carbon Dioxide (CO_2) has been recognized as a major greenhouse gas among many other heat-trapping gases due to its relative abundance in the atmosphere. The US National Oceanic and Atmospheric Administration (NOAA) stated that global average concentrations of CO_2 surpassed 400 ppm in March 2015 for the first time since they started tracking carbon dioxide in the atmosphere [1]. The human activity that pumped carbon dioxide into the atmosphere over the past 150 years raised its levels higher than during the industrial revolution, which began in around 1850, i.e. higher than 280 parts per million CO_2 [2].

Carbon capture and storage (CCS) is a promising option for CO_2 reduction. Numerous methods to capture carbon dioxide have been proposed, including post-combustion carbon capture processes focusing on advanced solid adsorbents and fluidization/membrane systems. The use of a fluidized bed reactor is one of the promising techniques for CO_2 capture in a post-combustion process. The main benefits from fluidization are high gas–solids contact efficiency and the continuous regeneration of adsorbents. Li et al. [3] studied CO_2 adsorption capacity over dry $K_2CO_3/MgO/Al_2O_3$ adsorbents in a fluidized bed reactor. Li and Lu et al. [4,5] utilized enhanced calcium-based adsorbents for CO_2 capture since calcium-oxide containing materials have high reactivity and adsorption capacity for CO_2 and low material cost. Valverde et al. [6] mixed silica and calcium hydroxide powder to enhance the CO_2 adsorption efficiency in the fluidized bed.

In this work, a microjet and vibration assisted (MVA) fluidized bed that is a patent-pending technology by the Andino research group at Arizona State University [7] was developed to enhance the fluidization quality of nanosized TiO_2 particles. The behavior of the fluidized bed of TiO_2 nanopowders was studied to optimize operating conditions in order to achieve a high CO_2 capture efficiency. It was hypothesized that the higher fluidized bed height achieved by the novel MVA system would correspond to improvements in the CO_2 capture efficiency. The actual CO_2 concentrations and breakthrough time were measured at the outflow of the fluidized bed system. Results from the MVA system were compared to a simple vibrating fluidized bed (VFB) to examine the impact of bed height on CO_2 capture.

References

1. Roberto M. C., 2015 State of the Climate: Carbon Dioxide (2016)
2. Lindsey, R., Climate Change: Atmospheric Carbon Dioxide (2018)
3. Lei Li, Yong Li, Xia Wen, Feng Wang, Ning Zhao, Fukui Xiao, Wei Wei, and Yuhan Sun, CO_2 Capture over K2CO3/MgO/Al2O3 Dry Sorbent in a Fluidized Bed, Energy Fuels 25 (2011) 3835–3842.
4. Li L, King DL, Nie Z, Li XS, Howard C. MgAl2O4 spinel-stabilized calcium oxide absorbents with improved durability for high-temperature CO_2 Energy Fuels 24 (2010) 3698–3703.
5. Lu H, Smirniotis PG, Ernst FO, Pratsinis SE. Nanostructured Ca-based sorbents with high CO_2uptake efficiency. Chem Eng Sci. 64 (2009) 1936–1943.
6. Valverde JM, Pontiga F, Soria-Hoyo C, Quintanilla MAS, Moreno H, Duran FJ, Espin MJ. Improving the gas-solids contact efficiency in a fluidized bed of CO_2adsorbent fine particles. Phys Chem Chem Phys. 13 (2011) 14906–14909.
7. An, K. and Andino*, J.M., Microjet and Vibration Assisted Fluidization of Nanoparticles, US Patent and Trademark Office Provisional Patent application #62/815,653 filed on 8 March 2019. Invention disclosure #D19-073/M19-140P, Filed with Skysong Innovations at Arizona State University on 12 December 2018.

577428 Computational Study on Biomass Fast Pyrolysis: Hydrodynamic Effects in a Laboratory-Scale Fluidized Bed

Fast pyrolysis is a leading candidate process for converting biomass to liquid fuels and high value chemicals. During fast pyrolysis in bubbling bed or circulating bed reactors, biomass particles are rapidly heated through contacting with hot gases and solids, and their constituent components decompose into volatiles, ash, and char. The product vapor/gas composition, which determines the yield of fuel and/or chemical feedstock-compatible molecules, is highly dependent on the bubbling intensity which promotes heat and mass transfer and also controls the particle residence time. As with other energy-intensive processes, accurate characterization of biomass pyrolysis reactors at smaller lab and pilot scales is a vital first step in developing successful commercial versions.

In this study, we simulate a lab-scale bubbling fluidized bed biomass fast pyrolysis reactor to explore the effects of hydrodynamics on biomass conversion as the gas flow is increased through the bubbling-to-slugging transition, with all the other operating variables held constant. We employ a 3D implementation of MFiX, which is an open-source software package supported by DOE that utilizes a continuum (two-fluid) approach for modeling the fluidzed reactor hydrodynamics. Bubbling intensity and dynamics and the resulting impact on pressure measurements have been reported for a similar reactor set up previously[http://dx.doi.org/10.1016/j.cej.2016.08.113].

We compare mixing and elutriation predictions from our computational model to experimental studies in the literature, as well as experimental pyrolysis yield measurements when reaction chemistry is coupled to the hydrodynamics. We also discuss implications from the results of this study for future numerical and experimental studies of similar reactors.

562346 Numerical Verification and Validation of the Biomass Pyrolysis and Gasification inside a Bench Scale Reactor with Mfix-TFM

Numerical verification and validation of the biomass pyrolysis and gasification inside a bench scale reactor with MFiX-TFM

Yupeng Xu[1,2], Mehrdad Shahnam[1], Andrea Porcu[3], Alberto Pettinau[3], Bhima S Sastri[4], William Rogers[1]

1. National Energy Technology Laboratory, Morgantown, WV 26505, USA

2. West Virginia University Research Corporation, Morgantown, WV 26506, USA

3. Sotacarbo Sustainable Energy Research Center, Carbonia, Italy

4. U.S. Department of Energy, Washington D.C., USA

Abstract

Biomass is a widely available renewable energy source, which could be an alternative fuel source to fossil fuels to alleviate environmental problems. Biomass gasification is a promising approach for combined heat and power generation and for the production of various products such as fuels and chemicals. Gas-solid fluidized-bed offers excellent mixing, heat and mass transfer between solid particles and fluidizing gas, thus is one of the most interesting technologies.

This work shows the result of an experimental and numerical comparison for biomass gasification as part of a research collaboration between Sotacarbo Sustainable Energy Research Center, Carbonia, Italy and National Energy Technology Laboratory (NETL), Morgantown, WV, USA. A bench-scale fluidized bed, shown in Figure 1, was used at Sotacarbo research center to gasify cypress biomass from Sardinian woods. The proximate, ultimate and calorimetric analyses of the fuel were carried out at Sotacarbo laboratories according to the international standards. A SympaTEC QICPIC particle analyzer was used to characterize the Cypress biomass and Olivine sand particles at NETL. Pyrolysis kinetics was measured through thermogravimetric analysis at Sotarcarbo. In the experimental activity, biomass feedstock is injected through a hopper via a screw feeder located at the lower side of the reactor. A small amount of N2 at room temperature is injected to aid the feeding process. The fluidizing gas (N2 or air) is injected through the bottom of the gasifier. The bed material consists of Olivine sand particles, to help with the fluidization of biomass particles and heat distribution in the bed. Various conditions are tested and the complete and accurate experimental data provide a benchmark dataset for CFD verification and validation.

Simulations are conducted using the U.S. Department of Energy's Multiphase Flow with Interphase eXchanges code, MFIX. The Eulerian-Eulerian, Two Fluid Model in MFiX, (MFiX-TFM) has been used to simulate the drying, pyrolysis, combustion and gasification of the biomass particles. Additionally, homogenous oxidation reactions such as CO, H2 and CH4 oxidation reactions and water gas shift reactions are included. Syngas composition at the outlet of the gasifier is compared with the experimental results. Figure 2 shows the comparison between the simulation results and the measurements for one of the conditions under consideration, where very good agreement is observed.

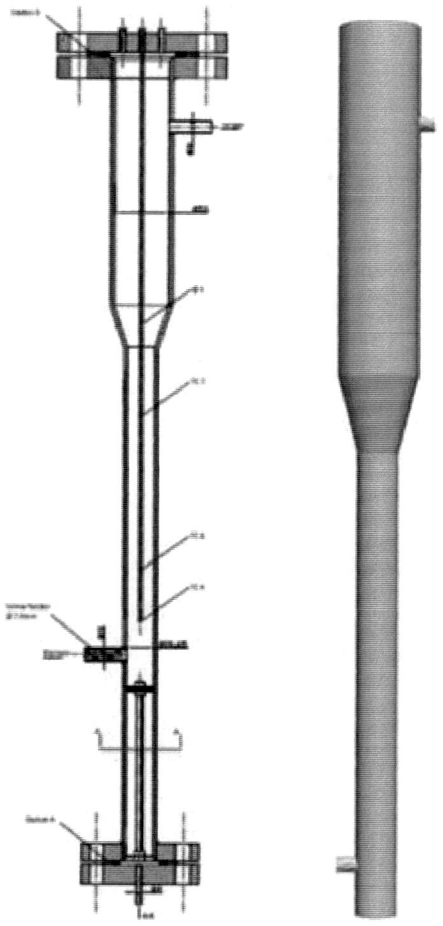

Fig.1. Direct comparison between simulation and experiments

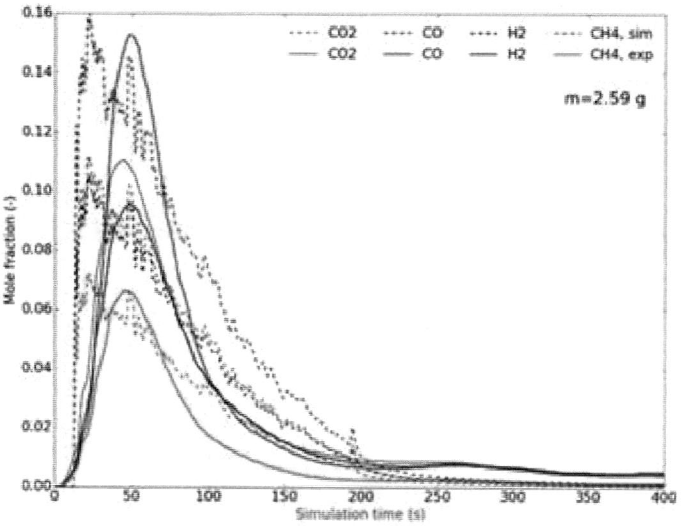

Fig.2. Detailed comparison between the instantaneous pyrolysis vapor between experiment and simulation

575673 CFD Simulations of Integrated Waste Treatment Unit

In a cost-saving computer modeling effort, the National Energy Technology Laboratory (NETL) assisted in solving a critical technical issue at the U.S. Department of Energy (DOE) Office of Environmental Management (EM) Integrated Waste Treatment Unit (IWTU), preventing a long-term delay of start-up operations. IWTU is an Energy Department facility designed to treat 900,000 gallons of radioactive liquid waste by heating and essentially drying it into a solid granular material for long-term storage. The heat required in this process is created by a piece of equipment called a denitration mineralization reformer (DMR), in which coal, steam, air and oxygen interact. Because this mixture contains multiple phases of matter (i.e., solids and gases), an understanding of multiphase flow is critical for design and troubleshooting. CFD simulations using the NETL's Multiphase Flow with Interphase eXchanges (MFiX) software suite, demonstrated that the problem — solids accumulating or agglomerating in the lower regions of the DMR — was inherent to the DMR's design and could not be overcome under normal operating conditions. NETL researchers then began a highly successful collaboration with EM support contractor Flour Idaho and the Idaho National Laboratory to create MFiX simulations that screened three alternative reactor designs under a range of conditions. This partnership was an exceptional example of teamwork in which a DOE contractor leveraged a variety of powerful resources, including cutting-edge modeling tools, the expertise of two national labs and thousands of computational cores, to efficiently solve a difficult technical issue. The simulations were pivotal in the design selection and guided both sub-scale and full-scale experiments. This computational approach eliminated the need to test each design under a range of operating conditions at full scale. IWTU made the selected design change to the DMR, which was successfully tested with no evidence of further accumulation of solids.

568494 Single Bubble Dynamics in a Dense Phase Fluidized Bed Gasification Environment in Presence of Biomass Pellets

Biomass gasification in fluidized beds is a process of important commercial value [1]–[3]. Simulation of these fluidized bed units strongly depends on establishing bubble dynamics in dense phase sand fluidized beds in the presence of biomass[4]. In the present study, a single bubble model is studied, and this to provide a phenomenological based framework for fully fluidized beds sand beds with several coexisting bubbles.

Studies have been made measuring bubbles with the CREC Optiprobe system [5] in the presence of biomass [6]. To improve bubble dynamics studies, a combination of CREC Optical Probe system, and a video camera is employed. Bubble velocity, and bubble dimensions, both vertical (bubble axial chord) and horizontal (bubble frontal radius) were measured. The effects of biomass pellet concentration on bubble rise velocity, bubble size and shape are evaluated at conditions close to minimum fluidization.

On this basis, two theoretical bubble models were considered. The first one is based on a spherical cap shaped bubble with a flat bubble-wake interphase, being this consistent with classic works in the matter [7] The second bubble shape model considered, has an irregular bubble-wake interphase, consistent with recently obtained bubble images [8]. These phenomenologically based models use the wake fraction to relate the bubble and the wake sizes, while additionally, including a wake angle parameter. These two models allow us to provide a theoretical based bubble axial cord and bubble frontal radius predictions with air volume as only input. Finally, bubble rising velocity as predicted in previous studies is compared with single bubble rise velocity in the presence of biomass.

The experimental data was collected using a 44 cm diameter and 40 cm of height fluidized bed, loaded with sand with a particle size distribution from 200 μm to 900 μm centered at 580 μm, and a biomass pellet with dimensions 2.7 cm in length and 0.8 cm in diameter. This is an anticipated reactor geometry and biomass pellet geometry in biomass gasifiers.

The measurements with out any treatment show an increment on the noise and a divergence from the theoretical values that grows with the injected volume. A data treatment had to be designed to improve the experimental data and close the bubble volume balance.

Figure 1 shows a comparation between the measured Bubble Axial Chord (BAC) and the theoretical calculated from the model. Furthermore, a comparation between different models of bubble rise velocity [4] are plotted together for comparation next to the experimental results for the experimental bubbles is shown in Figure 2 (a).

The effect of 5% vol. of biomass can be seen in Figure 2 (b). It is shown that the biomass makes the bubbles both smaller and slower. It is also shown that smaller bubbles are closer to the lines of the theoretical models than the bigger bubbles that tend to be more scatter and far from their predicted position in the plot.

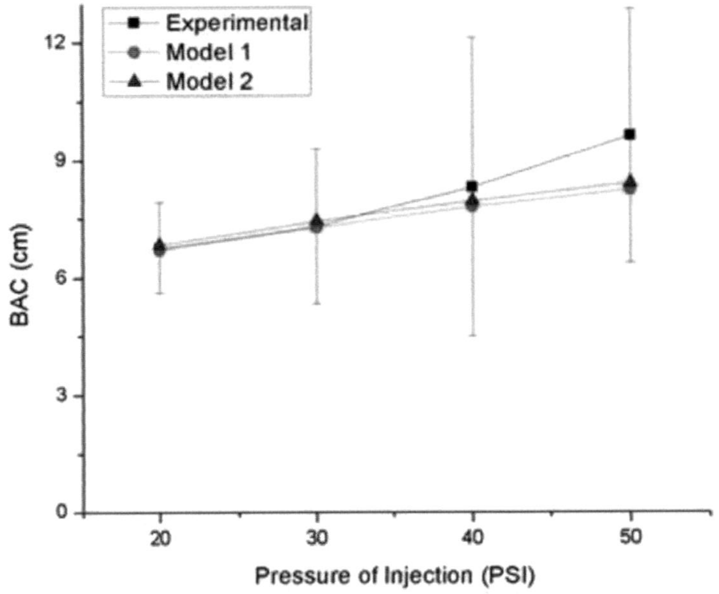

Figure 1. Comparation between the two model theoretical results and experimental measurements after data treatment.

a) b)

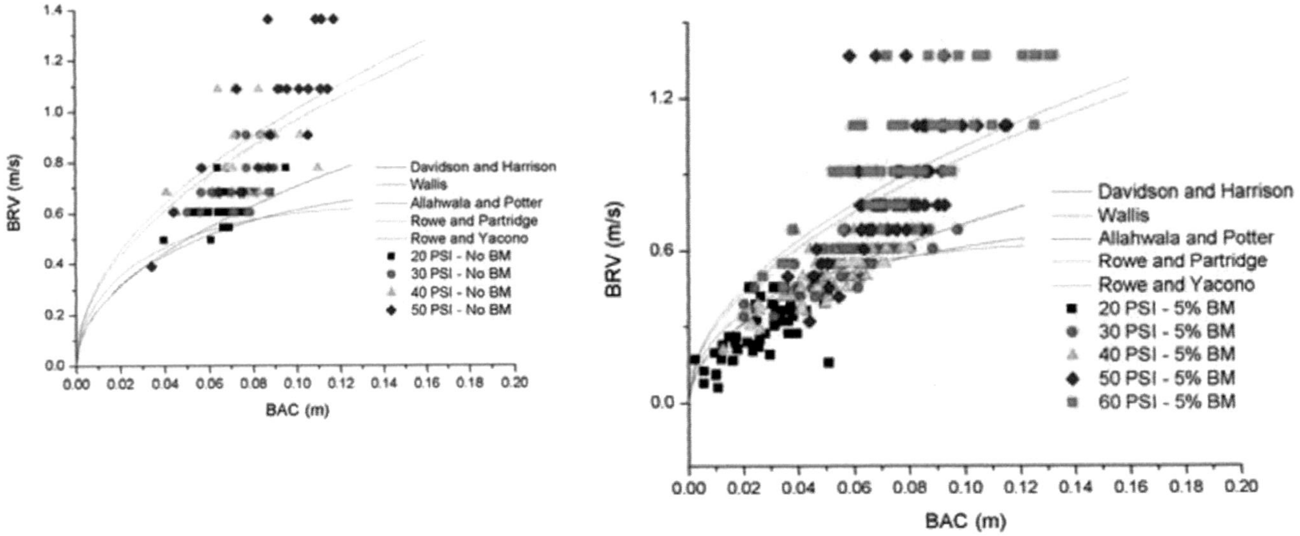

Figure 2. Comparation between various available models for bubble rise velocity (BRV) and bubble diameter, a) Bubbles with out biomass and b) Bubbles with 5% vol. biomass.

References

[1] L.-S. Fan, *Syngas Generation*. Cambridge University Press, 2017.

[2] J. D. Holladay, J. Hu, D. L. King, and Y. Wang, "An overview of hydrogen production technologies," *Catal. Today*, vol. 139, no. 4, pp. 244–260, 2009.

[3] H. I. de Lasa, E. Salaices, J. Mazumder, and R. Lucky, "Catalytic steam gasification of biomass: Catalysts, thermodynamics and kinetics," *Chem. Rev.*, vol. 111, no. 9, pp. 5404–5433, 2011.

[4] S. Karimipour and T. Pugsley, "A critical evaluation of literature correlations for predicting bubble size and velocity in gas-solid fluidized beds," *Powder Technol.*, vol. 205, no. 1–3, pp. 1–14, 2011.

[5] S. R. Nova, S. Krol, and H. I. de Lasa, "CREC Fiber-Optic Probes for Particle Velocity and Particle Clustering: Design Principles and Calibration Procedures," *Ind. Eng. Chem. Res.*, vol. 43, no. 18, pp. 5620–5631, 2004.

[6] F. Fotovat, A. Abbasi, R. J. Spiteri, H. I. de Lasa, and J. Chaouki, "A CPFD model for a bubbly biomass-sand fluidized bed," *Powder Technol.*, vol. 275, pp. 39–50, 2015.

[7] J. R. Grace and D. Harrison, "The influence of bubble shape on the rising velocities of large bubbles," *Chem. Eng. Sci.*, vol. 22, no. 10, pp. 1337–1347, 1967.

[8] C. M. Boyce, A. Penn, M. Lehnert, K. P. Pruessmann, and C. R. Müller, "Magnetic resonance imaging of single bubbles injected into incipiently fluidized beds," *Chem. Eng. Sci.*, vol. 200, pp. 147–166, 2019.

577009 Mfix-Exa: Adapting CFD-DEM to Exascale Computing

The Exascale Computing Project (ECP) is a U.S. Department of Energy's Office of Science and National Nuclear Security Administration with the goal of ensuring that the U.S. has a capable exascale ecosystem in 2021. Here, we describe progress on one of the software projects funded by this program. Specifically, the CFD-DEM capability of the open-source MFiX code, developed by NETL, is being rebuilt as a separate, stand-alone code. The goal of this project is to simulate a 50kW chemical looping reactor with about 1 billion particles in 1 day of simulation time. The framework of this new code will be described as well as preliminary results toward the goal.

572906 A Penalization Method for the Direct Numerical Simulation of Weakly Compressible Reacting Gas-Solid Flows

A penalization method for the direct numerical simulation of weakly compressible reacting gas-solid flows

1. **Hardy[1]**, **J. De Wilde[2]**, **G. Winckelmans[3]**

[1]*Institute of Mechanics, Materials and Civil Engineering (iMMC), Université Catholique de Louvain, baptiste.hardy@uclouvain.be*

[2]*Institute of Mechanics, Materials and Civil Engineering (iMMC), Université Catholique de Louvain*
[3]*Institute of Mechanics, Materials and Civil Engineering (iMMC), Université Catholique de Louvain*

Gas-solid flows are encountered in many natural and industrial phenomena. Fluidized beds are the most well-known application of gas-solid reactors in the chemical industry where they are valued for their drastic improvement of mass and heat transfer rates. Yet, the simulation of such equipment at industrial scale is still a challenge due to the tracking of billions of solid particles. In the past decades, different approaches have been addressed to model complex multi-scale and multi-phase reacting flows. The Two-Fluid Model (TFM) describes both the gas and the solid phases as continua while the Discrete Element Model (CFD-DEM) tracks each particle (or group of particles) in a Lagrangian manner. However, these large-scale models rely on closure relations to characterize the impact of the unresolved scales of the flow onto the large resolved scales. Direct Numerical Simulation (DNS) turns out to be a powerful tool in order to extract closure laws from fundamental principles for mass, momentum and heat transfer rates which can be subsequently inserted into the TFM or CFD-DEM models.

In the recent years, different DNS methodologies for particle-resolved simulation have been investigated, mostly based on the Immersed Boundary Method (IBM) originally developed by Peskin [1]to study elastic boundaries. The direct forcing method introduced by Mohd-Yusof [2] and later improved by Uhlmann [3]allows a better numerical treatment of rigid body problems and is therefore very popular in the fields of particulate flows. DNS-IBM was then applied by several research groups to predict fluid-particle mass, momentum and heat transfer rates [4]–[9].

Among immersed boundary methods, the penalization method developed by Arquis and Caltagirone [10]models solid obstacles as porous media with close to zero porosity. Originally used for solving external flows past bluff bodies, this method has been scarcely investigated for the study of heat and mass transfer problems in reactive gas-solid flows.

The present study aims at combining the penalization method to account for the presence of the solid phase with the low-Mach number assumption for the gas phase. Indeed, strong thermal effects induced by chemical reactions can induce non-negligible density gradients at the surface of solid particles and affect interfacial transfer laws. The low-Mach number assumption is of high interest for gas-solid reactive flows in that it allows density fluctuations while removing the constraint on the time step imposed by the speed of sound in fully compressible flows. Here, we extend the methodology of Lessani et al. [11]for low-Mach number flows in order to incorporate the penalization of the solid phase for momentum, heat and species transport. The proposed method allows the treatment of general boundary conditions for the transported scalars: Dirichlet-type boundary conditions for infinitely fast surface reaction and very rapid solid phase temperature homogenization, Neumann-type boundary conditions for finite rate surface reaction and coupled heat release and finally an intra-particle description with coupled heat and mass transfer between the solid and the fluid phase. Different reaction scenarios are investigated: heat consumption or release and gas expansion or compression. Finally, a comparison is established with the incompressible version of the penalization method to assess the impact of density fluctuations in view of building new closure laws for dense gas-solid flows.

References

[1] C. S. Peskin, "Flow patterns around heart valves: A numerical method," *J. Comput. Phys.*, vol. 10, no. 2, pp. 252–271, Oct. 1972.

[2] J. Mohd-Yusof, "Combined immersed-boundary/B-spline methods for simulations of fow in complex geometries," *Cent. Turbul. Res. Annu. Res. Briefs*, 1997.

[3] M. Uhlmann, "An immersed boundary method with direct forcing for the simulation of particulate flows," *J. Comput. Phys.*, 2005.

[4] N. G. Deen and J. A. M. Kuipers, "Direct Numerical Simulation of Fluid Flow and Mass Transfer in Dense Fluid–Particle Systems," *Ind. Eng. Chem. Res.*, 2013.

[5] Y. Tang, S. H. L. Kriebitzsch, E. A. J. F. Peters, M. A. van der Hoef, and J. A. M. Kuipers, "A methodology for highly accurate results of direct numerical simulations: Drag force in dense gas-solid flows at intermediate Reynolds number," *Int. J. Multiph. Flow*, vol. 62, pp. 73–86, 2014.

[6] N. G. Deen, S. H. L. Kriebitzsch, M. A. van der Hoef, and J. A. M. Kuipers, "Direct numerical simulation of flow and heat transfer in dense fluid-particle systems," *Chem. Eng. Sci.*, vol. 81, pp. 329–344, 2012.

[7] F. Municchi and S. Radl, "Consistent closures for Euler-Lagrange models of bi-disperse gas-particle suspensions derived from particle-resolved direct numerical simulations," *Int. J. Heat Mass Transf.*, vol. 111, pp. 171–190, 2017.

[8] B. Sun, S. Tenneti, and S. Subramaniam, "Modeling average gas-solid heat transfer using particle-resolved direct numerical simulation," *Int. J. Heat Mass Transf.*, vol. 86, pp. 898–913, 2015.

[9] M. Mehrabadi, E. Murphy, and S. Subramaniam, "Development of a gas-solid drag law for clustered particles using particle-resolved direct numerical simulation," *Chem. Eng. Sci.*, vol. 152, pp. 199–212, 2016.

[10] E. Arquis and J. P. Caltagirone, "Sur les conditions hydrodynamiques au voisinage d'une interface milieu fluide-milieu poreux: application à la convection naturelle," *CR Acad. Sci. Paris II*, vol. 299, pp. 1–4, 1984.

[11] B. Lessani and M. V Papalexandris, "Time-accurate calculation of variable density flows with strong temperature gradients and combustion," 2005.

Acknowledgements

Baptiste Hardy is a research fellow of the F.N.R.S-Fonds de la Recherche Scientifique, under grant n° 1.A.700.18F. This work was completed with the help of the computational resources provided by the Consortium des Equipements de Calcul Intensif (CÉCI).

581802 Finite Element Method (FEM) Modeling of Hopper Flow

Finite element method (FEM) modeling is often used to model the flow of particulate materials. FEM modeling provides a computationally inexpensive alternative to the popular approach of discrete element modeling (DEM) and is more accurate than the approximate analytical theories. This work aims to clearly define the applicability of FEM in predicting various hopper flow characteristics such as the stress and velocity profiles, the mass flow rate, the mass/funnel flow region, the critical outlet opening for cohesive particulate material to ensure continuous discharge, and also the applicability to modeling eccentric hopper systems. The validation of these various characteristics is done by comparing the FEM results to their experimental counterparts whenever available, or by comparison to the widely accepted approximate analytical theories. The critical outlet opening, the velocity profiles, and the mass flow rates are experimentally validated. Applicability of the FEM modeling to eccentric hopper system is demonstrated through experimental comparison of velocity profiles. Investigation of all these flow characteristics demonstrate that the FEM modeling based on simple constitutive models such as the Mohr-Coulomb or the Drucker-Prager model is reasonably accurate for modeling hopper flow of particulate materials exhibiting negligible compressibility. FEM approach proves advantageous over the DEM approach and the analytical theories for modeling large industrial scale complex hopper systems and materials exhibiting properties dependent on the consolidation stresses.

Research Interests: Finite Element Method (FEM), Hopper flow, Continuum modeling, Cohesive powders, Cohesionless particulate materials

Teaching Interests: Fluid Mechanics, Gas Dynamics, Finite Element Analysis, Elasto-plastic constitutive models

ChemPren: A new technology for the conversion of mixed municipal waste to basic chemicals and fuels

Dr. Piyush Sabharwall, Dr. Lynn Wendt, Ms. Roxanne Pinsky and Dr. Anne M. Gaffney
Idaho National Laboratory
Idaho Falls, ID 83404

Dr. Gennaro J Maffia, Amanda E. Peterman and Xufang (Katie) Liu
Manhattan College
Riverdale, NY 10471

Currently, over 90% of used plastics are not directly recycled and end up in U.S. landfills or incinerators. As more composite plastic materials are developed, this problem will only get worse. Technologies to up convert waste plastics to petrochemicals that can be used in the production of new plastics and materials, have been developed in the U.S., Japan, and Korea [1-4]. Catalytic conversion of waste plastics to Benzene, Toluene and Xylenes (BTX) was initially developed by ARCO chemical in the early to mid-1990s [1,2]. The BTX mixture is a versatile feedstock and can be used for the production of polyethylene terephthalate (PET), polycarbonates, polyurethanes, nylons, SBR, ABS and other polymers which are in turn used to produce beverage bottles, clothing, carpeting, automotive components and a broad range of other products. Intellectual property for improvements in the process and catalyst to up convert waste plastics to BTX was recently signed over to INL [5]. This technology capitalizes on INL owned intellectual property and advances the technical readiness level from the current level of 4 to a level of 7 with the project culminating in demonstrating at least 1000 hours of pilot plant operation.

References:

1. Jones, A.C., Gaffney, A.M., Leyshon, D.W., Cozzone, G.E., Sofranko, J.A., Process for reclaiming a thermoset Polymer, U.S. Patent No. 5,192,809, March 9, 1993.
2. Gaffney, A.M., Jones, A., Process for treating polyether polyols, U.S. Patent No. 5,462,971, October 31, 1995.
3. Nishino, J., Itoh, M., Fujiyoshi, H., Uemichi, Y, Catalytic degradation of plastic waste into petrochemicals using Ga-ZSM-5, Fuel 87 (2008)3681-3686.
4. Jung, S., Cho, M., Kang, B., Kim, J., Pyrolysis of a fraction of waste polypropylene and polyethylene for the recovery of BTX aromatics using a fluidized bed reactor, Fuel Proc. Tech. 91 (2010) 277-284.
5. Gaffney, A.M., Alloyed zeolite catalyst component, Method for making and catalytic application thereof, U.S. Patent Application 0215043 A1, August 23, 2013.

559476 Anaerobic Digestion of Food Wastes into Energy and Fertilizer

Anaerobic digestion (AD) has been evaluated as one of the most cost-efficient and environmentally friendly biotechnologies for renewable energy production, in the form of heat and power, organic waste management, and recovery of valuable nutrient rich digestate [1]. The latter has the potential to be used as fertilizer for agricultural utilization, landscaping, and soil amelioration since it contains essential nutrients such as nitrogen (N), phosphorus (P), and potassium (K) [2]. Digestate is produced from anaerobic degradation of organic wastes, mainly derived from food waste, crops and sewage sludge from wastewater treatment processes. However, the use of digestate has become a crucial challenge due to the limitation of cost and quality for the production of fertilizers as well as the application location. This study examined the optimization of an anaerobic digestion waste-to-energy system for food waste management and conversion of digestate to fertilizers. Following which, how the use of digestate for the cultivation of vegetables was investigated in comparison with chemical fertilizers. The objectives of this study are: 1. to optimize the operational parameters of a distributed anaerobic digestion waste-to-energy system for food waste management. 2. to study the efficiency of various fertilizers extracted from digestate and chemical fertilizers by cultivating green vegetables and comparing the major indexes for the growth of vegetables.

571999 Subcritical Water Upgrading of Coal Refuse with Food Waste

The goal of this research was to improve the fuel characteristics of coal refuse through co-hydrothermal carbonization (co-HTC) with food waste. Coal refuse, also known as coal waste and gob, are by-products resulted from coal mining. Coal processing plants reject 30-60% materials from the mined coal during the cleaning procedure and it contains 40-70% energy compared to coal. However, toxic elements (sulfur, heavy metals, and ash) present in coal refuse possess serious environmental threat and that is why it is not possible to combust coal refuse without prior treatment. Co-HTC of coal refuse along with food waste can be an efficient option to utilize this huge amount of available fuel resource. Co-HTC is a thermochemical process, where two feedstocks are treated at high temperature (180-280 °C) and corresponding vapor pressure (5-60 bars) for a short period of time. Coal refuse from local coal mine and food wastes from Ohio University Dining Hall were co-hydrothermally treated at three different temperatures (180, 230, and 280 °C) for 30 min. The acidic medium created by the hydrothermal carbonization of food waste catalyzed the upgrading of coal refuse. The elemental carbon (%) increased to a maximum value of 49.4 % for co-HTC solids from 18.3 % for raw coal refuses. Additionally, the ash percentage decreased from 66.4 % to 27.4 % for co-HTC solids. Moreover, the van-Krevelen diagram showed that the co-HTC solids have fuel characteristics similar to bituminous and sub-bituminous coals. The total sulfur concentration also decreased to less than 1.4 % from 9.9 %. The raw coal refuse samples and treated samples were further analyzed using the standard test method (ASTM D2492-02) and an induced coupled plasma optical emission spectroscopy (ICP-OES) to determine sulfate sulfur and pyritic sulfur concentration, respectively. The organic sulfur was calculated by subtracting the summation of sulfate and pyritic sulfur from the total sulfur concentration. The study also evaluated the distribution of sulfur among the co-HTC products. Later, an economic analysis was performed to check the economic feasibility of large scale hydrochar production. The economic study revealed that the production cost of co-HTC solids to be $39 per ton in the basis of a 110 MWe power plant.

556642 Polar Solvents Facilitate Efficient Microwave Pretreatment of Woody Biomass

Microwave pretreatment of woody biomass may be the key to unlocking an energy efficient process for recalcitrant waste stream for the production of renewable chemicals and fuels. Sugars derived from non-food sources such as woody biomass have the potential to be converted into a wide range of chemicals. However, woody biomass requires pretreatment to open the structure to access cellulose. Steam explosion, organosolv and sulphite pretreatment are currently used to remove lignin from wood [1]. However, these methods require substantial energy to be effective [1]. Development of an alternative, more efficient pretreatment processes for the release of sugars and other useful fractions from renewable biomass sources is necessary for a sustainable future.

Ionic liquids have been extensively researched for woody biomass pretreatment as they have proven to yield exceptional results via ionic conduction mechanism. However, limited research has been reported on the intentional exploitation of polar solvents, which increase the microwave absorption capacity via dipolar polarisation mechanism, for the pretreatment of recalcitrant biomass. Microwave absorption can vary significantly depending upon the ratio of effective dielectric loss factor (") to the dielectric constant (), of which the ratio is commonly known as the loss tangent (tan).

In this research, we investigated the use of solvents with high tan value to increase the efficiency of microwave-assisted solvolysis of woody biomass through an increase in the heating rates, and hence reduction of the energy consumption. To the best of our knowledge formaldehyde pretreatment of wood has not yet been reported under microwave irradiation. We studied the microwave irradiation of *Eucalyptus Regnans* and *Pinus Radiata* sawdust (~0.5 mm) in the presence of formalin solution containing 1.8, 3.7 and 10 wt% formaldehyde and water as the control at 180°C for 10 minutes.

It was found that 3.7 wt% formaldehyde (FA) solution is more efficient than water only for the pretreatment of wood sawdust at 180 °C and 10 min microwave exposure. The presence of formaldehyde in the solution accelerated the heating rate by 50% in the microwave. Furthermore, >75% of the hemicellulose sugars were released under mild microwave treatment conditions for both the wood species. Whereas in water-only pretreatment 32% and 55% of the hemicellulose was released from *E. regnans* and *P. radiata*, respectively.

A proposed mechanism for formaldehyde-mediated deconstruction of hemicellulose was developed. The lignin in the remaining sawdust from the FA pretreatment was analysed by 2D NMR, which detected the presence of stabilised lignin monomers. Whereas, with water-only pretreatment this does not occur. This stabilised platform may open-up a more advanced continuous process which could allow near theoretical lignin monomer yields through the subsequent hydorgenolysis.

576948 Comparative Leading Pretreatments for Improving Lignin Waste Valorization to Biobased Products

Biorefineries offer a potential solution to the current energy and environmental challenges facing our generation. However, there are still many technical barriers in developing a sustainable biorefinery. Due to the intrinsic recalcitrance of lignocellulosic biomass (LCB), the inherent value of LCB is highly dependent on deconstructing and valorizing lignin polymer in a biorefinery concept, and not just exploiting the more uniform carbohydrates. Lignin is the largest renewable aromatic carbon source and the second most abundant terrestrial polymer after cellulose. Valorizing lignin solid waste to value-added products shows the potential to make a sustainability of biorefineries. Despite these advantages and the significant progress, lignin waste remains as an untapped resource and the application of lignin valorization is relatively unexplored due to the heterogeneity of diverse lignin substrates. We hereby compared the leading pretreatments including sulfuric acid, steam explosion, liquid hot water, AFEX, and alkaline pretreatment for tuning lignin chemistry and improving lignin waste valorization to biobased products. Lignin characterization results showed that the lignin chemistry and reactivity obviously depended on the pretreatment employed. After the delignification of pretreated solid, the lignin stream was used as carbon source for polyhydroxyalkanoates (PHA) production by engineered *Pseudomonas putida* KT2440. Results showed that multilevel fractionation and utilization process can achieve high-value conversion of lignin for PHA with high yield. Based on the further characterization of lignin form different fractionation streams, the molecular mechanism has been proposed to unveil the improved processibility of lignin from multilevel utilization process. Overall, improving lignin waste valorization to biobased products by employing leading pretreatments provided a unique set of features to deliver a sustainable biorefinery process.

571984 Waste to Resource: Synthesis of Electrolyte and Electrode Material in Vanadium Redox Flow Battery from Leachate of Industrial Waste

Fly ash, a typical waste driven out of boilers in combustion plants, contains a large range of metal oxides, and leads to environmental consequence without proper treatment. In this study, we focus on a fly ash generated from combustion of oil refinery waste, which includes substantial amount of metal oxides (mainly vanadium oxide and nickel oxide). Instead of landfill, chemical leaching was applied to extract and separate metallic components in the fly ash. Because of the high concentration of vanadium ion inside, the leachate can be used as the electrolyte in vanadium redox flow battery (VRFB). Unlike vanadium-related product synthesized from leachate, this application avoids solid-liquid separation and minimizes the wastewater generation. Meanwhile, as the secondary metallic composition, nickel ions become a low-cost resource for synthesis of nickel-dope nanomaterial that can be applied in VRFB as electrode materials to enhance V^{3+}/V^{2+} and VO_2^+/VO^{2+} reactions.

567686 Synthesis of $LiNi_{0.6}Co_{0.2}Mn_{0.2}O_2$ Cathode Material By a Carbonate Co-Precipitation from Spent Lithium-Ion Batteries

Synthesis of $LiNi_{0.6}Co_{0.2}Mn_{0.2}O_2$ cathode material by a carbonate co-precipitation from spent lithium-ion batteries

*Lei Xing (speaker)[1, 2], Shuying Sun[1,2], Sen Lin[1,2], Ping Li[1,2], Jianguo Yu[1,2]**

1. *National Engineering Research Center for Integrated Utilization of Salt Lake Resources, East China University of Science and Technology, Shanghai, China, 200237.*

2. *State Key laboratory of Chemical Engineering, East China University of Science and Technology, Shanghai, China, 200237.*

**Speaker and corresponding authors: E-mail: xinglei8406@126.com, jgyu@ecust.edu.cn*

Abstract:

The spent lithium ion batteries (LIBs) are classified as hazardous solid wastes because they contain toxic heavy metals and corrosive electrolytes. However, they contain various valuable metals such as copper, aluminum, magnesium, nickel, cobalt and lithium. Recycling of valuable metals from spent LIBs mitigate both resource shortage and environmental contamination problems [1]. So, the aim of this work is to present a promising approach for recycling high value-added metals from the cathode material of spent LIBs.

In this work, $LiNi_{0.6}Co_{0.2}Mn_{0.2}O_2$ cathode material is synthetized by a carbonate co-precipitation using the leaching liquid of spent LIBs as raw stocks. Firstly, the spent LIBs are manually dismantled to remove both the metal cases and plastic, and the cathodic material is separated from the Al foil by ultrasonic cleaning in the solvent 1-methyl-2-pyrrolidinone [2]. The valuable elements such as nickel, manganese and cobalt are obtained by leaching in sulfuric acid, and $NiSO_4$¡¤$6H_2O$, $CoSO_4$¡¤$7H_2O$ and $MnSO_4$¡¤H_2O are added into leaching solution to obtain the raw stocks for the $Ni_{0.6}Co_{0.2}Mn_{0.2}CO_3$ precursor through adjusting the molar radio of Ni: Co: Mn = 6: 2: 2. Secondly, the $Ni_{0.6}Co_{0.2}Mn_{0.2}CO_3$ precursor is synthesized by a carbonate co-precipitation method using the prepared raw stocks from leaching solution of spent LIBs. Thirdly, the $Ni_{0.6}Co_{0.2}Mn_{0.2}CO_3$ precursor is mixed with Li_2CO_3 at a Li/M (M=sum of Ni, Co and Mn) molar ratio 1.06, and the mixture is calcinated at 800-850¡æ under oxygen atmosphere to fabricate $LiNi_{0.6}Co_{0.2}Mn_{0.2}O_2$ cathode material. Finally, the advanced characterization methods (XRD, ICP-AES TG-DSC SEM-EDS) are used to demonstrate the synthetized $LiNi_{0.6}Co_{0.2}Mn_{0.2}O_2$ cathode material having a well-ordered layered structure and a low degree of cation mixing.

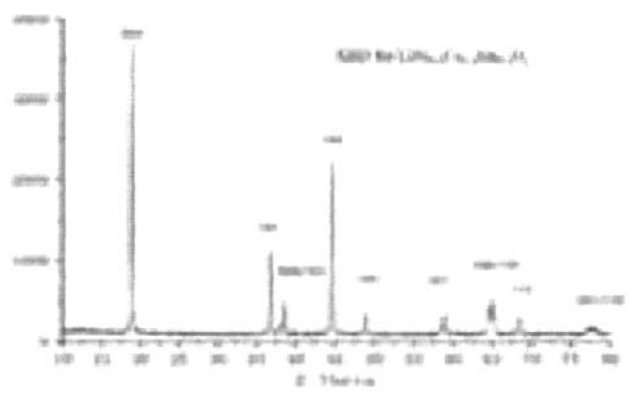

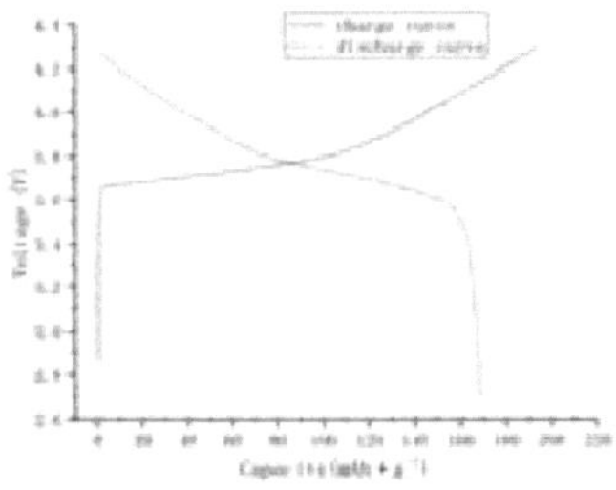

Furthermore, the electrochemical performance of the synthesized

$LiNi_{0.6}Co_{0.2}Mn_{0.2}O_2$ cathode material is tested in CR2025 coin cells. Electrochemical test methods include initial charge-discharge, continuous charge¨Cdischarge cycling, cyclic voltammetry and electrochemical impedance spectroscopy. It is found that the initial charge and discharge capacities of the $LiNi_{0.6}Co_{0.2}Mn_{0.2}O_2$ cathode material is 194.5mAh/g and 169.3mAh/g, and the initial coulombic efficiency is 87.05% when the initial charge-discharge curves are at a constant current density of 0.1C between 2.7V and 4.3V. The electrochemical performance of the prepared cathode material using the leaching liquid of spent LIBs is consistent with those of the cathode material synthetized from fresh materials. So, the proposed synthesis process has great potential for industrial-scale recycling from spent LIBs.

References:

[1] D.H.P. Kang, M. Chen, O.A.Ogunseitan, Potential environmental and human health impacts of rechargeable lithium batteries in electronic waste, Environ. Sci. Technol. 47 (2013) 5495-5503.

[2] L.H. He, S.Y. Sun, X.F. Song, J.G. Yu, Recovery of cathode materials and Al from spent lithium-ion batteries by ultrasonic cleaning , Waste Manag. 46 (2015) 523-528.

Keywords: lithium-ion batteries, cathode material, carbonate co-precipitation

557995 Sustainable Process for the Recycling of Polyethylene Phthalate

Polyethylene terephthalate (PET) is a thermoplastic polyester with excellent thermal and mechanical properties. PET is the third-most produced polymer (18% of world polymer production) and widely used in textile industry, X-ray films, food packaging particularly for water and soft-drink bottles.

Majority of plastic waste is going to land field. To recycle PET, chemical recycling has advantages over mechanical approaches in terms of better adaptability on heterogeneity of the solid wastes and no degradation of the product properties. In this paper, we will present a novel approach for chemical recycling of PET. This novel approach is featured by a simulated moving bed chromatography (SMBC) in the purification of Bis(2-Hydroxyethyl) terephthalate (BHET), which is the deploymerization product of PET. This novel approach can ensure a high purity of BHET that is suitable for reproducing chemical grade PET even with a low quality PET feedstock.

576903 Recycling Fe-P Slag As Energy Materials Via a Green Technics

Recycling waste is significant for the sustainable development of resource and environment. As one of the typical waste slags from phosphorus **industries, Fe-P generally has high melting point, plentiful iron and** phosphorus **elements and other impurities according to their raw materials and reaction technics. In order to recycle Fe-P slag as energy materials like $LiFePO_4$ and $FePO_4$, we have developed various routes including solid-state reaction and hydrothermal methods. The thermodynamics of different reaction processes are analyzed to obtain the possible routes with low energy consumption. Green technics with only water byproduct have been designed using the combined raw materials. $LiFePO_4$ and $FePO_4$ can be successfully obtained using $Fe_{1.5}P$ slag with ~1400°C melting point at no higher than 200°C. The purity and electrochemical performance of the as-synthesized samples are evaluated with various techniques. The conversion processes and key impact factors are discussed in detail. $FePO_4$ with good crystal structure and high purity can be obtained by adjusting the pH values of the product solution from the reaction of $Fe_{1.5}P$, H_3PO_4 and H_2O_2, which can meet the requirements for the production of $LiFePO_4$. On the other hand, $LiFePO_4$ can also be synthesized directly by using the raw materials of $Fe_{1.5}P$, H_3PO_4, H_2O_2 and LiOH. After being further coated with organic carbons using waste** sugarcane **slag, the as-obtained $LiFePO_4$ exhibits comparative electrochemical performance with that of current commercial samples. The discharge capacity can reach 150 mAh/g, and the capacity fade ratio is less than 1% after 100 cycles at 1.0C. The technics and equipment have also been proposed for mass production of $LiFePO_4$ and $FePO_4$ using Fe-P slag at low temperature. The green technics is interesting to recycle the waste with high melting point as function materials** through low-energy consumption way, **which is beneficial for the social sustainability with low environmental impact.**

574718 Predicting Critical Outlet Width of Hopper Systems Using Finite Element Analysis

Guidelines provided by Jenike (1961) are widely used for designing industrial scale hopper systems. One component of these guidelines is the minimum outlet width of the hopper to ensure continuous flow of the stored bulk solid. It has been experimentally observed that in most cases this critical outlet width estimate is conservative. In this work, critical outlet widths predicted by Jenike's theory, the empirical correlation by McLean (1986), Enstad's (1971) theory, and predictions from Finite Element Analysis (FEA) are compared. The FEA predictions use the Mohr-Coulomb elasto-plastic constitutive model provided by the ABAQUS commercial software package. Unlike Jenike's theory and the empirical correlation, Enstad's theory and the FEA model show that the critical outlet width depends on the material height. The effects of material parameters (internal friction angle, flow function), wall friction angle, and hopper angle on the critical outlet width predicted by the FEA are explored. In addition, comparisons to experimentally-observed critical outlet widths available in the literature are made (Eckhoff and Leversen, 1974). The primary advantage of using an FEA model for critical outlet predictions are that complex hopper systems may be simulated.

Jenike, A. W., "Gravity flow of bulk solids," *Bull. Univ. Utah*, vol. 52, no. 29, 1961.

McLean, A., "Empirical critical flow factor equations.," *Bulk Solids Handl.*, vol. 6, pp. 779–782, 1986.

Enstad, G., "On the theory of arching in mass flow hoppers," *Chem. Eng. Sci.*, vol. 30, pp. 1273–1283, 1975.

Eckhoff, R. K. and Leversen, P. G., "A further contribution to the evaluation of the Jenike method for design of mass flow hoppers," *Powder Technol.*, vol. 10, pp. 51–58, 1974.

572051 Computational and Experimental Studies of Mechanical Compression of Flexible Fiber Packings

A numerical flexible fiber model is developed based on the Discrete Element Method (DEM). The model is thereafter applied to simulate the compressions of flexible fiber packings. The normal and tangential fiber-fiber contact force models for the simulations are determined and verified by comparing with experimental results of uniaxial compression in a fixed container. To achieve a good prediction, the normal contact force needs to account for the various contact types between the sphero-cylinder elements, and the tangential contact force model has to take into account the process of static friction before sliding friction occurs. Simulations and experiments are also performed to investigate triaxial compressions of flexible fiber packings. It is observed that instead of yielding, long fibers exhibit a hardening behavior, where the stress increases rapidly with increasing strain at large strains and the packing density continuously increases. Thus, phase diagrams for classifying the bulk mechanical response as yielding, hardening, or a transition regime are generated as a function of the fiber aspect ratio, fiber-fiber friction coefficient, and confining pressure. The hardening packings can support much larger loads than the yielding packings, but larger internal axial forces within fibers and larger fiber-fiber contact forces occur.

566891 Modeling Particle-Induced Turbulence Using Sparse Regression with Embedded Invariance

Chemical conversion of solids into useful byproducts typically occurs in reactors that are highly turbulent and multiphase. Strong coupling between gas and solid phases can augment or obstruct thermochemical processes at play. Given current computational capabilities and the wide range of time- and length-scales of industrial systems, direct numerical simulation (DNS) is prohibitively costly. Thus, some degree of modeling must be employed. Current state-of-the-art modeling for turbulent multiphase flows is predominantly based on extensions to single-phase models, making it largely unsuccessful beyond the dilute limit. In this talk, we demonstrate the promise of using sparse regression with embedded invariance to determine tractable, compact models for such challenging turbulent flows. Specifically, we present the formulation of models using this technique for three classes of flow with increasing difficulty: (1) single-phase free shear turbulence, (2) turbulent flow over periodic hills and (3) gas-solid sedimenting flow.

- Abstract Text

572002 The Discrete Element Modeling (DEM) of Eccentric Weights Based Vibratory Dryer

Vibratory dryers are used in various industries, viz. agriculture, oil & gas, pharmaceutical, for filtration, washing, and vacuum drying. The drying of the powder is based on three-dimensional vibratory and/or rotary motion of material within the dryer. Multiple rotating eccentric weights at different angles relative (lead angle) to each other (known as weight setting) is used to create the complex three-dimensional vibratory motion of the dryer and control the motion of the material within the dryer. However, the effect of different weight settings on the material motion and the subsequent steady state flow-pattern is not well understood. Very few experimental and computational studies investigating vibratory dryers are available in literature. In this work, an experimentally validated mathematical framework is developed to describe the motion of a dryer. The motion of the drier was then implemented in LIGGGHTS v4.0, an open-source discrete element method (DEM) package. Subsequently, a parametric study using DEM was completed to understand the impact of both weight settings and material properties on the motion of material in the dryer, the distribution of material in the dryer and particle-particle collision behavior.

563167 Modeling Cohesive Powder Flow in a Rotating Drum Using the Finite Element Method and a Mohr-Coulomb Material Model

A Coupled Eulerian-Lagrangian finite element method (FEM) model assuming Mohr-Coulomb (MC) material behavior is developed to simulate the flow of cohesive powder in a rotating, circular drum. The MC properties used in the model are obtained from shear cell tests of microcrystalline cellulose. Corresponding rotating drum experiments using microcrystalline cellulose are also performed. The same binary image processing algorithm is used to capture the free surface changes from experiments and simulations. The free surface changes are characterized by the avalanche frequency for a constant rotation speed of the drum. The two sets of results are compared for various operating conditions.

A CFD-DEM Based Model for Wet Granulation Process in a Rotary Drum

Dr. Amit Patil, John Morud, Stein.T. Johansen, SINTEF Industry, Trondheim, Norway

Introduction

Industrial granulators are utilized widely in chemical process industries to produce various types of granular products. One of the main products produced by this method are granular fertilizer products. The fertlizer industry is known to produces a wide range of fertilizer types or grades using this method. This work focuses on manufacturing of fertilizers using a looped rotary drum granulator.

Most fertilizer grades are initially in a hot melt condition (150 ^{0}C) in the presence of small amount of water (about 4-10 %). In a rotary drum granulation process, this hot melt is sprayed onto a recycle stream of fine fertilizer granules. The liquid spray breaks into fine droplets that instantly begin to dry by loosing water by vaporization. These drying fertilizer melt droplets impinge on recycle granules to form liquid coats on them. The coated granules continue with the drying process where the melt looses moisture and solidifies untill all the water is lost. Besides, if the coated layer thickness is large enough on individual colliding particles, then they can further agglomerate together.

The granulation drying process is enhanced by injection of a co-current hot flue gas. The flue gas is injected alongside the recycle granules and liquid melt spray on the inlet side of the drum. The granulated particles overflow over a weir on the downstream side of the drum. The flue gas with collected moisture from the drying also escape from the drum outlet.

The drum produces a particle size distribution which has off spec (smaller or larger size for recycle) production alongside the desired product (medium size for output). Thus, the granules are fed onto a series of seives to seperate the product and recycle the remaining. The oversize particles are crushed in a crusher and turned into fines. They are gain combined with the seive fines and fed back into the drum.

In a drum operation the granular particles are in a constant motion due to the rotation of the drum. Most granulation drum processes in the chemical industry are operated in the rolling regime. In this regime, the particles are carried up by the rotating boundary wall by friction. They then come down due to gravity overflowing on the particulate mass. This cyclic motion regime is preferred due to higher heat and mass transfer characteristics with enhanced dispersion of melt.

This work concentrates on development of a CFD model for granulation drum process. Due to granulation being a complex process most CFD methods developed in the past have been simplified and limited in nature. Here the granulation mechanism is modeled using an Euler-Lagrange approach. The Euler-Lagrange model for granulation has always been a challenging method due to some inherent limitations. However, in this work, using some reasonable assumptions and model simplification utilizing such a method has been made possible.

The granulation mechanism is thermo-chemical process where the particulate mass exchanges heat and mass with the gas phase in an inter-dependent manner. This complex interplays of momentum, heat and mass transfer is dependent on the driving force created by melt thermodynamics equilibrium. Depending on the rate of heat and mass transfer the granulation rate is controlled. In this work a CFD model that accommodates all the mechanisms involved has been developed and presented.

Modeling method

Model background

This proposed model is fundamentally an Euler-Lagrange based method for particulate granulation and drying. In this method, gases are modeled as a single continuous phase fluid mixture or Eulerian phase. The injected melt droplets or liquid sprays and granular particles are modeled as Lagrangian discrete phases. This means in this method, the melt consisting of fertilizer in liquid state and solidified fertilizer in solid which are both at high density are modeled as Lagrangian and low-density compressible gas is Eulerian.

The continuous fluid phase consists of 2 species represented by water vapor (Wv) and dry flue gas (a). Generally, industrial flue gases are wet i.e. they contain some moisture. Typically composition of flue gas mixture is Nitrogen (80 %), Carbon dioxide (15 %) and Water vapor (5 %). Thus, the mixture of Nitrogen and Carbon dioxide are treated as dry flue gas species and water vapor as the other.

The Lagrangian droplet or particle phase consists of 2 species which are the fertilizer mixture and water liquid. Depending on the water liquid (Wl) content, the fertilizer exists as either fertilizer liquid (Fl) or fertilizer solid (Fs). The fertilizer liquid exists as a mixture with the water content in the particle as an outer core layer and the fertilizer solid is completely dried (or water less) solidified core of the particle. The fertilizer liquid and water liquid in the melt form a proportional composition.

Both Lagrangian particles and droplets are treated as discrete elements. All Lagrangian discrete elements here are characterized by a solidified fertilizer core and a fertilizer melt layer with water on the outside. The droplets sprayed or injected start as a 100 % fertilizer melt with no solidified core. However, as the droplets loose moisture a solidified core develops. Similarly, particles may also possess a melt layer due to melt droplet deposition. Melt layer on particles are much thinner compared to that of injected droplets. However, if particles have a large enough melt layer thickness, they may also undergo coalescence with other wet or dry particles.

Thus, in this model 5 types of coalescence can be modelled which are: droplet-droplet, droplet-dry particle, droplet-wet particle, wet-particle-wet particle and wet-particle-dry particle. Any Lagrangian discrete element pair can undergo coalescence depending on enough melt layer thickness on either or both particles. Many works in the past have developed criteria for coalescence mechanism for droplets and wetted particles [1]. According to these works, coalescence of droplets and wet particles depend on several parameters like Webber number, impact inclination, melt liquid layer thickness, etc. These parameters effect the particle effective restitution coefficient leading to coalescence on satisfying of criteria.

In the current work a threshold thickness has been utilized for model validation and implementation. Further, a more complex criteria of coalescence has been implemented and drum operations have been simulated. The present model is developed or implemented in ANSYS Fluent 19.2 using user defined functions (UDF). All the source terms relating to mass, momentum, species and heat exchange between phases are managed by the UDF.

Continuous phase flow model

The mass continuity equation that represents for all fluid species is given by;

$$\frac{\partial}{\partial t}(\varepsilon_m \rho_m) + \nabla \cdot (\varepsilon_m \rho_m \vec{u_m}) = S_m, \tag{1}$$

Where, the suffix 'm' denotes the mixture of all fluid species in the system which are modeled as fertilizer, water liquid, water vapor and air. The source term S_m represents the exchange of mass with the Lagrangian particulate phase. The fluid phase exchanges mass with particle phase by water vaporization. The exchange is mapped between the Eulerian and Lagrangian grid using a mapping function. This relationship is given by;

$$S_m = -\sum_{p\in cell} S_p\delta(\vec{r}-\vec{r_p}), \qquad (2)$$

Where, S_p is the source term for individual particles belonging to the given cell. This includes the mapping function that collects all Lagrangian particles belonging to the given Eulerian grid cell.

The momentum equation representing for all fluid species is given by;

$$\frac{\partial}{\partial t}(\varepsilon_m\rho_m\vec{u})+\nabla\cdot(\varepsilon_m\rho_m\vec{u}\vec{u}) = -\varepsilon_m\nabla p-\nabla\cdot\left[\varepsilon_m\mu_m\left(\nabla\vec{u}+\nabla\vec{u}^T\right)\right]+\varepsilon_m\rho_m\vec{g}+S_{m,d}. \qquad (3)$$

Where; $S_{m,d}$ is the source term for the drag force interaction between particle and fluid phase. The mass source term for the drag force acting on the fluid $S_{m,d}$ due to particles is given by;

$$\vec{S}_{m,d} = \sum_{p\in cell} \vec{F}_{p,drag}\delta(\vec{r}-\vec{r_d}) \qquad (4)$$

Where, the drag force on individual particles $F_{p,drag}$ are accumulated onto the discretized Eulerian grid cell points with a smoothed Diriac-delta function. The ρ_m and μ_m in Eq. (3) represent the mixture density and viscosity for the discretized cell. The mixture density ρ_m is given by;

$$\rho_m = \frac{p}{RT\sum\frac{Y_i}{M_i}} = \frac{p}{RT(\frac{Y_{Wv}}{Mw_v}+\frac{Y_a}{M_a})}, \qquad (5)$$

Continuous phase species and energy model

The individual fluid species modeled as scalar transport are given by;

$$\frac{\partial}{\partial t}(Y_i\varepsilon_m\rho_m)+\nabla\cdot(Y_i\varepsilon_m\rho_m\vec{u_m}) = S_{m,i}, \qquad (6)$$

The species mass fraction denoted by Y_i sum up to 1.0 for the fluid phase. The gas phase species include the flue gas 'a' and water vapor 'wv'. The flue gas specie is not exchanged with the particulate phase. Thus, it has no source term. The water vapor is not a defined specie for particle phase as discussed earlier. However, it receives a source supply by water liquid vaporization from particle surface. Therefore, for the fluid phase, the water vapor source is given by;

$$S_{m,Wv} = \left\{-K_{w,p}(Y_{Wv}-Y_{Wv}^{eq})|_{\varepsilon_p>0}\right\}\delta(\vec{r}-\vec{r_p}). \qquad (7)$$

Where; $K_{w,p}$ is the mass transfer by vaporization coefficient for water at the particle surface. This is same for the source term for the mass equation S_m. Y_{Wv}^{eq} equilibrium vapor composition at the gas-particle interface. This is function of the vapor pressure of water in the given gas phase at the prevailing temperature. This term is evaluated similar to several previous works [2].

The energy equation for fluid mixture is given by;

$$\frac{\partial}{\partial t}(\varepsilon_m\rho_m C_{p,m}T_m)+\nabla\cdot(\varepsilon_m\rho_m\vec{u_m}C_{p,m}T_m) = \nabla\cdot(\varepsilon_m k_m\nabla T_m)-S_{m,E}. \qquad (8)$$

Where; $C_{p,m}$ is the mean heat capacity and k_m is the mean thermal conductivity of species belonging to the Eulerian cell. $S_{m,E}$ is the energy source term for the fluid phase.

Discrete phase flow model

The motion of individual particles are tracked as Lagrangian points. The dynamic particle motion is simply a balance of drag force, gravity and contact force given by;

$$m_p \frac{d^2 \vec{r}_p}{dt^2} = F_{p,drag} + F_{p,gravity} + F_{p,contact} + F_{p,agg}. \quad (9)$$

Where, $F_{p,drag}$ is particle drag force due to interaction with gas, $F_{p,gravity}$ is force due to gravity, $F_{p,contact}$ is contact force and $F_{p,agg}$ is the agglomeration force which is active when an agglomeration is under way.

Discrete phase species and energy equation

The total mass of a discrete Lagrangian particles consisting of fertilizer liquid, solid and water liquid is given by;

$$m_p = m_{p,Fl} + m_{p,Fs} + m_{p,Wl}, \quad (10)$$

Therefore, the total fraction of the components sum to 1.0 given by;

$$x_{p,Fl} + x_{p,Fs} + x_{p,Wl} = 1.0 \quad (11)$$

The fertilizer melt is a mixture of water liquid and fertilizer liquid that exists in a solute equilibrium. This equilibrium is defined by;

$$R_m = f(T, P) = \frac{x_{p,Wl}}{x_{p,Fl}} = \frac{m_{p,Wl}}{m_{p,Fl}} \quad (12)$$

Where, R_m is a function of temperature and pressure that defines the proportion in which the two components are in equilibrium. The fertilizer exists in the particle either in melt state or solidified state. In this work, water liquid fraction is assumed to be the basis that defines the proportion of fertilizer in melt state and solidified state. Depending on the temperature and water liquid content in the particle, this proportion is fixed. Thus, the solidified fertilizer mass in particle is given as;

$$m_{p,Fs} = m_p - (1 + \frac{1}{R_m}) m_{p,Wl}, \quad (13)$$

The discrete particles loose water liquid by vaporization. Thus, the rate of change of particle mass (same as rate of change of water liquid mass) is also the mass exchanged source term S_p with the fluid phase is given by;

$$\frac{dm_p}{dt} = \frac{dm_{p,Wl}}{dt} = S_{p,Wl} = K_{w,p}(Y_{W_v} - Y_{W_v}^{eq}). \quad (15)$$

The total fertilizer mass in the particle is given by;

$$m_{p,F} = m_{p,Fl} + m_{p,Fs}, \quad (16)$$

Discretizing this with respect to time, substituting and re-arranging, we get;

$$\frac{dm_{p,Fs}}{dt} = \frac{dm_{p,F}}{dt} - \frac{1}{R_m} \frac{dm_{p,Wl}}{dt}. \quad (17)$$

The dynamic heat balance for Lagrangian particles is given by;

$$\frac{d\left[(m_{p,F} C_{p,F} + m_{p,Wl} C_{p,Wl}) T_p\right]}{dt} = S_{p,E} = h_p A_p (T_m - T_p) + S_{p,Wl,vap} \lambda_{W,vap} \quad (18)$$

Where; h_p is the heat transfer coefficient for particle, $\lambda_{W,vap}$ is the latent heat of vaporization. $S_{p,Wl,vap}$ is vaporization mass which here is same as $S_{p,Wl}$ in Eq. 15.

Discrete phase coalescence model

The DEM model in this work was introduced with a coalescence model for accommodating melt physics. The melt layer on individual particles or droplets make them adhere together during collisions and become a single discrete element. This is a complex physics where substantial work has been done experimentally and with resolved CFD methods [3]. In this work, simplified closures have been adapted for performing large scale unresolved CFD-DEM simulations.

Consider 2 Lagrangian particles 'a' and 'b' undergoing coalescence where 'a' is bigger in mass than 'b'. In this model, if coalescence criteria is satisfied at the beginning of the collision, discrete entity 'b' is removed instantly and the mass of 'b' is added to 'a' in a slow discrete manner. This is because dense DEM simulations are often closely packed where individual particles are surrounded by many neighboring particles. If particle 'a' is added with mass of 'b' instantly the resulting new size of particle can cause large overlaps with other surrounding particles. This can cause abrupt unnatural increase in collision forces leading to sudden high particle velocities. This makes the simulations unstable and unreal leading to simulation failure.

Note that the removal of discrete element 'b' is instantaneous at the beginning of the first time step of coalescence. The discrete element 'b' could have also been treated with shrinking model. However, this was not done as this discrete element will any way disappear and tracking smaller particles can bring numerical instability.

The time step size used in DEM simulations varies from 10^{-6} s to 10^{-3} s depending on the particle size. Coalescence is not an instant process and it takes more time than the DEM timestep size. Therefore, for smooth functioning of DEM simulations the particles should slowly grow in size over the agglomeration time. Here we have used a linear integration of particle mass to its final agglomerated size for the particle 'a' which is given by;

$$m_{p,a,t} = m_{p,a,0} + \int_0^t \frac{m_{p,b}}{T_{agg}} dt, \qquad (19)$$

Where; T_{agg} is the agglomeration time. The agglomeration time depend on various collision impact parameters such as relative velocity, melt layer thickness and collision inclination. The agglomeration time varies depending on the wetting, size and relative velocity of particles. While undergoing coalescence particles need to traverse through a part of other particle radius. Several previous works have experimentally measured wet particle deceleration and adhesion during collision [3]. Depending on the relative velocity, the coalescence time can vary from within a microsecond to several milliseconds. In this work, we perform some test simulations to demonstrate this where the agglomeration time T_{agg} is set as a fixed value.

The liquid melt layer from both entities are added into the outer layer mass and the solidified core are added together in the resulting mass of particle 'a'. An Equation same as Eq. (19) is utilized for book keeping of melt layer fertilizer, melt layer water and solidified fertilizer mass. At the end of time step after the incremental mass is updated all other physical particle parameters are updated. The resultant velocity and temperature are set to the mass weighted resultant of the two particles.

Results and discussion

Agglomeration test

In this test discrete particles of size d_p = 0.005 mm were injected in a rectangular cuboid domain and allowed to fall under gravity into the smallest cross-section corner. These particles collect at the bottom of the domain. The agglomeration criteria were set to 2 different particle relative velocity at the time of collision (0.8 m/s and 3.0 m/s) in two simulation tests. The results are shown in Fig. 1, where particles have been colored based on particle size and particle velocity.

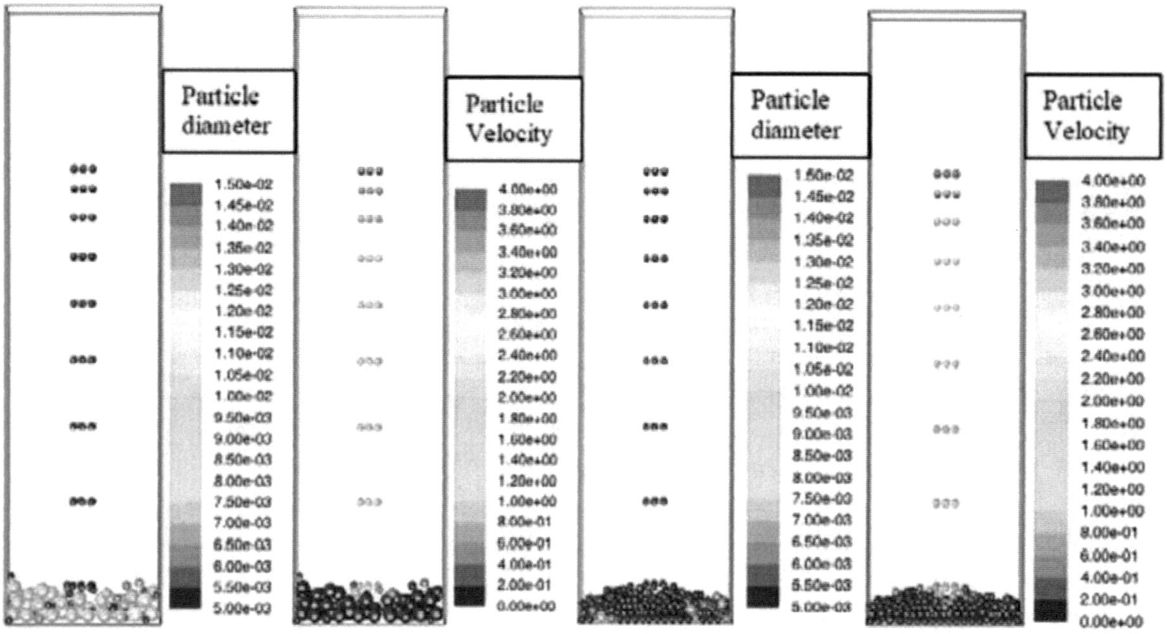

(a): Threshold relative velocity v_{th}=0.8m/s (b): Threshold relative velocity v_{th}=3.0m/s

Fig. 1: Simulation test showing percolation of particles with different thereshold agglomeration criteria.

The agglomeration time T_{agg} described in Eq. (19) had been set to 0.05s in the following simulation. It should be noted that using instant agglomeration can lead to numerical issues relating to particle collision forces in the DEM. As can be seen, for lower particle relative velocity (Fig. 1(a)), larger agglomerates are formed relative to higher particle relative velocity (Fig. 1(b)).

Heat and mass transfer test

In this test case, a 3D array of particles, of size 0.005 mm, was positioned and fixed in a rectangular domain towards an exit boundary. At the inlet side hot flue gas at inlet velocity 0.1 m/s and temperature 700K was blown into the domain. An array of 9 droplets of size 0.0025 mm were continuously injected at the inlet side with an initialized velocity of 0.25 m/s. The particles were initialized at a water fraction of 0.1 % and 380 K temperature and the droplets at 0.08 % and 423 K, respectively. The R_m factor in Eq. (12) was set to be 0.1 which is a typical range for the fertilizer industry. Thus, for droplets fertilizer melt fraction was nearly 100 % and for particles it was small.

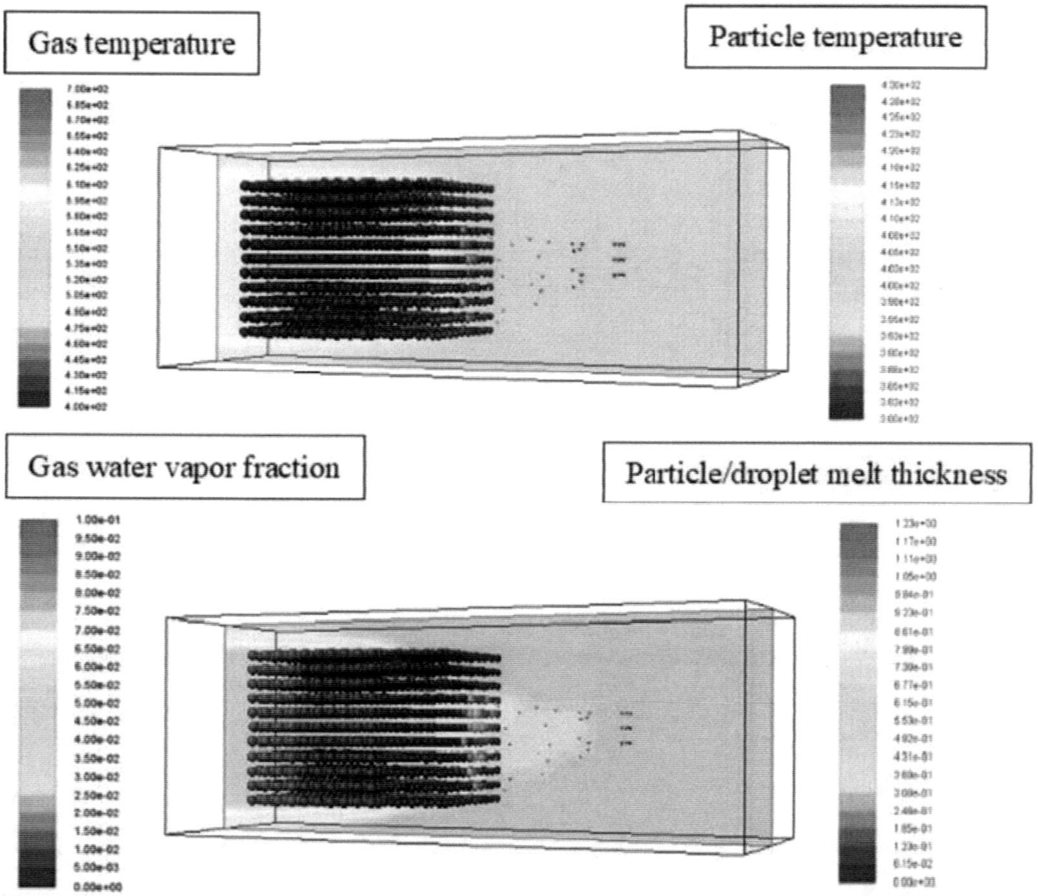

Fig. 2: Droplet spraying on an array of fixed particles test with simultaneous heat and mass transfer.

The injected droplets get sprayed onto the stationary particles. This test involved simultaneous heat and mass transfer where water present in the droplet melt vaporized. The stationary particles and droplets loose moisture in them to the hot gas blowing in the surroundings. At the same time the loose heat to the surrounding due to water vaporization and gain heat by heat transfer. When droplets are impinging onto particles and agglomerating, they are equilibrating in temperature and moisture content. All of these observations can be made in Fig. (2). This demonstrates the successful implementation of granulation model into a CFD-DEM.

Conclusion

This work demonstrates a first of its kind granulation model that includes most of the prevailing mechanism. This model is suitable to run simulations on granulation drums and provide an insight into drum operation process and mechanism. In this model, an agglomeration mechanism was developed to simulate DEM coalescence for dense systems using an agglomeration time. This model was successfully demonstrated with an agglomeration time. Further, the drying process of particles with simultaneous heat and mass transfer was demonstrated.

Acknowledgement

The author would like to thank the Norwegian Research Council and YARA International ASA for supporting this work through project no.: 269507/O20 (HEFTY). The authors would also like to thank Jakub Bujalski and Bjørn Glemmestad from YARA for providing suggestions to the work.

References

1. Ko, Gwon. H. and Ryou, Hong. S. (2005), "Droplet collision processes in an inter-spray impingement system," *Journal of aerosol science* 36, pp. 1300-1321.
2. Finotello, Giulia. Padding, Johan T. Buist, Kay A. Schijve, Annelien. Jongsma, Alfred. Innings Fredrik. and Kuipers, Johan. A. M. (2019), "Numerical investigation of droplet-dropletollisions in a water and milk spray with coupled heat and mass transfer," Drying technology, pp. 1-23.
3. Antonyuk, Sergiy. Heinrich, Stefan. Deen, Neils. Kuipers, Hans (2009), "Influence of liquid layers on energy absorption during particle impact," *Particuology* 7, pp. 245-259.

562296 Reactive Composites of Boron and Bismuth Fluoride with Varied Fluoride Concentrations

Boron is a thermodynamically attractive fuel that suffers from long ignition delays and burn times. The primary reason for the ignition delay is the nascent oxide/hydroxide layer on the particle that acts as a diffusion barrier at low temperatures. At higher temperatures, the combustion of boron in aerobic environment is kinetically slow. The approach previously found to be effective in addressing these shortcomings was to modify the chemistry of boron oxidation by introducing metal fluorides as oxidizers. Boron-fluoride composites were prepared by arrested reactive milling with 50 wt. % of the fluorides: CoF_2 and BiF_3. The composites exhibited very low ignition temperatures (~ 275 °C at a heating rate of 2500 K/s). Ignition was driven by fluorination reactions involving the boron hydroxide layer. With the active removal of condensed phase oxides, the composite B·BiF_3 was found to burn much faster than pure boron in aerobic conditions while generating favorable gas phase combustion products: BF_3 and BOF

The current research aims to extend the benefits of fluoride oxidizer BiF_3 while lowering its weight fraction in the composite to improve the calorific content for the reactive material combusting in an aerobic environment. The compositions prepared and characterized included boron with 5, 10, 20, 30, and 40 wt. % of BiF_3. The samples with higher weight percentages of BiF_3, 10-40 wt. %, were prepared by milling. Chemical deposition was employed to obtain BiF_3 as a surface coating on boron for lower weight percentages of 5 and 10 wt. %. All the prepared samples were found to be reactive when initiated in small quantities by a burning filter paper. A systematic reduction of BiF_3 in the composite resulted in the elevation of ignition temperatures with the 5 wt. % sample having the highest ignition temperature of ~750 °C at a heating rate of 2500 K/s. The ignition and combustion characteristics of these samples in air will be presented. Results of thermo-analytical studies for these materials heated in an oxidizing gas will also be discussed. The plausible reaction pathways in the experimental conditions will be explored.

564265 Aluminum-Based Thermites Prepared By Staged Milling with Different Process Control Agents

Process control agents (PCAs) serve as an efficient tool for tuning thermite particle morphology and size distribution during mechanical milling. It has been observed that upon interaction with aluminum, hydrocarbons may form carbide and nitride layers in addition to the naturally occurring oxide. These chemical and morphological changes in the surface layers separating aluminum from oxidizers are expected to affect ignition mechanisms of respective thermites. In this study, aluminum-rich thermites (Al·CuO, Al·Fe_2O_3) were prepared by arrested reactive milling in presence of different polar (acetonitrile) and non-polar (hexane) hydrocarbon PCAs. Several samples were prepared and evaluated with different PCA or PCA combinations. In one set of experiments, Al·CuO thermites were milled in different volume of acetonitrile, and very fine, equiaxial composite powders were formed. In the next set of experiments, the thermite powder milled in acetonitrile was recovered and milled in hexane in the second stage. This did not cause agglomeration of the formed powders; instead, further refinement of the produced thermite particles was observed. In another set of experiments, powders of Al and CuO were pre-milled in acetonitrile separately and then blended and milled in hexane. Milling Al in acetonitrile yielded flake-like particles. The flakes of Al milled with CuO during the second stage formed relatively coarse, rounded composite powders. For every combination of process control agents and milling stages, significant differences in particle morphology and size have been observed. It is hypothesized that such differences affect the interfacial contacts between the fuel and the oxidizer, and thus affect kinetics of reactions governing ignition in these materials. Further experiments, including ignition of the prepared powders using an electrically heated filament as well as thermo-analytical experiments will also be presented and discussed.

565335 Electrostatic and Steric Stabilization of Nanoenergetics Suspensions

The US Army has an interest in developing Insensitive Munitions (IM) that can replace munitions in current use without jeopardizing performance and cost. Recently, methods that have been developed to decrease the sensitivity of energetic materials while keeping their performance involve reducing the size of the energetic material to nano scale. In the present work, one such method selected to produce nanoenergetic materials is bead milling, as it is highly scalable and easily controlled. During the milling process, the behavior of the product suspensions is increasingly influenced by particle-particle interactions, which can lead to undesirable outcomes such as agglomeration. Therefore if these interactions are better understood and controlled, higher nanoenergetics stability and quality could be achieved. Consequently, the main focus of this work relates to understanding the interaction forces of stabilization mechanisms of nanonitramines using electrostatic, steric and depletion techniques.

The RDX electrostatic stabilization mechanism was the aim of the first part of this work. The maximum Zeta potential is reached under alkaline conditions of Ph 11. The results indicate that the ionic strength, even at low concentrations of NaCl (0.002 mol/L), can increase the Zeta potential of the nanoRDX by 37%. The DLVO theory predicts that 300 nm RDX particles are stable at a low salt concentration of 0.002 mol/L and unstable at 0.1 mol/L. The turbidity test qualitatively confirms the results predicted by DLVO theory which shows that the nanoRDX dispersion is stabilized with the addition of electrolyte in an alkaline solution.

In the second part of this work, the polymeric stabilization of HMX using PVP polymer is investigated by zeta potential and turbidity measurements. This study showed that the solvent and the molecular weight affects the PVP adsorption. The turbidity test indicates that the stability is enhanced by increasing the ethanol concentration. Finally, an HMX milling study indicated that PVP in 50% water/ 50% ethanol was able to stabilize HMX by depletion and reduce the size to 180 nm within 10 minutes of milling. Strong aggregation was observed for HMX milled in the absence of PVP.

The present study will play an important role in the stabilization of nanoenergetics and will facilitate the manufacturing of nanoenergetics in a more effective and efficient manner at the production scale.

568304 Heterogeneous Oxidation of Boron Powders and Its Role in Ignition and Combustion

Despite significant thermodynamic benefits, applications of boron combustion are limited, primarily, due to its low bulk burn rates. It has been suggested that the rate limitations of boron ignition and combustion are due to relatively slow kinetics of the associated heterogeneous reactions. However, such reactions presently are poorly characterized. Experimentally, they are hard to quantify because of the complex surface morphology of boron powders containing agglomerates of partially fused nano-sized primary particles. In this study, an approach enabling quantitative description of boron oxidation is proposed, presenting each boron agglomerate as a spherical particle with a porous reactive shell and impenetrable inert core. This approach is validated experimentally, enabling one to use this reaction geometry to develop an explicit model for boron oxidation. Kinetic parameters of this model are determined using low-temperature, low-heating rate thermo-analytical experiments. For this interpretation, the reactive shell is assumed to consist of packed, spherical, mono-sized primary particles. The primary particle size is determined based on thermo-gravimetric measurements combined with analyses of specific surface areas of the powders tested. Using the particle size and thickness of the reactive shell enables one to assess the total reactive surface area, where oxidation occurs so that the rate of oxidation per unit of boron surface is obtained. The model is valid until a critical oxide layer thickness forms on the surface of the reacting boron nano-particles that changes the distribution of oxide, total reacting surface area and particle geometry. It is thus suitable to describe reactions leading to boron ignition. Interestingly, at high temperatures typical of full-fledged boron combustion, oxide layers are expected to volatilize, once again exposing boron surface to external gaseous oxidizer. Thus, the same oxidation process may be governing both low- and high-temperature reactions of boron. At high temperatures, however, it is projected that the complex morphology of boron agglomerates is lost and the molten boron is expected to form spherical droplets. The surface area of the molten burning droplets can therefore be estimated accounting for the size of the original agglomerates. The kinetics of boron oxidation recovered from the thermo-analytical measurements is then used to predict the reaction rates observed in combustion experiments. Comparison of these predictions and measured particle burn rates for boron will be presented and discussed.

573945 Designing Composite Energetic Nanomaterials with Tunable Interfacial Activities Via Laser Ablation Synthesis in Solution

Energetic nanomaterials (ENMs) find applications in solid-state propellants and explosives. Yet, the large heat release in the Al nanoparticle (NP)-based first-generation ENMs were offset by hindered detonation rates due to the fuel-oxidizer diffusion lengths and rates being compromised by excessive oxide shell formations and NP aggregations. Efforts have been made tune their energetic behaviors by tailoring their interfacial structures that can control the oxide shell formation while leading to excessive internal stresses within the metallic cores. Yet, there exists weak fundamental understanding and considerable challenges in the rational design and synthesis of such nanostructured architectures. To this end, carbon (C) coatings on Al NPs facilitate safe handing while promoting enhanced activities due to the added advantages of the coating itself oxidizing into gaseous products (CO_2, CO etc.) without any residual ash formation, while allowing the C shell to retard NP aggregations. But, the challenge remains in the facile yet, chemically clean synthesis of these encapsulated NPs without contaminating and/or oxidizing the metal cores. Here, we address this challenge through rational design and structure-property characterizations of graphitic shell coated Al NPs (< 20 nm sizes) dispersed in pyrolyzed C matrices via laser ablation synthesis in solution (LASiS) to preserve high surface areas and interfacial properties of Al NPs. Such nanostructures allow tailored design of interfacial structures that can either lead to strain energy manifestation or, rate-controlled release of solid propellants under high pressure/temperature to prevent oxide shell-mediated surface passivation. Energetic activities of the C/Al composite NPs were tested via Laser-induced Air Shock from Energetic Materials (LASEM) technique at the US Army Research Laboratory, Aberdeen Proving Ground, MD. We demonstrate that synthesis parameters such as organic solvents, laser flux and ablation times can be tuned to provide superior control on NP sizes/aggregation with the aid of the C shell formations and, in turn, their energetic behavior. The study unveils synthesis-structure-property relations in LASiS-based manufacturing of ENMs capsuled in graphitic shells that are safe to handle and undergo kinetically controlled energy release under desired conditions. Such Al/C-based ENMs can be immediately employed for high energy density munitions in defense applications.

575299 Boron-Based Analogs of Graphene Exfoliated from Layered Metal Diborides As Catalytic Additives to Improve the Performance of Energetic Materials

Several research groups are investigating the effect of nanoscale boron as a catalytic additive for improving the energetics of the system. Nanoscale boron grabs the attention of the scientific community because of the scope of enhanced accessibility of high energy density of elemental boron. The primary challenge in utilizing the potential of elemental boron is the formation of an inert boron oxide layer. We can overcome this challenge by combining boron with energetic metals, and therefore metal borides are the best alternative as they inherently possess both boron and metal in their native chemical structure. However, the knowledge on utilizing the potential of metal borides is in its incipient stages. In this talk, we present that layered metal diborides (LMDBs) of the form MB_2 are excellent catalytic additives in improving the energetics of AP. Our group has developed the science of exfoliating LMDB crystals to obtain two-dimensional (2D) boron-based nanostructures.[1–3] These Boron-based nanostructures possess enhanced surface area with pronounced accessibility to the active sites. We demonstrate that the boron-based nanostructures or boron-based analogs of graphene obtained by exfoliation of LMDBs show superior catalytic activity in tuning the energetics of AP. Here, dry ball milling approach is used to exfoliate LMDBs to yield large-scale production of boron-based nanostructures. The physicochemical properties of these boron-based nanostructures were obtained through XRD, FE-SEM, TEM, AFM, FTIR, PSA (particle size analysis), and BET surface area. The catalytic activity of these boron-based nanostructures on thermal decomposition of AP was studied using DSC and TGA. We found that the catalytic activity of these nanostructures increases with an increase in milling time. Addition of a small fraction of these nanostructures to AP depicts a remarkable catalytic activity in the system. This study presents a new perspective to utilize the potential of boron-based nanostructures obtained from LMDBs as a super active nanocatalyst and adds an immense prospect to the field of energetic materials.

References

1. Das, S. K., Bedar, A., Kannan, A. & Jasuja, K. Aqueous dispersions of few-layer-thick chemically modified magnesium diboride nanosheets by ultrasonication assisted exfoliation. *Sci. Rep.* **5,** 10522 (2015).
2. James, A. L. & Jasuja, K. Chelation assisted exfoliation of layered borides towards synthesizing boron based nanosheets. *RSC Adv.* **7,** 1905–1914 (2017).
3. Gunda, H., Das, S. K. & Jasuja, K. Simple, Green, and High-Yield Production of Boron-Based Nanostructures with Diverse Morphologies by Dissolution and Recrystallization of Layered Magnesium Diboride Crystals in Water. *ChemPhysChem* **19,** 880–891 (2018).

576486 Hot-Spot Formation in Composite Energetic Materials Due to Ultrasonic and Impact Excitation As a Function of Young's Modulus and the Work of Adhesion

Energetic materials such as propellants and explosives fulfill a variety of roles in addressing modern engineering and scientific challenges. Of primary interest for the purposes of increasing the safety of handling these materials as well as improving performance is the response of the energetic material to a wide variety of mechanical insults. It is widely accepted throughout the energetics community that the initiation of energetic materials due to mechanical insults such as impact or shock is primarily due to thermomechanical processes. In particular, the localization of energy within the material leads to the creation of small regions of greatly increased temperatures which can subsequently grow due to chemical decomposition of the material. These localized temperature fields or "hot-spots" allow for ignition or initiation of energetic materials at energies far below what would be required to bring the bulk material to temperatures sufficient for thermal decomposition. Consequently, the susceptibility to various mechanical insults or sensitivity can be understood to be a function of chemical composition, intensive properties such as thermal conductivity, and physical parameters including heterogeneity of the material and the inclusion of air bubbles. The dependence on the exact nature of the mechanical insult further complicates a full description of the sensitivity of a material. Traditionally, sensitivity of a given material has been characterized by the response to friction, impact, and electrostatic shock. Further tests in more specialized setups may quantify the response to mechanical shock, but these three tests have been the standard for determining the safety of a material. Recent developments including the expectation of operating at more extreme conditions and demands for performance have led to the discovery of a sensitivity to acoustic excitation or vibration. This new excitation mode has been proven to be sufficient to force energetic materials to violently decompose over relatively short timeframes demonstrating an immediate need for the reevaluation of sensitivity testing and the incorporation of designs to mitigate the possibility of rapid unplanned decomposition of energetic materials. This study aimed to explore the role of adhesion and stiffness as two material properties that might substantially contribute to the heating observed during acoustic excitation. With early promising results, further impact excitation tests were carried out in order to explore how changing one material property may result in different outcomes in the sensitivity response.

High frequency acoustic excitation was provided in the form of ultrasonic excitation and non-shock mechanical excitation was applied via drop weight impact. In order to examine the possible mechanisms of hot-spot formation for ultrasonic excitation, a mechanically-delaminated inclusion of HMX in polymer binder was subjected to ultrasonic excitation utilizing an ultrasonic transducer. With weak coupling provided via ultrasound gel, these experiments allowed for exploration of a low incident energy regime unexplored by other works in the field. For the investigation of the role of adhesion and in order to control the mechanical properties of the materials, various modifications were made to the polymer components of the samples. The ultrasonic excitation experiments utilized samples consisting of a block of polymer binder enclosing a crystal at a set distance from the top surface of the sample. The thermal properties of each polymer material were also evaluated in order to evaluate the energy dissipation at the embedded crystal. Hydroxyl terminated polybutadiene (HTPB) and polydimethylsiloxane (PDMS) polymer binders were utilized with particular additives or modifications to the curative ratios in order to provide a range of adhesive and mechanical properties for testing. As current state-of-the-field experimental methods are insufficient for evaluating the true mechanical and adhesive properties of composite polymer materials under high frequency excitation, the mechanical and adhesive properties at static conditions and room temperature were utilized as figures of merit to determine if any overt relationship existed between these material properties and the energy dissipation due to ultrasonic excitation.

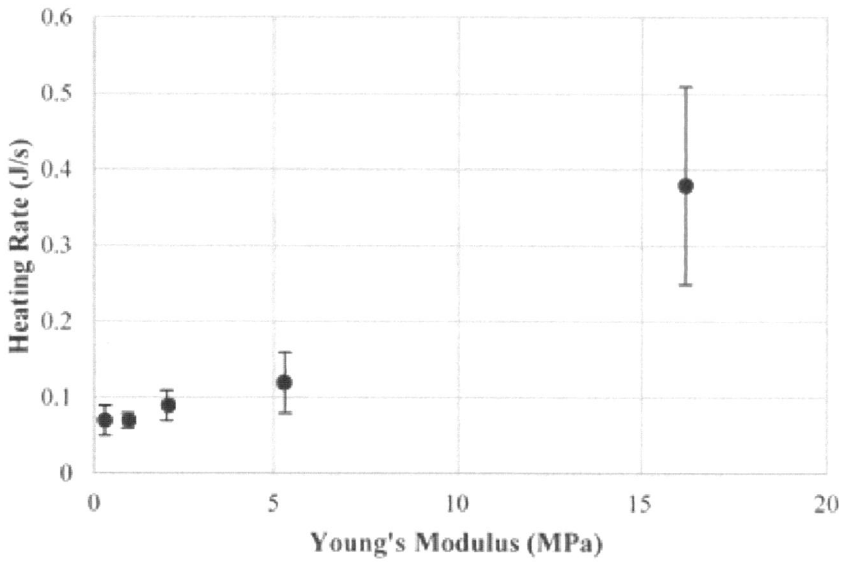

Figure 1: Heating Rate at Particle Due to Ultrasonic Excitation as a Function of Young's Modulus

While the ultrasonic excitation experimental results suggest that the work of adhesion has no effect on the heating rate at the inclusion, a positive correlation exists between the Young's modulus of the polymer material and the heating rate at the inclusion. The variation of the heating rate with the Young's modulus is shown in **Figure 1**. While slight, the positive correlation may suggest either less energy is lost due to viscoelastic effects or that the energy dissipation mechanism at the internal interface is modulated by the stiffness of the binder.

Drop weight impact tests were performed on samples containing 85% HMX crystals by mass in order to investigate if a similar relationship existed between the studied material properties and non-shock impact sensitivity. Specific heights were determined through utilization of the Neyer Sentest software which uses a D-optimal approach to quickly determine the drop height as well as the shape of the distribution assuming that the sample responses are normally distributed. The L50 or 50% probability of explosion of the polymer-HMX systems tested so far are displayed in **Table 1**. There is some indication of possible bimodal behavior in the results, but further experimentation is necessary in order to determine if this is simply an artifact of inherent sample variability. While such a possibility would certainly be an interesting indicator of two distinct mechanisms being responsible for the behavior, further refinement of the statistical method used would be required to account for the novel shape of the distribution. Regardless, early results generated from this study suggests that stiffer composite energetic materials are more sensitive to impact than more pliable materials which superficially corresponds to results from the ultrasonic heating experiments but is likely driven by a different mechanism.

These results suggest that the stiffness of the material plays an important role in the energy dissipation mechanisms responsible for hot-spot formation in these materials and that more compliant materials exhibit a lesser degree of sensitivity. Further investigations will be necessary to fully map the parameter space and determine if the behavior becomes dominated by different properties for greater excitation parameters. Modification of the drop weight impactor could also allow for high speed imaging of hot-spot formation providing insight into possible bimodal behavior.

Binder	Young's Modulus (MPa)	L50 (cm)	L50 (J)
Sylgard 184/527 25%	0.3	63.8±2.3	31.3±1.1
Sylgard 184/527 41%	0.6	43.0±4.2	21.1±2.0
Sylgard 5:1	2.0	37.5±8.9	18.4±4.3
HTPB	5.3	32.7±2.4	16.0±1.2

Table 1: L50 as a Function of Young's Modulus

576701 Microwave Enhancement of Boron Combustion for Solid Fuel Ramjet Applications

Solid fuel ramjet (SFRJ) propulsion systems will enable the next generation of supersonic and hypersonic transport. SFRJ systems provide a high degree of range, specific impulse, safety, and simplicity and boron-based fuels are poised to enable this technology with boron's high volumetric theoretical combustion enthalpy, 1.5 times that of aluminum and 3 times that of hydrocarbon fuels. Unfortunately, difficulties created by the oxide layer surrounding the boron particles, such as fuel particle agglomeration, oxygen and boron species transport barriers, and evaporative cooling, make swift ignition and combustion of boron particles difficult to achieve within the timescales of ramjet combustors. Microwave radiation has been found to couple with both gas-phase flame regions and condensed-phase energetic materials. Recently, a 60% increase in the burning rate of an alkali-doped, aluminized, ammonium perchlorate composite propellant was reported. This burning rate enhancement was hypothesized to be, in part, the result of dielectric absorption by high temperature aluminum oxide species present in the flame, as the loss tangent of many metal oxides exhibit exponential temperature dependence. This suggests that microwave coupling with other metal flame structures, such as those in boron-based fuels and propellants, could significantly enhance combustion efficiencies in solid fuel ramjet applications. This research explores the microwave enhancement of boron and boron carbide combustion, both within a composite propellant and within particle seeded burners. S-band, 2.46 GHz microwave energy is applied within a single-mode, impedance-matched, resonant microwave cavity during combustion. Time-resolved temperature measurements of propellant and seeded burner flames are made using single-point and spatially-resolved methods. Microwave irradiation effects on the burning rates of boron/boron carbide propellant are measured using high-speed video techniques, while its effects on the burning time of boron/boron carbide particles are determined using long-exposure photography.

577775 Microwave Controlled Ignition and Combustion of an AP Composite Solid Propellant Using Graphene Oxide-Wrapped Nanoscale Thermites

This effort reports on the development of AP composite propellants that, in small scale, high thermal loss environments, have ignition, combustion, and extinguishment that can be controlled by application of microwave irradiation. In order to localize microwave ignition at the burning surface of the propellant, we incorporate into the propellants recently developed graphene oxide (GO)-wrapped nanoscale Al/Fe2O3 thermites, whose GO coatings, upon heating, undergo reduction to form reduced graphene oxide (rGO) coated thermites. In this effort we cast small ~1 mm diameter wires of propellant supported on quartz and PTFE substrates that are ignited within a 2.46 GHz resonant microwave cavity with a microwave power of ~300 to 1000 W. Microwave ignition is achieved on one end of the wire by first igniting and quenching the propellant strand to convert GO near the burning surface to rGO. Modulation of the microwave field demonstrates that ignition with delays of ~0.4 s are possible, and upon removal of the microwave field, the burning front is extinguished due to the high thermal loss of the propellant size and configuration. We demonstrate the ability to successfully ignite, extinguish, and reignite propellant many times in succession. We explore the effects of microwave power (field strength) on ignition delay and burning rate, report the effects of GO and rGO on the dielectric properties of the propellant, and investigate the composition of quenched burning surfaces in order to understand mechanisms responsible for the microwave reignition capability.

578039 The Microstructural Evolution and Kinetics of Thermal Decomposition of Selected Energetic Cocrystals

The Microstructural Evolution and Kinetics of Thermal Decomposition of Selected Energetic Cocrystals

J.T. Dean[1], M. Örnek[1], T.D. Manship[2], S.F. Son[1,2]

[1]School of Mechanical Engineering, Purdue University, West Lafayette, IN.

[2]Purdue Energetics Research Center, Purdue University, West Lafayette, IN.

Abstract:

Energetic materials are a class of materials which possess significant amounts of stored chemical energy, which can be released upon stimulation from heat, friction, or impact. These materials are commonly used in explosives, propellants, and pyrotechnics, with a wide variety of applications spanning from military ordnance and aerospace to mining and beyond. The improvement of energetic material performance and reduction of their sensitivity have been the utmost importance and often are the driving force behind ongoing research into energetic materials.

Cocrystallization, in contrast to its longstanding use in the pharmaceutical industry, has recently been adopted to energetic materials [Landenberger 2010, 2012; Vuppuluri 2019]. Energetic cocrystals are mixtures of two or more types of material, known as coformers, with a unique crystal arrangement that are held in suspension by weak intermolecular forces [Landenberger 2010]. Because of this unique crystal arrangement, the cocrystal has the propensity to possess properties or characteristics which differ from its energetic coformers. While in the past increased energetic safety led to a decrease in performance, the intent behind energetic cocrystallization is to optimize both safety and energetic performance beyond what is currently feasible through the coformers or even physical mixes. Energetic cocrystals are a relatively new topic when compared to other aspects of energetic material research, therefore the properties and behavior of these energetic cocrystals are fairly uncharted.

The kinetic parameters for thermal decomposition of energetic materials are critical to understand the properties of an energetic material and to determine their usage, transportation, and storage. With accurate kinetic parameters, the risk of violent reaction or thermal runaway can be greatly reduced. This knowledge is of particular importance when considering energetics with military applications, as energetics will be subjected to a wide variety of conditions, circumstances, and scenarios.

In this study, we investigated the microstructural evolution and the kinetics of the thermal decomposition of selected cocrystals by following two primary pathways: Simultaneous differential-scanning calorimetry/thermogravimetry analysis (DSC/TGA) coupled with mass spectroscopy (MS) using an open alumina pan under argon flow and hot-stage microscopy with interspersed micro-computed tomography (Micro-CT) imaging. Thermal decomposition of four cocrystals (2,4,6-trinitrotoluene (TNT)/ 2,4,6,8,10,12-hexanitro-2,4,6,8,10,12-hexaazaisowurtzitane (CL20) in 1:1 molar ratio; 1,3,5,7-tetranitro-1,3,5,7-tetrazacyclooctane (HMX)/CL20 in 1:2 molar ratio; 1-methyl-3,5-dinitro-1,2,4-triazole (MDNT)/CL20 in 1:1 molar ratio and 4-amino-3,5-dinitropyrazole (ADNP)/diaminofurazan (DAF) in 1:1 molar ratio), their physical mixtures in the same molar ratios, and their individual coformers were studied by MS/DSC/TGA.

The kinetic parameters including activation energies (E_a) and pre-exponential factor (A) have been determined using Kissinger and Ozawa methods across different heating rates from the DSC/TGA analysis. Figure 1 shows the representative DSC traces of TNT/CL20 cocrystals as a function of different heating rates. For all the heating rates, an endotherm is visible around 140 ^{o}C; this correlates to the delayed melting of TNT from its usual melting point of approximately 80 °C. This delay in the melting of TNT demonstrates the potency of cocrystallization to alter the thermal behavior of energetic materials. The behavior displayed on this figure agrees with the literature [Zhang 2018; Jia 2019].

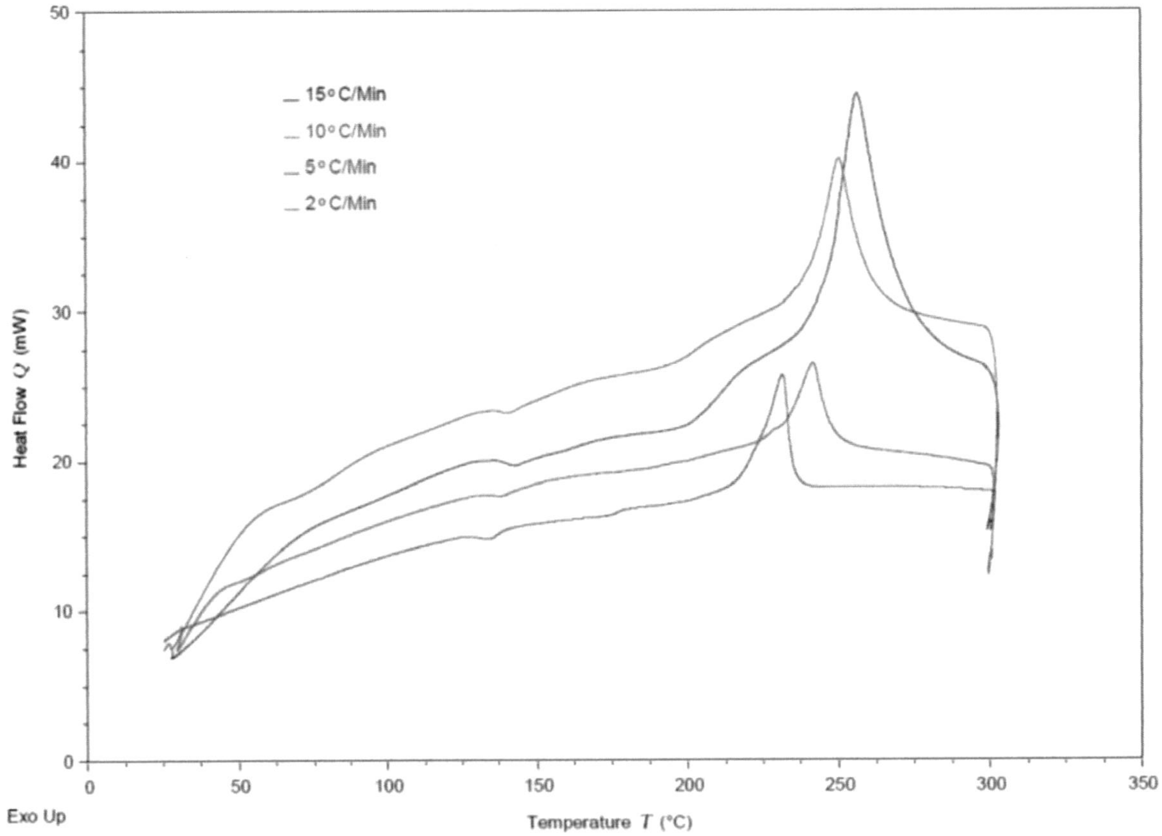

Figure 1: DSC Data of TNT/CL20 Cocrystals at Selected Heating Rates

The activation energies determined through Kissinger's and Ozawa's kinetics analysis methods for the TNT/CL20 cocrystal are listed below in Table 1. While these values differ from the literature [Zhang 2018; Jia 2019], this difference can be attributed to different testing conditions; specifically, Zhang et al [2018] and Jia et al [2019] used a different purge gas and sealed aluminum pans.

Table 1: Activation energies calculated using Kissinger and Ozawa methods for TNT/CL20 cocrystal

Heating Rates (°C/Min)	Exotherm Peak Temperature (°C)	Activation Energy, (Kissinger' method)(kJ/mol)	Activation Energy, (Ozawa's method) (kJ/mol)	Activation Energy [Zhang 2018] (kJ/mol)	Activation Energy [Jia 2019] (kJ/mol)
2	231.41	179.63	179.25	Kissinger	Kissinger
5	241.63			239.40	215.45
10	250.61			Ozawa	Ozawa
15	254.78			235.45	211.79

Using the insight gained from the DSC/TGA regarding the decomposition reaction, we then visually observed the material at critical points in the decomposition process, such as phase changes, melting points, and decomposition onset, using hot-stage microscopy and micro-CT imaging. A series of images taken from the heating of HMX/CL20 cocrystals at a heating rate of 10 °C/min to 300 °C are presented below (Figure 3). The cocrystals closely resemble the structure and size of HMX, which is unique considering the molar ratio is skewed towards CL20. During the decomposition, cocrystals were observed to change color and begin to form agglomerates. Towards the end of the decomposition, the camera was obscured as molten cocrystal surged over and covered the viewing window.

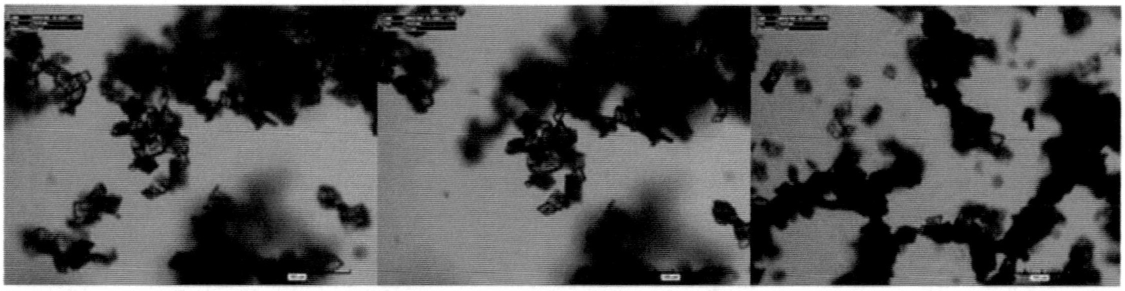

Figure 2: Microscopy Images of HMX/CL20 Cocrystal; (Left) Prior to heating, (Center) During heating, (Right) Moments before molten cocrystal obscured the microscope

The more time spent on the study of energetic cocrystals only discovers there is much more to learn. There are unique characteristics of energetic cocrystals which do not align with current mental models regarding energetic materials, such as the HMX/CL20 cocrystal's appearance resembling that of HMX, despite HMX being the minority coformer. More research and work must be done before the potential held by energetic cocrystallization is unlocked.

Keywords: Cocrystal; Thermal decomposition; Energetic materials

References:

K.B. Landenberger, A.J. Matzger. "Cocrystal Engineering of a Prototype Energetic Material: Supramolecular Chemistry of 2,4,6-Trinitrotoluene." *Crystal Growth and Design* 10 (2010) 5341-5347.

K.B. Landenberger, A.J. Matzger. "Cocrystals of 1,3,5,7-Tetranitro-1,3,5,7-Tetrazacyclooctane (HMX)." *Crystal Growth and Design* 12 (2012) 3603-3609.

V.S. Vuppuluri, J.C. Bennion, R.A. Wiscons, I.E. Gunduz, A.J. Matzger, S.F. Son. "Detonation Velocity Measurements of a Hydrogen Peroxide Solvate of CL-20." *Propellants, Explosives, Pyrotechnics* 44.3 (2019) 313-318.

J.Q. Zhang, Y.L. Xu, Q. Jia, S.J. Zhang, N. Liu, H.X. Gao, and R.Z. Hu. "Nonisothermal Decomposition and Safety Parameters of HNIW/TNT Cocrystal." *Royal Society of Chemistry Advances* 8 (2018): 31028-31036.

Q. Jia, J.Q. Zhang, K.C. Kou, S.J. Zhang, and Y.L. Xu. "Preparation, Characterization, and the Thermodynamic Properties of HNIW-TNT Cocrystal." *Propellants, Explosives, and Pyrotechnics* 44 (2019): 1-10.

559758 Thermal Behavior of CL-20 at Ultra-High Heating and Cooling Rates Using Flash DSC

First synthesized in the late 1980s, Hexanitrohexaazaisowurzitane (CL-20) has been an energetic material of significant interest due to its exceptionally high molecular density. Conventional differential scanning calorimetry (DSC) has been used extensively in past literature to characterize the thermal transition behaviors of CL-20's various polymorphs. However, the conventional DSC work cannot determine the melting temperature for CL-20 due to thermal degradation prior to melting due to the relatively slow heating of the DSC ($\leq$ 200 K min^{-1}; $\leq$ 3.3 K s^{-1}). In the current work, a Flash DSC was used to characterize the thermal behavior of CL-20 using ultra-high rates of heating and cooling. In Flash DSC, thermal rates can be as high as 40,000 K s^{-1} (2,400,000 K min^{-1}) for heating and 4,000 K s^{-1} (240,000 K min^{-1}) for cooling. The chip calorimeter uses a sample mass on the order of 10-1000 ng, as compared to a few milligrams in conventional DSC. Here, the thermal behavior of CL-20 was determined using heating rates up to 1000 K s^{-1} (60,000 K min^{-1}) and allowed for several experimental repetitions reaching temperatures up to 300 °C, which is well above the typical point of decomposition of CL-20. At high heating rates, little to no sample degradation was observed. Preliminary results suggest the appearance of a melting peak just beyond 200 °C (473 K). As was previously demonstrated for TNT using the same technique [1], the goal is to produce an amorphous glass by quenching a melt at a fast enough rate to avoid crystallization. We aim to similarly investigate the thermal behavior of different material compounds with CL-20 (e.g. HMX, polyvinyl acetate) in the form of co-crystals and spray-dried material, some of which have been shown to form amorphous structures [2].

[1] N. Shamim, Y.P. Koh, S.L. Simon, and G.B. McKenna, "The glass transition of trinitrotoluene (TNT) by flash DSC," Thermochimica Acta, **620**, 36-39 (2015).

[2] V. Stepanov, R. B. Patel, R. Mudryy, and H. Qiu, "Investigation of Nitramine-Based Amorphous Energetics," Propellants Explos. Pyrotech. **41**, 142-147 (2016).

568596 Cerium Oxide Promoted Nickel Nanoparticles Deposited By Atomic Layer Deposition for Dry Reforming of Methane

Cerium oxide (CeO_2) was used to promote the catalytic performance of nickel (Ni) nanoparticles supported on four-channel structured alpha-Al_2O_3 hollow fibers for dry reforming of methane. Highly dispersed Ni nanoparticles were deposited on the hollow fibers by atomic layer deposition (ALD) using bis(cyclopentadienyl)nickel and hydrogen as precursors and CeO_2 was loaded by liquid phase impregnation method. X-ray diffraction, transmission electron microscopy, temperature program reduction, X-ray photoelectron spectroscopy, and temperature program surface reaction were conducted to reveal the properties of the catalysts. It was found that CeO_2 interacted with NiO and improved the catalytic activity. The CeO_2 promoted catalyst also exhibited an excellent performance after regeneration. In addition, CeO_2 can inhibit coke formation.

568416 The Impact of Support in Syngas Conversion over Rh-Mn Particles on WC

The conversion of synthesis gas, a mixture of CO/CO_2 and H_2, is a promising technology to transform carbon-containing feedstocks to higher value platform chemicals. Depending on the catalyst used, a variety of different products can be obtained, ranging from saturated hydrocarbons to methanol and higher alcohols. The synthesis of higher alcohols is challenging, since a catalyst that can activate the appropriate percentage of CO bonds is required. Specifically, Rh shows selectivity towards higher alcohols in synthesis gas conversion and the selectivity can be improved when Mn is used as a promoter. On a silica support, Mn suppresses the formation of methane by passivating cluster edges of Rh nanoparticles[1]. However, on a WC support, formation of methanol is suppressed [2], which indicates a different role for Mn in this case.

Here we use first principles modeling to understand the role of Mn in the conversion of synthesis gas over WC supported Rh clusters. In a first step, we construct 22 different Rh_xMn_y clusters containing either 31 or 37 metal atoms and model their stability on a WC support. We find that a distribution of layered clusters with Mn in direct contact with the support is thermodynamically stable. We then calculate C and O binding energies to different high symmetry sites on these clusters. Subsequently, we use selected binding energies as descriptors to assign specific activities and selectivity to each of the active sites on the different clusters. We find that sites at the bottom of the supported clusters, which are mainly selective to methanol for monometallic Rh clusters, become passivated when Mn is introduced to the system. Additionally we find that a core-shell type cluster with Mn in the core and Rh in the shell shows a significantly higher activity in this catalysis and might be the most active part of the cluster distribution.

[1] Y. Liu, F. Göltl, I. Ro, M.R. Ball, C. Sener, I.B. Aragao, D. Zanchet, G.W. Huber, M. Mavrikakis, J.A. Dumesic, ACS Catal. 7, 4550-4563, 2017

[2] Y. Liu, L. Zhang, F. Göltl, M.R. Ball, I. Hermans, T.F. Kuech, M. Mavrikakis, J.A. Dumesic, ACS Catal. 8, 10707-10720, 2018

572482 Modular Self-Assembly of Metal/Polymer Nanoreactors

Self-assembled structures such as micelles and vesicles that confine catalyzed reactions to the hydrophobic core of micellar nanoreactors can facilitate organic phase reactions in a bulk aqueous environment. Flash NanoPrecipitation is a rapid, scalable, and modular method capable of continuously producing such structures. We have explored the use of Flash NanoPrecipitation (FNP) as a platform technology to rapidly self-assemble kinetically trapped core-shell polymer nanoreactors that incorporate catalytic gold nanoparticles within tunable hydrophobic microenvironments. Specifically, nanoreactors using polystyrene, dodecane, dodecanethiol, or dodecylamine have been achieved to investigate the effect of core-catalyst interactions as well as the phase (liquid vs. solid) of the microenvironment on reaction kinetics using reduction of 4-nitrophenol as a model reaction. Both dodecanethiol and dodecylamine nanoreactors did not catalyze the 4-nitrophenol reaction. This is likely due to the interaction of the amine and thiol moieties with the gold nanoparticle surface, effectively blocking catalytic active sites. Dodecane nanoreactors catalyzed the reaction with an apparent rate constant of $0.049 \pm 0.012\ s^{-1}\ cm^{-2}$. Interestingly, nanoreactors with a dodecane core showed 10-fold shorter induction time and 2-fold faster kinetics than polystyrene cores. These results demonstrate that the microenvironment of the nanoreactor can have a dramatic impact on reaction kinetics, possibly due to reagent solubility. Thus, the effect of reactant permeability on overall nanoreactor performance will be discussed.

572578 Gold Bimetallic Nanoparticles for Applications in Selective Oxidation Catalysis

Gold nanoparticles have demonstrated catalytic activity and excellent selectivity for various oxidation reactions such as carbon monoxide oxidation, benzyl alcohol oxidation, and propene epoxidation. The activity and selectivity of gold nanoparticles can be further tuned by adding a second metal to the system creating a bimetallic nanoparticle. As an example, recent work from Hutchings et al. has demonstrated the activity of bimetallic AuPd in direct methane partial oxidation.[1, 2]

Building off of successes in partial oxidation reactions with gold and gold-alloy nanoparticles,[1-4]this work demonstrates new synthetic methods for producing small bimetallic clusters with gold. In small particles, there are a large fraction of surface atoms that are undercoordinated, and it is believed in many systems that the undercoordinated sites are the most catalytically active sites. Our new method uses a facile one-pot synthesis where copper and nickel have been used to alloy with gold in our materials to produce bimetallic clusters with triphenylphosphine as the bound ligand. These particles also stable when deposited onto support materials such as titania and silica. As a comparison, this work also synthesizes supported nanoparticles consisting gold alloyed with copper or nickel synthesized using strong electrostatic adsorption (SEA) methods. While the synthesis of some bimetallic nanoparticles using SEA methods has been demonstrated, catalytic testing of these materials largely has not been investigated. This work takes the SEA method-synthesized gold alloy bimetallic nanoparticles and uses them as catalysts for oxidation reactions such as benzyl alcohol oxidation and CO oxidation.

HAADF-STEM confirms that the alloyed nanoparticles are sub-3 nm synthesized via strong electrostatic adsorption and sub-1 nm for the clusters synthesized in solution. Liquid phase batch reactions analyzed with were conducted for benzyl alcohol oxidation with aliquots removed and analyzed via GC. Gas phase CO oxidation activity was measured using an in-line GC-MS. XPS and ICP-MS were utilized to determine the composition of bimetallic nanoparticles and result in the prescribed ratio of metals. Combining the characterization data and the catalysis data, relationships between the catalyst structure and composition have been developed for these oxidation reactions and will be presented. These results help to inform the design of bimetallic catalysts for important these and other important oxidation reaction systems.

References

1. C. Williams, J. H. Carter, N. F. Dummer, Y. K. Chow, D. J. Morgan, S. Yacob, P. Serna, D. J. Willock, R. J. Meyer, S. H. Taylor and G. J. Hutchings, *ACS Catalysis*, 2018, **8**, 2567-2576.
2. N. Agarwal, S. J. Freakley, R. U. McVicker, S. M. Althahban, N. Dimitratos, Q. He, D. J. Morgan, R. L. Jenkins, D. J. Willock, S. H. Taylor, C. J. Kiely and G. J. Hutchings, *Science*, 2017, **358**, 223-227.
3. A. Maclennan, A. Banerjee, Y. Hu, J. T. Miller and R. W. J. Scott, *ACS Catalysis*, 2013, **3**, 1411-1419.
4. M. O. Nutt, J. B. Hughes and M. S. Wong, *Environmental Science & Technology*, 2005, **39**, 1346-1353.

576536 Role of Poly(*N*-vinylpyrrolidone) and Its Influence on Ag^+ Reduction Kinetics during Formation of Ag Nanostructures

Poly(*N*-vinylpyrrolidone) (PVP) is ubiquitously used in shape-controlled polyol synthesis of metal nanoparticles (NPs). Among the various systems using PVP, polyol synthesis of Ag nanocubes (NCs) has emerged as one of the most robust systems, where PVP is considered as the structure-directing agent, and glycolaldehyde (GA), an oxidation product of ethylene glycol (EG) as the reducing agent for Ag^+ at elevated temperature. While several reports indicate molecular weight (M_w) and monomer concentration (C_m) of PVP impact the final shape of NPs, a universal agreement on the role of PVP is thus far lacking. Recent experimental studies from our group indicate the differential heat of adsorption of PVP to (100) versus (111) facets is too small to predict a cube over an octahedron by thermodynamic considerations. We further demonstrate chloride (Cl^-) added as HCl is the structure-directing agent in the synthesis with about half monolayer of Cl present on the Ag NCs. However, it has been observed PVP can impact the final shape of Ag NPs as its C_m or M_w is changed, suggesting PVP may influence the rate of reduction leading to kinetically-preferred shapes even in the presence of Cl^-. We demonstrate PVP acts both as the dominant reducing agent, where the end groups are responsible for influencing Ag^+ reduction kinetics, and the pyrrolidone ring is unique for stabilizing the Ag NPs. We constructed an experimental phase diagram for the formation of Ag NCs by varying the PVP C_m and M_w at constant temperature and Cl^- concentration. Between the upper boundary (related to the role of PVP in Ag^+ reduction) and the lower boundary (related to a required minimum amount of PVP for stabilization) of the phase diagram, any combination of PVP C_m and M_w, can yield uniform Ag NCs. Ag nanowires formed above the high boundary were found to lack the Cl^- monolayer necessary for shape control in cubes. Optical studies of Ag^+ reduction at different PVP C_m and M_w show direct dependence of Ag^+ reduction on the molar ratio of PVP/Ag and a decreased rate with higher M_w at constant C_m. These combined results suggest PVP rather than EG or GA plays a dominant role in the reduction of Ag^+. Most of the GA (boiling point lower than the synthesis temperature) formed *in situ* from EG oxidation, is lost during the initial 20 h of Ag NC synthesis. Colorimetric test with 2,4-DNPH confirmed concentration of GA is much lower than the total Ag^+ present and the concentration varies very slightly over the period of 20 hours. Although, the PVP end-group/Ag^+ ratio is also well below stoichiometry, we suspect Ag^+ reduces auto-catalytically after the initial reduction by PVP end-groups. In order to further investigate the complex mechanism of Ag^+ reduction by PVP, we synthesized end-groups specific PVP polymers with a precise control over the average M_w. Experimental results indicate end groups are indeed responsible for influencing Ag^+ reduction kinetics; PVP with –OH end groups generate kinetically-preferred Ag NPs (nanorods) lacking adsorbed Cl^- while those with –CHO end groups form Ag NCs covered with a half monolayer Cl^-. This work provides key insights into understanding the role of PVP in influencing reduction kinetics of Ag^+, and thereby the shape of Ag NPs during polyol synthesis.

559920 Enhancement of CO_2 Desorption Performances in CO_2 -Loaded MEA Solution By Using CMK-3 Bifunctional Catalysts

Abstract

Reduction of the required heat duty and enhancing the desorption rate for amine solvent regeneration in stripping carbon dioxide (CO_2) from a CO_2-rich monoethanolamine (5M MEA) solution have been identified as the major challenges in the CO_2 desorption and amine regeneration process. In this study, to reduce the energy penalty four novel bifunctional heterogeneous catalysts CMK-3-SiO_2 CMK-3-MCM-41 and CMK-3-SBA-15 were synthesized and introduced for the first time to study their catalytic effectiveness on CO_2 desorption and monoethanolamine (MEA) solvent regeneration with an initial CO_2-loading of 0.5±0.01 mol CO_2/mol MEA at 97°C for 3hours. The synthesized catalysts were characterized by Fourier transform-infrared spectrometry, pyridine-infrared spectroscopy, X-ray diffraction, N_2-adsorption-desorption, and ammonia-temperature programmed desorption. Based on the obtained results the newly CMK-3 mesoprose catalyst can significantly reduce the energy required for CO_2 desorption and the CMK-3-SiO_2 catalyst presented the highest catalytic efficiency compared to the other CMK-3 catalysts and the solid acid catalyst (SZ), 36.5% of energy required was reduced compared to 5M monoethanolamine (MEA) without catalyst and enhance the desorption rate by 73.45%. Additionally, the CMK-3-SiO_2 catalyst exhibited the best catalytic performance compare to the literature with an excellent stability and had no unfavorable effects on the CO_2 absorption activity by the regenerated MEA solvent.

568086 Identification of Stable Bimetallic Nanoclusters Via a Mathematical Optimization Framework

Metallic nanoparticles (NPs) have distinct properties from bulk metals and show potential for use in a wide range of applications, including catalysis, electronics, and optics [1]. While at the nanoscale there are many possible structures that can be imagined, it is not obvious which is the most stable for a given particle size and composition. This work proposes a mathematical optimization-based design framework to address this combinatorial problem.

The cohesive energy of an NP accounts for the strength of its inter-metallic bonds [2]. Therefore, we can deduce the structure of the most stable nanocluster of a given mass by maximizing its cohesive energy. Our work is based on a bond-centric cohesive energy model for bimetallic nanoclusters proposed by Zihao et al [3]. This model relates a particle's cohesive energy with the constituent atoms' coordination numbers. In this work, we formalize the search for the most cohesive structure via a mathematical optimization model by introducing binary variables to indicate the presence or absence of certain type of atoms in design space. This leads to a Mixed-Integer Linear Programming (MILP) optimization model that can be solved to identify optimal structures using well-established numerical optimization techniques.

In order to improve the tractability of this design framework and be in position to address practically relevant sized NPs in reasonable computational time, we also propose a two-step "structure-then-order" heuristic approach. First, we identify stable nanoparticle geometries using a monometallic nanocluster geometry optimization model that we have developed previously [4], and which results in a pool of nanocluster geometries that may be good candidates for bimetallic NPs. The pool can be further extended by local search based on a metaheuristic search framework such as simulated annealing and genetic algorithms. Then, we use a second MILP model for searching over the space of alloy compositions on a particular geometry. This second model possesses significantly reduced combinatorial complexity compared to the fully-flexible bimetallic structure optimization counterpart, and can thus be efficiently solved to guaranteed optimality for any particle size of interest. By iteratively solving many such alloy composition optimization models, we can obtain good solutions to the fully-flexible counterpart.

We have used this optimization framework to study a wide range of transition bimetallic systems. In many cases, our results reveal segregation patterns such as well-known core-shell structures, which are supported in the literature by both experimental and computational results. However, in some cases, unintuitive trends can be observed for certain bimetallic systems. These results will help us efficiently explore the bimetallic design space and gain a deeper understanding of relationships between structure, order and stability.

[1] Ferrando, Riccardo, Julius Jellinek, and Roy L. Johnston. "Nanoalloys: from theory to applications of alloy clusters and nanoparticles." *Chemical reviews* 108, no. 3 (2008): 845-910.

[2] Kittel, Charles and Paul McEuen. Introduction to solid state physics. Vol. 8. New York: Wiley, 1976.

[3] Zihao Yan, Michael G. Taylor, Ashley Mascareno, and Giannis Mpourmpakis. "Size-, Shape-, and Composition-Dependent Model for Metal Nanoparticle Stability Prediction." *Nano letters* 18, no. 4 (2018): 2696-2704.

[4] Natalie M. Isenberg, Michael G. Taylor, Zihao Yan, Christopher L. Hanselman, Giannis Mpourmpakis, Chrysanthos. E. Gounaris, "Identification of Optimally Stable Nanocluster Geometries via Mathematical Optimization and Density-Functional Theory", *Under review*

568804 Oxidative Dehydrogenation of Ethane to Ethylene with Fe_2O_3@SiO_2 Core@Shell Catalysts

The engineering of materials on the nanoscale enables precise tailoring of materials' functionality. Core@shell materials are a widely studied class of engineered nanomaterials with application in various technologies. From a reaction engineering perspective, these nanocatalysts can be considered "nano-reactors" with porous membrane walls for preferential diffusion of molecules, which enables tuning of selectivity by tailoring the porosity of the shell material. Our previous studies have demonstrated the preferential diffusion and conversion of H_2 in H_2-CH_4 mixtures with lattice oxygen in NiO@SiO_2 due to a "sieving effect" by the porous SiO_2 matrix, confirming the potential of core@shell nanocatalysts towards tuning selectivity.

In the present work, we are extending this approach towards preferential oxidation of H_2 in H_2-C_2H_4/C_2H_6 mixtures to demonstrate the potential of such engineered core@shell materials for oxidative dehydrogenation of ethane to ethylene. Fe_2O_3@SiO_2 core@shell catalysts were synthesized, achieving with fine control of SiO_2 shell thickness with near nanometer precision by adjusting synthesis parameters, including SiO_2 precursor concentration and reaction time. This allows control of the degree of preferential diffusion and thus evaluation of the targeted selective conversion of hydrogen. The reactivity of these core@shell catalysts was evaluated by C_2H_6 pulse experiments at 800°C to observe the gradual consumption of lattice oxygen in Fe_2O_3@SiO_2 in comparison to a conventional Fe_2O_3/SiO_2 catalyst as a reference. We observe noticeably reduced H_2 and CO_2 formation over the core@shell Fe_2O_3@SiO_2 catalyst, and a dependence of this effect on the shell thickness of the catalyst, strongly supporting our concept of selective H_2 oxidation due to "sieving" by the porous SiO_2 shell. Such core@shell materials hence open a relatively straightforward and rationally predictable approach towards designing selective catalysis based on well-established nanostructuring.

563576 Packing Prediction of Pharmaceutical Powders Via Bond Number Estimation

The packing of pharmaceutical powders plays an important role in various unit operations such as die filling, compaction, and feeder performance. In the case of pharmaceutical tablets, the final tablet porosity can dictate critical tablet properties such as tensile strength and dissolution kinetics, and is dependent on blend porosity. Therefore, pharmaceutical powder or blend porosity is an important property that requires better understanding of how it is impacted by both mixing process parameters and constituent material properties. Recent work has introduced mechanistic models to predict the porosity of pharmaceutical powders, which employs a semi-empirical approximation using commonly measured particulate properties such as particle size, particle density, and surface energy. Here, this initial work has been extended to include a wider range of active pharmaceutical ingredients (APIs) and high functional pharmaceutical excipients, both before and after dry coating with nano-silica. Results show that the recently introduced porosity prediction model can predict most pharmaceutical powder porosities fairly accurately, but such is not the case for certain types of materials including functional excipients such as Avicel grade microcrystalline cellulose (MCC) excipients. These functional excipients have a distinctly rougher particle surface, compared to smooth particle surfaces for most other pharmaceutical powders. This distinction could be the key to improving the proposed porosity prediction model, and such a topic has been discussed in previous literature. In this work, we also examine how a prevalent approximation of natural surface asperity size of 200 nm diameter requires re-examination. The specific surface area, internal void volume and particle shape descriptors of the pharmaceutical powders were experimentally measured, and the dependence of powder packing on these characteristics was investigated. Additionally, binary blends of APIs and commonly used excipients were tested for their bulk density (packing). Results show that API dry coating significantly alters the interparticle cohesiveness of these binary blends, and their packing behavior is more similar to non-cohesive glass beads. Such processing significantly improves the design space for micronized, cohesive APIs for use in processes such as high speed direct compression. However, extensions of the current models are also necessary for accurate blend porosity prediction if dry coating processing was used for one or more blend constituents. The experimental work presented here illustrates the need to improve the current semi-empirical, mechanistic porosity prediction model and may help assess if our initial attempts are adequate.

571265 Theoretical and Experimental Investigation into the Impact of Tribo-Charging on the Performance of Dry Powder Blends

Introduction

Powder charging is known to greatly affect the industrial processing and performance of particulate and granular materials. The underlying process of contact charging is referred to as tribo-electrification[1]. Pharmaceutical materials are usually insulators and often accumulate electrostatic charges that can detrimentally affect their processability[1]. Thus, tribo-electric charging can be expected to provide an important contribution to particle-particle and particle-wall interactions during handling and transport of powders of dielectric materials. It would therefore be desirable to systematically determine the factors that contribute to powder tribo-charging. Generally, the extent of tribo-charging is expected to be related to the difference between the work function (WF) values of the two contacting surfaces[2]. Other material properties relevant to tribo-charging include surface energy, hygroscopicity/wettability, surface roughness, and particle shape[1]. Therefore, it is anticipated that different materials as well as particulate properties will be crucial in the tribo-charging of different pharmaceutical blends. Finally, also, the effect of unit operations involved in the manufacturing of pharmaceutical products must be taken into account. Thus, it was the aim of our work to systematically understand the factors critically affecting this phenomenon, and its respective impact on formulation performance.

Design of the studies

Model pharmaceutical materials were selected and their material and particle properties characterized. Tribo-charging acquired during the flow over a stainless steel tube was investigated in order to obtain relevant information of their behavior during unit operations such as feeding, transport, blending and capsule filling etc. The results of relative charging in the materials were interpreted considering their distinct chemical and physical properties. More concretely, evaluation of the net charge of the different materials was carried out using Granucharge™ (Granutools). Measurements were performed under defined environmental conditions. First, initial charge to mass ratio (q0) was recorded by dispensing the powder directly into a Faraday pail. In a consecutive step, the powder was transferred to a vibrating stainless-steel V-tube and the charge to mass ratio (q1) was recorded again after the powder was transported through the tube. The acquired charge during flow, thus charging tendency, was determined as $\Delta q = q1 - q0$.

Tribo-charging trends predicted from first principles

The tribo-charging of five different pharmaceutical materials acetylic salicylic acid (Merck KGaA), Ibuprofen (Merck KGaA), a-lactose monohydrate (LH100, Meggle), trehalose (Cargill Deutschland GmbH), and D-Mannitol (P160C, Roquette) were experimentally investigated. In order to concentrate on the effect of electronic properties of the materials, other characteristics and experimental parameters, in particular the particle size (distribution) of the used powders, were kept identical for the five investigated materials. The measured tribo-charging trends were compared with the energies from electronic structure calculations.

It was found that varying the level of theory (from semi-empirical/PM6, over DFT/GGA, to DFT/meta-GGA) affects the correlation between tribo-charges and calculated energies only weakly. However, of the calculated values (highest occupied molecular orbital (HOMO), energy levels, vertical and adiabatic ionization potentials (IP), and work functions (WF)) only adiabatic IP values showed a useful correlation with tribo-chage, with a pearson correlation around $r = 0.9$. The calculated WF values however, the measure that is traditionally used for prediction of contact-electrification, showed a poor correlations with our experimental data. This unexpected result is presumably due to a combination of two facts: 1) experimental IP and WF values usually show a good correlation, and 2) the lowest unoccupied molecular orbital (LUMO) energy, which is used here to estimate electron affinity and for the calculation of the WF, is notoriously difficult to calculate accurately, this carrying over to the calculated WF values. We conclude that since the IP can be calculated more accurately, and it correlates with the WF, the IP can be used as a replacement for the WF, in particular if one is primarily interested in relative values. Although limited in number, our data suggests that electronic structure calculations can be used to estimate relative tribo-charging trends for pharmaceutical materials, provided that other material properties are comparable.

Tribo-charging trends of different particulate materials

Although prediction of tribo-charging trends using first principles is desirable to determine *a priori* the molecular properties affecting the contact of different materials, its applicability is still limited as it does not discriminate between different solid and particle states of materials (as discussed in the previous section). Therefore, the tribo-charging of different salbutamol sulphate (SBS) (Selectchemie) particles engineered using jet-milling (Spiral air Jet Mill 50 AS, Hosokawa Alpine AG) and spray-drying (Nano spray-dryer B-90, Büchi) was experimentally investigated and compared[4].

It was demonstrated that different surface chemistries arising from distinct solid-states of the same material, different contact areas due to particle properties like shape, size and morphology, as well as variable water contents can critically impact the polarity and magnitude of tribo-charging. Amorphous, spherical shaped SBS particles with a high water content (generated via spray-drying) attained positive charge of a higher magnitude. In contrast, the milled fine particles with irregular shape and lower water content tended to charge negatively and to a lower extent[4].

Tribo-charging trend of different material blends

To understand how different particulate characteristics can affect the tribo-charging of pharmaceutical powder blends, the distinctly engineered fine particles of SBS (1-5 μm) were mixed with coarse D-mannitol (about 100 μm) (MAN) at 2 drug loadings (2 wt% and 5 wt% of SBS). Likewise, the tribo-charging of the obtained powder blends was evaluated[4].

Comparison of the two crystalline powders (jet-milled SBS and MAN) revealed that theoretically calculated WFs can provide a rank order of the sign of charge when the materials were in contact with metal, supporting our previous conclusions that electronic structure calculations can be used to estimate relative tribo-charging trend. Concerning the charging tendency of the powders blends, an increased charge mitigation effect was observed when a higher concentration of fine SBS particles were blended with the coarse MAN. Likewise, it was found that the charging tendency of the coarse particles is less relevant when blended with fine materials. Consequently, it was possible to conclude that evolution of charge in the blend seemed to be dominated by the nature of fines and their concentrations contributing to inter-particle contacts[4].

Tribo-charging trends arising from distinct process parameters

In a final step, powder blends have to be processed into a final dosage form to be administered to the patient. Capsules are one of the most common solid oral and inhalation dosage forms. Therefore, the capsule filling process was selected to investigate the impact that processability can have on the tribo-charging. For that, size 3 gelatin (Quali-G™, Qualicaps) as well as thermally (Vcaps® Plus, Capsugel) and chemically gelled HPMC capsules (Quali-V®-I, Qualicaps) were filled (Labby, MG2) targeting distinct fill weights (25 and 75 mg) and the effect on tribo-charging was evaluated and compared[5].

Comparing the charge tendency of the different capsules, it was seen that gelatin (Gel-Cap) and the thermally gelled HPMC (TG-HPMC-Cap) capsules showed a higher tendency to acquire charge during contact with stainless steel than chemically gelled HPMC (CG-HPMC-Cap). When different fill weights were used, it was observed that the Gel-Cap and TG-HPMC-Cap had a decreased tendency to acquire charge as the powder fill level increased. In contrast, the powder fill level had negligible impact on the charging properties of CG-HPMC-Cap capsules. This revealed the similar charging tendencies of capsules manufactured with different materials and the dissimilarity between capsules of the similar material[5].

Conclusion

Tribo-charging is a complex and very relevant phenomenon that is still largely unpredictable. In this work, we propose a bottom-up approach to the understanding of tribo-charging trends of pharmaceutical materials. From molecular properties to particulate and powder bulk characteristics to the influence of process conditions on tribo-charging, providing a systematic understanding the factors critically affecting this phenomenon, and its respective impact on dry powders performance.

References

[1] Wong J, Kwok PCL, Chan HK. Electrostatics in pharmaceutical solids. *Chem Eng Sci*. 2015; 125:225-237.

[2] Helmholtz H. Studien über electrische Grenzschichten. *Ann der Phys und Chemie*. 1879; 243(7):337-382.

[3] Brunsteiner M, Zellnitz S, Pinto JT, Karrer J, Paudel A. Can we predict trends in tribo-charging of pharmaceutical materials from first principles?. *Powder Tech.* 2019; Submitted.

[4] Zellnitz S, Pinto JT, Brunsteiner M, Schroettner H, Khinast J, Paudel A. Tribo-charging behaviour of inhalable mannitol blends with salbutamol sulphate. *Pharm Res*. 2019; Accepted.

[5] Wutscher T, Zellnitz S, Kobler M, Buttini F, Andrade L, Daza V, Mercandelli A, Biserni S, Probst SE, Khinast J, Paudel A. Influence of Storage Humidity and Powder Fill Level on Charging Behaviour of Different Capsules Types. RDD Europe 2019; Lisbon, Portugal.

571333 Influence of an External Lubricant on the Tribo-Charging Behaviour of Capsules

Introduction

During the manufacture of empty two-piece hard capsules, an external lubricant is applied to the outer surface of the capsules to prevent sticking to each other after packaging and to improve transfer of the empty capsule during filling operations. This lubricant also helps the capsule to minimize friction in a capsule filling machine and therefore is needed for a smooth processing [1, 2]. For the external lubricant, the same substances are used for capsules comprised of either gelatin or hydroxypropyl methylcellulose (HPMC). These lubricants, for example, are magnesium stearate (MgSt), sodium lauryl sulphate (SLS) and carnauba wax (CW). The structural differences among lubricant can lead to a different work function of the material (the minimum energy needed to withdraw an electron of the surface of a solid), which places them at different positions of the tribo-electric series. When materials with different work functions brought into contact and are separated, electrons are transferred and tribo-electric charges are created. Tribo-charges are often responsible for many manufacturing issues and can influence the quality, safety and efficiency of pharmaceutical products [3] or for example the aerodynamic performance of a dry powder inhaler (DPI). Tribo-charges can also cause problems in terms of powder mixing homogeneity and dosing accuracy [4], and can lead to particle adherence on surfaces and agglomeration of particles during manufacturing or powder release from the capsule [5].

This study examines the impact of the charging behaviour of capsules that are externally lubricated with different materials, before and after transport over stainless steel and PVC. The transport over steel was intended represent the transfer from the bulk container to a filling machine hopper-feeder, whereas the transport over PVC could give an idea of how the capsules behave/or will be charged within a packaging material.

Material and Methods

Different types of size 3 gelatin capsules (Quali-G®) and chemically gelled HPMC capsules (Quali-V®-I) were received from Qualicaps (Qualicaps Europe, S.A.U., Spain) with various variations of colorant and external lubricant. Gelatin capsules were supplied with the external standard lubricant SLS, unlubricated, MgSt, and CW. Batches with 1g and 2g of MgSt on 10kg of capsules were received, to study the influence of the amount of external lubricant. Gelatin capsules with MgSt or CW as external lubricant were further divided into two groups, in which different colorants were used. HPMC capsules were also used in an unlubricated state, as well as with CW, which is the standard lubricant for HPMC capsules, SLS, and MgSt.

Charge measurements were performed in triplicate using 2.0g ±0.15g capsules on the GranuCharge™ (Granutools, Belgium) instrument. The room was conditioned to a relative humidity (RH) 50% ±3% at 23.5°C ± 0.6°C, whereas the conditions inside the device remained constant at 50% ±2% at 23.5°C ± 0.5°C. Charge measurements started with the measurement of the initial charge to mass ratio (q0) by pouring the capsules directly out of the bag into the inbuilt Faraday cup of the device. Afterwards, the capsules were transferred to the grounded vibrational feeder of the GranuCharge and fed into a V-shaped steel or PVC tube to drop into the Faraday cup again after transport. Then, the charge to mass ratio was calculated for a second time (q1). By subtracting the initial charge to mass ratio from that obtained after sliding through the V-tube yielded the charge to mass ratio (Δq = q1-q0) that describes the charge gained.

To differentiate the surface energy (SE) of the capsules with different external lubricants, contact angle measurements were made using the pendant drop method (Easydrop system, Krüss GmbH, Germany) at 22°C ± 0.3°C. For this purpose, the contact angle of water and diiodomethan was measured on the outer surfaces of the capsules (as received). Surface energy was calculated via Owens-Wendt-Rabel-Kaelble method (OWRK) [6].

Results and Discussion

Overall, the charge to mass ratio (Δq) of capsules after the transport over steel and PVC was positive, although a higher charge to mass ratio was observed when capsules were in contact with PVC tube walls. Gelatin capsules on steel charged in a range of 0.63 (± 0.05) to 2.87 (± 0.07) nC/g whereas HPMC charged from 0.43 (± 0.06) to 1.45 (± 0.06) nC/g. Charges of gelatin and HPMC capsules after passing over PVC surface were from 0.72 (± 0.27) to 2.62 (± 0.71) nC/g and 1.61 (± 0.05) to 3.05 (± 0.13) nC/g, respectively. Capsules with MgSt showed an exception because they were charged to the same amount on both materials. As Staniforth and Rees [7] stated, MgSt is an electropositive and therefore able to lower the charging effect of the capsule material and wall interactions. Within one type (gelatin or HPMC) of capsules, the un-lubricated capsules showed the highest charging and those lubricated with SLS showed the lowest charging. It can be considered, that the electron transfer of the HPMC and gelatin is hindered due to the work function of the contacting surface of the capsule material or the external lubricant. The batches with two different amounts of MgSt showed that a lower amount of lubricant gave higher charging. This seems to fit to the theory that lubricants sometimes work as antistatic agents as it is also reported for coated particles [8]. A part of the lower charging tendency of HPMC capsules compared to gelatin capsules can be explained by the antistatic properties of HPMC itself [9] although other ingredients of the capsule formulation like the gelling agent or different colorants also play a role . The effect of a lower charging tendency due to lubricants is supported by the fact that a lower amount of external lubricant lead to an increase in charge. In order to assess the batch to batch variability, two batches of gelatin capsules without lubricant were measured. The two batches gave the same results in charging, although the surface energy was slightly different. Overall, SE showed results between 3.18 to 8.08 mN/m for HPMC capsules and 26.6 to 38.59 mN/m for gelatin capsules, respectively. The HPMC capsules with the highest SE showed the lowest charging tendency within the group of HPMC capsules, whereas in gelatin the highest SE also gave the highest charging tendency. It can be assumed that the lubricants cover the main part of the capsule surface but do not fill all cavities of the underlying material and therefore the differences between HPMC and gelatin occur.

Conclusion

In summary both, the external lubricant and the capsule materials themselves have an influence on the tribo-charging behaviour of the capsules. The low amount of lubricant on a single capsule (only 2% of the average capsule weight), seems to smoothen the surface of the capsule. Since it is not known how uniform the distribution of the external lubricant is, it is possible that some regions of the capsule surface are not covered enough to protect the capsule material from directly interacting with the environment. Hence, the results for charging can be a mixture of the interactions of the capsule material and the external lubricant. The work functions of HPMC and gelatin on one hand and the work functions of the lubricants on the other hand do not explain the differences on their own. Calculated work function values depend on well-defined molecular structures, therefore real work function values are hard to define for HPMC and gelatin as impurities in the material and the orientation of functional groups on the surface also have an impact. Overall, tribo-charging effects of capsules seem to be an interplay of capsule materials, parameters like surface roughness of the capsule and the interacting machine, water content and lubricant attributes as well as the lubrication level. It could be shown, that lubrication of capsules plays a role in tribo-electrification.

579032 Investigation into the Effects of Roller Compaction on Granule and API Properties

Solid oral dosage forms are the most popular drug delivery system. The pharmaceutical industry is a very highly regulated one and as such the processes involved in the manufacture should be carefully controlled. One key aspect to control is the particle size and shape of the Active Pharmaceutical Ingredient (API), as this has been shown to have significant effects on prcoessibility (e.g. punch sticking) and the performance of the final product (e.g. dissolution rate) (Leane et al. 2015).

However, this can prove challenging as the API is subjected to harsh environments in various unit operations, such as roller compaction, which can cause attrition of the particles (Gamble et al. 2014). For control purposes it is important to understand what is occurring to the API within these processing steps, however, tracking changes in the API's properties once it has been incorporated into a blend is challenging. Consequently the performance of the end product are often related to the input properties, which are likely to have changed during processing.

A modern approach combining static image analysis (SIA) and Raman Spectroscopy can be utilised to measure the particle size and shape of a single component within a multi-component system (Gamble et al. 2015) thus enabling changes in API particle size during processing to be measured. This approach has been applied to investigate the effects of roller compaction process parameters on API properties. The results of this study provide an insight into the dry granulation processing route and the effects of processing conditions on the properties of API particles and the subsequent granule. The presentation will also address topics including the nature of the API within granulate, both in terms of size and location, and the possible impact on to downstream processing such as tabletting.

References:

Gamble, J. F., Hoffmann, M., Hughes, H., Hutchins, P. and Tobyn, M. (2014). "Monitoring process induced attrition of drug substance particles within formulated blends." International Journal of Pharmaceutics **470**(1): 77-87.

Gamble, J. F., Tobyn, M. and Hamey, R. (2015). "Application of Image-Based Particle Size and Shape Characterization Systems in the Development of Small Molecule Pharmaceuticals." Journal of Pharmaceutical Sciences **104**(5): 1563-1574.

Leane, M., Pitt, K. and Reynolds, G. (2015). "A proposal for a drug product Manufacturing Classification System (MCS) for oral solid dosage forms." Pharmaceutical Development & Technology **20**(1): 12-21.

568348 In-Depth Experimental Analysis of Pharmaceutical Continuous Twin-Screw Wet Granulation in View of the Development of a Generic Compartmental PBM

INTRODUCTION

Recently, a mechanistic Population Balanced Model (PBM model) has been developed and calibrated for twin-screw wet granulation for a few specific formulations [1]. In-depth data collection and analysis was performed to investigate the influence of process parameters on granule formation mechanisms along the twin-screw granulator barrel and resulting granule Critical Quality Attributes (CQAs) [2]. A logic next step is to make this model applicable for any formulation by developing a generic PBM for continuous twin-screw wet granulation, predictive for granule size distribution (GSD) in function of both process parameters and raw material properties. Given the potential importance of granule porosity on granule formation and final drug product, information with regard to this property was collected per granule size fraction along the length of the granulator barrel. This information will serve for the development of a generic 2D-PBM model.

MATERIALS AND METHODS

Different commercially available Active Pharmaceutical Ingredients (APIs) with major differences in physicochemical properties, primarily solubility and wettability, were selected. Milled and non-milled APIs were utilized to investigate the effect of API particle size on granule formation and final granule CQAs.

Blends were prepared using a fixed formulation base consisting of microcrystalline cellulose, HPMC, lactose, and 5% or 50% API for the low and high dose formulations, respectively.

APIs, excipients and blends were thoroughly characterized for particle size distribution, bulk and tapped density, dissolution rate, water holding capacity, solubility, residual moisture content, contact angle, ring shear test, true density, angle of repose, dispersion charge, and specific surface area.

Granules were produced using the high shear twin-screw wet

granulation module of the ConSigma™-25 system (GEA Pharma systems, Collette, Wommelgem, Belgium).

A fixed screw configuration was used with 2 kneading compartments, each consisting of 6 kneading elements arranged at 60° stagger angle. Both kneading zones were separated by a conveying element and at the end of the granulator barrel, an additional conveying element was added together with three size-control elements.

In order to characterize the granule size distribution and granule porosity along the length of the twin-screw granulator barrel, samples were collected in four well-defined compartments: (i) the wetting zone, (ii) after the first kneading zone, (iii) after the second kneading zone and (iv) at the granulator outlet (after the size control elements) using the method described by Verstraeten et al. [2].

A 5-level Rotatable Central Composite Circumscribed Design of Experiments (CCC) (MODDE 12, MKS Umetrics, Umeå, Sweden) was performed to evaluate the effect of process parameters liquid-to-solid ratio (L/S), material throughput and screw speed on local and final granule size distribution and porosity. The granulator barrel temperature was kept constant at 25 °C. Granules collected at the different locations were dried at ambient temperature and relative humidity (25 °C, 40% RH) for at least 24 hours and granule size distribution (Partan^{3D}, Microtrac Inc, Montgomeryville PA, USA) and granule porosity by mercury intrusion porosimetry (Porotec, Thermo Finnigan, California, US) were measured. The flowability of the granules at the end of the granulator was evaluated by the Hausner ratio. Also, friability was measured at the end of the barrel to evaluate the effect of granulation process parameters and raw material characteristics on granule strength.

RESULTS AND DISCUSSION

The experimental results not only showed prominent similarities but also differences with what has been observed before due to the different starting material properties [2]. The effect of TSG process parameters material throughput, screw speed and L/S ratio on granule size distribution along the length of the barrel was highly dependent on formulation properties. Furthermore, L/S ratio remained the most important factor dictating the size distribution of granules collected at the outlet of the twin-screw granulator.

However, barrel fill level, dictated by screw speed and material throughput, also had a relevant effect on granule size distribution. While the impact of the narrow chopper section on the granule particle size has been shown to be insignificant at low L/S ratios [2], in this study, the size control elements at the end of the granulator barrel appears to have a major effect on granule size, both at low and high L/S ratios.

CONCLUSION

This experimental study provided valuable information on the granulation mechanisms in function of process parameters and raw material attributes. Interestingly, the results indicate that the size-control elements and raw material properties have an important effect on granule formation and final granule CQAs. This extensive data collection and analysis is a first crucial step in the development of a generic 1D as well as a 2D PBM model for twin-screw wet granulation.

References

[1] Van Hauwermeiren, D., Verstraeten, M., Doshi, P., am Ende, M. T., Turnbull, N., Lee, K., ... Nopens, I. (2018). On the modelling of granule size distributions in twin-screw wet granulation: Calibration of a novel compartmental population balance model. *Powder Technology*, *341*, 116–125. https://doi.org/10.1016/j.powtec.2018.05.025

[2] Verstraeten, M., Van Hauwermeiren, D., Lee, K., Turnbull, N., Wilsdon, D., am Ende, M., De Beer, T. . (2017). In-depth experimental analysis of pharmaceutical twin-screw wet granulation in view of detailed process understanding. *International Journal of Pharmaceutics*, *529*(1–2), 678–693. https://doi.org/10.1016/j.ijpharm.2017.07.045

567489 The Use of T-PLS to Investigate the IMPACT of Raw Material Properties on Granule Quality Attributes Obtained after Continuous Twin-SCREW WET Granulation

INTRODUCTION

Since 15 years, the interest for continuous manufacturing has steadily increased in the pharmaceutical industry. The advantages offered in terms of process efficiency, costs reduction, flexibility and environmental footprint reduction have been explained in multiple articles and reviews. Among the different continuous drug product manufacturing processes currently available, twin-screw granulation (TSG) is the most evaluated continuous wet granulation process. However, a predictive platform linking the influence of raw material properties, the ratio's in which the raw materials are mixed, and the applied process settings with the granule quality attributes is not yet available.

PURPOSE

In this study, T-shaped partial least squares (T-PLS) was used to study the combined influence of raw materials properties, formulation composition and process settings on granule quality attributes.

MATERIALS AND METHODS

A set of 25 powders consisting of 13 active product ingredients (APIs), 9 fillers and 3 binders for immediate release was characterized using 15 different techniques covering dry powder characteristics (particle size, compressibility, flow properties, ...) and wet powder characteristics (solubility, contact angle, dissolution rate, ...). Principal Component Analysis (PCA) was then performed on the API and filler raw material characteristics subsets to elucidate correlations between the powders and the raw material properties. Based on this analysis, formulation blends were selected, ensuring that a maximal area of the material variability space was covered. Each formulation consisted of an API, a filler and a binder. Herewith, three different blend ratios were used for each formulation, corresponding with an API content of 10, 40 or 70%. As the binders were intended for immediate release, their content was varied from 2 to 5%. Each formulation blend was granulated using the twin-screw wet granulation unit of the ConsiGma™ 25 system (GEA Pharma systems, Wommelgem, Belgium). A 2-level full factorial Design of Experiments (DoE) was performed on each formulation blend to investigate the influence of the granulation process parameters mass flow rate (MFR) and liquid to solid ratio (L/S ratio) on the resulting granule quality attributes. The MFR was varied from 10 to 20 kg/h, while the applicable L/S ratio ranges were experimentally determined for each formulation blend, because this depended on the properties of the selected raw materials. Granule size and shape (PartAn^{3D}, Microtrac, Montgomeryville, United States), friability (Friabilator, PTFE Pharma Test, Hainburg, Germany) and flow properties (GranuDrum, GranuTools, Awans, Belgium) were measured for the obtained granules after each granulation experiment.

RESULTS AND DISCUSSION

The T-PLS model was able to indicate which raw material properties affected the most the granulation process. It was seen that dissolution rate, solubility, porosity, specific surface area and water binding capacity had the highest impact on the variability in granule quality attributes. Furthermore, it was also possible to highlight the most influencing raw materials for each granule quality attribute. For example, the granule size fraction that remained ungranulated (i.e. granule size fraction <150 μm) was correlated with water binding capacity, while it was anti-correlated with solubility, dissolution rate and compressibility. The T-PLS model indicated as well that L/S ratio affected more the granulation process than MFR. But, in general, the variation in granule quality was more influenced by raw material properties than by process parameters.

CONCLUSION

Besides an enhanced process understanding of the impact of raw material properties on granule quality attributes, the developed T-PLS model has also the potential to predict a reasonable starting point for formulation and process parameters for new API's. This approach can reduce the amount of API and the time needed to find optimal process parameters during product development.

567670 Visualization of Granule Temperature Along the Length of the Barrel Using Thermal Imaging to Improve Process Understanding of Pharmaceutical Twin-Screw Melt Granulation

INTRODUCTION

The pharmaceutical industry has been showing an increasing interest in continuous manufacturing for the past 15 years. In order to improve process efficiency, reduce costs and decrease the environmental footprint, innovative continuous manufacturing processes are gaining more interest. Twin-screw melt granulation (TSMG) is an interesting technique to perform agglomeration due to its multiple advantages. For instance, the use of a solid meltable binder to achieve agglomeration instead of using a solvent like in wet granulation enables the use of moisture-sensitive drugs and the exclusion of a drying step. Additionally, the high-dosed formulations with up to 95% of API and the ability to modulate release kinetics can be added to the extensive list of advantages.

In batch melt processes, the heat transfer on powder particles is not uniformly distributed and is uncontrolled. In combination with a long residence time, batch melt processes are not desirable to be used in pharmaceutical industry to obtain uniform quality attributes of granules. On the contrary, in continuous TSMG heat transfer is uniformly distributed and therefore also granule growth can be controlled because every powder particle is exposed to an equal amount of heat and mechanical stress. Furthermore, the residence time in continuous melt granulation is much lower than in batch melt granulation.

The aim of this study is to increase process understanding of the granulation mechanism in twin-screw melt granulation by evaluating the influence of different screw configurations on granule formation, granule temperature and the relation between this.

MATERIALS AND METHODS

Soluplus was used as a thermoplastic binder in the 2 formulations selected for this study. Metoprolol tartrate (MPT) and Caffeine anhydrous (CAF) were chosen as model drugs due to their different miscibility with Soluplus [1]. The former is miscible, while the latter is immiscible with Soluplus. Physical mixtures (85% API/binder w/w) were processed using a twin-screw extruder (Prism Eurolab 16, Thermo Fisher Scientific) with a screw configuration containing 2 kneading zones of each five kneading elements and a screw mixing element at the end of the screws. A D-optimal screening design was performed with screw speed (150-500 rpm), material throughput (0.4kg/h-0.8kg/h), barrel temperature (60°-80°C for MPT and 90°-110°C for CAF), stagger angle (30°, 60° and 90°) and direction of the kneading zone (forward and reversed) as factors.

Granule and process properties were evaluated as responses like friability, granule size, shape, density, porosity and torque. These properties were evaluated for samples collected at 4 different locations along the length of the granulation barrel to visualize and understand the granule formation along the length of the barrel [2]. Samples were collected after only conveying elements, after the first kneading zone, after the second kneading zone and after the screw mixing element at the end of the barrel.

Furthermore, the granule temperature and barrel temperature were monitored at these 4 locations. Therefore, a FLIR A655sc Infrared Camera (FLIR systems, Frankfurt am Main, Germany) was used. Heat can be transferred onto the granules from the heated barrel wall, but heat can also be generated due to the shear created by the friction forces of the kneading elements. The temperature of the barrel wall was monitored with the data logging system of the granulator, but this does not represent the actual temperature of the granules. The maximum and the mean granule temperature was calculated and the difference between the programmed barrel temperature and the measured granule temperature was calculated for the different runs of the DoE to evaluate the influence of the different factors on the granule temperature. Granule formation, granule temperature and resulting granule properties are closely related. Therefore, the link between granule and process properties and granule temperature was also investigated in this study.

RESULTS AND DISCUSSION

The most influencing factors of the DoE on the granule temperature increase is the reversed configuration with a small stagger angle and as to be expected the programmed barrel temperature. The higher granule temperature for this screw configuration can be linked with an increased torque value and therefore a higher amount of shear. Because of this, the binder can be melted sufficiently to attain granule formation. Granules that are obtained with this kneading zone configuration also have the best quality attributes, e.g. the granules produced are the largest and strongest and have a monomodal granule size distribution.

It can be observed that granule formation occurs in the zones along the barrel that create higher shear and subsequently higher temperature which ensures enough melting of the binder. Therefore, no granule formation is occurring in the conveying zones or in the kneading zones with forward configuration. The torque value for the forward configuration is not increasing sufficiently and hence not enough shear is created to form granules with good quality attributes. This can be countered by changing the screw configuration or by elevating the barrel fill level. For these two options there will be an increase of the torque value to ensure enough shear at the kneading zones to obtain granules with good quality attributes.

CONCLUSION

Heat transfer in continuous melt granulation occurs in a more controllable and uniform way compared to batch melt granulation, resulting in granules with more uniform quality attributes. By applying different amounts of shear and barrel temperature, different granule properties can be obtained. The smallest stagger angle in reversed configuration creates the highest shear and the largest granule temperature increase and therefore, more melting of the binder and densification occurs and better quality attributes of granules are obtained. Because the solid state of various active pharmaceutical ingredients are not resistant for all levels of shear and temperature, the screw configuration can be optimized for every formulation so that optimal quality attributes can be obtained.

REFERENCES

[1] T. Monteyne *et al.*, "The use of rheology to elucidate the granulation mechanisms of a miscible and immiscible system during continuous twin-screw melt granulation," *Int. J. Pharm.*, vol. 510, no. 1, pp. 271–284, 2016.

[2] M. Verstraeten *et al.*, "In-depth experimental analysis of pharmaceutical twin-screw wet granulation in view of detailed process understanding," *Int. J. Pharm.*, vol. 529, no. 1–2, pp. 678–693, 2017.

562508 3D Printing: A New Era of Precise Manufacturing of Individually Developed Pharmaceuticals

The growing demand for customized medicines increases the impact of 3D printing as one of the emerging technologies for precise manufacturing of individually developed dosage forms. In all 3D printing technologies, large number of materials, geometrical and process factors affect the quality of these computationally designed dosage forms and safety of their use. In addition, for printing an identical dosage form, the quality may vary when built using different 3D printers, even when the same model of the printer, printing parameters, process steps, and raw materials are used. Therefore, knowledge of how the variability of each input parameter and processing step affects the final finished product is critical to ensuring its quality. There is no specific guidance concerning the design, manufacturing process and quality testing considerations for 3D printed solid dosage forms. This presentation will then highlight some of specific methodologies and control strategies considerations over the material, geometrical and process parameters for 3D printing of solid dosage forms. A case study of employing 3D pneumatic assisted microextrusion based 3D printing to manufacture modified release tablets and polypills will be presented. considerations for process validation will be also discussed so that quality for all printlets is maintained when built in a single cycle, between cycles, and by different printers. The issue of identifying and linking the critical factors to the critical quality attributes of the printlets will be also highlighted with an emphasis on the regulatory considerations involved.

561294 Smart Design and Virtual Product Analysis for 3D Printing Pharmaceutical Tablets for Patient Centric Product Development

The Additive Manufacturing is an innovative and promising method that is suitable for the fabrication of personalized dosage forms of tablets. The formulation selection and geometry design of the tablets are key elements for product design. Tuning the formulations and geometry and design of tablet would be based on personalized dose and release profile, and tablets can be smartly designed to provide adequate physical strength and desired drug release profile.

In this work, two mathematical models were developed and validated by experimental data:

The first model is a solid mechanics model by Finite Element Method technique. The model studies the various internal geometry and design of the tablet on tablet axial and radial strength. The sensitivity analysis demonstrates how changing the printing parameters and internal structure of the tablet will impact the tablet hardness. Experimental data for tablet axial and radial strength are in confirmation with the model's results.

The second model is a dissolution and drug release profile simulation. Effects of formulation variabilities (soluble to insoluble ingredients ratio) and tablet geometry on drug release profile have been studied. The model is developed based on unsteady diffusion of water to the solid tablet and swelling (moving boundary PDEs) of the tablet, and unsteady diffusion of drug from the tablet to the dissolution medium.

The models demonstrate how the smart design and virtual product analysis for the 3D printed tablets could be used for patient centric product development.

Disclaimer: This scientific publication reflects the views of the authors and should not be construed to represent FDA's views or policies.

573787 Rapid on-Site Manufacturing Approach with 3DP Capsules: Delayed Burst Release for Regional Absorption

A critical gap existing in our first-to-human strategy is the ability to evaluate complex dosage forms under the on-site formulation (OSF) strategy. We propose to use 3D printing (3DP) to fill this gap by enabling OSF with custom dosage forms using this flexible, high quality, highly reproducible, rapid, and material sparing approach. The 3DP platform has the potential to enable project teams to gain a better understanding of the complex in vitro-in vivo relationships across different drug release mechanisms through accelerated pharmacokinetic screening clinical trials during or immediately following Phase I, and potentially diminish a portion of the decade-long timing from molecule patent to commercialization. Specifically, human regional absorption and permeation is difficult to determine in early clinical phases. The models we use today for calculating regional permeation in the GI tract are largely based on datasets collected in the last few decades using one of two approaches: 1) a double lumen or triple lumen catheter system, or 2) either a single or double tube-balloon system, where the tube would be fed through the intestinal tract and the balloons would be inflated to isolate a portion of the intestines during the study. Both methods are invasive and not feasible as a routine study in humans. An oral delivery system has been developed by Medimetrics called IntelliCap, which is capable of investigating regional absorption; however, this technology is proprietary and therefore may not be cost effective to use for routine studies.

A Hyrel 3D System 30M printer along with various modular heads were purchased from Hyrel 3D, whose heads were modified or custom designed to be able to print new polymer filaments that are semi-brittle, soft, or flexible, or a combination thereof, or to be capable of injecting fluids with viscosities between 20-3000 cP for a single-click print-and-fill dosage form. In this presentation, the development of a set of capsule formulations for delayed burst release with controlled timing is shown to facilitate regional absorption studies on new active pharmaceutical ingredients. By manipulating the thickness of the capsule walls, the onset of release can be controlled. Due to the swelling behavior of PVA, it was discovered that the mechanical properties were critical for robust formulations due to delamination failure that would occur if the layer-layer weld during the printing process was not sufficiently strong. It was also found that fillings containing a large fraction of swellable materials, such as water, affected the physical stability of the formulations. As a first step to enable robust formulations, finite element analysis was used to understand the effect of water-containing drug fillings on the dissolvable capsule shell, and the shell geometry was optimized to reduce swelling stresses that may cause delamination in- vitro or vivo. Second, software techniques were developed, called zoning, to print capsules with no functional defects. Additionally, due to the nature of the 3DP process, in vitro analysis revealed a unique anisotropic dissolution behavior. Custom USP2 hanging baskets were designed and manufactured via 3DP for orientation control during dissolution. Lastly, an in vivo (beagle dogs) study was run using two 3DP capsule formulations and an immediate release compacted tablet as a control, and found a good in vitro – in vivo relationship.

557916 Production of Small-Batches of 3D Printed Multi-Compartment Capsular Devices for Oral Drug Delivery Using High-Throughput Robotic Manufacturing Technologies

In this talk, we present an automated production process enabling the manufacturing of multi-compartment capsular devices for oral drug delivery. This process includes *i)* the production of capsule parts and *ii)* filling the capsules with formulations containing active pharmaceutical ingredients. In this respect, our technology allows fast and efficient fabrication of fixed-dose combination products in small, personalized batches.

The capsular devices manufactured with this process have a structure that is different from the traditional capsules that are already on the market – generally made either of gelatin or hydroxypropyl methylcellulose. Our capsular devices have a modular structure that consists of multiple inner compartments, which are independent of one another. Each compartment can be filled with different active ingredients and with a range of doses, as well as formulations, of the same drug.

Moreover, these capsular devices transform the polymeric barriers of traditionally coated drug delivery systems into self-operating shells. Therefore, they represent a step forward in controlled release – separating the container (i.e. the capsule shell) from the content (i.e. the drug formulations contained in each compartment). The release performance of a compartment is determined by the composition and design features (i.e. morphology and thickness) of its walls. By varying the composition and thickness of each compartment, its release profile can be modified without affecting the properties of the other compartments. In other words, these multi-compartment capsules products yield multiple release kinetics in a single unit.

Multi-compartment capsular devices are especially advantageous for therapies involving multiple drugs that are mutually incompatible or interact in the gastrointestinal tract. Using multiple compartments simplifies the dosing schedule and has a positive impact on overall patient compliance. Moreover, these capsular devices enable the independent development of the drug formulation and the capsule shell, reducing time-to-market and development cost for new drug delivery systems.

The talk will present an example of this new class of capsular devices. The system consists of a two-compartment capsule shell composed of three parts: two hollow halves and a middle part, acting both as a joint and a partition. The joint allows the hollow parts to be assembled into a closed capsular device while dividing the internal cavity into two separate compartments. The hollow parts can have different shapes and thicknesses, resulting in compartments with different release properties and internal volume.

As a proof of concept, we will discuss the fabrication of capsule shells composed by 400 µm-thick and 800 µm-thick compartments using hydroxyl propyl cellulose (Klucel™ LF). We selected 3D printing by fused deposition modeling (FDM) as a suitable manufacturing technique for small-scale batches. This required the development of an industrial-grade FDM system able to realize pharmaceutical products. The talk will discuss in detail the development steps required to manufacture both the filament (through hot melt extrusion) and the capsule shells (through FDM 3D printing).

The FDM process requires as input material thermoplastic polymers supplied in the form of filaments – with suitable diameter and mechanical characteristics. We designed custom extrusion and 3D printing systems, and validated them to fulfill strict pharmaceutical quality requirements. In both cases, the study of the process was aimed at the identification of the critical operating parameters that affect the quality attributes of the final product. We assessed the stability of hydroxyl propyl cellulose after hot-processing and 3D printing, ruling out contamination and the formation of any hazardous degradation product. This led to the formation of an internal protocol for evaluation of by-products. These results demonstrate that the filaments and prototype capsule parts meet the target specifications – in terms of composition, absence of microbiological contamination, and absence of degradation by-products.

Once filled and assembled, the two-compartment capsular devices showed the expected two-pulse release behavior, in agreement with the swellable/erodible nature of the polymer. Furthermore, we compared the dimensional and release characteristics of multi-compartment capsules prepared by fused deposition modeling with those of analogous devices obtained by injection molding (IM). The real-time prototyping capabilities of 3D printing can significantly speed up the R&D process for IM. Both represent viable manufacturing techniques. 3D printing enables the rapid and inexpensive realization of small on-demand batches. Then, once the prototype capsules have been tested and refined, IM allows the large-scale production of capsule shells with the desired behavior.

After the capsule shells are produced, a proprietary robotic system fills each compartment with specific drug formulations. This part of the process relies on a set of automated filling modules working simultaneously, in parallel. The flexibility of this configuration is particularly suitable for the efficient production of small batches of fixed-dose combination products. The automated deposition stage allows the rapid and accurate adjustment of the combination and dosage strength of the active ingredients. Cross-contamination is avoided by enclosing each filling station in a dedicated clean room. A robotic arm (designed to safely collaborate with human operators) moves the capsular devices form one filling station to the next. The capsule filling systems were validated with preliminarily filling trials, which used powders with different particle size distribution and bulk density. We considered 25 batches of 50 capsule each for each powder formulation. In all cases, the amount of active ingredient deposited by the robots had an average coefficient of weight variation (i.e. the ratio between the standard deviation and the average value of the weight of the deposited powder) below 3%. These results demonstrate the accuracy and the flexibility of the powder deposition systems.

This robotic capsule filling system enables the realization of batches that are much smaller than the ones typical of traditional large-scale production processes. This level of personalization requires innovative in-process, non-destructive controls in order to guarantee the quality of the products in a cost-effective manner. We developed quality control machines that are able to check the thickness and weight of the capsule parts. Additionally, we use spectroscopic techniques to check the weight uniformity and identity of the pharmaceutical formulations.

566475 Pharmaceutical Application of Inkjet Based 3D Printing Process

Pharmaceutical Application of Inkjet Based 3D Printing Process

Koyel Sen [1], Sameera Sansare [1], Si Wan Li[2,3], Bodhi Chaudhuri [1,2,3], Anson Ma [2,3]

[1]Department of Pharmaceutical Sciences, University of Connecticut [2]Department of Chemical and Biomolecular Engineering, University of Connecticut [3]Institute of Material Sciences, University of Connecticut

Correspondence: koyel.sen@uconn.edu

Background:

There has been a growing interest in using 3D printing (3DP) for drug tablet manufacturing. This is motivated mainly by 3DP's potential in achieving different dosage geometries, attaining complex drug release kinetics, and combining multiple drugs efficiently for personalized medicine applications. The aims of this study are to: (i) experimentally evaluate an inkjet-based 3D printing method for manufacturing of tablets, (ii) understand the underlying physics, and (iii) to compare against tablets produced by the conventional tableting process.

Methods:

In an inkjet-based 3DP process, a liquid binder is printed onto a layer of excipient powder to bind the powder in a pre-determined pattern. Then, a fresh layer of powder is then spread evenly on top, followed by another round of binder printing. The process is repeated until the construction of the 3D object, or drug tablet in this case, is finished. The surface tension, viscosity, and density of the binder solutions were measured to predict their jettability. This was achieved using dimensionless groups that compare the relative importance of inertia, surface forces, and viscous forces. Two different types of powders and three different types of binder solutions were studied. Successfully printed tablets were further characterized for their weight variation, hardness, disintegration, true density, crystallinity, and physical stability.

Results and Conclusions:

Inkjet-based 3DP was performed using the selected pharmaceutical excipients and binder solutions. The final attributes of the printed tablets were clearly linked to the physical properties of the powder and binder solutions. Such physical understanding was used to create a "design space" in terms of material selection for successful 3D printing of drug tablets. This research lays the foundation for improving the precision and accuracy of the 3DP process for personalized medicine and controllable drug release applications, which will in turn improve patient compliance, lower drug price, and fulfill public health mission.

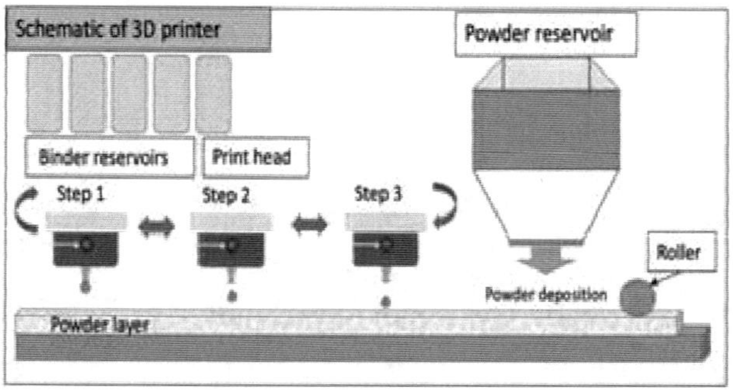

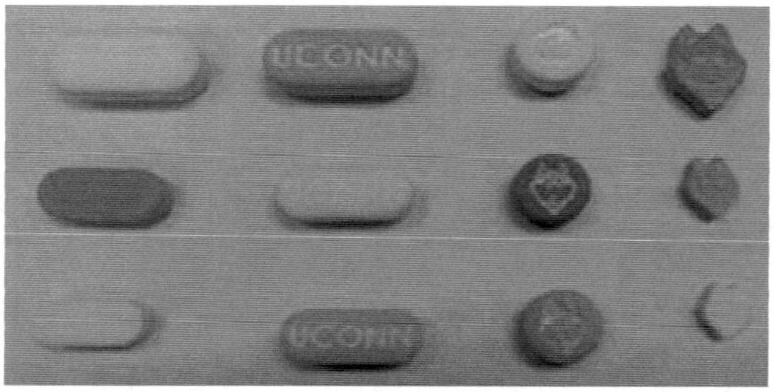

Figure 1- a) Schematic diagram of the inkjet based 3D printer used for the study (left) b) Non-pharmaceutical grade 3D printed tablets using inkjet based printing method

570318 3D Printed Tablets Using Binder Jetting and Pharmaceutical Excipients

Binder jetting is an additive manufacturing (AM) technology used to create three-dimensional model constructs. The materials used in binder jetting are relatively limited generally due to insufficient knowledge surrounding powder technology. The ability to self-formulate powder materials for use in binder jetting machines is not common practice today as not enough rigorous knowledge exists for the requirements of a 3D printable powder.

In this study we characterised several commercial 3D printable powders used in a 3DS Projet460 binder jetting printer, using techniques central to powder flow and powder wetting. The goal was to determine the key powder parameters that are likely to result in a successful binder jetting process, and then to use these parameters to screen standard pharmaceutical excipients for their likelihood to be suitable for 3D printing in a standard commercial printer.

Five different commercial binder jetting powders of various generations from different manufacturers were used as the benchmarks for a binder-jettable powder. Particle size and shape was measured using a Camsizer X2 and the powder flow (bulk and dynamic) was characterised using a Freeman FT4 powder rheometer. All five commercial powders had similar d_{10} and d_{50} values and similar shape distribution profiles. The FT4 showed that all powders fell within a narrow range for the parameters comprising basic flowable energy, specific energy, flow rate index and compressibility.

Drop penetration tests were used to determine the likely construct quality prior to the powder being 3D printed. The dried nuclei from these tests were analysed for spread, depth and hardness, which correlates to the dimensional accuracy of the powder after 3D printing. Drop penetration time seemed to have a wide range of acceptable values, but nuclei construct strength was important. A preliminary framework highlighting the quantitative parameters for binder jetting powders was created.

Finally, to test the framework approach, a simple formulation of 20% VA64 Kollidon and 80% Microcrystalline cellulose was used to successfully 3D print a range of basic tablet shapes in a Projet460 binder jet printer. The use of these commercial powders and characterisation techniques ultimately provide a useful benchmark for determining the suitability of standard pharmaceutical powder APIs and excipients to achieve good quality printed constructs, and sets out the considerations for specially engineered pharmaceutical particles and formulations that would need to be developed in future.

575111 Potential of 3D Printing in Precision Combination Therapy: A Case Study of Cardiovascular Polypills

3D printing is one of the emerging manufacturing technologies that is well suited to produce precision (personalized) medicine. The unique versatility of 3D printing to engineer desired designs with specific geometry allows for achieving targeted release profiles of printed drugs. It has a high potential for individualization of drug therapy combination through the so called "Polypill printlets" by which all the drugs required for the therapy can be printed into a single dosage unit. In this regard, this presentation will highlight a case study of engineering and manufacturing a cardiovascular polypill of Metoprolol (MT) and Atorvastatin (AT) using 3D pneumatic assisted micro-extrusion (PAM) technique. The polypills were designed to provide extended release properties of MT and immediate release properties of AT. The effect of printing, geometrical and formulation parameters on breaking strength and disintegration of the polypills and release properties of the drugs will be discussed. Various designs of the polypills will be demonstrated to modulate the release rate of the drugs from the polypills. The design spaces for formulating these polypills will be discussed in terms of identifying the critical material attributes and key geometrical parameters that achieve independent release properties of both drugs from the polypills. In addition, the regulatory considerations and failure modes of the polypill printlets will be also shared.

DISCLAIMER

This abstract reflects the views of the presenter and should not be construed to represent FDA's views or policies.

560253 Bringing Particle Scale Properties into Descriptions of Energetic Powder Behavior Via the Enhanced Centrifuge Technique

Energetic materials powder technology plays a significant role in defense and security applications[1,2]. Most solid explosives and propellants are processed at some point in the form of powders. The behavior of such powders is impacted by the behavior of the individual particles which comprise them[3], especially their adhesion. While computational models have been developed to predict the behavior of energetic materials and propellants[4], these cannot accurately account for the effects of particle-scale properties on the behavior of the energetic powder as a whole. In turn, the ability to model and optimize the manufacturing and handling of energetic materials is limited.

This work focuses on developing an experimental and modeling framework that maps particle-scale properties onto experimentally-validated 'effective' Hamaker constant distributions that describe van der Waals adhesion forces between particles in powders of energetic material. These distributions represent an engineering approach that allows powders comprised of particles of complex shape and roughness, which are challenging to model, to be described as if they were perfect, smooth spheres, which are comparatively simple to model. The complexity associated with the shape and size distributions of the individual particles is captured by the effective Hamaker constants[5]. This prior work is extended here to measure the adhesion force distribution of explosive powders against different surfaces and to characterize the explosive powder behavior. The resulting size-dependent 'effective' Hamaker constant distributions guide the selection of binders, surfaces, and excipient materials used in explosives and propellants.

References

(1) Yu, H. A.; Becker, T.; Nic Daeid, N.; Lewis, S. W. Fundamental Studies of the Adhesion of Explosives to Textile and Non-Textile Surfaces. *Forensic Sci. Int.* **2017.**

(2) Chaffee-Cipich, M. N.; Sturtevant, B. D.; Beaudoin, S. P. Adhesion of Explosives. *Anal. Chem.* **2013**.

(3) Litster, Jim Ennis, Bryan Liu, L. *The Science and Engineering of Granulation Processes*. **2004**.

(4) Walters, D. J.; Luscher, D. J.; Yeager, J. D.; Patterson, B. M. Cohesive Finite Element Modeling of the Delamination of HTPB Binder and HMX Crystals under Tensile Loading. *Int. J. Mech. Sci.* **2018**.

(5) Thomas, M. C.; Beaudoin, S. P. An Enhanced Centrifuge-Based Approach to Powder Characterization: Experimental and Theoretical Determination of a Size-Dependent Effective Hamaker Constant Distribution. *Powder Technol.* **2017**.

NUMERICAL SIMULATION OF MUNITION RESPONSE FOR FRAGMENT IMPACT

Ruslan S. Mudryy

U.S. ARMY Combat Capabilities Development Command - Armaments Center, Picatinny, NJ

Abstract

The primary purpose of full-scale Insensitive Munitions (IM) testing is to establish the response of a munition to a variety of stimuli under specified conditions. This information can be used to determine compliance with the national IM policy, which is a strict requirement in many NATO countries. This ensures improved weapon safety and decreased risk for the occurrence of unplanned hazardous events. Numerous standardization agreements (STANAGs) [1] outline the various Insensitive Munitions (IM) threats simulating the hazards that munitions are commonly exposed to on the modern battlefield. These threats include: sympathetic detonation, shaped charge jet impact, bullet and fragment impact, and slow and fast cook-off. Advances in the state-of-the-art physics-based models have created an opportunity for the weapon development community to use these tools effectively to screen design candidates and to predict full scale test outcome.

The overall modeling and simulation capability addresses a wide range of physics phenomena that relate to the energetic material, its behavior when exposed to threat stimuli, and ultimately its final response that determines the outcome of an IM event. The objective of this work is to develop a numerical model using the Arbitrary Lagrange/Eulerian hydro code (ALE3D) that accurately simulates a fragment impact event for tested projectile. The model will predict the shock pressure that the round undergoes due to the impact insult. Various fragment velocities were investigated in order to estimate at which velocity the shock pressure is above the threshold value for detonation obtained from the LSGT (Large Scale Gap Test). This result will be classified as a "fail" and it will be assumed the projectile will not survive the shock from fragment stimuli that will mark the threshold value of fragment velocity. The effect of aim point variation on the shock pressures transmitted to the explosive has also been analyzed computationally. The planar impacts at two different aim points were modeled, one at the main fill and other at the booster area.

Introduction

Safety related hazards fall into two distinct categories from a physics appearance. The first is mechanical insult from kinetic energy deposition on a weapon that results from a bullet, fragment or shaped charge jet impacts offer an extensive spectrum of conditions for an analysis of weapon response. The second is a thermal insult that transfers the heat within a weapon and has the potential for numerous scenarios that encompass slow heating (slow cook-off), fast heating (fast cook-off) and heating rates that fall in between.

In recent years, modeling and simulation technology is more widely used in predicting explosive detonation behavior in numerous tests outlined by standardization agreements (STANAGs), which simulate the mechanical [1] and thermal insults. Advanced hydro codes, based primarily upon deterministic continuum mechanics methods, have been used by thousands of munitions developers to study and improve the performance of systems undergoing a variety of hydrodynamic events [2]. These types of scenarios require solution of the conservation equations, mass, momentum, and energy, and that some type of constitutive and failure models exist. The overall M&S capability addresses a wide range of physics phenomena that relate to the energetic material, its behavior when exposed to threat stimuli and ultimately its final response that determines the outcome of an IM event. Recent advances in the computational models have also created an opportunity for the weapon development community to use these tools effectively to screen design candidates and to predict lab scale or full scale test outcomes.

Two of the codes most commonly used by DOD researchers are CTH code produced by Sandia National Labs (SNL) and ALE3D code produced by Lawrence Livermore National Labs (LLNL). CTH is explicit Eulerian finite-volume formulation hydrocode, where the solution scheme uses the mesh fixed in space and the material flows through the mesh. It has models appropriate for strong shock wave, high explosive detonation and large deformation calculations. CTH has many attractive features such as adaptive mesh refinement and uses second-order accurate numerical methods to reduce dispersion and dissipation to produce accurate results efficiently. However, it is an explicit hydrocode, so it cannot efficiently solve problems that must be solved over a much longer time frame. ALE3D is the most popular code among DOD participants. It is very robust, efficient, and constantly being updated. ALE3D incorporates a variety of multi-physics and multi-time capabilities and uses arbitrary Lagrangian-Eulerian (ALE) techniques. The ALE and mesh relaxation capability broadens the scope of applications in comparison to tools restricted to Lagrangian or Eulerian (advection) only approaches, while maintaining accuracy and efficiency for large, multi- physics and complex geometry simulations. As a result, ALE3D hydrocode was employed in numerous DOD projects to simulate the behavior of energetic material when exposed to threat stimuli and ultimately its final response. Several successful numerical studies have been conducted using ALE3D to model thermal, mechanical, and chemical behavior of heating, ignition, and explosion in cookoff tests [3-4]. Additional work has been done to model bullet or fragment impact against confined energetics. Simulations of impact events involves large deformations and fracture, mesh elements undergo extreme distortion; and in these situations the advection capabilities of ALE3D are very helpful to recover from element deformation [5]. Significant work has been done to develop fragment impact barrier concept for different munition systems in order to mitigate disruptive threat energy. It was demonstrated experimentally that computationally modeled barriers can significantly reduce the violence associated with fragment impact scenarios [6-7]. Fragment velocity variation and projectile tilt upon impact have been investigated recently using ALE3D simulations and successfully validated by experimental data [8].

The objective of the present work is to develop an integrated ALE3D computational model to analyze fragment impact penetration of the M72 LAW system. To simplify the geometry and meshing only impact on the warhead is considered. Calculations will predict the maximum pressure transferred to the explosive during the impact event. And pass or fail fragment impact sensitivity test can be judged based on comparison of calculated maximum pressure with experimental data from LSGT (Large Scale Gap Test). As another practical application of the developed numerical model, the use of liners placed around the warhead body were investigated to help reduce violent response from fragment impact. The effect of aim point variation on the shock pressures transmitted to the explosive has also been analyzed computationally.

Numerical Model

A meshed domain was created for this problem, which consists of fragment surrounded by the air and a warhead body accepting the impact. The fragment material, weight, diameter, and velocity at the impact are specified in NATO STANAG4496 procedure for fragment impact testing. A schematic diagram of a simulated domain is presented in Figure 1. The warhead main fill is PBXN-9 explosive material, and the booster is filled with PBXN-5. These energetic materials are widely used and well characterized so it makes a model development process less complicated. It is a common practice in fragment impact modeling to monitor shock pressures in other specific locations – at explosive critical diameter pathway and at fragment diameter pathway (Fig.1). Predicting the reaction of an explosive fill based on reaching the LSGT threshold means there is a 50% chance that the given stimuli will result in a transition to a detonation. There are a number of other factors that can affect the reaction of the explosive to shock stimuli. Each explosive fill has a critical diameter below at which a detonation wave cannot be sustained even if an overdriving shock is applied. Therefore, knowing the shock pressure history at the explosive critical diameter locations is important for better prediction of the reaction outcome. And another crucial location is fragment diameter. Due to the shear effects on the fragment edges shock pressure can be increased at these locations sometimes. ALE3D axisymmetric model has been established with sufficiently high mesh density in order to capture all the warhead features. Maximum pressure in the explosive versus time will be calculated. Also pressure–time histories will be collected along pathway of the flying projectile.

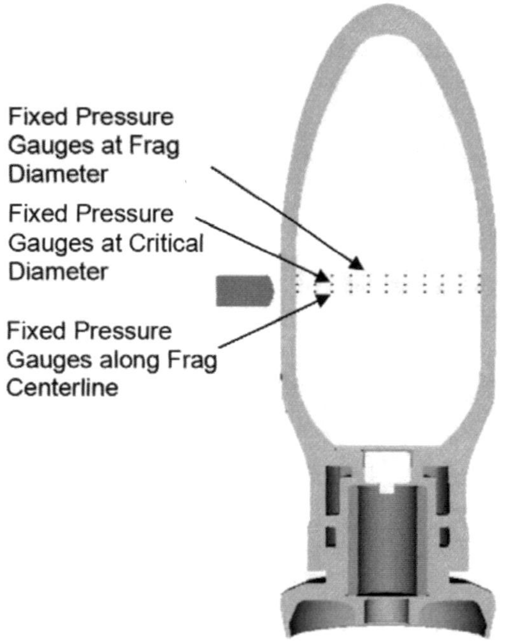

Figure 1. Material plot of fragment impact event in the main fill.

Pressure gauges will be fixed along fragment diameter locations, along fragment centerline locations, and along explosive critical diameter locations. The volumetric response for all materials was modeled using the Mie Gruneisen equation of state. The strength was modeled using the Johnson-Cook constitutive model for parts made of steel and the Steinberg-Guinan constitutive model for parts made of aluminum. The fragment was modeled using mild carbon steel properties and the target warhead body was modeled as 7075 aluminum. The strength and equation of state parameters for these materials were taken from the ALE3D Material Properties Database (MATROP) [9]. The planar impacts at two different aim points were modeled, one at the main fill and other at the booster area. Corresponding explosives were modeled as inerts during the impact and after. The PBXN-9 and PBXN-5 are similar explosives, so Hugoniot data available only for PBXN-9 was used for both of them [10]. And for modeling the strength of explosives, the materials were treated as isotropic elastic perfectly plastic. The model was set up with a symmetry boundary condition along the axis of the projectile. All other boundaries were modeled as flow out boundaries.

Results and Discussion

Predictions by the model shock pressure will be compared with the threshold pressure for shock to detonation transition obtained by the NOL Large Scale Gap Test (LSGT) for PBXN-9 material. Results reported by NSWC Indian Head indicate a 50 % threshold of detonation of 179 cards which corresponds to 26.7 kBars [11]. The explosive was pressed to the density of 1.743 g/cc so the same density was used for our computations. Note that the analysis will only predict whether or not the threshold will be reached. No prediction for the type of the reaction will be given as once the threshold is reached the explosive could react anywhere from a slow burn to a full detonation. Figure 2 presents ALE3D output for a fragment impacting the main fill of the warhead at 5, 28, 50, and 72 microseconds. Simulation analysis predicts that the fragment will penetrate through the entire warhead, and it will take about 72 microseconds. The run also predicts fragment breakup upon steel impact. Shock pressure distribution inside the warhead is shown on the set of snapshots in Fig. 3. For a more distinctive view a logarithmic scale was selected. Maximum pressure was observed at the moment of impact and then the shock wave propagates ahead of the fragment and sideways, with some reflections and decays.

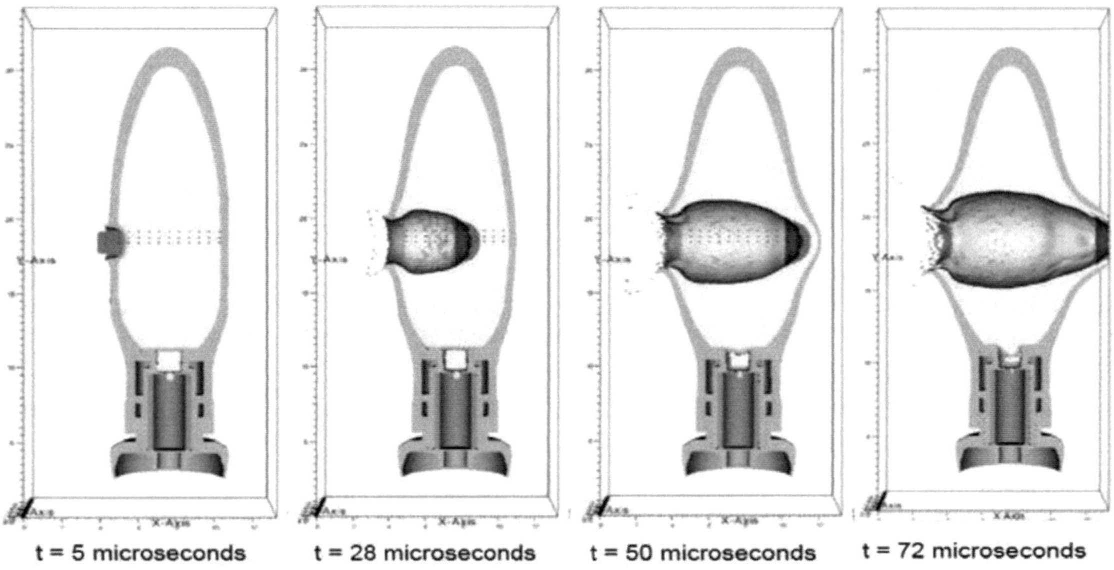

Figure 2. Material plot of fragment impact event in the main fill.

Tracers at the locations specified earlier collected pressure histories and the results are shown in Figures 4, 5 and 6. The highest pressure was predicted and recorded by the first tracer (#0) placed in the main fill along fragment centerline just behind the warhead body. It's equal approximately 160 kbars as the initial shock pressure, with some additional reflections as the pressure decays. Obviously, the fragment is losing kinetic energy penetrating the warhead further, so the next tracers recorded lower pressure values. However, the first five tracers (#0, #1, #2, #3, #4, #5) still show higher shock pressure values when compared to the initiation pressure from LSGT test for the PBXN-9 material.

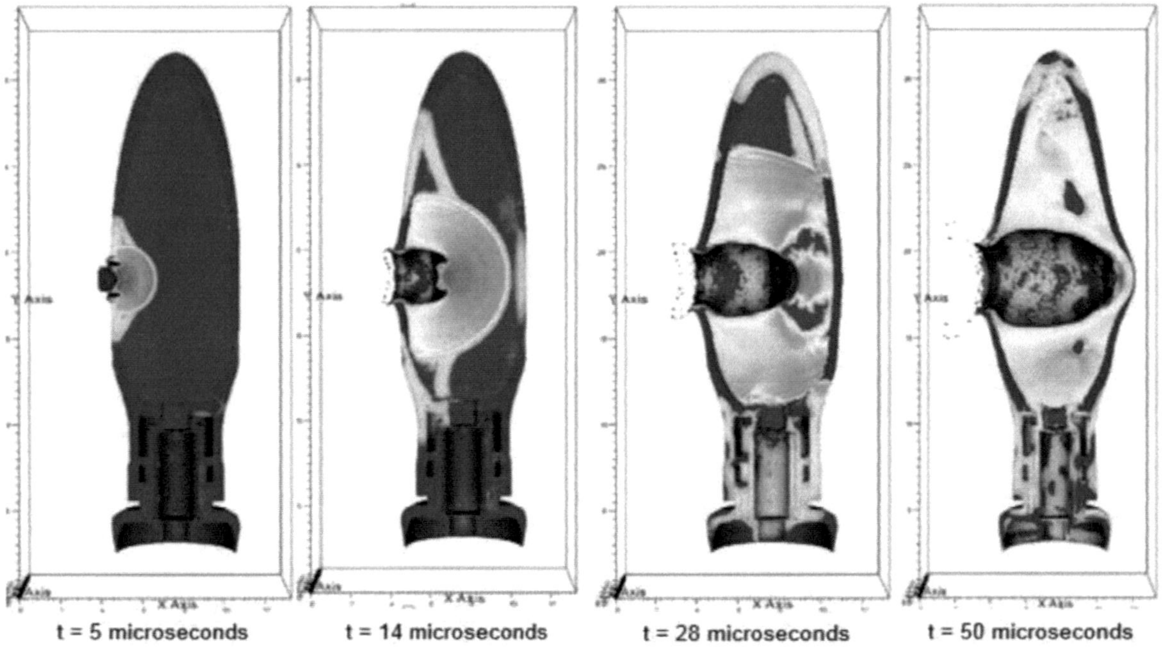

Figure 3. Shock pressure plot in a logarithmic scale.

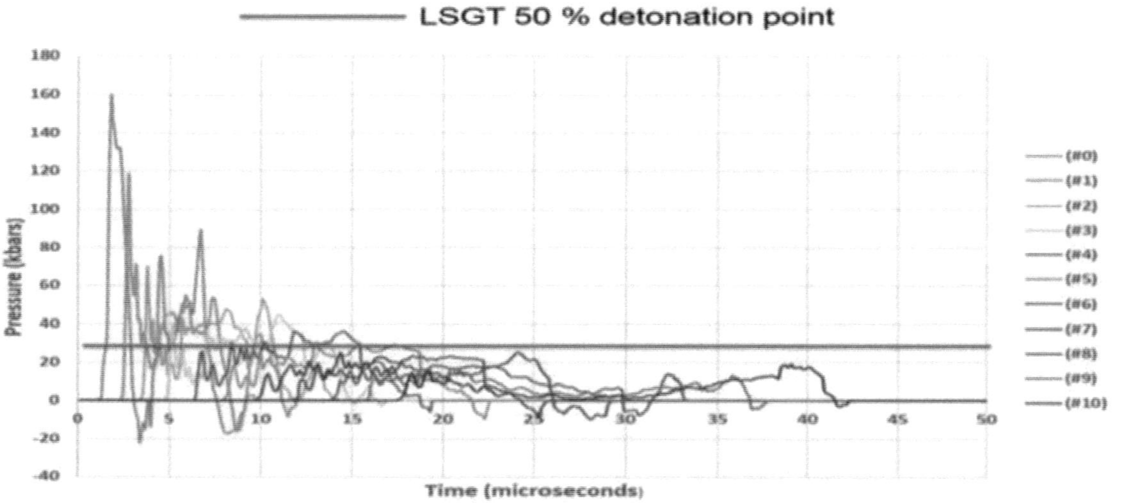

Fig. 4. Pressure-Time histories along fragment centerline in the main fill (inert PBXN-9).

Figures 5 and 6 demonstrate shock pressure time histories collected by the tracers located at other specific points - the explosive critical diameter and the fragment's diameter locations. The initial shock pressures are lower at these locations compared to the data collected from the fragment centerline. However, all of these tracers show that shock pressures are in excess of the minimum threshold for reaction obtained from LSGT test. This result is classified as a "fail", the warhead will not survive the shock from fragment stimuli according to the numerical predictions. In some apparent "pass"cases, where impact pressure is slightly below the LSGT threshold may be misleading. Due to reflections, the pressure may actually be higher at locations other than the tracers. In order to observe such an important scenario peak pressure in the entire region was collected as a function of time. A shapshot of the peak pressure in the entire region filled with the PBXN-9 is shown in Fig.7. The peak pressure is also significantly higher than the minimum theshold.

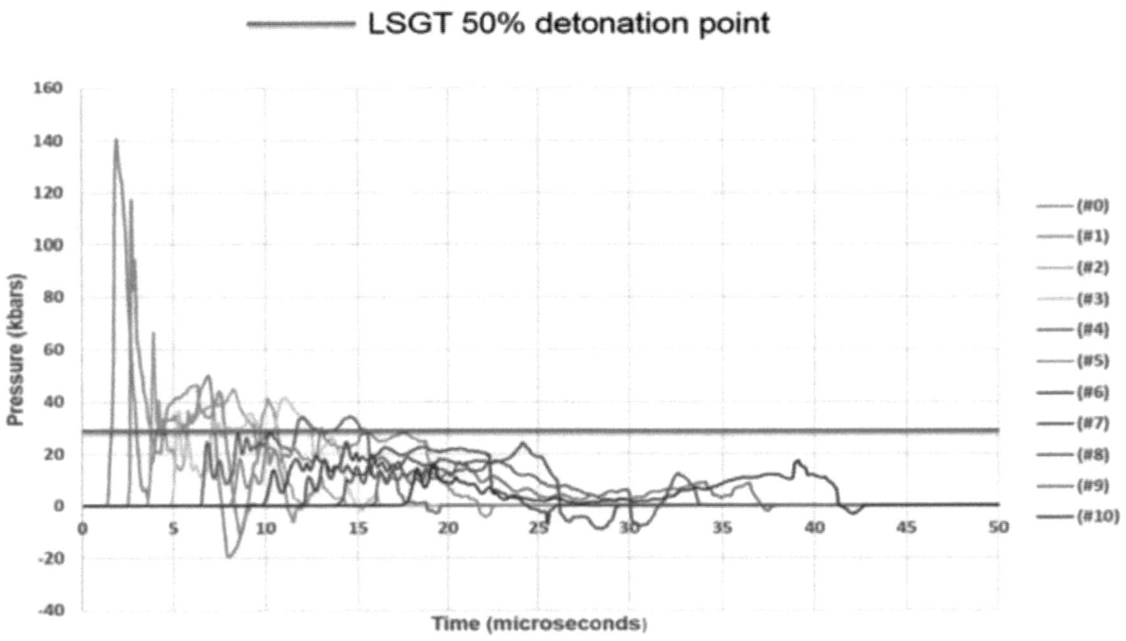

Fig. 5. Pressures-time histories along axis of critical diameter in the main fill (inert PBXN-9).

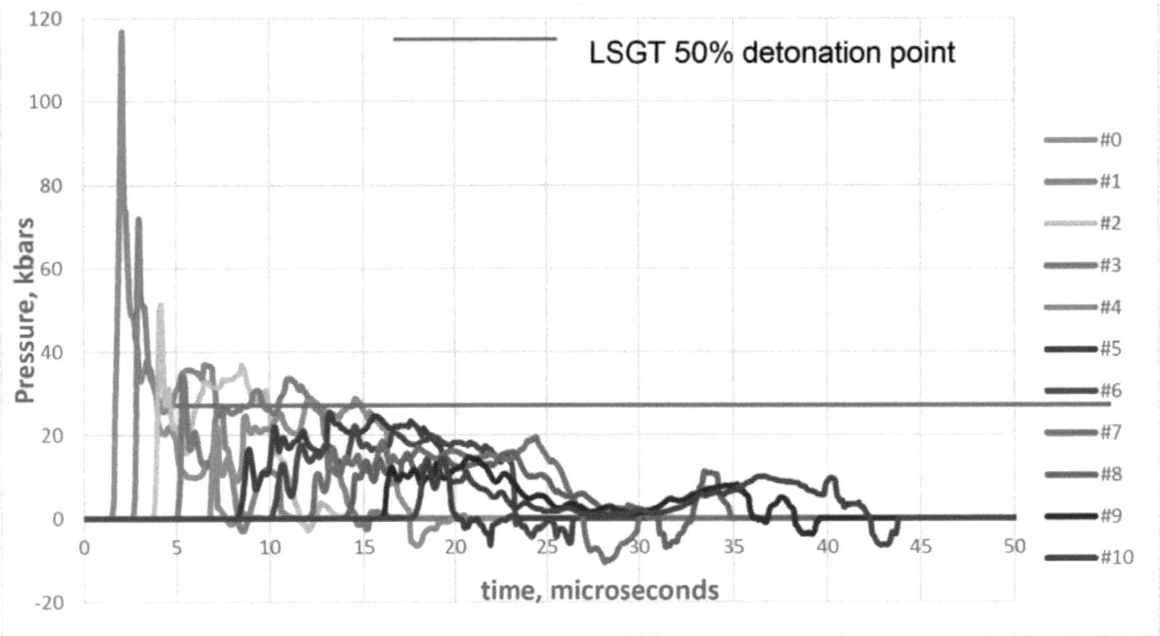

Fig. 6. Pressures-time histories along fragment diameter in the main fill (inert PBXN-9).

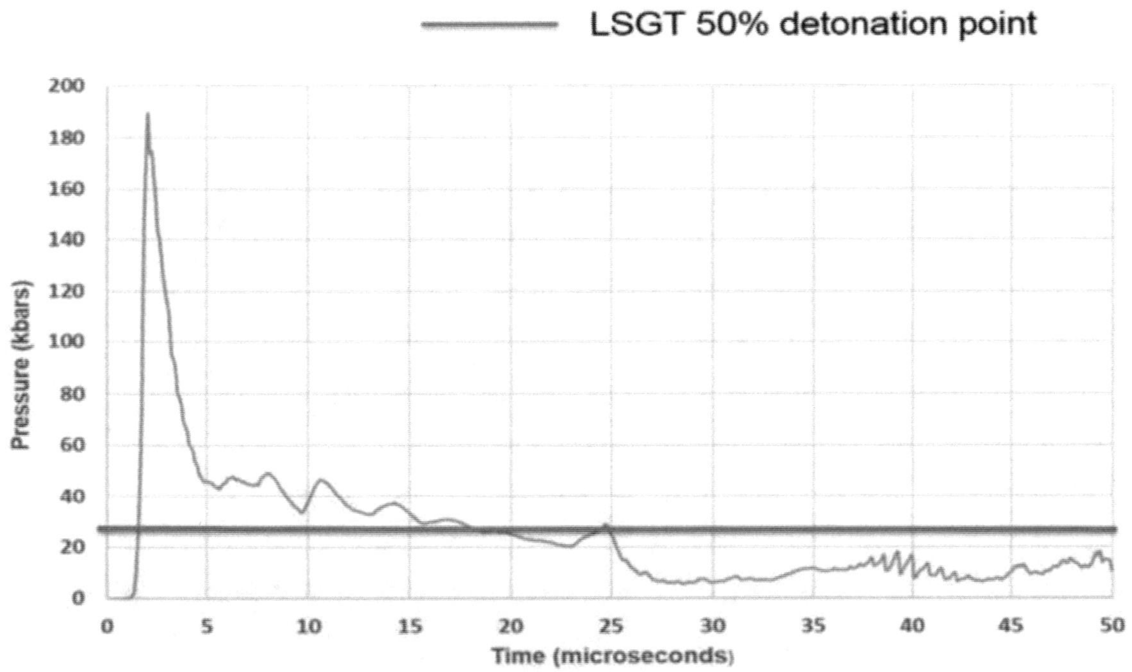

Fig. 7. Maximum Pressure in the inert PBXN-9 versus Time.

Impact on other shock sensitive booster regions was simulated. The booster is filled with the PBXN-5 material that has slighly different charcteristics. It is higher density material, and more sensitive compared to the PBXN-9. Experimental data collected from NOL LSGT test for PBXN-5 material shows that a 50% threshold for detonation requires 206 cards, which corresponds to 19.8 kBars of initiation pressure [11]. Numerical modeling and simulation analysis was conducted for a booster impact event. A snapshot of the analysis is presented in Fig.8. a,b. Due to the limition on space, only impacts at two different times are illustrated. The model accurately represents the fragment impact event, showing the dynamics of fragment penetration.It takes 51 microseconds for the fragment to penetrate through completely. Shock wave propagation inside the warhead is also illustrated at the right. Pressure-time histories at the specified locations were collected during the computational run for comparison with the a 50% threshold from NOL LSGT test for the PBXN-5 material. Numerical calculations showed that shock pressure excess is more than10 times the threshold initiation pressure. Based on this analysis, one can say that with given shock stimuli there is a high chance of explosive reaction.

Another particular interest of modeling application was to design a barrier between the fragment and the warhead body that will mitigate the impact. A new configuration of warhead body protected by one, two and three liners made of steel with 3mm thickness of each were studied numerically (Fig.9. a). The analysis showed that the fragment will still penetrate the warhead entirely, but it takes longer time (82 mcseconds). However, the liners mitigate the shock initially introduced by the fragment. Shock pressure distribution inside the warhead at different times is shown in Fig. 9b.

Compared to results of pressure–time histories, along fragment demonstrated shock pressure reduced with the usage of a single liner (Figures 10 and 11). Having two and then three liners further reduces the maximum pressure transferred to the explosive during the impact event. Although it is not low enough to meet the IM criteria, it provides understanding and identifies a direction for future warhead design. Thicker or stronger liners made of metallic alloys could be sufficient for preventing a violent reaction in the M72 LAW system.

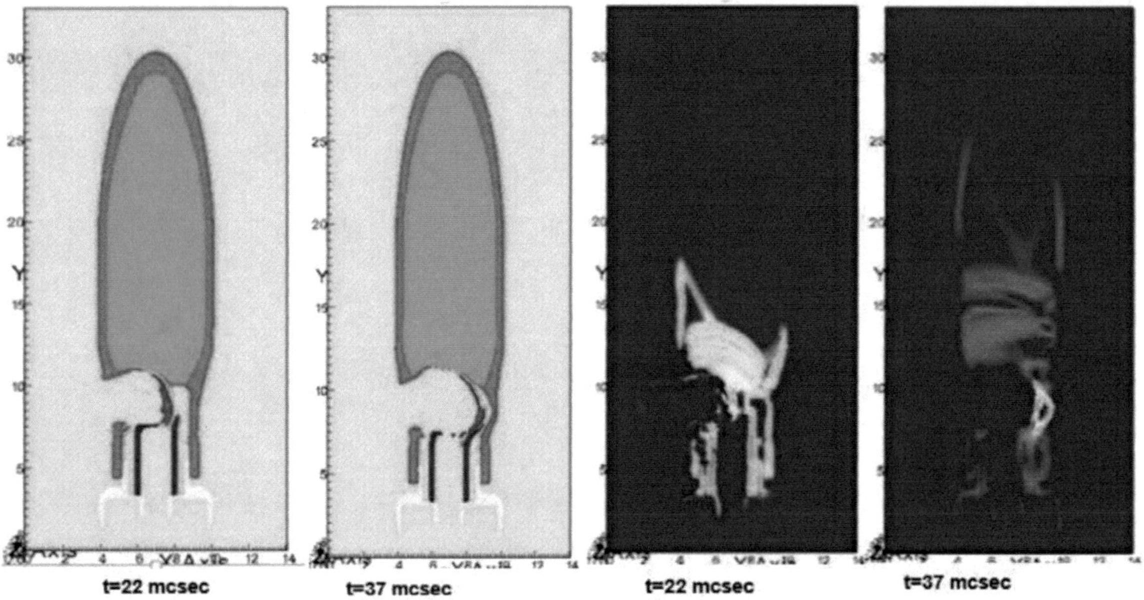

Figure 8. a. Material plot and b. Shock Pressure plot (logarithmic scale) for fragment impact in the booster.

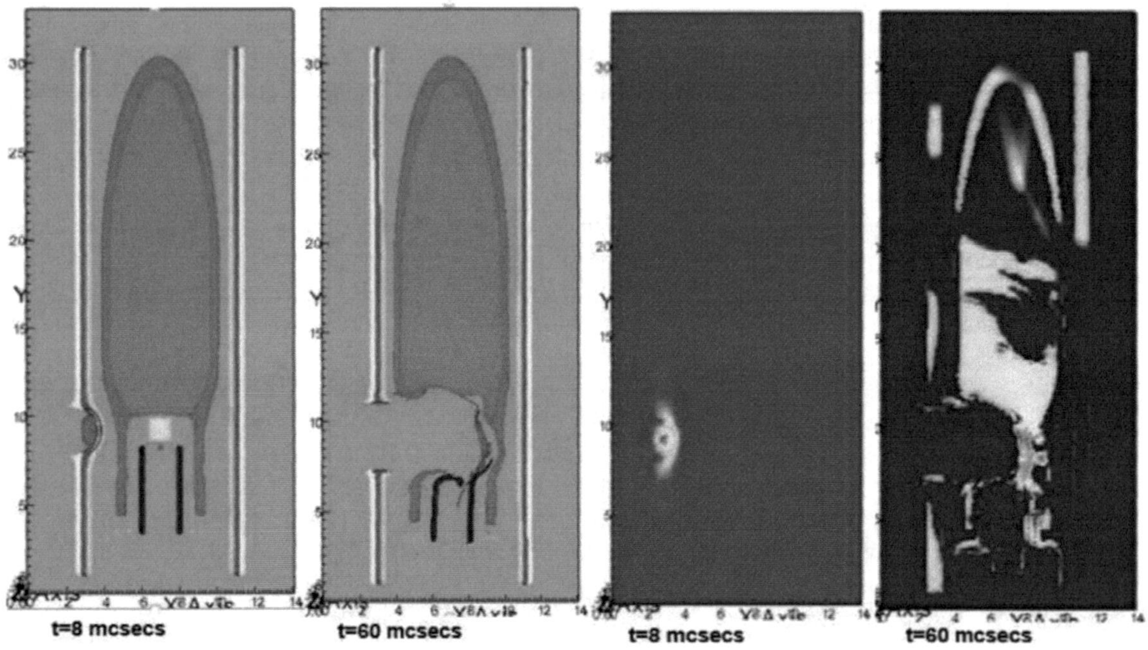

Figure 9. a. Material plot and b. Shock Pressure plot (logarithmic scale) for fragment impact in the booster. Three liners are used for mitigation effect.

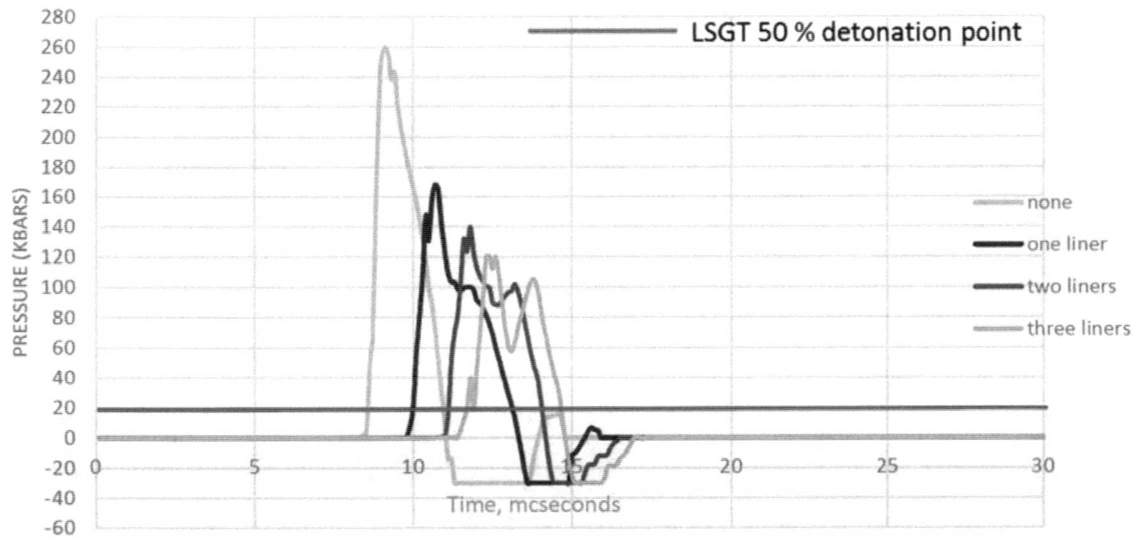

Figure 10. Pressure-Time histories along fragment centerline in the inert PBXN-5. Mitigation effect of various number of liners on shock pressure.

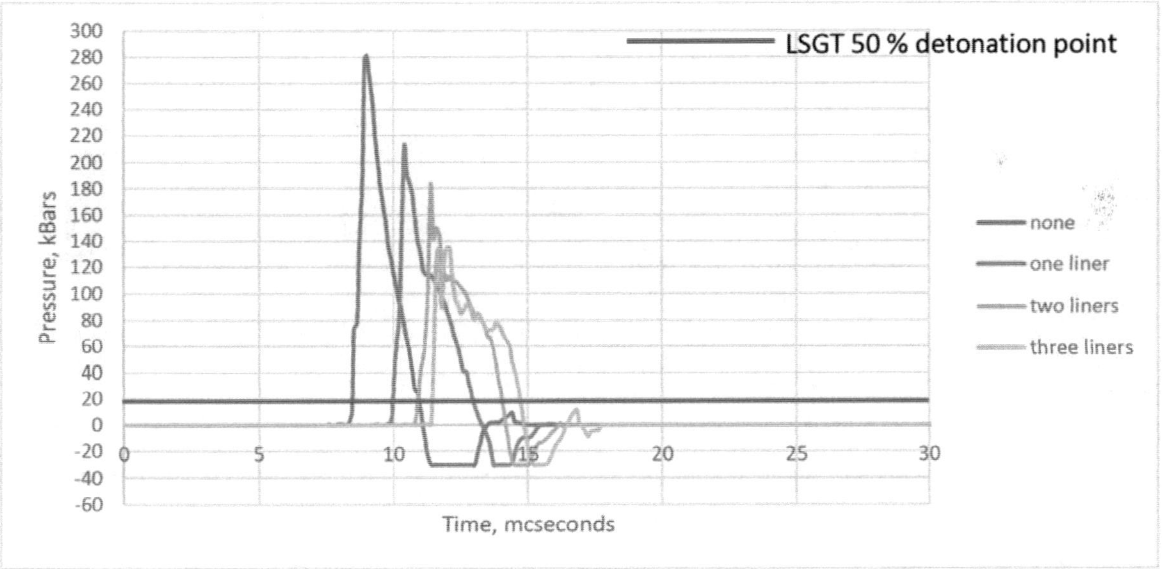

Figure 11. Pressure-Time histories along axis of PBXN-5 critical diameter. Mitigation effect of various number of liners on shock pressure.

Conclusions

A comprehensive numerical investigation was conducted to explore the fragment impact event for M72 LAW warhead. An advanced ALE3d model was developed to compute the shock pressures transmitted to the main fill (PBXN-9) and the booster fill (PBXN-5). Simulated results were compared with the NOL Large Scale Gap Test (LSGT) data to predict the IM warhead performance. It was demonstrated that the shock pressures are much higher than the initiation shock pressure for the corresponding explosive material. The usage of 3mm steel liners was investigated as a mitigating effect. Significant pressure reduction was observed, but the IM requirements were not met with the current design. The usage of thicker and/or stronger liners is recommended. The model can be used in the future as a tool for selection the optimum composition and thickness of the liner material and designing the protective system that will help to survive the impact.

References

1. Fragment Impact, Munitions Test Procedure, STANAG 4496 Ed. 1. NATO Standardization Agency, Brussels, Belgium, 2006.
2. Zukas, Jonas A., Introduction to Hydrocodes, Elsevier, New York, 2004.
3. McClelland M.A., Tran T. D., Cunningham B.J., Weese R.K., and Maienschein J.L., Cookoff Response of PBXN-109: Material Characterization and ALE3D Thermal Predictions, Report UCRL-JC-144009-REV-1, Lawrence Livermore National Laboratory, 2001.
4. Nichols, A. L. III, Couch, R., McCallen, R. C., Otero, I., and Sharp, R., "Modeling thermally driven energetic response of high explosives", Eleventh International Detonation Symposium, Office of Naval Research ONR 33300-5, Snowmass, CO, 1998, p. 862.
5. L.D. Leininger, S.A. Minkoff, R. J. Dorgan, S.E. DeFisher, H.K. Springer, R.C. McCallen, "Capability Improvements for Modeling Fragment Impact in ALE3D", 26th International Symposium on Ballistics, Miami, FL, 12-16 September, 2011.
6. A. Daniels, J. Pham, K. Ng, and D. Pfau, "Development of Particle Impact Mitigation Sleeves to Reduced IM Response", 2007 Insensitive Munitions & Energetic Materials Technical Symposium, Miami, FL, October 15-17, 2007.
7. M. Steinberg, J. Stanfield, "Fragment Impact Mitigation Through Shock-Induced Treat Damage", 2016 Insensitive Munitions & Energetic Materials Technology Symposium, Nashville, Tennessee, 12-15 September 2016.
8. E. Baker, N. Al-Shehab, K. Miers, and D. Pudlak, "Insensitive Munitions Fragment Impact Gun Testing Technology Challenges", Propellants, Explosives, Pyrotechnics, 41, p.572-579, 2016.
9. J.K. Durrenberger, R.C. Becker, D.M. Goto, J. R. Neely, B.K. Wallin, "MatProps: Material Properties Database and Associated Access Library", Report UCRL-TR-233699, Lawrence Livermore National Laboratory, August 15, 2007.
10. N.J.Sanchez, R.L.Gustavsen, D.E. Hooks,"Shock Initiation Behavior of PBXN-9 Determined by Gas Gun Experiments, Report LA-UR-09-04889, Los Alamos National Laboratory, Los Alamos, NM, USA, 2009.
11. L.J. Montesi, K. Burrows, M. Beyard, "Large Scale Gap Test (LSGT) Sensitivity of Explosives: Complication of Test Data, Vol. I, Report IHTR 2597, Indian Head Division, Naval Surface Warfare Center, 24 April 2006.

562777 Filled Spherical Composite Particles Prepared By Mechanical Milling

Spherical powders are deemed highly desirable owing to improved flowability and ease in their handling and processing in energetic, pharmaceutical and other formulations exploiting composite materials and structures. Spherical micro particles are most commonly prepared by cooling droplets of melts. This method however cannot be employed for materials susceptible to heating, such as most energetic materials, for which the structure may be compromised during the processing. Other existing techniques include extrusion, spheronization, and emulsion polymerization, applicable to a relatively narrow range of soft materials. Hollow particles, such as colloidosomes, can be prepared from Pickering emulsions. However, filling them to achieve high density is difficult and involves preparing complex double emulsions. In this study, we introduce a simple, versatile, cost effective and readily scalable technique of synthesizing densely filled spherical powders by mechanical milling. Powders of different materials, including elemental Al, B, and Ti, oxide Fe_2O_3, as well as reactive composites Al·CuO, Al·Fe_2O_3, and Al·B were ball milled in presence of two immiscible fluids (hexane and acetonitrile) serving as a liquid process control agent. Emulsification of acetonitrile takes place during milling. The emulsion is stabilized by particles adsorbing at the liquid interface, so that Pickering emulsion forms. Most of the solid remains suspended in the continuous phase, hexane. It is hypothesized that upon high shear and stress interaction between suspended particles and Pickering emulsion droplets caused by the milling media, particles from the suspension penetrate and get trapped inside the emulsion droplets. Eventually, densely filled spherical particles form upon going through repeated destabilization and repairing of the droplet/continuous fluid interfaces laden with particles. Effects of liquid to solid ratio and milling time have been explored in this study thus far. Upon recovery of the powders, the surface morphologies and particle sizes are characterized using scanning electron microscopy (SEM). The sizes of the forming spheres are observed to decrease with the increasing milling time. The study aims at elucidating the mechanism of spherical particle formation during high-energy milling. In future, effect of different milling parameters, such as milling media size and type, energy introduced during milling, and presence of surfactants on the particles shapes and density will be investigated. The applicability of this milling technique to preparing spherical powders of a wide range of materials involving different immiscible fluid systems will also be explored.

564558 Influence of Nanoconfinement on the Degradation of CL-20 and HMX

Nanoconfinement refers to a state in which materials are confined to regions below 100 nm in a single dimension, resulting in dramatic deviations from their typical bulk behavior. In particular, the melting point of materials under nanoconfinement are typically reduced. A common approach to investigate nanoconfinement is to load materials into a controlled pore glass (CPG), usually via melting. For example, the melting point depression of pentaerythritoletranitrate (PETN) and trinitrotoluene (TNT) was shown by McKenna and co-workers [1, 2] to decrease by ~22 and 27 ^{o}C, respectively, upon nanoconfinement within CPGs whose interior pore diameter was 8.1 nm. Similar results, although to a lesser extent, were observed for CPGs with larger pore diameters up to 70 nm. The goal of the current work is to perform a thermal analysis of hexanitrohexaazaisowurtzitane (CL-20) and cyclotetramethylenetetranitramine (HMX) upon nanoconfinement within CPGs. The samples were loaded into two different CPG nanoporous materials for the calorimetric study. Currently, the melting temperature of CL-20, even under nanoconfinement, remains elusive due to decomposition. We had initially hypothesized that nanoconfined CL-20 would yield a reduction in its melting point to a temperature before the onset of decomposition. However, nanoconfined CL-20 is found to decompose at lower temperatures than in the bulk material resulting in the onset temperature for decomposition decreasing with decreasing pore size. The results suggest an increase in CL-20's sensitivity upon nanoconfinement. Future works aim to characterize the degradation behavior of CL-20, HMX, and a mixture of CL-20 and HMX in nanoconfinement.

[1] B. Xu, X. Di and G.B. McKenna, "Melting of pentaerythritol tetranitrate (PETN) nanoconfined in controlled pore glasses (CPG)," J. Thermal Analysis and Calorimetry, **113**, 539-543 (2013).

[2] X. Di, B. Xu and G.B. McKenna, "The melting behavior of trinitrotoluene nanoconfined in controlled pore glasses," J. Thermal Analysis and Calorimetry, **113**, 533-537 (2013).

565693 Influence of Catalyst on the Mechanochemical Nitration of Toluene

Recently, it was shown that toluene can be nitrated mechanochemically employing a solvent-less process with a metal oxide catalyst and sodium nitrate as the nitronium source. This nitration method is attractive because it alleviates environmental concerns associated with the currently used techniques involving, most commonly, fuming nitric acid combined with sulphuric acid. Mechanochemical nitration is also scalable and the approach can be extended to nitrate other organic compounds. This effort is aimed to compare different metal oxide catalysts and clarify the mechanism of mechanochemical nitration of toluene. Metal oxides with different Lewis and Bronsted acidities, densities and oxidation states were tested as in experiments. The oxides considered included MoO_3, WO_3, V_2O_5, TiO_2, Bi_2O_3, MnO_2, Fe_2O_3, γ-Al_2O_3, and zeolites HZSM-5(Si/Al-39) and HZSM-5(Si/Al-350). Experiments were performed in a planetary mill. Detectable nitration of toluene occurred only with MoO_3, WO_3, and V_2O_5. For these oxides, the specific surface areas were measured after milling runs using BET method. The results suggest correlations between Bronsted acidity, surface area and yield of nitro-products. The experiments with varied milling times, altered concentration of H-bonds on the catalyst surfaces, and FTIR characterization of the evolving catalyst surfaces will also be presented and discussed.

566159 Ignition of Reactive Material Powders in Vicinity of an Electrostatic Discharge

Electrostatic discharge (ESD) is a common ignition stimulus for powders of reactive and energetic materials. It is a common safety hazard when handling such materials; it can also be used to initiate pyrotechnic formulations. In laboratory experiments, ESD is produced by discharging a high-voltage capacitor via a gap between two electrodes. Produced electric spark typically lasts for several microseconds. In all reported studies, the powder being tested is placed in a holder, which serves as one of the ESD electrodes. It was shown that the ignition is caused by Joule heating, with the main heat release occurring at the interfaces between powder particles through which the current flows. However, ignition may occur in situations when the ESD does not pass directly through the powder. Indeed, ESD generates a shock wave, which can lift the powder from a nearby substrate, even if this substrate was not an electrode. ESD also generates an expanding kernel of plasma heated to ca. 10,000 K, which can exist much longer than the ESD current between electrodes. The effect of such a shock and plasma on a reactive material powder remains unexplored. Such effects are important for developing appropriate safety protocols for handling reactive materials. They can also be of interest for understanding the ignition mechanisms for different materials, which can be exposed to shock and plasma. In this study, a reactive material powder is placed on a substrate located in vicinity of an ESD initiated between two pin electrodes. The electric current and voltage of ESD are monitored using induction coils. The pressure generated by the ESD is recorded using a piezoelectric transducer. Experiments are performed with both inert and reactive material powders. For inert powders, the space above the substrate is illuminated using a laser sheet so that the particles lifted in air are visible using a high-speed video camera. For reactive powders, emission from the ignited particles is sufficient to observe them in similar high-speed videos. The ESD energy and distance from ESD electrode gap and powder are varied. Ignition is observed to occur for a range of distances between the ESD gap and the substrate holding the powders. Experiments are performed with powders of flammable metals, Magnesium and Zirconium, as well as with powders of nanocomposite thermites prepared by Arrested Reactive Milling. The results are interpreted considering the size and lifetime of the plasma kernel generated by ESD and also visualized by the high-speed video. A correlation between ignition of powders exposed to ESD plasma and shock and their thermally initiated ignition observed in different experiments is considered.

557599 Particle ALD of Tungsten Nitride As a Hydrogen Environmental Barrier Coating for Nuclear Space Propulsion

Nuclear thermal propulsion is an attractive in-space propulsion technology due to its large specific impulse about twice that of a chemical combustion engine. This specific impulse is achieved by using energy released from uranium fission to heat hydrogen gas to over 2400°C before expelling it out a supersonic nozzle. However, hydrogen diffuses into the nuclear fuel elements and embrittles them, hindering the use of nuclear thermal engines. Here, we investigate the use of nanoscale tungsten nitride (WN) films as hydrogen environmental barrier coatings to address this issue.
Ultrathin WN films were deposited by atomic layer deposition (ALD) on zirconia nanoparticles and yttria stabilized zirconia (YSZ) micropowders using a fluidized bed reactor. The as-deposited film contained tungsten (W) in a variety of oxidation states, but was shown to dissociate nitrogen and form a crystallized W film upon annealing. The film's intended use as a hydrogen barrier coating motivated the investigation of hydrogen diffusion in W using density functional theory (DFT) and differential thermal analysis (DTA). The diffusion pathway was determined from the lowest energy adatom site and the hydrogen charge state at various sites along the pathway were calculated. DTA was used to investigate the efficacy of the W/WN films in preventing hydrogen attack experimentally. The temperature at which hydrogen reacted with the sample increased with film thickness, thereby indicating that the film inhibited this reaction. However, the existence of a hydrogen reaction peak in the thicker film indicated a material with greater than 2.0 eV energy barrier is necessary for a substantial decrease in hydrogen interaction with the substrate. This is the first study for ALD of WN on particles from bis(t-butylimido)bis(dimethylamino)tungsten(VI) and the use of WN ALD films as environmental barrier coatings (EBCs) for high-temperature H2 above 1000oC.

557075 Preparation of the Core-Shell Aluminum Nanoparticle with a Magnetic Nickel Shell

Aluminum nanoparticle has many advantages as the metal fuel in solid propellants for rockets, such as fast burning rate, high burning efficiency and no agglomeration in the combustion process. Because aluminum nanoparticles are very oxidative to form alumina, it is necessary to coat their surface with inert materials to prepare the core-shell aluminum nanoparticles. In this work, we proposed a novel process to prepare nanoparticles with an aluminum core and a magnetic shell. The preparation process consists of three steps as the purification of aluminum nanoparticles by differential centrifugation, the corrosion reaction of alumina layer, and the in situ displacement reaction of nickel ion with aluminum to form a magnetic nickel shell. The magnetic shell makes the nanoparticles can be easily separated from the liquid phase by the magnetic force. The nickel shell tightly bonds with the aluminum core, and protects it from the oxidation. The monodisperse nanoparticles with uniform shells can be produced using this new process. The methodology for regulating the size, structure and performance of the nanoparticle was investigated as well as the mechanism involved.

564800 Internal Structure of a Drying Nanoparticle-Polymer Film

Drying process is essential to various industrial applications including pharmaceutical drug-release films for dermal and oral drug delivery. One of the primary concerns is the distribution of active ingredients within dry films. In colloid-polymer mixtures, the drying mechanism is complicated and its study is limited. Prior studies show the drying behavior of the colloid-polymer mixtures at various pH, including particle dispersion and flocculation due to depletion interactions between colloidal particles and non-adsorbing polymers by using small-angle X-ray scattering (Kim et al., Macromolecules, 49, 2016) and stress development (Kim et al., Langmuir, 25, 2009). Also, recent simulation suggests stratifications in drying colloid-polymer mixtures depending upon polymer chain length and colloid diameter (Michael et al., Langmuir, 33, 2017). This work is an experimental in-situ investigation of the drying mechanism and film formation of colloid-polymer mixtures using confocal laser scanning microscopy to visualize the internal structural evolution of surrogate "active" ingredients. The mixtures prepared at high and low pH start as colloidal liquids or gels and are investigated over various colloid and polymer concentrations. Suspensions are drawn into thin films on glass substrates and dried at room temperature. During drying process, the film is evaluated by monitoring the dynamic evolution of the distribution and 3D structure of colloidal particles by analyzing the concentration profile, the number of nearest neighbors, the radial distribution function, and the Voronoi polyhedra distributions. In between these 3D scans, 2D scanning to determine the microrheology and internal flow is also obtained. The concentration and gel formation lead to the differences in particle distribution of the dry films. In gel samples, the time evolution of the structure is highly complex, sometimes experiencing periodic gel formation, gel breaking, flow and restructuring before the final structure is permanently immobilized. Moreover, the particle aggregation between liquid and gel mixtures during drying shows distinguishing patterns which are more comprehensible and useful for the industrial process and determining the final structure that dictates drug delivery.

570932 Nanocoating of Metals Oxides on Nanoscale Substrates in Organic Solvents

A new coating technique in a liquid phase is reported to coat carbon nanotubes (CNTs), carbon black (CB), and iron oxide (IO) as a substrate by titanium dioxide (TiO_2), niobium oxide (Nb_2O_5), and aluminum oxide (Al_2O_3). The coating process was carried out under an inert atmosphere by using a specially-designed set up. Heptane as an organic solvent was used to disperse the substrates with a deposited water film. The metal oxides precursors are also dissolved in the organic solvent. The thickness of the metal oxide coatings can be controlled from 5 to 80 nm. In addition, Nb_2O_5 Nano shells and porous Al_2O_3 nanotubes with a high surface area are produced by burning off the carbon. Transmission electron microscopy (TEM), electron low loss energy spectrum (ELLS), and thermo gravimetric analysis (TGA) were carried out to get deep insight for the coatings. In this presentation, we will demonstrate the nanocatings of various metal oxides on different substrates and demonstrate their advanced applications as catalyst supports and absorbents.

572495 Fabrication and Characterization of Superparamagnetic Hybrid Cellulose Nanocrystals (CNC)/Iron Oxide (Fe_3O_4) Nanoparticles

Cellulose nanocrystals (CNC) were coated with iron oxide (Fe_3O_4) nanoparticles by a single-one step co-precipitation method. In this project, two types of CNCs were used: (a) longer bacterial cellulose nanocrystals (BCNC) isolated in-house from nata de coco with a length of 505.06 ± 200 nm and a height of 7.66 ± 4.5 nm and (b) shorter wood pulp CNC obtained from southern bleached softwood kraft (SBSK) pulp with a length of 122 ± 70 nm long and a height 6.1 ± 3 nm. In both cases, CNCs were isolated by sulfuric acid hydrolysis. The sulfur content in CNC was measured by conductometric titration with sodium hydroxide solution, and it was found to be 0.41 ± 0.5 wt% in wood pulp CNC and 0.24 ± 0.4 wt% in BCNC.

Each CNC was then coated with Fe_3O_4 by in-situ co-precipitation of $FeCl_2$ and $FeCl_3$ in the presence of ammonium hydroxide. The reactions were done with theoretical mass compositions of CNC/Fe_3O_4 and BCNC/Fe_3O_4 of 1:1, 1:2, and 1:4. The resultant magnetic cellulose nanocrystals (MagCNC) were found to have moderate dispersibility in water and can be easily recovered by a magnetic field. Transmission Electron Microscopy (TEM), Thermal Gravimetric Analysis (TGA), and Fourier Transform Infrared (FTIR) spectroscopy experiments confirmed the successful assembly of CNC with magnetite. TEM images showed that increasing the magnetite concentration increased the coated area of CNC and apparently more aggregation between the hybrid MagCNC. The actual amounts of CNC and Fe_3O_4 in MagCNC were calculated by TGA. The actual amounts of CNC in the MagCNC were lower than the theoretical values due to some loss of CNC not attached to magnetite during the purification steps. Increasing the magnetite concentration decreased the loss of CNC indicating a greater number of coated CNC. The FTIR spectroscopy results indicated the interaction of magnetite nanoparticles with the hydroxyl group of CNC surface. Moreover, the colloidal stability of MagCNC in water was tested by Dynamic Light Scattering (DLS), UV-Vis spectroscopy and zeta potential measurements. The UV-vis spectroscopy data suggest that particles with higher ratios of Fe_3O_4/CNC are less colloidally stable than neat CNC at neutral pH. The hydrodynamic radius of neat CNC, neat Fe_3O_4, and MagCNC were determined by DLS. Increasing the magnetite composition in the MagCNC increased the hydrodynamic radius, providing additional evidence for aggregation among the particles. Zeta potential measurements showed that MagCNCs were moderately stable in water from pH 5 to pH 9 (zeta potential values ˜ -30 mV). Furthermore, the magnetic properties of MagCNC were measured by Vibrating-Sample Magnetometer (VSM). The normalized saturation magnetizations of MagCNCs were 31.79 emu/g, 45.36 emu/g and 45.40 emu/g for CNC/Fe_3O_4-1:1, CNC/Fe_3O_4-1:2, CNC/Fe_3O_4-1:4, respectively while for neat Fe_3O_4 nanoparticles it was 77.37 emu/g.

Many papers reported various methods for the modification of CNC and their applications but only a few have been done with metal oxide particles. The prepared magnetic cellulose nanocrystals can be used to stabilize magnetically-controlled emulsions and foams for applications in drug delivery or low energy separations.

557755 Particle Atomic Layer Deposition of Alumina for Flash Sintering Yttria-Stabilized Zirconia

Yttria-stabilized zirconia (YSZ) is the most-common electrolyte material for solid oxide fuel cells due to its reasonable oxygen-ion conductivity and chemical stability. To achieve suitable ionic conductivities, YSZ ceramics must be near theoretical density, requiring sintering temperatures around 1450°C. Alternatively, flash sintering has been shown to densify YSZ ceramics at ~750°C in just a few seconds. During flash sintering, an electric field is applied across the sample and once a threshold temperature is reached, the sample conductivity and power dissipation rapidly increase causing densification. Since densification occurs quickly, grain growth can be difficult to control. The addition of a small amount of aluminum oxide (Al_2O_3) has been shown to hinder grain growth, enhance densification, and increase YSZ ionic conductivity during conventional sintering. Typically, secondary phase particles (Al_2O_3) are added to primary ceramic particles (YSZ) by mechanical mixing which can lead inhomogeneities in the final microstructure. In this work, particle atomic layer deposition (ALD) was used to conformally coat each YSZ particle with a thin film of amorphous Al_2O_3, homogeneously dispersing the Al_2O_3 prior to flash sintering. Transmission electron microscopy and scanning electron microscopy were used to characterize the as-deposited Al_2O_3 films and to determine the effect of the Al_2O_3 films on grain growth and microstructural homogeneity. Density measurements showed that a small amount of amorphous Al_2O_3 added by particle ALD increased the final density of YSZ after flash sintering. Particle ALD homogeneously adds amorphous Al_2O_3 to YSZ particles, resulting in higher final densities and greater microstructural control to enhance properties of interest (mechanical strength, ionic conductivity, etc.) after flash sintering.

577167 Modifying the Photocatalytic and Dispersion Properties of TiO_2 P25 Nanoparticles By Scalable Molecular Layer Deposition

Titanium dioxide (TiO_2) is used for a range of applications, with a global market of over 6 million metric tonnes per annum. Almost 2/3 of this demand stems from the paints and coatings industries, where the high efficiency with which TiO_2 scatters visible light imparts highly favourable characteristics for its use as a bright white pigment. TiO_2 also acts as an effective UV-screening agent, and nanoscale TiO_2 is an increasingly important constituent of various cosmetic products, most notably as a UV blocker in sunscreens. The excellent performance of TiO_2 as a UV-screening agent stems in part from its semi-conductive properties, which makes it absorptive in the UV spectrum. This property is of particular interest for photocatalytic applications, where nanoscale TiO_2 has been studied as a potential water cleaning agent, degrading organic pollutants through the photocatalytic generation of reactive oxygen species (ROS). Although this generation of free radical species is utilised in certain self-cleaning coatings, for other applications where TiO_2 is used as a white pigment it is highly unfavourable, reducing product life-time through accelerated degradation. The high-molecular weight polymers in the organic binders found in coating applications are particularly susceptible to photochemically induced chain-scission reactions. The production of these oxidative species is also undesirable in cosmetic products, where the use of TiO_2 in sunscreen has raised concerns of potential carcinogenic, and mutagenic effects.

Due to these problems, considerable attention has been placed on finding approaches by which the photocatalytic activity of TiO_2 P25 nanoparticles can be suppressed, preferably without compromising its white pigmentation. Most TiO_2 commercially produced undergoes some form of surface treatment, not only to reduce photo-reduction of final use products, but also to improve the dispersibility of the TiO_2 particles in water and a range of organic liquids. Inorganic coatings are typically used for this purpose, with alumina and silica films the most commonly utilised metal oxide films. Alumina coatings as thin as 1 nm were found to effectively suppress the photodegrading properties of TiO_2. Although less used, organic coatings are of increasing interest for this photo-suppressive effect, often providing advantages in biocompatibility, lower raw material costs, and environmental considerations. The biocompatibility of polymeric films is of particular interest to the cosmetics industry, and also promises applicability to fields such as nanomedicine. A major advantage of using organic rather than inorganic coatings is the opportunity to improve the dispersibility of the underlying particle substrate in organic media. Not only is this promising for the aforementioned applications of TiO_2 to organic-based paints and cosmetics, but it may also be of interest for the production of polymer nanocomposites. Certain industrial polymers such as PVC are UV-sensitive, and various photo-stabilising systems have been investigated as a means to reduce their degradation rates. Two key issues preventing TiO_2 from fulfilling this role are its photocatalytic activity, which can enhance degradation rates, and difficulties in adequately dispersing the nanoparticles throughout the polymer. Therefore a reliable, scalable approach to produce organic coatings on TiO_2 may be of some value to polymer research and manufacture.

Typically, the production of these organic coatings on the particle substrates has been achieved through liquid-phase techniques. Liquid approaches are well established, providing versatility and a large degree of control over experimental conditions, however they provide non-conformal coverage and produce excessive amounts of waste. The requirement for multiple, at times complex synthesis steps, is also undesirable, as is the frequent need for post-processing separation and drying stages. Vapour phase approaches remove these steps, and prevent the undesirable effects of wetting and surface tension which can arise when using liquid-phase approaches. Chemical vapour deposition (CVD) is the most widespread vapour-phase coating process, and the introduction of both monomers in a single step allows simultaneous polymerisation and thin film deposition. However the coatings produced from this method can be non-conformal, reducing their capacity for surface passivation. To combat these issues, thicker films than desirable are typically produced to ensure complete surface coverage, however this is wasteful and expensive. Alternating or sequential vapour deposition polymerisation (VDP) was developed to provide greater control over film growth, and the formation of highly orientated polymer thin films is attributed to the interchanging reaction and deposition process achieved by introducing alternate monomers to the vacuum chamber. Although the control provided by alternating VDP is superior to that of CVD, monomers which remain present in the vapour phase following their introduction will react with sequentially introduced monomers, reducing the extent to which film growth can be quantifiably predicted.

The failings inherent to the above processes led to the development of Molecular layer deposition (MLD) to obtain the controlled growth of ultrathin organic films. Like sequential VDP, MLD is a vapour phase process which involves the cyclical introduction of organic precursors into the reaction chamber, however in MLD each pulse of precursor is separated by a purge step with an inert gas, minimising uncontrolled CVD growth. The consecutive, self-limiting chemical reactions between the substrate surface and the gaseous precursors enables the controlled, layer-by-layer growth of conformal polymeric coatings on a range of substrates.

In this work, we deposit ultrathin PET films on gram-scale batches of TiO_2 P25 nanoparticles (NPs) using terephthaloyl chloride and ethylene glycol as precursors. The MLD process is carried out in an atmospheric-pressure fluidized bed reactor at 150 °C for a broad range of growth cycles (i.e., from 10 to 50). Ex-situ diffuse reflectance infrared fourier transform spectroscopy (DRIFTS-FTIR) shows the presence of the characteristic C=O stretch of ester groups in the range 1730-1750 cm^{-1}, thus demonstrating the successful MLD reaction. DRIFTS-FTIR and thermogravimetric analysis (TGA) confirm the self-saturating behaviour of the precursors, and the linear increase in mass with the number of cycles. Moreover, TGA highlights the good thermal stability of the PET films, which are stable up to ~225 °C. Transmission electron microscopy (TEM) enables the visualization of uniform PET films on the surface of TiO_2 NPs. TEM observations suggest a low PET growth per cycle (GPC), around 0.06 nm. This value is also in agreement with the GPC estimated by the amount of deposited material on TiO_2 NPs measured by TGA as well as with the GPC of PET on SiO_2 wafers measured by spectroscopic ellipsometry. UV photocatalytic dye degradation tests demonstrate the efficacy of the produced PET films in suppressing the photocatalytic activity of the underlying TiO_2 P25 NPs. Still, UV-Vis diffuse reflectance spectroscopy measurements show that the absorption spectra of the PET-coated TiO_2 are virtually unaffected by the PET coating, indicating that the bulk optical properties (e.g., white color and brightness) of TiO_2 are retained. Furthermore, zeta potential measurements in aqueous solutions indicate greater stability of the suspensions with PET-coated TiO_2 than with uncoated TiO_2. Finally, to demonstrate the improvement of PET-coated TiO_2 in organic media, we disperse both the uncoated and PET-coated powders in mixtures of an apolar organic solvent (diethyl ether) and water. While uncoated TiO_2, being hydrophilic, stays in the aqueous phase, PET-coated TiO_2 stays in diethyl ether, thus indicating clear hydrophobization of the particles upon the MLD process.

TiO_2 P25 is a well-known photo-catalyst, serving as an ideal model substrate to assess the capability of MLD produced organic coatings to suppress the generation of free radical species. PET, a mass produced plastic, was selected on the grounds of its ideal characteristics of good chemical stability, relatively high thermal stability, and biocompatibility. The precursors used to produce the PET coating (terephthaloyl chloride and ethylene glycol) are also widespread in manufacturing, further demonstrating the industrial relevance of this study. Finally, by utilising a fluidized bed reactor operating at atmospheric pressure, we demonstrate a scalable process currently capable of depositing organic coatings on gram-scale quantities of nanopowder. This contrasts with the approach of most MLD studies, utilising highly controlled, low-pressure systems which are both complex and expensive. We therefore believe that this study is among the first demonstrations of a commercially viable MLD process with real world applicability.

An Approach of Coupled KTGF of Euler Granular Phase and DEM of Lagrangian Discrete Particles Applied to Gas-Solid Fluidized Beds

Zhang Qinghong, Jiang Xiaoxue, Lu Huilin [1], Cai Wenjian

School of Energy Science and Engineering, Harbin Institute of Technology, Harbin, 150001, China

1 E-mail address: huilin@hit.edu.cn

Abstract: An approach of coupled kinetic theory of granular flow (KTGF) for Euler granular phase, discrete element method (DEM) for Lagrangian discrete particles phase and turbulent model for Euler gas phase is proposed in a gas-particles fluidized bed. The coefficients of restitution (CoR) of KTGF are determined from the motion of discrete particles using DEM. The velocities of Euler gas phase, Euler granular phase and Lagrangian discrete particles phase are predicted in a fluidized bed. The simulated axial velocities of Euler granular pages and Lagrangian discrete particles phase are in agreement with experiments in a center gas jet fluidized bed. A cluster structure-dependent (DCSD) drag model is proposed to improve drag predictions for heterogeneous gas-solid flows in risers. The DCSD drag model is verified by CFD simulations in a riser.

1. Introduction

Computational fluid dynamics (CFD) is an effective tool for understanding the fundamental hydrodynamics of gas and particles in risers. However, a major challenge for CFD models of risers is to realistically resolve the effect of clusters on the momentum exchange between the gas phase and solids phase. The numerical simulations of the bed materials of gas-solid fluidized beds can be described by the Euler-Euler two-fluid models (TFM) using the kinetic theory of granular flow (KTGF). The viscosity and pressure of Euler granular phase depends on granular temperature due to the collisions of particles which relate to the coefficient of restitution e_s (i.e., p-p CoR). On the other hand, the flow behavior of the bed materials of the gas-solid fluidized bed is modeled by means of Lagrangian approach to track the discrete particles in space and time using discrete element method (DEM). Hence, the main feature of CFD–DEM is that it can generate detailed particle-scale information, such as the trajectories, forces and impact velocity of individual particles in gas-solid fluidized beds.

The drag force acting on particles is represented by the product of the drag coefficient β_{gs} and the slip velocity between the two phases. The energy minimization multiscale approach (EMMS) developed by Li et al. (1999) has been used to predict steady flow in circulating fluidized beds. However, the validation of EMMS drag model is still a problem because the model equations stem from a global fluidized bed system, not for grid cells of numerical simulations. Also the equation for predicting cluster diameter is not valid to use in the counter-current flow with gas flow-up and clusters flow-down. On the other hand, the filtered drag models proposed by Sundaresan group (2011) are obtained using finely resolved simulations. These filtered models are the analogue of large-eddy simulation of single phase turbulent flow, where one simulates spatio-temporal patterns occurring at the macro-scale using the conservations of mass and momentum, but accounts for the effects of meso-scale structures occurring at a scale smaller than the grid size through additional closure relations.

2. A model of Euler gas phase-Euler granular phase-Lagrangian discrete particle phase

2.1. Conservation equations for Euler granular phase with KTGF

The momentum conservation of the Euler granular phase in the absence of phase change and chemical reactions is:

$$\frac{\partial(\rho_s\alpha_s)}{\partial t}+\nabla\cdot(\rho_s\alpha_s\vec{u}_s)=0 \tag{1}$$

$$\frac{\partial}{\partial t}(\alpha_s\rho_s\vec{u}_s)+\nabla\cdot(\alpha_s\rho_s\vec{u}_s\vec{u}_s)=-\alpha_s\nabla p-\nabla p_s+\nabla\cdot\bar{\bar{\tau}}_s+\vec{f}_{gs}+\alpha_s\rho_s\vec{g} \tag{2}$$

where the Euler granular phase pressure p_s are expressed as follows (Gidaspow, 1994)

$$p_s=\alpha_s\rho_s\theta+2\alpha_s^2\rho_s g_o(1+e_s)\theta \tag{3}$$

where e_s is the p-p CoR of Euler granular phase. θ is the granular temperature, and the balance equation for the granular temperature is

$$\frac{3}{2}[\frac{\partial}{\partial t}(\alpha_s\rho_s\theta)+\nabla\cdot(\alpha_s\rho_s\vec{u}_s\theta)]=\nabla\cdot(k_s\nabla\theta)+(\nabla p_s\mathbf{I}+\bar{\bar{\tau}}_s):\nabla\vec{u}_s+\gamma_s-3\beta_{gs}\theta \tag{4}$$

2.2 Conservation equations for Lagrangian discrete particle phase

The displacement of an individual discrete particle is expressed as follows

$$m_i\frac{d\vec{v}_{pi}}{dt}=\sum_{j=1,i\neq j}^{N}\vec{f}_{c,ij}+\vec{f}_{d,i}+\vec{f}_{pf,i}+m_i\vec{g} \tag{5}$$

$$I_{pi}\frac{d\vec{\omega}_{pi}}{dt}=\sum_{j=1,i\neq j}^{N}\vec{T}_{p,ij} \tag{6}$$

2.3 Conservation equations for Euler gas phase

The equations of mass and momentum conservation for Euler gas phase are (Gidaspow, 1994):

$$\frac{\partial(\rho_g\alpha_g)}{\partial t}+\nabla\cdot(\rho_g\alpha_g\vec{u}_g)=0 \tag{7}$$

$$\frac{\partial}{\partial t}(\alpha_g\rho_g\vec{u}_g)+\nabla\cdot(\alpha_g\rho_g\vec{u}_g\vec{u}_g)=-\alpha_g\nabla p+\nabla\cdot\bar{\tau}_g-\vec{f}_{gs}+\alpha_g\rho_g\vec{g} \tag{8}$$

2.4. Interactions of Euler granular phase, Euler gas phase and Lagrangian discrete particles phase

The drag forces term f_{gd} appears in the momentum equations of the Euler gas phase and Euler granular phase as well as the motion equation of discrete particles. The coupling of Euler gas phase and Euler granular phase is achieved by applying Newton's third law to account for the forces generated on the Euler gas phase.

$$\vec{f}_{gs} = \beta_{gs}(\vec{u}_g - \vec{u}_s) \tag{9}$$

The drag force acting on the discrete particles is calculated according to the momentum exchange coefficient β_{gs}.

$$\vec{f}_d = \frac{\beta_{gs} V_p}{\alpha_s}(\vec{u}_g - \vec{v}_p) \tag{10}$$

2.5. Coupling of CoR between Euler granular phase and Lagrangian discrete particle phase

The p-p CoR and p-w CoR measure the fraction of kinetic energy recovered during a collision. From simulated e_{ij} between the discrete particle i and j and e_{iw} between the wall and discrete particle i, the p-p CoR and p-w CoR of Lagrangian discrete particles in the computational cell are:

$$e_{sd} = \frac{1}{N}\sum_{j=1, i\neq j}^{N} e_{ij} \quad \text{and} \quad e_{wd} = \frac{1}{N}\sum_{i=1}^{N} e_{iw} \tag{11}$$

These equations give the relations of p-p CoR and p-w CoR of Euler granular phase used in the KTGF and the e_{sd} and e_{wd} of Lagrangian discrete particles phase determined from DEM.

2.6 Comparison of experiments and simulations in a spout fluidized bed

The flow behavior of bed materials is measured using a particle image velocimetry and positron emission particle tracking in a spout fluidized bed. The pseudo-2D spout fluidized bed has a width, depth and height of 145 mm, 20 mm and 2500 mm with a spout jet at the center of the distributor, as shown in Fig. 1. The diameter and density of particles are 3.0 mm and 2505 kg/m^3 with the minimum fluidization velocity of 1.9 m/s. Gas is supplied though the spout jet with spouting velocity u_{sp} and the distributor with background velocity u_{bg}.

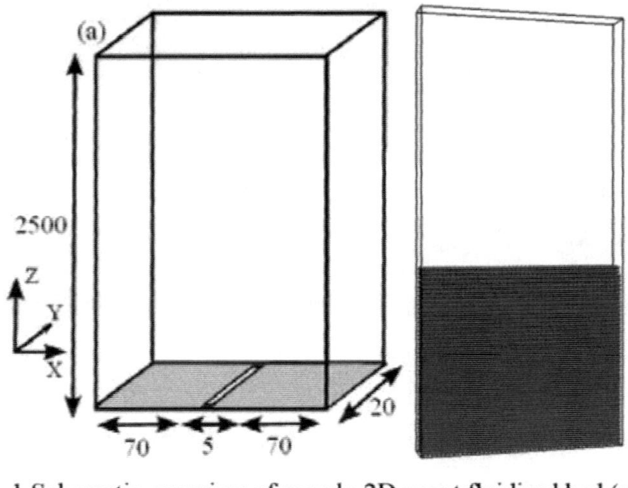

Fig. 1 Schematic overview of pseudo-2D spout fluidized bed (mm)

The instantaneous volume fraction contour of Euler gas phase, velocity vector of Euler granular phase and velocity color of discrete particles are shown in Fig. 2 at the inlet spouting velocity u_{sp} and background velocity u_{bg} of 43.5 m/s and 2.4 m/s. The colors of particles represent the velocity magnitude of discrete particles. The numerical simulations show that the dense annulus region near the walls and the dilute spout region at the center exist in the spout fluidized bed. The inlet background velocity is around the minimum fluidization velocity and the inlet spout velocity is high to enable the spout channel to form. This leads to the dilute spout channel with the upward flow of Euler granular phase and discrete particles in the center. However, the down-flow of Euler granular phase and discrete particles and the up-flow of Euler gas phase exist in the annular region. The circulation of Euler granular phase and discrete particles is formed in the bed. It is also found that the spout jet region is unstable, and moves the spout channel from side to side due to the competition between the collisional interactions and the hydrodynamic interaction of the spout jet.

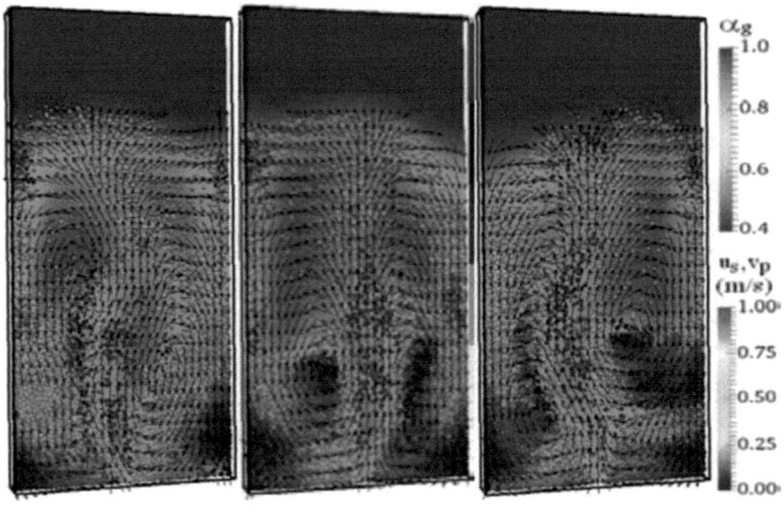

Fig. 2 Instantaneous volume fraction contour of Euler gas phase, velocity vector of Euler granular phase and velocity color of discrete particles

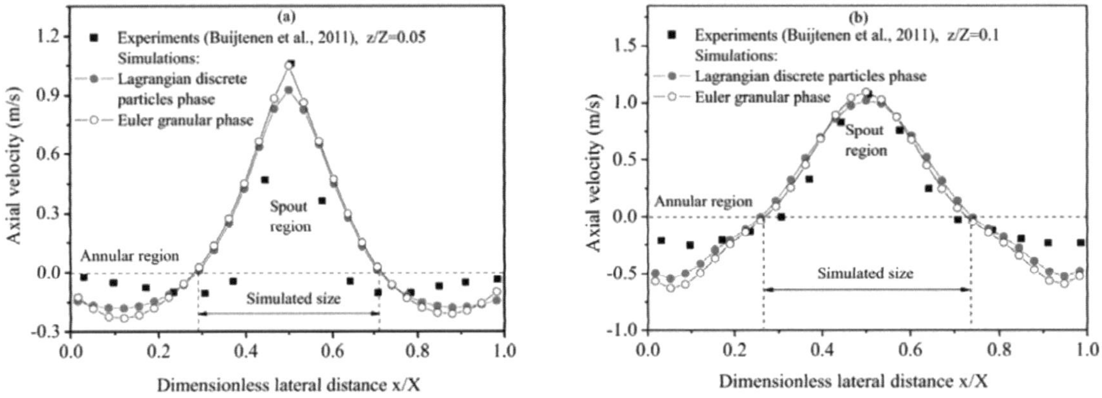

Fig. 3 Comparison of simulated and measured axial velocities at two bed heights

The simulated and measured axial velocities of bed materials are shown in Fig. 3 along bed width. The simulated axial velocities of Euler granular phase and Lagrangian discrete particles phase match with experimental measurements. Both simulations and measurements show the axial velocities are positive at the spout region and negative at the annular region. However, the difference of axial velocities exists between simulations and experiments at two heights. The axial velocities of Lagrangian discrete particles phase are smaller than that of Euler granular phase at the dilute spout region. However, it is reverse, and the axial velocities of Lagrangian discrete particles phase are larger than that of Euler granular phase in the annular region. The simulated axial velocities are larger than that of measurements at the interface between the spout region and the annular region, indicating that the numerical simulations over-predict the spout size in comparison with measurements in the spout fluidized bed.

Fig. 4 shows the distributions of p-p CoRs and p-w CoRs as a function of volume fractions of Euler granular phase in the bed. Both decrease with the increase of granular volume fractions. The mean values of p-p and p-w CoRs are 0.971 and 0.890, respectively. We also found that the p-p and p-w CoRs are low at high granular volume fractions because of the effect of multiple collisions of discrete particles. When the granular volume fractions are high, the correlation in particle velocity increases, indicating the binary collision model might not be accurate to model dynamics of Euler granular phase in fluidized beds.

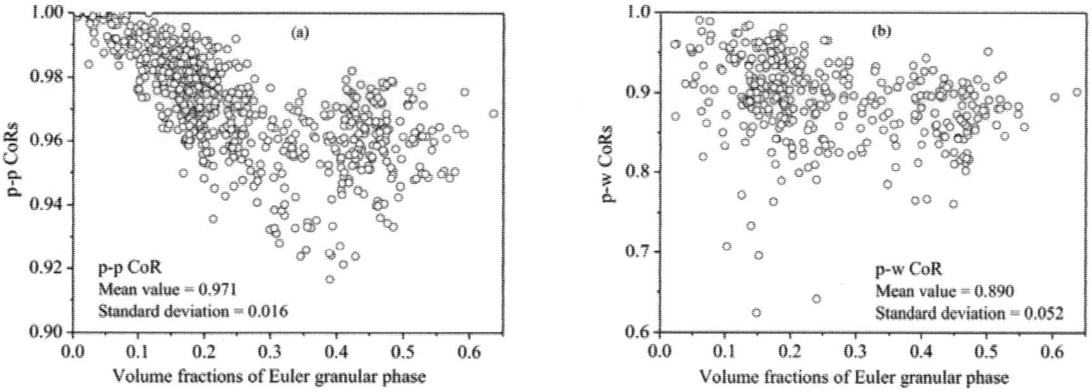

Fig. 4 Distributions of p-p CoRs and p-w CoRs as a function of granular volume fractions

3. Gas-solids two-fluid model (TFM) with modified DCSD drag model

3.1 Governing equations with cluster structure-dependent (DCSD) drag model

The flow structure of particles is shown in Fig. 5 in a grid cell in which the motion of particles is in the form of clusters in the dense phase and in the form of a dispersed particle in the dilute phase. The heterogeneous structure of gas-particles mixture is described by ten independent variables. That is, gas volume fractions of the dilute phase and the dense phase (α_d, α_c), gas superficial velocity of the dilute and the dense phases ($U_{g,d}$, $U_{g,c}$), superficial velocity of particles in the dilute phase and the dense phase ($U_{s,d}$, $U_{s,c}$, granular temperatures of particles of the dilute and dense phases (θ_d, θ_c), cluster size (d_c) and volume fraction of dense phase (f). On the other hand, the parameters of a grid cell are velocities of gas and solids phases (u_g, u_s), gas volume fraction (α_g), granular temperature (θ) and gas pressure gradient ($\partial p/\partial z$) along flow direction from KTGF-based TFM.

The description of hydrodynamics of gas and particles as fully interpenetrating continua includes the mass and momentum conservation equations with KTGF. The momentum equations of gas and solid phases are

$$\frac{\partial}{\partial t}(\alpha_g \rho_g \vec{u}_g)+\nabla\cdot(\alpha_g \rho_g \vec{u}_g \vec{u}_g)=-\alpha_g \nabla p+\nabla\cdot\bar{\tau}_g-\beta_{gs}(\vec{u}_g-\vec{u}_s)+\alpha_g \rho_g \vec{g} \tag{12}$$

$$\frac{\partial}{\partial t}(\alpha_s \rho_s \vec{u}_s)+\nabla\cdot(\alpha_s \rho_s \vec{u}_s \vec{u}_s)=-\alpha_s \nabla p-\nabla p_s+\nabla\cdot\bar{\bar{\tau}}_s+\beta_{gs}(\vec{u}_g-\vec{u}_s)+\alpha_s \rho_s \vec{g} \tag{13}$$

The stress-strain tensor of particles is as a function of shear viscosity and bulk viscosity of particles. Both solid pressure and viscosity are modeled as a function of granular temperature from KTGF. Hence, we find that one additional constitutive equation is needed to calculate drag coefficient β_{gs} to solve gas and solid momentum equations with the effect of clusters.

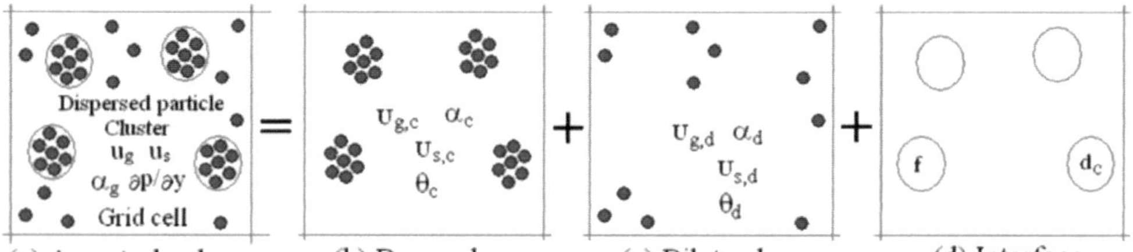

(a) A control volume (b) Dense phase (c) Dilute phase (d) Interface

Fig. 5 Flow structure of particles in a grid cell

3.2 Equations of kinetic energy balance and minimization of energy dissipation rate

The kinetic energy equation of solid phase was derived according to the kinetic-theory-based model (Gidaspow, 1994; Fox, 2014). For flow of heterogeneous structures, the total kinetic energy of gas and solid phase consists of the kinetic energy components of gas and dispersed particles in the dilute phase ($E_{g,d}$ and $E_{s,d}$) and the kinetic energy components of gas and clusters in the dense phase ($E_{g,c}$ and $E_{s,c}$). The kinetic energy balance of gas-particles suspension is

$$\begin{aligned}
&\frac{\partial}{\partial t}\{[f(1-\alpha_c)\rho_s E_{s,c}+(1-f)(1-\alpha_d)\rho_s E_{s,d}]+[f\alpha_c \rho_g E_{g,c}+(1-f)\alpha_g E_{g,d}]\}\\
&+\frac{\partial}{\partial z}[f^2(\rho_s U_{s,c}E_{s,c}+\rho_g U_{g,c}E_{g,c})+(1-f)^2(\rho_s U_{s,d}E_{s,d}+\rho_g U_{g,d}E_{g,d})]\\
&+(\alpha_g u_g+\alpha_s u_s)\frac{\partial p}{\partial z}+\frac{\partial}{\partial z}[(\frac{fU_{s,c}}{1-\alpha_c})p_{s,c}+(\frac{(1-f)U_{s,d}}{1-\alpha_d})p_{s,d}]+\frac{\partial}{\partial z}(q_{s,d}+q_{s,c})\\
&+\frac{\partial}{\partial z}(q_{g,d}+q_{g,c})+[f^2\rho_s U_{s,c}+(1-f)^2\rho_s U_{s,d}]g+[f\rho_g U_{g,c}+(1-f)\rho_g U_{g,d}]g\\
&=[n_d F_d U_d+n_c F_c U_c+n_i(1-f)F_i U_i]+(\Gamma_{s,c}+\Gamma_{s,c})+(\Gamma_{g,d}+\Gamma_{g,c})
\end{aligned} \tag{14}$$

We observe that the terms at the right hand side represent the rate of energy dissipations of gas-particles suspension through three contributions by hydrodynamic forces, interactions of collisions of particles and gas viscous forces of heterogeneous structures.

$$\Omega=[n_d F_d U_d+n_c F_c U_c+n_i F_i U_i(1-f)]+(\Gamma_{s,d}+\Gamma_{s,d})+(\Gamma_{g,d}+\Gamma_{g,c}) \tag{15}$$

To keep the energy dissipation rate as low as possible, particles tend to aggregate into clusters to reduce the gas flow resistance and the collisional energy dissipations, while gas tends to bypass the clusters instead of penetrating through clusters (Gidaspow, 1994; Huilin, 2017). Hence, the criterion of the existence of heterogeneous structures is the kinetic energy dissipation rate per unit mass of gas-particles mixture is minimum:

$$N_{df} = \frac{[n_d F_d U_d + n_c F_c U_c + n_i F_i U_i (1-f)] + (\Gamma_{s,d} + \Gamma_{s,d}) + (\Gamma_{g,d} + \Gamma_{g,c})}{\varepsilon_g \rho_g + (1-\varepsilon_g)\rho_s} = \Omega_{dr} + \Omega_{co} + \Omega_{vi} = \text{minimum} \quad (16)$$

This is an additional constitutive equation which is an extreme value equation of the rate of kinetic energy dissipation including the energy dissipation component of drag forces Ω_{dr}, the collisional energy dissipation component Ω_{co} and gas viscous energy dissipation component Ω_{vi}.

3.3 Dynamic cluster structure-dependent (DCSD) drag model

From gas momentum balance equation with no gas acceleration and gravity in the dense phase and dilute phase (Huilin, 2017), the DCSD drag coefficient is expressed as follows:

$$\beta_{DCSD} = \frac{n_c F_c + n_d F_d + n_i F_i}{|u_g - u_s|} \qquad N_{df} = \text{minimum} \quad (17)$$

On the other hand, the homogeneous drag coefficient is calculated using Huilin-Gidaspow drag correlation β_{HG} with the switch function φ_{gs}:

$$\beta_{HG} = \varphi_{gs}\left(\frac{150\alpha_s(1-\alpha_g)\mu_g}{\alpha_g d_s^2} + \frac{1.75\alpha_s \rho_g |u_s - u_g|}{d_s}\right) + (1-\varphi_{gs})\frac{3C_D \alpha_s \alpha_g \rho_g |u_s - u_g|}{4d_s}\alpha_g^{-2.65} \quad N_{df} \neq \text{minimum} \quad (18)$$

From Eq. (17) and (18), the drag coefficient in the computational cell is modeled as follows:

$$\beta_{gs} = \begin{cases} \beta_{DCSD} & N_{df} = \text{minimum} \\ \beta_{HG} & N_{df} \neq \text{minimum} \end{cases} \quad (19)$$

We see that the extreme value equation (16) takes as a closure equation for the determination of drag coefficient in flow of heterogeneous structures.

3.4 Numerical results and discussion

Simulations are conducted using the DCSD drag model and Huilin-Gidaspow drag models and results are compared with the experimental data in a 30.5 mm vertical pipe using a laser-Doppler velocimeter. The diameter and density of particles are 54 μm and 1398 kg/m^3, respectively. Fig. 6 shows the distribution of drag coefficient as a function of solids volume fraction. Roughly, the drag coefficient predicted by DCSD drag model decreases, reaches minimum, and then increases with the increase of solids volume fraction. The difference of drag coefficients between the DCSD drag model and Huilin-Gidaspow drag model is obvious. We see that the heterogeneity reduces drag instead of increasing it at the low solids volume fraction. The reason is that when the flow is dilute, the particle collision probability is small, and results in decreased resistance with more gas flow through the dilute phase. So that the dilute phase occupies more space, and there are few opportunities to form clusters and the flow tends to be homogeneous. As a result, the drag reaches a minimum. As the solids volume fraction increases beyond some critical value, the interaction between the dense phase and the dilute phase increases. Therefore, clusters can only form within a certain range of conditions.

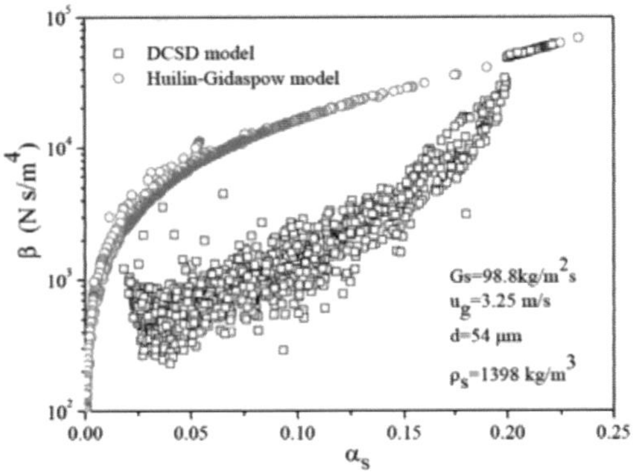

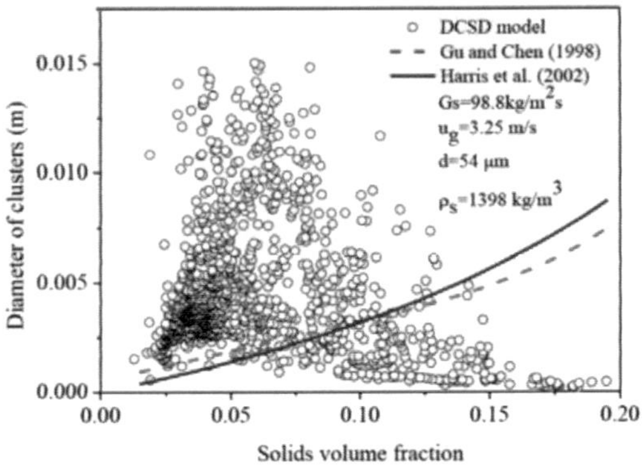

Fig. 6 Drag coefficient as a function of volume fractions Fig. 7 Cluster diameter as a function of volume fractions

Fig. 7 shows the distribution of predicted cluster diameter as a function of solids volume fraction. As the amount of solids increases, particles tend to aggregate to achieve less resistance. This leads in general to larger diameter, and until the cluster size reaches a turning point. Beyond this turning point, the cluster diameters decrease, hinting at a possible change in the flow patterns over the region originally thought to be completely homogeneous. Harris et al. (2002) presented correlations for predicting the properties of cluster of particles traveling near the riser wall. However, the cluster diameter correlation proposed by Harris et al. (2002) is actually the mean vertical cluster length which is quite different from the surface volume mean diameter of an equivalent spherical cluster in the DCSD drag model. The cluster sizes in experimental investigations usually refer to the mean vertical cluster length because of the considerable variation of cluster shapes.

Fig. 8 shows the profile of predicted volume fraction of dense phase (*f*) and solids volume fraction of clusters ($\alpha_{s,c}$) as a function of solids volume fraction. Generally speaking, the predicted volume fractions of dense phase, f, increase, reach maximum, and then decrease with the increase of solids volume fraction. The calculated solids volume fraction of clusters using the correlation proposed by Gu and Chen (1998) is given as a function of solids volume fraction. Both numerical simulations and correlation give the solids volume fractions of dense phase increase with the increase of solids volume fraction. However, the difference between them is obvious. Examination of the correlation proposed by Harris et al (2002) concerning solids volume fraction of clusters suggests that the volume fraction of cluster requires the local instantaneous solid concentration greater than the time-mean solid concentration by at least n times the standard deviation. A cluster would thus be identified if the instantaneous solid concentration exceeds this threshold.

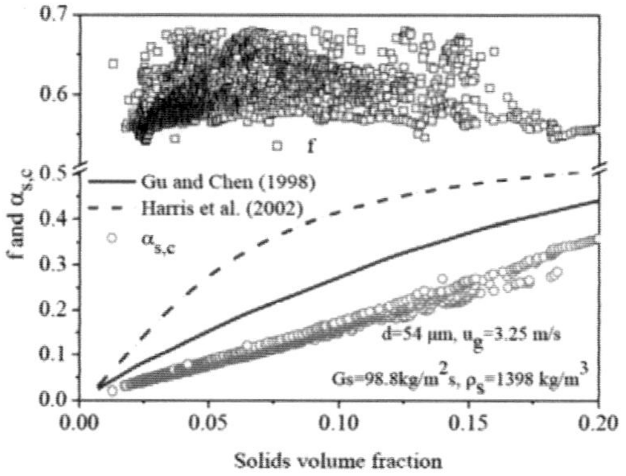

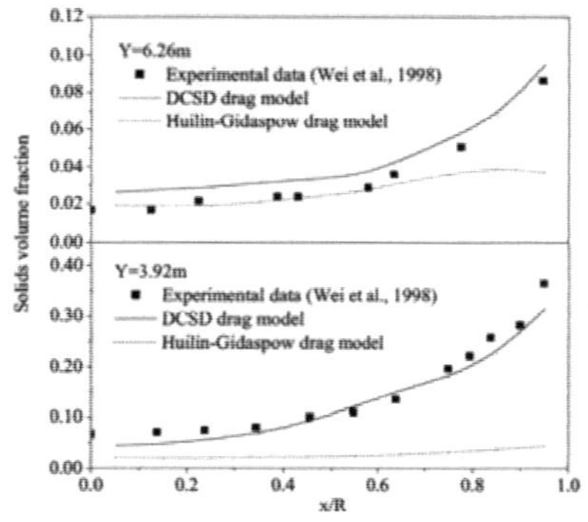

Fig. 8 Volume fraction and solids volume fractions   Fig. 9 Experimental and computed solids volume fractions

Fig. 9 shows a comparison of experimental and computed solids volume fraction profiles at two heights. Both drag models give a high solids volume fraction near the wall, and a low volume fraction of particles at the center regime. Comparing to simulations by means of Huilin-Gidaspow drag model, the predictions using DCSD drag model give a reasonable quantitative agreement with experiments. It should also be noted that the predictions by means of DCSD drag model show a profile of the experimental data more accurately than those using Huilin-Gidaspow drag model in the riser.

4. Conclusion

A numerical simulation approach coupled with Euler gas phase-Euler granular phase-Lagrangian discrete particles is proposed with KTGF and DEM to model gas-particles fluidized beds. The CoRs of Euler granular phase of KTGF is determined by means of DEM. The simulated axial velocity and solids volume fractions are close to experimental measurements in a spouted fluidized bed.

The performance and applicability of DCSD drag model are evaluated by comparing with experimental data in risers. The DCSD drag model and gas-solids Euler approach with kinetic theory of granular flow is combined in the numerical simulation of heterogeneous gas-solid two-phase flow in risers. The comparison of simulations shows that the DCSD drag model gives a reasonable quantitative agreement with the experimental data.

Acknowledgment

This work is supported by the Natural Science Foundation of China through Grant Nos. 51776059 and 51752115.

Reference

[1]. Agrawal K, Loezos PN, Syamlal M, Sundaresan S., The role of meso-scale structures in rapid gas-solid flows. J. Fluid Mechanics, 445, 151-185, 2011.

[2]. Ansys-Fluent Inc., Fluent users manual Version 14.0, New Hampshire, USA, 2012.

[3]. Fox R. O., On multiphase turbulence models for collisional fluid-particles flows, J. Fluid Mech., 742, 368-424, 2014

[4]. Gidaspow D., Multiphase flow and fluidization: continuum and kinetic theory descriptions. Academic Press, New York, 1994.

[5]. Gu W.K., Chen J.C., A model for solid concentration in circulating fluidized beds. In: Fan, L.S., Knowlton, T.M. (Eds.), Fluidization IX. Engineering Foundation, Durago, Colorado, 501-508, 1998.

[6]. Harris A.T., Davidson J.F., Thorpe R.B., The prediction of particle cluster properties in the near wall region of a vertical riser, Powder Technology, 127, 128-143, 2002.

[7]. Huilin L., Kinetic theory for dense fluid-solid flow, Science Press, 2017.

[8]. Li J., Chen C.L., Zhang Z.D., Yuan J., Nemet A., Fett F. N., The EMMS model-its application, development and updated concepts. Chem. Eng. Sci. 54, 5409- 5425, 1999.

[9]. O'Brien T.J., Syamlal M., Particle cluster effects in the numerical simulation of a circulating fluidized bed, 4th International Conference on CFB, Somerset, USA, 430-443, 1993.

574182 3D Printed Static Mixers for Continuous Reactive Crystallization

Crystallization is an important unit operation in the pharmaceutical industry for separation and purification of intermediate chemical compounds and active pharmaceutical ingredients. The driving force for crystallization is the supersaturation of the solution, which is defined by the difference between the solute concentration and its solubility. There are various methods to create supersaturation. In anti-solvent or reactive crystallization, different process streams are mixed together to generate supersaturation. Consequently, mixing plays an important role in the performance of anti-solvent or reactive crystallization. Initially, mixing is required to obtain a homogeneous solution to obtain predictable bulk nucleation and to prevent local nucleation, which may occur due to concentration gradients[1]. Once crystals are formed, mixing is needed to maintain the crystals in suspension. Plug-flow behavior is typically desired to provide a uniform residence time for each crystal and hence deliver uniform product quality attributes such as crystal shape and size. Tubular crystallizers can provide plug-flow behavior if mixing is fast in radial direction but slow in axial direction. One of the methods to approach such mixing patterns involves static mixers. A kenics-type static mixer has previously been demonstrated to be useful for achieving plug-flow conditions for anti-solvent crystallization of ketoconazole.[2] However, other types of static mixers such has Komax, SMX, or low pressure drop (LPD) mixers have not been studied for their application as a tubular crystallizer.[3] Additionally, the mixing elements of these static mixers can be optimized for a given application.[4] Finally, the mixing requirements may change when the process material flows from the entrance to the outlet of the tubular crystallizer. In particular, in the first part of the crystallizer, rapid mixing of liquid process streams needs to be achieved, whereas large crystals need to be kept in suspension further downstream. Fouling and plugging should be avoided when the crystal mass increases. Therefore, when using a static mixer as a tubular crystallizer, the structure and dimensions of the internals may need to change for optimal operation. These requirements for customization and the absence of any moving parts call for 3D printing to be used as the preferred fabrication method of tubular crystallizers based on static mixers.

The objective of this work is to design, fabricate and characterize tubular crystallizers with different designs for static mixing for reactive crystallization. Stereolithography is identified as the most suitable 3D printing technique among the four major techniques, because it allows for fabrication of leak-proof parts and easy removal of uncured resin from the flow channel.[5] Reactive crystallization of barium sulphate is conducted in a Y-mixer and komax-type mixer fabricated using a commonly available desktop stereolithography 3D printer (Form 2, FormLabs) with clear resin. The results demonstrate practical feasibility of using 3D printing to fabricate innovative tubular crystallizers and a reduced agglomeration is observed in the komax mixer compared to a conventional Y-mixer. Subsequently, reactive crystallization of an active pharmaceutical ingredient, salicylic acid from sodium salicylate and an acid, will be investigated. COMSOL simulations show that a kenics-type static mixer with a standard design provides faster mixing compared to a Y-mixer, komax or a LPD mixer. COMSOL simulations will be used to optimize the different static mixers for the rate of mixing and to minimize zones with low liquid velocity where crystals might settle. Finally, the crystallization performance in terms of yield, crystal size distribution and degree of agglomeration will be characterized experimentally in the 3D printed tubular crystallizer. This work demonstrates a novel type of application of 3D printing in which there is a clear benefit of customization and fabrication speed.

Acknowledgement: The work described in this abstract was supported by a grant from the Research Grants Council of the Hong Kong Special Administrative Region, China, Project No. 16214418.

References

(1) Hacherl, J. M.; Paul, E. L.; Buettner, H. M. Investigation of Impinging-Jet Crystallization with a Calcium Oxalate Model System. *AIChE J.* **2003**, *49* (9), 2352–2362.

(2) Alvarez, A. J.; Myerson, A. S. Continuous Plug Flow Crystallization of Pharmaceutical Compounds. *Cryst. Growth Des.* **2010**, *10* (5), 2219–2228.

(3) Ghanem, A.; Lemenand, T.; Della Valle, D.; Peerhossaini, H. Static Mixers: Mechanisms, Applications, and Characterization Methods - A Review. *Chem. Eng. Res. Des.* **2014**, *92* (2), 205–228.

(4) Singh, M. K.; Anderson, P. D.; Meijer, H. E. H. Understanding and Optimizing the SMX Static Mixer. *Macromol. Rapid Commun.* **2009**, *30* (4–5), 362–376.

(5) Bhattacharjee, N.; Urrios, A.; Kang, S.; Folch, A. The Upcoming 3D-Printing Revolution in Microfluidics. *Lab Chip* **2016**, *16* (10), 1720–1742.

576831 Ultra-Low-Cost Continuous Manufacturing Platforms through 3D Printing and Telescoped Isolation Free Purification

Chemically resistant parts for continuous flow chemistry have been produced using the 3D printing process fused deposition modelling (FDM), from poly(etheretherketone) (PEEK). PEEK has greater chemical resistance than common FDM materials such as acrylonitrile butadiene styrene (ABS), polypropylene (PP), or even high-performance plastics like poly(etherimide) (PEI), and also has good thermal resistance and excellent mechanical strength. These printed parts have been demonstrated to be capable of withstanding high pressures, allowing the superheating of solvents for optimal reaction conditions. The ability to use FDM for these reactors allows highly customisable, cost effective flow reactors to be produced directly in the laboratory and integrated into equipment already available.

The integration of complicated internal structures, difficult to produce with powder bed fusion or resin photo-curing techniques such as selective laser melting (SLM) or stereo lithography (SL), was also possible with FDM, and allowed structures as small as 0.5 mm to be produced reliably within the reactor internal channels. These integrated elements were shown to enhance reaction mixing and the efficiency of liquid-liquid extractions, improving reaction throughput. The printed reactors were used for flow chemistry and then integrated with continuous liquid-liquid extraction followed by continuous phase separation or isolation free continuous crystallisation within an agitated bed crystalliser.

The phase separator is a compact and easily scalable setup, at a fraction of the cost of current commercial systems, requiring only simple back pressure control. The agitated bed allows continuous purification *via* crystallisation on a suspended seed bed, with the option to perform solvent swap as a final step if required. When combined with the printed reactors a customisable synthesis path results, allowing low cost integrated isolation free purification with continuous flow chemistry.

Scheme 1. (Top) 3DP reactor followed by 3DP workup stage and continuous flow phase separation in phase separator. (Bottom) 3DP reactor followed by confined suspension crystalliser

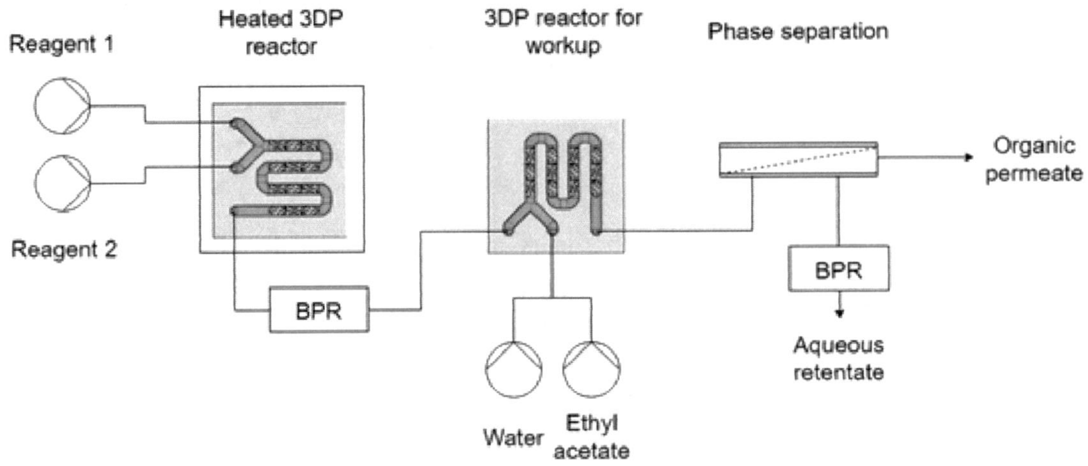
Reagent 1
Heated 3DP reactor
3DP reactor for workup
Phase separation
Organic permeate
Reagent 2
BPR
BPR
Aqueous retentate
Water
Ethyl acetate

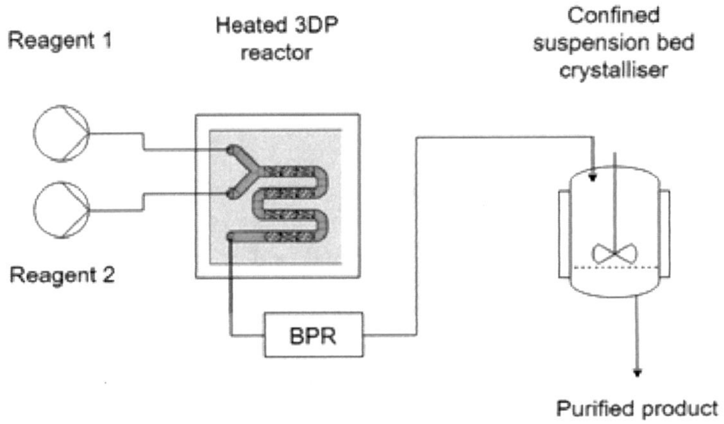
Reagent 1
Heated 3DP reactor
Confined suspension bed crystalliser
Reagent 2
BPR
Purified product

560518 Modeling of Poly(ether imide) in Material Extrusion Additive Manufacturing

Additive manufacturing (AM), is a manufacturing modality that allows the production of small scale, customized parts quickly and cheaply. Material extrusion (MatEx) AM is one of the most popular types of AM. This process uses a filament feedstock to produce a part by depositing thin roads of material in 2D layers which are then stacked in the desired geometry to create a 3D shape. However, this method of manufacturing is currently limited primarily to the rapid production of prototypes and physical models. The ultimate goal of MatEx is to produce parts for the end-use market that will experience practical use. The current state of the process prevents this due to a lack of robust mechanical properties and material versatility. The lack of versatility stems from a highly limited material catalog. MatEx is primarily commercially limited to polylactic acid (PLA) and acrylonitrile butadiene styrene (ABS). This small catalog of materials also limits the mechanical properties of the parts produced with this technique, as the mechanical properties of these two materials are much less than the common engineering thermoplastics such as polyetherimide (PEI) and polyether ether ketone (PEEK). The weakness of the produced parts also stems from a lack of z-axis, interlayer, adhesion. This problem is inherent to the process due to a low amount of time above the glass transition temperature of the material during printing, preventing the diffusion of polymer chains across the interface formed where two roads meet.

To address these issues, we propose a more thorough understanding of MatEx is required in order to more efficiently identify novel materials as well as their optimal printing conditions. Because in-situ measurements of the process are difficult, often resulting in unforeseen changes in the system that render any information gained with the measurements incorrect, we suggest utilizing multiphysics based process models to examine the effect of various process parameters on the quality of the final product. In this study, we developed multiple predictive models based on thermal transport, diffusion, and stress development. By examining the effect caused by modifying various parameters in these models, we can gain a detailed understanding of the process and how it can be modified to allow the printing of novel materials as well as identifying the optimal printing parameters for the production of parts with near bulk properties. These predictive models were then compared to empirical measurements.

Harnessing a High Energy, Superconducting Electron Beam for Additive and Far-from-Equilibrium Manufacturing

Adam Duzik, Ryan Reeves, Justin J. Hill†
Mainstream Engineering Corporation, Rockledge, FL, USA
†Corresponding author: jhill@mainstream-engr.com

Superconducting linear accelerators that generate electron beams open up new opportunities for advanced materials processing since they can continuously deliver extremely high energy electrons. The increased energy delivery rate of these systems enables the production of far-from-equilibrium materials and selective processing within a material surface and subsurface. High energy electrons penetrate deeper within materials and can increase the processing depth and rate. With the support of the Office of Naval Research, NASA, and DoE, Mainstream Engineering commissioned a superconducting linear electron accelerator system specifically designed to investigate advanced materials processing, production, and additive manufacturing. The goal of Mainstream's Electron Beam Enabled Advanced Manufacturing (EBEAM) center is to develop materials that are uniquely produced with high energy electron beam systems and alleviate significant technological deficiencies. We will present results of irradiating various metals and their alloys and the unique material properties that result. Monte Carlo simulations of the interaction of the electron beam with the sample were used to optimize the processing parameters and study far-from-equilibrium processing conditions. Among the applications that will be discussed are additive manufacturing of stainless steels for small modular nuclear reactors and corrosion-resistant cladding via bonding of dissimilar metals for next-generation molten salt reactors.

Introduction

Current manufacturing is subtractive in nature, either removing undesired material from a block of material or joining of sheet metal or simpler components into a more complex whole. These methods result in large amounts of waste and leaving weak weld joints components. Additive manufacturing is an up and coming alternative to subtractive manufacturing. Material is selectively added in 2D layers, resulting in a 3D part. Polymer-based 3D printing as a subset of additive manufacturing is now well-established. Metals and ceramics, however, melt at much higher temperatures, requiring an alternate approach. Metal AM is maturing, with established methods in material feeding and selectively bonding material only in the desired location. Wire-fed systems inject material out of a nozzle, much like 3D printing, while powder bed systems involve spreading successive thin layers of fine powder to construct a 3D build volume.[1] In either case, a metal AM system requires a high intensity, localized heat source to fuse the wire or powder in the desired spots in the final part. Lasers and electron beams are the primary heating methods. Lasers can operate in air, but lack penetration depth, slowing processing speed, and increasing stresses and microfractures.[2,3] Electron beams require vacuum, but can penetrate deeper, melt thicker layers, and can achieve higher power than laser systems. Small parts can be produced in short times, but multiple centimeter-scale parts take several hours, too long for industrial use.

Faster AM times require thicker melt layers without compromising part quality. This in turn requires much faster heating and cooling rates to melt and solidify the powder to solid material in less time. Flash melting to much higher temperatures in turn causes flash cooling and solidification when the beam moves. Rapid cooling creates a strong driving force for nucleation, producing a nanocrystalline structure. As such, higher power localized heating results in faster processing times. To reach more industrially useful times, much higher than current 50-100 kV laser or electron beam systems. Electron beam systems are more amenable to high energy as electrons can reach any energy given enough acceleration. Superconductive resonance cavity linear accelerators are capable of MeV level electron energies, 2 orders of magnitude higher than conventional electron beam AM systems. To this end, Mainstream Engineering Corp. has acquired the Electron Beam Enabled Advanced Manufacturing (EBEAM) system. The superconducting cavity is designed to accelerate an electron beam up to 2 MeV at 4 kW average power, resulting in 10^6 K/s heating and 10^4 K/s cooling rates. Nanocrystalline grain microstructures are accessible at fast raster speeds, enabling larger and stronger parts produced in less time with less waste than conventional manufacturing. Such microstructures are stronger, allowing smaller parts with less material to supports the same stresses as larger parts made with traditional manufacturing. Thus, miniaturization efforts will benefit from such technology.

EBEAM Overview

The EBEAM electron linear accelerator system is shown in Figure 1. The EBEAM system consists of a 60 keV Schottky electron gun and a superconducting linear electron accelerator and currently is capable of producing an electron beam up to 1.4 MeV at 1 kW. Past the accelerator cavity, the electron beam is directed down into a processing chamber; inside, the electron beam strikes component powder material placed onto a build plate. The powder coalesces into a melt pool under electron beam impact. As the beam rasters over the build plate area, the melt pool moves, melting more powder material while the previous material rapidly cools and recrystallizes into solid material.

Additive manufacturing part quality depends on many parameters including: melt pool depth, heating rate, and cooling rate and their roles in controlling solidification and crystallization. Electron beams possess an important advantage over lasers in depth of penetration; this gives electron beams the ability to re-melt and recrystallize larger volumes of already solid parts and materials. Figure 2 shows the recrystallization of existing coarse-grained material into ultra-fine grain material, producing material property advantages, including mechanical yield strength, surface hardness, radiation tolerance and magnetic permeability.[4-13] Such a high powered electron beam also enables new claddings consisting of two dissimilar metals bonded together, enabling high strength and corrosion-resistance even on irregular parts. Such advances are promising for modular nuclear reactor components.

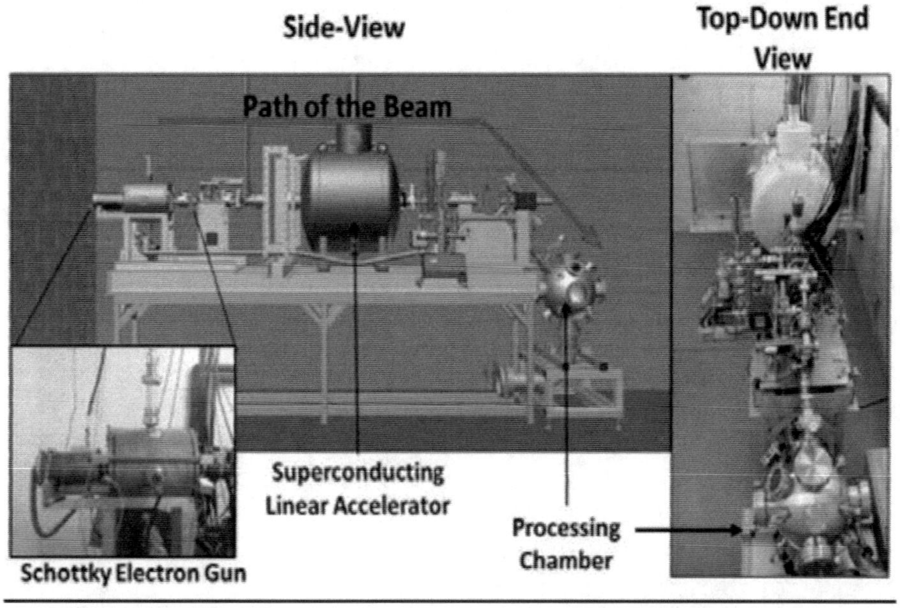

Figure 1. Electron Beam Enabled Advanced Manufacturing (EBEAM) electron linear accelerator.

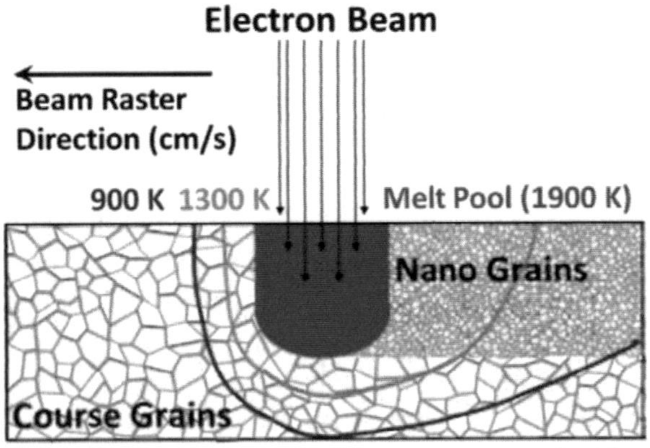

Figure 2. Nanocrystallization process under the higher electron energy possible in the EBEAM system.

Stainless Steel 316L Manufacturing for Small Modular Nuclear Reactors

As developing nations construct a nuclear power infrastructure and aging continues in developed nations' nuclear infrastructure, new generations of nuclear reactors have been proposed. One such concept is the small modular reactor (SMR) that requires far less space than conventional nuclear plants. As a modular design, it is possible to scale these reactors up to meet higher demand and locate the reactors within the consumer site. They can also eliminate power transmission losses and large capital investments needed to build new conventional nuclear plants. Response to rising demand can be met more quickly and economically as an alternative to fossil fuel sources.

SMRs require miniaturization of large nuclear reactor components such as fuel support assemblies, spray nozzles, cooling channels, steam separators, and instrument brackets.[14] Some components are difficult to produce with conventional manufacturing, raising capital costs to more than $1 billion and construction time to 4.5 years.[15] AM stands as a means to cut 6 months off this schedule at a 20% lower cost.[16] Maintenance costs also decline as parts are more easily replaced. However, several SMR parts are too large for effective production with current AM rates. The faster rates possible with the EBEAM system can meet this need, but much microstructure and mechanical property characterization remains to be done. Effects of raster speed, beam power, powder size, and porosity from incomplete sintering all need to be established. This work is an effort to answer these questions for 316L stainless steel powder, as commissioned by the DoE under the SBIR program, Phase II (DE-SC0017685).

Experimental Setup

Additive manufacturing was conducted using 316L stainless steel secured from LPW Technology as the raw feedstock. The powder consisted of spherical, gas atomized particles with a diameter of 44–106 μm and an average Hall flow rate of 15s/50g. The powder bed was held under vacuum at $\approx 10^{-7}$ mbar while the beam was magnetically rastered in a snake pattern, tracing in a single direction then turning around at the ends to begin the next row in the opposite direction. This rastering pattern leads to the linear ridging and troughs are apparent in the resulting solid coupons shown below. The EBEAM operating parameters are given in Table 1. Characterization included Rockwell hardness testing, optical microscopy (OM) and scanning electron microscopy (SEM).

Table 1. Electron Beam Processing Parameters for 316L Powder in EBEAM System.

Fixed Parameters	
Beam Energy (MeV)	1.16 ± 0.10
Beam Spot Diameter (mm)	3 ± 1
Powder Packing Fraction	57%
Varied Parameters	
Emitted Gun Current (mA)	1.5–2.25
Collected Beam Current (mA)	0.12–0.18
Beam Scan Rate (mm/s)	0.2–20
Raster Line Overlap	0–50%
Powder Layer Thickness (mm)	1–3

Results

Macroscale material morphology after EBEAM processing is shown in Figure 3. This fabricated from 316L stainless steel powder showing fusion of a rectangular area of material into a single piece is possible. However, the coupon is riddled with hole and ridges from uneven heating and cooling effects. Beam energy and spot size fluctuations (8.6% and 33%, respectively) are the most likely culprit. Efforts are ongoing to improve the beam control to tighten the spread. Another consideration is 316L stainless steel is slightly paramagnetic.[17,18] The local magnetic and electric fields generated from the electron beam is quite large given the large electron velocities, causing a Lorentzian force on the molten pool. This is not verifiable at present as these forces are tied to the beam fluctuations. As beam control improves, the magnetic effects will become consistent.

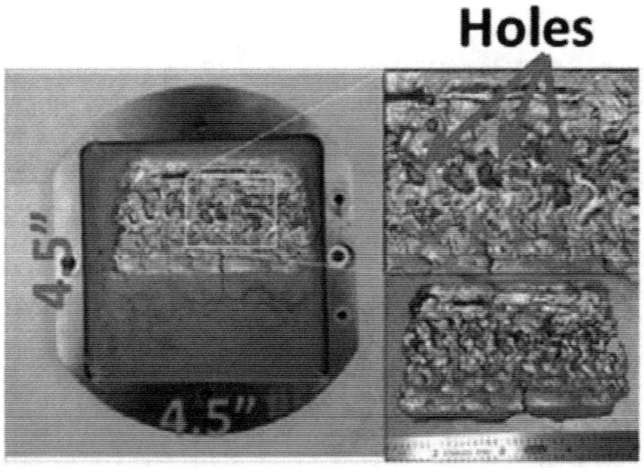

Figure 3. EBEAM test coupon fabricated out stainless steel powder. While forming a complete, solid piece, many holes and ridges exist from uneven heating and wide beam power variation. Efforts are underway to reduce such variation in process conditions.

This uneven morphology has prevented any tensile tests for direct yield strength comparison to the conventional hot-rolled, annealed and pickled (HRAP) 316L stainless steel cladding. Improved control of the process is necessary, including tighter control of the electron beam profile and energy, raster speed, pattern, and dwell time. Figure 4 shows that tightening the beam spot size variation and energy has improved the uniformity in powder melting, eliminating the holes and reducing height variation. Much work remains to be done on raster parameter optimization to eliminate the height variation. Once flat enough samples are fabricated with the AM process, tensile test samples can be cut out for a more direct comparison.

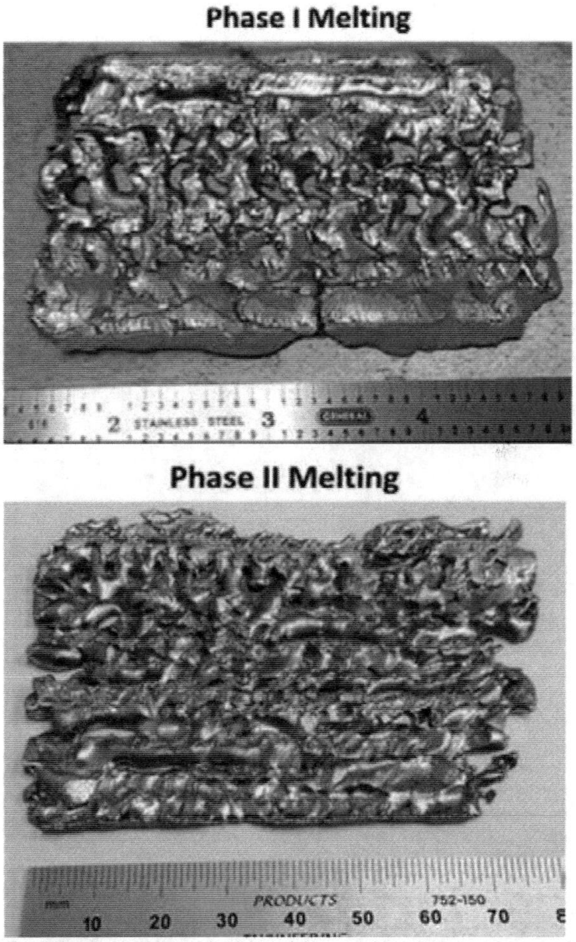

Figure 4. SBIR Phase I coupon vs. the improved version in Phase II. Improvements to electron gun control and tighter energy dispersion shows no holes and smaller ridges, indicating more uniform heating and cooling.

On the other hand, properties such as porosity, hardness, density and grain size in the test coupons can be compared to conventional HRAP plate. Microstructures as seen in OM and SEM are shown in Figure 5. Images of the HRAP plate microstructure in Figure 5a, c are consistent with equiaxial austenitic grains common to 316L.[19] Measured properties after both processing methods are given in Table 2.

Grain sizes in the AM processed material were 200-700 nm, less than 15% of the HRAP grain size. Significant porosity and incomplete sintering was present in the AM coupons, resulting in a low density, only 57% of the HRAP material. Local hardness was still measurable on the AM coupon in well-sintered regions. Hardness values were comparable for both methods, but the AM method is not useful until the porosity is eliminated. As with the large scale mounds and valleys, evening out the melting pool and thermal profile will make for more uniform solidification, reducing porosity.

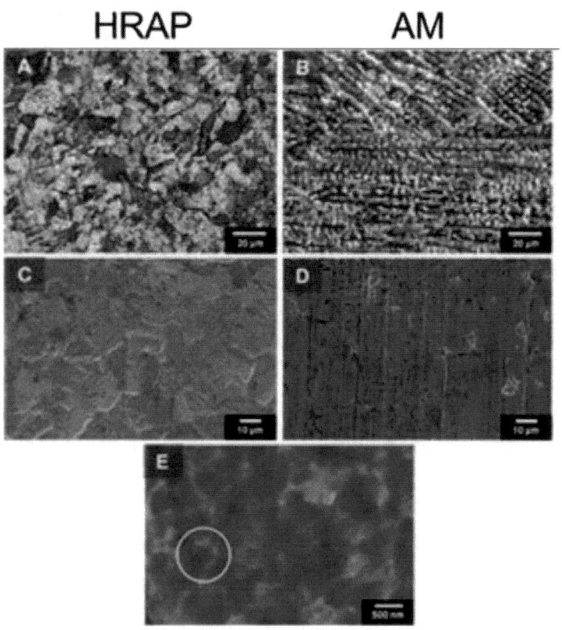

Figure 5. Microstructure after conventional HRAP processing in 316L stainless steel (a, c) and EBEAM powder fusion (b, d, e). The circle in (e) notes a potential grain in the EBEAM-processed material. HRAP produced a uniformly equiaxed structure while EBEAM processing produced a mix of equiaxed and dendritic grains.

Despite the clear uniformity problems in the AM coupons, the technique is still quite promising. This approach is applicable to any part shape, whereas HRAP processing is limited to plate and cylindrical shapes. The rapid heating and cooling rates produce smaller nanocrystalline grain sizes than HRAP plate, resulting in higher strength and hardness. Refinement of the EBEAM approach will reduce porosity and void volume to match or exceed HRAP part properties.

Table 2. 316L stainless steel properties comparison after EBEAM vs. HRAP processing.

	AM 316L Sample Coupon	HRAP 316L Plate	AM vs. HRAP
Hardness (RHC)	33.4 ± 2.5	33.7 ± 0.9	~100%
Density (g/cm^3)	4.6	8.0	57 %
Grain Structure	Dendritic + Equiaxial grain structure	Equiaxial austenite	—
Avg. Grain Size (μm)	0.2–0.7	4.8	< 15 %

EBEAM Simulation

Experimental test coupon production is instructive but suffers from limited throughput and high production cost. A large amount of liquid helium is necessary to maintain the accelerator cavity at superconductive temperatures, limiting the number of runs. Computational methods are necessary to weed out undesirable electron beam conditions. To this end, Monte Carlo simulations have been carried out to track electron motion within 316L stainless steel, as shown in Figure 6. From this, the melt pool is predicted along with the temperature profile tail following the rastering direction. Under the beam conditions of 1.16 MeV accelerating voltage, 3 mm beam diameter, 0.12 mA collected current, and 1.9 mm/s raster scan rate, we expected the melt pool to be hemi-ovoidal in shape with 0.82 mm diameter at the plate surface and 0.59 mm depth (Figure 7). Along with incorporating latent heat of melting into the model, we accounted for the fact that 316L does not have a single melting point but rather solidus and liquidus transition temperatures. This is because 316L is an alloy, and therefore, individual grains have their own melting points depending on the local, elemental grain composition. These simulations are helpful for estimating the effects of melt pool size and depth on the cooling time and thus material tensions that result in ridges and uneven height. On the other hand, reducing the melt pool size reduces processing speed. Hence, a balance exists for melt pool size, and work is ongoing to balance experimental beam parameters to obtain a fast speed while keeping height and width variations within acceptable limits requiring a minimum of post-AM machining. Still, a simple model of the melt pool guides powder layer thickness for each spread. A bigger melt pool enables thicker layers faster layer completion, substantially reducing part production time.

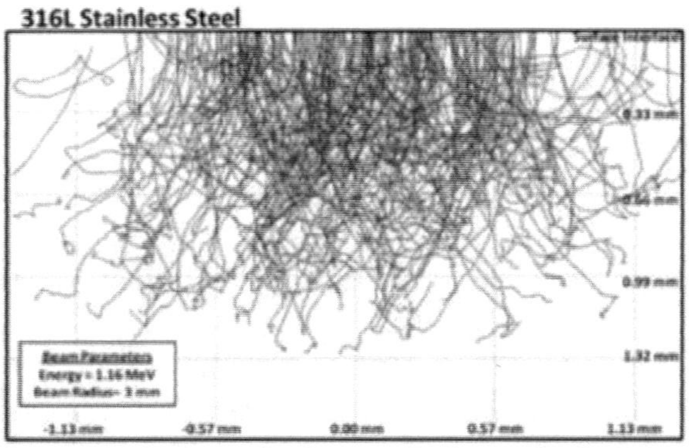

Figure 6. Monte Carlo prediction of electron paths within 316L stainless steel. Blue lines are electron paths, red lines represent electrons reflected back out of the simulated coupon.

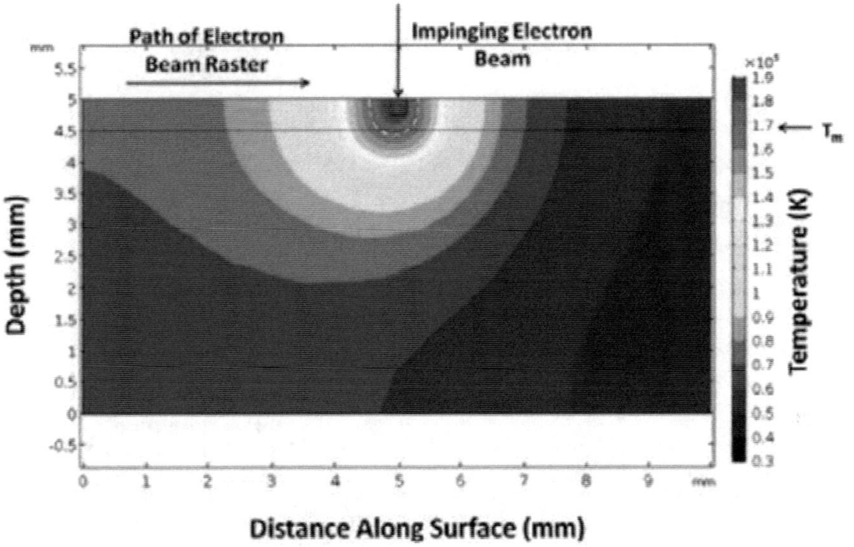

Figure 7. Melt pool prediction from the Monte Carlo simulation. Melt pool size, depth and the cooling of the tail predict heating and cooling rates. This in turn informs how to reduce residual stresses and liquid metal migration into ridges. This also shows the need for low variation in beam parameters.

Conclusions

Mainstream has commissioned an electron accelerator additive manufacturing system that enables many new applications in additive manufacturing. The two-order of magnitude higher electron beam power enables larger, deeper melt pools with 10^6 K/s heating rate and 10^4 K/s cooling rate. Larger parts, faster process speeds and stronger nanocrystalline structures all become possible under these conditions. This equipment was applied to the problem of making larger scale (multiple centimeter) parts for SMRs for upcoming nuclear power applications. 316L stainless steel powder was fused into test coupons and compared to conventional manufacturing methods. The AM coupons exhibited high porosity but similar hardness as conventional manufacturing. AM coupons were riddled with holes, but these have since disappeared with improved control. We expect the ridge height variation to diminish further with improved beam control techniques, leading to a viable large-scale 3D AM process. Once perfected, this technique has many profound implications beyond AM including applying corrosion resistant claddings to irregular parts, leveraging nanograin structure for higher strength parts, ceramic AM and laminar parts of strongly-bonded, alternating layers of material.

Acknowledgements

This material is based upon work supported by the U.S. Department of Energy, Office of Science, Office of Nuclear Energy, under Award Number(s) DE-SC0017685.

References

1 Gibson, I., Rosen, D. & Stucker, B. *Additive Manufacturing Technologies: Rapid Prototyping to Direct Digital Manufacturing*. (Springer, 2010).

2 Abe, F., Osakada, K., Shiomi, M., Uematsu, K. & Matsumoto, M. The manufacturing of hard tools from metallic powders by selective laser melting. *J. Mater. Process. Technol.* **111**, 210-213 (2001).

3 Facchini, L. *et al.* Ductility of a Ti-6Al-4V alloy produced by selective laser melting of prealloyed powders. *Rapid Prototyping Journal* **16**, 450-459 (2010).

4 Abramova, M. *et al.* Grain boundary segregation induced strengthening of an ultrafine-grained austenitic stainless steel. *Mater. Lett.* **136**, 349-352 (2014).

5 Huang, H., Wang, Z., Lu, J. & Lu, K. Fatigue behaviors of AISI 316L stainless steel with a gradient nanostructured surface layer. *Acta Mater.* **87**, 150-160 (2015).

6 Muley, S. V., Vidvans, A. N., Chaudhari, G. P. & Udainiya, S. An assessment of ultra fine grained 316L stainless steel for implant applications. *Acta Biomater.* **30**, 408-419 (2016).

7 Ueno, H., Kakihata, K., Kaneko, Y., Hashimoto, S. & Vinogradov, A. Enhanced fatigue properties of nanostructured austenitic SUS 316L stainless steel. *Acta Mater.* **59**, 7060-7069 (2011).

8 Hammond, V. H. Magnesium Nanocomposites: Current Status and Prospects for Army Applications (ARL-TR-5728). (U.S. Army Research Laboratory, Aberdeen Proving Ground, MD, Sept. 2011).

9 Walsh, S. M., Scott, B. R., Jones, T. L., Cho, K. & Wolbert, J. A materials approach in the development of multi-threat warfighter head protection. (U.S. Army Research Laboratory, Aberdeen Proving Ground, MD, Dec. 2008).

10 Chookajorn, T., Murdoch, H. A. & Schuh, C. A. Design of stable nanocrystalline alloys. *Science* **337**, 951-954 (2012).

11 Guo, X., Yang, G., Weng, G. & Zhu, L. Numerical simulation of ballistic performance of bimodal nanostructured metals. *Materials Science and Engineering: A* **630**, 13-26 (2015).

12 Huang, Y. & Langdon, T. G. Advances in ultrafine-grained materials. *Mater. Today* **16**, 85-93 (2013).

13 Tschopp, M. A., Murdoch, H. A., Kesckes, L. J. & Darling, K. A. "Bulk" Nanocrystalline Metals: Review of the Current State of the Art and Future Opportunities for Copper and Copper Alloys (ARL-TR-5728). *JOM* **66**, 1000-1019 (2014).

14 Babu, S., Love, L. J., Peter, W. & Dehoff, R. Report on Additive Manufacturing for Large-Scale Metals Workshop (ORNL/TM-2016/202. (Oak Ridge National Laboratory (ORNL), Oak Ridge, TN, May 2016).

15 Modro, S. *et al.* Multi-application small light water reactor final report. (Idaho National Engineering and Environmental Laboratory, Idaho Falls, ID ..., 2003).

16 NEUP. Fiscal Year 2017 Consolidated Innovative Nuclear Research - Appendix A: Workscopes for U.S. University-led Program and/or Mission Supporting R&D Projects. (Nuclear Energy University Program with

U.S. Department of Energy, Idaho Falls, ID, 2017).

17 Menéndez, E. *et al.* Controlled generation of ferromagnetic martensite from paramagnetic austenite in AISI 316L austenitic stainless steel. *J. Mater. Res.* **24**, 565-573 (2009).

18 AK_Steel. 316/316L Stainless Steel Product Data Sheet. (AK Steel Corporation, West Chester, OH, 2007).

19 Weidmann, E., Guesnier, A. & Taylor, B. Application Note: Metallographic preparation of stainless steel. (Struers A/S, Ballerup, Denmark, February 2016).

577916 Characterization of Spatial Variability in Material Properties in Echogenically-Segmented Metal Components Fabricated By Additive Manufacturing

Background, Motivation, and Objective

The difference in thermal history during layer-by-layer manufacturing of metal components by selective laser melting (SLM) and related additive techniques may result in microstructural anisotropy and spatial variability in strength, density, porosity, and other material properties. The difference is particularly significant between interior and surface regions and in the build direction. While the microstructural variation may be characterized noninvasively by X-ray computed tomography, the options for nondestructive assessment of spatial heterogeneity in end-use material properties are lacking. This paper describes our progress towards the development of a novel approach to the ultrasonic characterization of spatially varying material properties and its application in additive manufacturing.

Statement of Contribution/Method

The traditional nondestructive ultrasonic evaluation gives an aggregate characterization of mechanical properties of solids and is insensitive to their spatial variabilities. This paper adapts the previously developed method for the ultrasonic measurements of segmental temperature distributions in solids [1–2] to estimate the spatial variability of mechanical properties in SLM-manufactured parts. We build upon the earlier report on assessing heterogeneity in additively manufactured aluminum parts [3] and describe our progress in designing echogenic features segmenting 3D printed parts without interfering with end-use functionality while improving the accuracy in segmental characterization.

Results

The application of the developed approach is demonstrated by characterizing segmental variability in the titanium-alloy and stainless-steel parts fabricated by SLM. Three echogenic features incorporated during the fabrication produce the train of echoes, the time-of-flight of which provides information which we use to assess the segmental variability in material properties. The ultrasonic results were found to correlate with the destructive characterization of mechanical properties. Overall, the experimental results indicate the feasibility of detecting heterogeneity of the material properties in 3D-printed parts following the developed approach.

[1] Y. Jia & M. Skliar, *Energy & Fuels*, 2016. [2] Y. Jia et al., *Ultrasonics*, 2016. [3] M. Roy, et. al., *2018 IEEE IUS*, Kobe, Japan.

575361 Lab Enablement through 3D Printing - Nitrogen Inertion Funnel

Hydrogenation reactions pose a significant fire hazard due to the use of a mix of metal catalysts, reagents and solvents. Solvent flames are especially dangerous as they are barely visible and as such a present danger for burn injuries. Most metal catalysts supported on finely divided carbon powder are highly flammable and can ignite solvents and hydrogen. The catalyst material is especially dangerous after it has been used in a hydrogenation reaction leaving the catalyst metal soaked with adsorbed hydrogen. An inertion device has been developed that reduces fire risks associated with the filtering of metal catalysts. The inertion setup is composed of a 3D printed inertion ring that clips onto commercially available filter funnels. This has been tested at two different sized filter applications to determine how much inert gas, nitrogen in this case, was required to reduce the atmospheric oxygen below 12vol% ignition level and has also been tested to see at what flow rates actual ignition is suppressed when methanol is the solvent.

3D printing was a significant factor in not only the design and prototyping of the device but also in the manufacturing of the end use devices. The final product is topologically complex. Most methods of traditional manufacturing would have added significantly to the cost and complexity. This device required non linear internal passage ways and guide vanes to direct and smooth the flow of air. To produce these devices using more traditional means would have necessitated either manufacturing in multiple parts and heat welding or combining many methods, such as sheet metal forming, machining and welding which would have significantly increased cost and limited the design in ways that could have effected the efficacy of the function.

569117 Formulation of UV Curable Resins Utilized in Vat Photo Polymerization for the Additive Manufacturing of Gun Propulsion Charge in 3D Printers

Formulating resins for Additive Manufacturing (AM) utilizing UV laser stereolithography (SLA) is a new technique that makes it possible for the fabrication of complex geometries with high dimensional resolution. This layer by layer photopolymerization approach spans various industrial sectors from adhesives, inks and optical fibers to nanotechnology and biomaterials. UV curable resins such as epoxides, vinyl ethers and other acrylates are important monomers that offer effective mediums for energetic materials, and the potential exists to develop environmentally friendly formulations with suspended energetic materials at various solids loading levels.

Developing techniques for UV curing formulations of highly loaded energetic suspensions is a challenging feat that must satisfy several requirements and produce a high quality formulation with synergistic ingredient combinations to enhance propulsion phenomena. The candidate formulation must be able to operate in an (SLA) resin tank, meaning the suspension must be at least as fluidic as conventional SLA resins; and a thorough understanding of the polymer network structure and cure kinetics is essential for a resulting polymer that exhibits good mechanical properties.

569198 Viscoelastic Properties of Energetic Material Filled UV Photo Polymer Resins and Thermoplastic Feed Stock for the Additive Manufacturing of Gun Propulsion Charges

Additive manufacturing (AM), or commonly known as 3D printing, of various materials such as metals, thermoplastics and ceramics is growing in popularity as a practical and effective manufacturing method. This technology has blossomed over the past few decades and is starting to become cheaper and more available. Utilization of AM technology for the manufacture of gun propellants enables charge designers to realize very complex geometries in comparison to today's standards of gun propellants. The rheological properties of the materials for use in 3D printers is crucial for prototype development of propulsion charges. This paper will go over baseline rheological properties of both UV curable resins and thermoplastic binders for 3D printing. These materials will be loaded with nitramine filler in order to increase their energy content. The viscoelastic properties of the feed stock with energetic materials will drive the resolution and print quality of the prototype propulsion charges. This information will allow for material optimization related to diluents, viscosity modifiers while maintaining high energy density and the ability to 3D print samples.

577360 MgB_2/Fluoropolymer Filament and Pyrolant Combustion Studies

Lance N. Kotter[a] and Lori J. Groven[b,*]

[a]South Dakota School of Mines and Technology, Department of Material Science and Engineering, Rapid City, SD, US

[b]South Dakota School of Mines and Technology, Department of Chemical and Biological Engineering, Rapid City, SD, US

***corresponding author email: lori.groven@sdsmt.edu**

Magnesium diboride has been shown to be an excellent fuel in combination with sodium iodate for biocide formulations, but little work has been reported as a pyrolant and more specifically for the development of an energetic filament. Magnesium diboride is a hard ceramic fuel that is a viable replacement for metal fuels, such as aluminum, in energetic formulations. A fluoropolymer binder, THV, will be used to develop printed traces and structures for the combustion study. In this work, two methods were used to assess the role of MgB_2 interacting with the binder. First, a reactive filament at different ratios of MgB_2/THV was formulated using a Filabot extruder. Second, comparable mixtures were also made and printed using a solvent based ink approach. Fuel to oxidizer ratios of MgB_2 to THV binder (30/70, 40/60, 50/50, 60/40, and 70/30) are studied and compared. It is shown that MgB_2 performs on par with micron sized aluminum in these formulations and provides enhanced structural characteristics when used in printed structures. Printed structures with different geometries using a commercial Lulzbot Taz6 printer will also be discussed.

573726 Spherical Composite Powders As Feedstock for Additive Manufacturing

Our recent work showed that mechanical milling of powders of different materials in presence of two immiscible fluids as process control agent may yield composite spherical particles. It is hypothesized the fluids form a suspension so that spherical composite powders form after emulsified droplets capture and retain suspended particles. Because of their excellent flowability, such composite spheres are attractive as feedstock material for additive manufacturing. This effort is aimed to develop spherical powders with morphology and composition suitable for additive manufacturing of reactive materials. In one approach, metal powders are milled in presence of molten tin and wax as two immiscible fluids. The experiment is aimed to encapsulate metal particles inside tin droplets and recover solidified tin-bound spheres containing densely packed reactive metal particles. An experimental setup is developed including a miniature attritor mill combined with a heater and an enclosure to perform the milling in an inert gas environment. The inert gas is needed to prevent the possibility of ignition of evaporating wax. The experiments are performed with different metal powders, varying the milling time and milling intensity adjusted selecting different ball to powder mass ratios and rotation rate of the impeller. In a different approach, core-shell spheres are desired, with the core being a reactive material and shell being a low-melting metal. An example could be Ti-core, Al-shell particles. Preparing core-shell spheres relies on a staged milling in presence of two immiscible fluids, hexane and acetonitrile. The core powder is milled first and is expected to be encapsulated in emulsified acetonitrile droplets. The shell material is then added and milling continues. Initially, conditions are explored yielding spherical particles containing core material spheres. In follow-up experiments, preparation of core-shell spheres is targeted. This talk will present methodologies and results of the described effort.

580349 Solids Metrics for Successful 3D Printing of Energetic Feedstocks

Traditional manufacturing methods for producing warheads, propellants, and pyrotechnic systems are timely and costly. Additive manufacturing (AM) approaches offer one path of alternative manufacturing technologies. Among various technologies, powder bed fusion (PBF) is one promising approach. A PBF printer is essentially a miniature manufacturing facility in a box, with significant powder handling challenges that must be overcome for robust production performance and part quality [1, 2]. Feedstock handling in a powder bed fusion printer is no different than and faces the same challenges as full-scale solids handling processes.

For a feedstock to function properly in a powder bed fusion 3D printer, it must be designed to meet several material characteristics. Some of the most important material properties include: powder flowability & cohesion, powder and roll friction, permeability, particle size & shape distribution (PSSD), triboelectric charging propensity, wetting characteristics, compaction characteristics and thermal softening point.

This paper presents select studies of developing feedstocks for PBF printing of energetic and other materials. Key attributes benchmarked against included powder flowability, aeration attributes, wetting behavior, spreadability, and fusion, as well as final part performance (bulk density, detonation, sensitivity, and part tolerances.)

[1] Ennis, B.J. et al. (2008) Section 21: Solids-Solids Operations & Processing. Perry's Chemical Engineers' Handbook. Perry, R. H., & Green, D. W. (eds.), New York: McGraw-Hill.

[2] Ennis, B.J. et al. (2017), HPC4Mfg Modeling of Powder Dynamics in Metal Additive Manufacturing, Austin, TX. Link for presentation: https://hpc4mfg.llnl.gov/pdfs/events/powder2017/Powder_Dynamics_EG_17oct2017_FINAL.pdf

[3] Ennis, B.J. et al. (2017). Navy SBIR Phase I and Phase II Awards. Development of Explosive Feedstock for Commercial-off-the-Shelf (COTS) 3D Printers. Link for award:https://www.navysbirprogram.com/NavySearch/Award/award.aspx?pk=869CC83B-5945-4E75-A598- AA2527D51045

[4] "Chattanooga firm to help U.S. Navy figure out how to 3-D print explosives" (2017), T. Omarzu, Chattanooga Times Free Press. Story link:http://www.timesfreepress.com/news/business/aroundregion/story/2017/dec/10/chattanoogfirm- help-us-navy-figure-out-how-3-/458777/

570103 Performance of Different Types of Baffles Used in Gas-Solids Fluidized Bed Strippers

Fluidized bed strippers are essentially flowing fluidized beds where an up-flowing gas removes (strips out) product entrained in a down-flowing stream of the emulsion phase (denser solid phase). The stripper has to facilitate the transfer of the gas in the interstices of the emulsion phase and that adsorbed on the particles into the bubbles of the stripping gas. The counter-current contacting between the down-flowing emulsion phase stream and the up-flowing bubbles stream is enhanced by the use of an assembly of horizontal or inclined baffles. In unbaffled fluidized beds, the bubbles movement is unrestricted so, as they rise, interact, coalesce and grow in size they tend to move laterally toward the center. This would result in too low of a mass transfer to be an effective stripper.

Fluidized bed strippers play a major role in hydrocarbon processing with respect to fluid catalytic cracking (FCC) and fluid coking. In the FCC process, the catalyst collected by the reactor cyclones contains significant amounts of the product hydrocarbon vapors. Failure to remove these vapors from the catalyst results in a loss of a valuable product and excessively high temperatures in the regenerator. Thus, entrained and adsorbed hydrocarbon vapors are typically removed from the catalyst in a fluidized bed stripper using steam. In a fluid coker stripper, steam is used to remove entrained and adsorbed hydrocarbons from the coke particles descending from the fluid coker reactor thus minimizing carry-under of valuable hydrocarbon product. For both unit operations, a variety of proprietary and standard baffles are used with these fluidized bed strippers. The most common types are disk and donut trays, horizontal sheds and structured packings. Horizontal grating trays, sometimes referred to as subway gratings, are also used but to a much lesser extent. This paper reviews studies available in the open literature on fluidized bed strippers as applied to the FCC and fluid coker unit operations. The review has shown that experimental work has dealt with stripping efficiencies, stripper flooding and other flow dynamics issues. Fundamental studies on mass transfer in fluidized bed strippers are few and the application of computational fluid dynamic techniques to model fluid bed strippers appears to be a fairly recent addition. Overall, fluid catalytic cracking strippers, in particular disk and donut strippers, have received the most attention, undoubtedly, due to their much wider industrial usage.

568641 Dilute Phase Riser Dynamics – a Comparison of Scale

2019 AIChE Annual Meeting

Topic/Group: Fluidization and Fluid-Particle Systems

Tentative Session: Circulating Fluidized Beds

Title: Dilute Phase Riser Dynamics – A Comparison of Scale

Authors:

Ronald W. Breault: Ronald.Breault@netl.doe.gov

Steven L. Rowan: Steven.Rowan@netl.doe.gov

Jingsi Yang: Jingi.Yang@netl.doe.gov

Justin Weber: Justin.Weber@netl.doe.gov

Key Words: Fluidization, Circulating Fluidized Bed, Scaling, Cold Flow, Energy

Abstract:

Risers in gas-solids circulating systems are used extensively in the petrochemical and energy industries as well as being proposed for advanced technologies from chemical looping combustion/gasification. The performance of these units has been linked to the meso-scale and the dynamic behavior of the flow field. To examine the effects of riser scale on these dynamic behaviors, experimental data collected in a 0.1 m diameter riser with a 5.3 m height is compared to data collected under similar operating conditions in a 0.3m diameter riser with a 16m height. Experiments were conducted primarily in the dilute upflow regime prior to core-annular flow, with Gs/Gs* ratios between 0.2 and 0.3, and superficial gas velocities ranging between 5.5 – 6.0 m/s. Comparisons between riser pressure, solids fraction, voidage, and particle velocity profiles are considered for a range of bed materials, including 750 μm high density polyethylene (HDPE) and 180 μm glass beads. Additionally, where available, three-dimensional measurements of the solids volume fraction obtained via a Tech4Image Electrical Capacitance Volume Tomography (ECVT) system are discussed. The effect of system scale is analyzed and discussed.

575552 CFD Simulation of Carbon Capture Using an Amine Based Solid Sorbent

Carbon capture using of amine-based solid sorbents from existing flue gas of fossil fuel-fired power plants, has been considered as a promising method compared to the conventional liquid based systems. National Energy Technology Laboratory (NETL) designed an integrated adsorption and regeneration system of fluidized beds to investigate the amine-based sorbents capacity. In this study, the Eulerian-Eulerian two fluid model is used to conduct two dimensional Computational Fluid Dynamic (CFD) simulation of Carbon dioxide sorption and regeneration using amine-based solid sorbents in a circulating fluidized bed, based on NETL setup. kinetic theory was used as constitutive relation to analyze particulate phase flow model along with frictional stress and viscosity for the dense solid flow regimes. Stable operation of the system and continues circulation of the solids between carbonator and regenerator fluidized beds has been verified by comparing simulation results of solid circulation rate by experimental data.

Also, at the second stage of the study, carbonation and regeneration heterogeneous reactions was added to the CFD model. So, in addition to continuity and momentum conservation equations, species conservation and energy conservation equations were also solved to calculate the rate and extend of CO_2 sorption and temperature distribution in the system

574864 Coupled CFD-DEM Model of Fluidized Bed Drying of Pharmaceutical Wet Granules

Coupled CFD-DEM model of fluidized bed drying of pharmaceutical wet granules

Syed N. Ahsan

Pharmaceutical Sciences
University of Connecticut

Storrs, CT

/ed_nazmul.ahsan@ucon.edu

Bodhisattwa Chaudhuri

Pharmaceutical Sciences

Chemical and Biomolecular Engineering and Institute of Material Sciences (IMS)

University of Connecticut, Storrs, CT

bodhi.chaudhuri@ucon.edu

Yijie Gao

Takeda Pharmaceuticals
Cambridge, MA
yijie.gao@takeda.com

Here, we propose to present preliminary works to develop a model based on Computational Fluid Dynamics (CFD) and Discrete Element Method (DEM) coupling using opensource software for simulating gas-solid multiphase-multicomponent flow (MMF) associated to the drug manufacturing process. MMF flow is prevalent in a myriad of processes used in the pharmaceutical industry. Here, we seek to understand the processes through the application of the two numerical techniques with the overarching goal to the model the fluidized bed drying process. The experimentally validated fluidized bed dryer model will allow for both qualitative and quantitative study of fluidization and drying kinetics, that is heat and mass transfer during drying. In addition, we envision to investigate experimentally and numerically the different regimes of flow during the drying process by systematically changing the process parameters. The software's modular approach permits to incorporate new models and can simulate large scale models with MPI parallelization that would be beneficial for the engineers of Industry. The validated and fine-tuned CFD-DEM coupling model will be beneficial as it is cost-effective, amenable to changes in the boundary or the material properties of different phases and finally offers better control in the processing parameters. Further, the proposed process modeling will help reduce the drug development time thereby preventing product delays, improve scale-up optimization and ease the manufacturing process by serving as a priori predictive tool.

Keywords:

Multiphase-Multicomponent flow, Fluidized bed, Powdered Material Drying, Computational Fluid Dynamics, Discrete Element Method

Preliminary results:

The CFD-DEM coupled model of packed bed using the opensource software CFDEM to simulate multiphase-multicomponent flow pertaining to fluidization phenomena has been preliminarily implanted by PI's group. Such a model developed to assess the capability of the simulation tool is verified to be capable to replicate the fluidization process of

Figure: Effect of different level of cohesive force between particles in fluidization.

varying particulate system and produce comparable results with the empirical model available in the literature. The force model is implemented to take into account the interaction among particles and between particle and wall. The model used is the Gran-Hertz-History model where a non-linear relationship is assumed between overlap distance during the collision and normal contact force. The simplified fluid bed model has been used to assess different coupling models, fluid-particle interaction force models, and interdependency of mesh-particle size and volume fraction evaluation models. In addition, different level of cohesive forces are investigated based on simplified 'jkr' cohesion model to accommodate the fluidization process of pharmaceutical powdered material.

575108 Numerical and Experimental Investigation of a Lab Scale Wurster Coating Process

Numerical and Experimental Investigation of a Lab Scale Wurster Coating Process

Martina Trogrlic[1], Thomas Forgber[1], Sharareh Salar Behzadi[1], Johannes Khinast[1,3], Avik Sarkar[2], Dalibor Jajcevic[1]

[1] Research Center Pharmaceutical Engineering, Inffeldgasse 13, 8010 Graz, Austria

[2] Worldwide Research and Development, Pfizer Inc. Groton CT, USA

[3] Institute of Process and Particle Engineering, Inffeldgasse 13/III, 8010, Graz, Austria

Coating is one of the essential processes in the pharmaceutical, chemical, food and other industries. In the pharmaceutical industry the Wurster coater application is commonly used because of its wide range of application: e.g. coating, drying, pelletizing and granulating. The typical path of a particle begins in the Wurster tube, where the particles get accelerated upwards and enter the spray zone. After spraying, the particles move into the fountain region, and subsequently fall into the down bed region where they are fluidized. The liquid part of the deposited droplets evaporates from the particle surface either in the fountain or in the down bed region. From the down bed region the particles eventually get drawn back in the tube again, repeating the cycle.

The coating layer is formed by two events: depositing the spray droplets on the particles in the spray zone, and the evaporation of the liquid part of the deposited droplets. The formation and quality of the coating film are influenced by the conditions the particle was exposed to in the process equipment, and the time it was exposed to those conditions. As [1] showed in their experimental study, there are large regional differences in the coater with respect to temperature and relative humidity (RH). Furthermore these differences vary with the operating conditions, i.e. batch size, spray rate and fluidization air flow. However, the data obtained from the experiments is restricted with respect to spatial resolution, and it is very difficult to get the picture of the entire coating process relying only on this information.

Using the fully coupled (i.e. mass, energy and momentum) CFD-DEM model we are able to resolve the hydrodynamics of very large number of particles [2]. With the numerical model, not only can we extract trajectories, residence and cycle times in specific zones, but also individual particle temperatures, environment conditions and the rate of evaporation.

In this work we consider four different cases, where the fluidization air rate and the inlet temperature are varied. We validate our numerical model against the experimental data, and investigate the influence of the varied inlet parameters on the spray deposition and evaporation rate. We attempt to identify zones with specific environment conditions (i.e. temperature and humidity) and show how their size varies with the process parameters and investigate the residence time of particles in those zones.

This investigation is important because the environment conditions in the Wurster coater have an effect on the formation of the coating layer. This study attempts to investigate the particle history in the coater, to gain a better understanding of the Wurster coating process itself.

[1] S. J. Maronga and P. Wnukowski, "The use of humidity and temperature profiles in optimizing the size of fluidized bed in a coating process," *Chem. Eng. Process. Process Intensif.*, vol. 37, no. 5, pp. 423–432, 1998.

[2] D. Jajcevic, E. Siegmann, C. Radeke, and J. G. Khinast, "Large-scale CFD-DEM simulations of

fluidized granular systems," *Chem. Eng. Sci.*, vol. 98, pp. 298–310, 2013.

566857 3D CPFD Simulation of Circulating Fluidized Bed Riser and Downer: Comparisons of Flow Structure and Solids Back-Mixing Behavior

1. Introduction

Circulating Fluidized Bed (CFB) widely applied to industrial processes such as Fluidized Catalytic Cracking (FCC) and Coal Combustion is mainly composed of a riser in which the gas and solids move upward co-currently and a downer in which the two phase flow moves downward by the gravity [1]. CFB risers have a lot of advantages over conventional bubbling fluidized bed and turbulent fluidized bed because of its high gas-solids contact efficiency, high gas-solids throughput and reduced gas-solids axial dispersion [2]. However, CFB risers still meet with some problems such as severe solids back-mixing and clustering effects caused by the nonuniform gas-solids flow structure in axial and radial direction which will reduce the interphase contact efficiency and reaction selectivity [3]. Compared to the conventional riser, downer as a type of novel reactor showing a rather uniform flow structure, less axial and radial solids back-mixing which is beneficial to the reaction with short contact time at high temperature [4]. Therefore, downer had drawn much more attention in the past decade. In this study, a 3D CPFD model based on the modification of drag model considering the clustering effect is established to describe hydrodynamics of gas and solids and solids back-mixing behavior in CFB riser and downer. The model is validated by comparing the simulation results with the experimental data [5] under different operating conditions. The residence time distribution (RTD) of solids is introduced to describe the degree of solids back-mixing behavior.

2. Method

Computational particle fluid dynamics (CPFD) is a new form of Eulerian-Lagrangian approach which is based on the multi-phase-particle-in-cell (MP-PIC) method [6]. In CPFD scheme, interphase interactions are described as a spatial gradient and particles sharing the same properties such as size, density, velocity and so on are defined as a parcel [7]. Then throughout tracking the motion of each individual parcel describes the overall flow structure so that the CPFD approach can extremely reduce the computational cost. Therefore, the CPFD approach is chosen to simulate the circulating fluidized bed riser and downer. Meanwhile the EMMS Matrix drag model and a cluster-based drag model are adopted to simulate the riser and downer by separately considering the effects of cluster in both riser and downer.

3. Results and Discussion

The results demonstrate that the proposed model based on the modification of cluster offers more accurate prediction of solids holdup and solids vertical velocity than Gidaspow drag model. In addition, although the solids holdup in downer is much lower than that in riser under the same operating condition, the clustering phenomenon can be observed from the simulation results in downer. Because of the opposite flow direction in riser and downer, the formation of cluster will cause a higher solids vertical velocity in downer while a lower solids vertical velocity in riser. The cluster is evenly distributed in downer while it is concentrated near the wall in riser. This is caused by the different formation mechanism of cluster in riser and downer. The solids flow in downer has a self-adjustment mechanism which will restrain the formation of cluster so that the flow structure in downer is much more uniform. Furthermore, it is demonstrated that residence time distribution curve of solids is narrower with a higher peak in the downer compared to the riser which indicates that less solids back-mixing is existed in the downer and the flow structure is nearly approached to the ideal plug flow. The standard deviation of solids residence time is introduced to describe the solids back-mixing behavior quantitatively and when the value is high indicates a severe solids back-mixing. The value of solids residence time standard deviation in downer is 0.02 s while that in riser is 3.6 s at the height of 5 m in fully developed region which quantitatively indicates that the solids flow structure in downer is much more uniform than riser.

References

[1] Zhang M H , Chu K W , Wei F , et al. A CFD–DEM study of the cluster behavior in riser and downer reactors[J]. Powder Technology, 2008, 184(2):151-165.
[2] Zhang H , Huang W X , Zhu J X . Gas-solids flow behavior: CFB Riser vs. Downer[J]. AICHE Journal, 2001, 47(9):2000-2011.
[3] Wu B , J.-X. Zhu, Briens L . A Comparison of Flow Dynamics and Flow Structure in a Riser and a Downer[J]. Chemical Engineering & Technology, 2007, 30(4).
[4] Wang C , Barghi S , Zhu J . Hydrodynamics and reactor performance evaluation of a high flux gas-solids circulating fluidized bed downer: Experimental study[J]. AICHE Journal, 2015, 60(10):3412-3423.
[5] C. Wang, High-density Gas-Solids Circulating Fluidized Bed Riser and Downer Reactors Dissertation The University of Western Ontario, Canada, 2013.
[6] Shi X , Lan X , Liu F , et al. Effect of particle size distribution on hydrodynamics and solids back-mixing in CFB risers using CPFD simulation[J]. Powder Technology, 2014, 266:135-143.
[7] Shi X , Sun R , Lan X , et al. CPFD simulation of solids residence time and back-mixing in CFB risers[J]. Powder Technology, 2015, 271:16-25.

566906 Thermal Decomposition Synthesis of Monodisperse, Phase-Pure Iron Oxide Nanoparticles for Mpi and Magnetic Hyperthermia Applications

Thermal decomposition syntheses are often the method of choice to produce highly crystalline inorganic metal and metal oxide nanoparticles with tightly controlled physical size and shape distributions. A hallmark of the thermal decomposition synthesis is that it is typically performed in inert atmospheres, because the reaction temperature is often above the flash point of the organic solvent. However, such conditions may be unfavorable to the formation of phase pure metal oxide particles. For iron oxide nanoparticles, the anoxic conditions in the reactor favor the formation of nanoparticles with a mix of magnetic and non-magnetic domains or phases and the magnetic properties are poorly controlled and sub-par compared to bulk crystals. This poses a critical limitation, since the exact nature of the nanoparticles' response to externally applied static or time-varying magnetic fields forms the basis for their use in many applications, and this response is directly related to the nanoparticle's magnetic properties.

Methods to manipulate crystal phase and promote the formation of magnetite have been reported, either through post-synthesis oxidation or by tuning the electrochemical redox potential of the reaction mixture during synthesis. However, the reported syntheses still offer limited tunability in size. To overcome this limitation, our group has explored controlled addition of molecular oxygen to promote *in situ* oxidation of the nanoparticles during growth, and we have shown that phase-pure magnetite nanoparticles with magnetic properties approaching those of bulk magnetite could be synthesized reproducibly with fine control in size over the range of 10-20 nm. Here we report efforts to extend this size range to 10-40 nm and improve reproducibility of nanoparticle synthesis. First, we improved reproducibility of the iron oleate precursor synthesis, as we believe variability in the chemical structure of the oleate prevented us from obtaining particles larger than 20 nm. Next, we explore the reaction parameter space to identify conditions over which monodisperse iron oxide nanoparticles can be grown reproducibly, both in an inert atmosphere and with controlled molecular oxygen addition to promote the formation of phase-pure magnetite with magnetic properties approaching those of the bulk. Finally, we characterize these particles for their potential biomedical applications in magnetic hyperthermia and the new molecular imaging modality called magnetic particle imaging.

558787 Magnetic Nanoparticle Clusters for Efficacious Bilirubin Removal from Body Fluids

Accumulation of bilirubin in the blood of patients suffering from end-stage liver disease severely compromises their health and prospects for survival[1]. Hyperbilirubinemia - if left untreated - may cause hepatic encephalopathy leading eventually to irreversible damage to the brain[2]. Traditional extracorporeal blood purification techniques and liver support devices only poorly remove bilirubin and suffer from membrane saturation and clotting[3]. Magnetic blood purification based on the capturing of toxins and pathogens by freely diffusing magnetic nanoparticles may offer an attractive alternative[4]. Here, we report on the functionalization of 100 nm superparamagnetic magnetite cluster nanoparticles functionalized with cationic polymer brushes grown using Surface Initiated ATRP for bilirubin capturing from body fluids. We demonstrate that the bilirubin removal capacity depends on the brush length and composition. Batch extraction efficiencies of more than 99% were found for bead concentrations of 1mg/mL and extraction/separation times of less than two minutes. While the most efficacious capturing agents were also the most cytolytic ones, significant bilirubin removal efficiencies of 40% could be obtained at minimal bead concentrations of 25µg/mL in albumin rich solutions. At these low concentrations, cytolytic activity remained indistinguishable from untreated controls. Taken together, we demonstrate the feasibility of quantitative bilirubin capturing by polymer-functionalized magnetic nanoparticles with prospective implications for artificial liver support.

References

1. Yong, A.; Ying-Bin, X.; Qian-Jin, Z.; Hyperbilirubinemia after extracorporeal circulation surgery: A recent and prospective study World J Gastroenterol 2006; 12(41): 6722-6726
2. Wennberg, R.P.; The Blood–Brain Barrier and Bilirubin Encephalopathy Cell Mol Neurobiol 2000; 20: 97.
3. Stadlbauer, V; Krisper, P; Beuers, U; et al. Removal of bile acids by two different extracorporeal liver support systems in acute-on-chronic liver failure. ASAIO J 2007; 53: 187.
4. Herrmann, I. K.; How Nanotechnology-Enabled Concepts Could Contribute to the Prevention, Diagnosis and Therapy of Bacterial Infections. Crit Care 2015, 19, 239-243.

576603 Antioxidant Protection from Oxidative Stress Damage in Human Pulmonary Cells

Exposure to pollutants and particulate matter cause oxidative stress through the direct introduction of exogenous ROS and compounds that drive free radical reactions, or indirectly through the recruitment and activation of inflammatory cells which release free radicals. The large surface area of the lungs and direct contact with inhaled air make the respiratory system highly susceptible to oxidative stress related injury, including DNA damage, cellular barrier dysfunction, and direct tissue damage. More susceptible populations include immunodeficient individuals, infants, elderly and those with pulmonary diseases such as asthma.

Previous groups have investigated the acute and subchronic exposure responses of human pulmonary cell line A549 to particulate exposure, including production of reactive oxygen species and cell viability. Antioxidant supplementation with Trolox, a commercially available vitamin E analog, has been investigated by various groups to decrease DNA damage and cell cycle arrest in A549s when acutely exposed to oxidants, including particulate matter (PM10). The first goal of this work was to characterize acute and chronic oxidative stress response parameters of the A549 cell line by the ratio of reduced and oxidized glutathione, chemokine production, DNA integrity, and surfactant secretion. The second goal was the development of an antioxidant formulation to prevent oxidative stress damage to the pulmonary cells.

575981 Crosslinking of Ovalbumin Nanoparticles Affects Cellular and Humoral Immune Responses

Cancer immunotherapy takes advantage of the body's own immune system to fight cancer. In antigen-specific immunotherapy, cancer-specific antigens are delivered to dendritic cells (DCs), which can initiate an immune response against the tumor by processing and presenting captured antigens to T cells. Despite great progress in recent years, there are still challenges that need to be addressed, such as insufficient induction of immune responses and off-target side effects of immunotherapeutics.

Here, instead of decorating preexisting particles with peptide or protein antigens of interest, we have developed a new class of protein nanoparticles that are entirely made of the model protein antigen (protein antigen nanoparticles, PANPs).

We found that as the Young's modulus of PANPs increases, their interactions with immune cells can be tailored. In *in vivo* immunization studies that were conducted in a prime-boost regimen, PANPs with intermediate Young's modulus elicited 50-fold and 10-fold increase in serum IgG titers compared to soluble antigen. Bone marrow-derived dendritic cells (BMDCs) internalized more PANPs with lower elasticity. Overall, by tuning the elasticity of PANPs, we were able to alter the humoral and cellular responses in terms of their uptake by BMDCs, OT-1 CD8+ T cells proliferation and antibody production.

571964 Encapsulation of Poorly Water-Soluble Drugs into Yeast Glucan Particles By Spray Drying: Improvement of Dispersion and Dissolution Properties

Encapsulation of poorly water-soluble drugs into yeast glucan particles by spray drying: Improvement of dispersion and dissolution properties

Gabriela Ruphuy*, Ivan Salon, Jan Tomas, Petra Šalamunová, Jaroslav Hanus, František Štepánek

Department of Chemical Engineering, University of Chemistry and Technology Prague, Technická 3, 166 28 Prague 6, Czech Republic

*E-mail: ruphuycg@vscht.cz

Yeasts are single-celled microorganisms belonging to the fungus kindom. They are more commonly known for their use in food products and, in the biomedical field, they have been of great interest for drug delivery. The so-called yeast glucan particles (GP) are obtained after removing all the internal organelles and cell wall components, other than ß-glucans, from *Saccharomyces cerevisiae* (baker's yeast). The result are hollow and porous shells which are suitable for drug encapsulation. Since they are obtained from microorganisms, they activate pattern recognition receptors of host immune cells, triggering immune responses [1]. For this reason, most of the studies have been focused on the use of GP for the encapsulation and macrophage-targeted delivery of drugs.

Peptides [2], liposomes [3], siRNA [4], DNA [5], among others water-soluble payloads, have been successfully encapsulated into GP for macrophage-targeted delivery. Publications on the encapsulation of poorly-water soluble drugs, on the other hand, are scarse, constituting an area of study yet to explore. Low water-soluble resveratrol, for instance, was encapsulated in living yeast [6], and curcumin, also insoluble in water, was loaded in 1,3-ß-glucan isolated from mushrooms [7]. Given that GP are mainly composed of amorphous polysaccharides (>85% ß-glucans), they are suitable candidate-materials for the development of amorphous solid dispersions (ASDs). To the best of our knowledge, no studies are found either regarding the encapsulation of low-water soluble drugs in beta-glucan particles by spray drying, or regarding the use of beta-glucan particles for the amorphization of low water-soluble drugs.

In this work, low water-soluble drug models, such as ibuprofen (IBU) and curcumin (CC), were loaded into GP using spray drying. Microparticles with different IBU-to-GP mass ratios were prepared and characterized. Samples with low IBU content (20% or less) exhibited amorphous state, but crystallinity peaks appeared and increased with drug content. The presence of IBU crystals outside the GP was observed in the case of the samples with higher IBU/GP mass ratio (>20% IBU).

This was evident also in the XRD results.

Different sets of spray-drying parameters were tested to evaluate the influence of droplet size and initial solid content on encapsulation efficiency (EE). It was determined that both, droplet size and initial solid content, have an influence on EE, but the effect of droplet size seems to be more significant; larger droplet sizes lead to higher encapsulation efficiencies. For instance, approximately 100% EE was obtained for a CC/GP composite sample produced using an ultrasonic nozzle (USN), whereas a drug content of 61.5 % was obtained for the CC/GP sample prepared using the 2-fluid nozzle (2FN). With the USN, much larger droplets are produced than with the 2FN under the conditions used. Therefore, the difference in EE is attributed to losses of curcumin that precipitated outside of the glucan particles in the case of the 2FN-sample due to a smaller droplet size (see **Fig. 1**). Such curcumin particles are very small; therefore, there is a high possibility that they were not collected in the cyclone, causing the losses along the spray dryer.

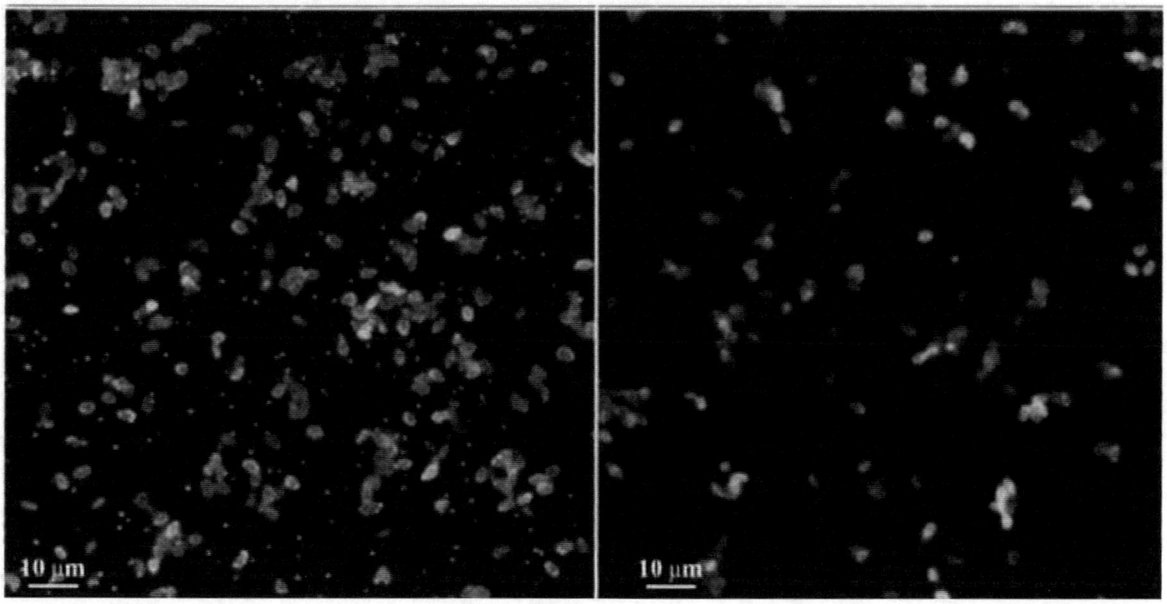

Figure 1. Confocal microscopy images of YGP/CC produced with 2-fluid nozzle (left side), and ultrasonic nozzle (right side).

Lastly, dissolution kinetics were evaluated for IBU/GP composites produced using the USN, and physical mixtures of the composites with crude crystalline ibuprofen. Progressively faster dissolution profiles (see **Fig. 2**) were obtained with increasing mass fraction of the composites, until the solubility limit was reached. In fact, the fast release of the IBU/GP composites lead to supersaturation, a typical characteristic of ASD. This is an important property because,if the supersaturated state is maintained for sufficient time, the *in-vivo* drug absorption may increase, resulting in an overall improved oral bioavailability [8-10]. Finally, besides fast dissolution, these

IBU/GP composites exhibited outstanding wettability and dispersion properties.

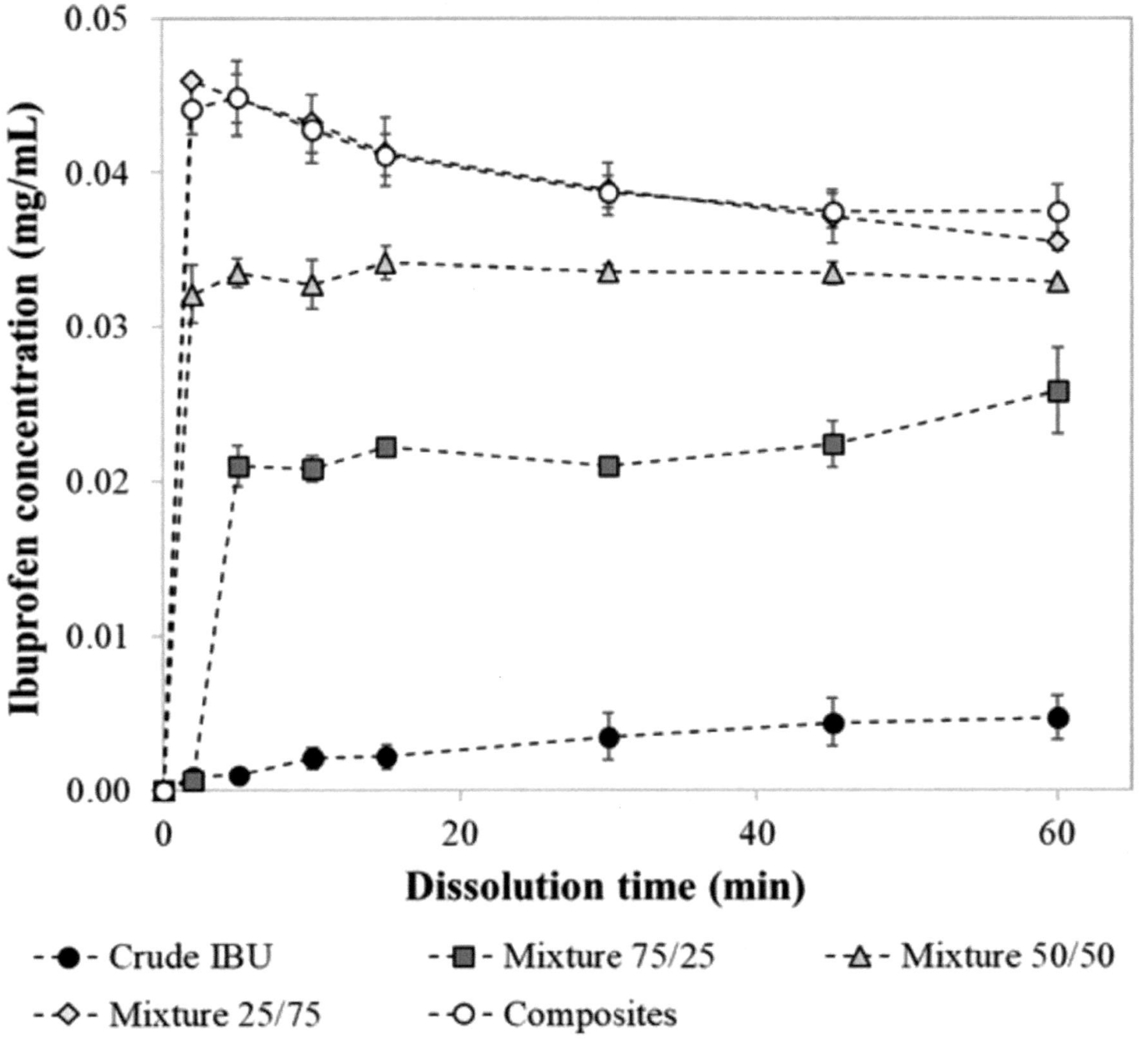

Figure 2. Dissolution kinetics of crude ibuprofen, IBU/GP composites prepared with ultrasonic nozzle, and physical mixtures of them (e.g. mixture 75/25 = 75 wt.% of crude ibuprofen and 25 wt.% of IBU/GP composites).

Acknowledgements. Authors acknowledge financial support by the European Structural and Investment Funds, OP RDE-funded project 'CHEMFELLS4UCTP' (No. CZ.02.2.69/0.0/0.0/17_050/0008485), and by the Agency for Healthcare research, project no. 16-27522A.

References

[1] A.B.C. Samuelsen, J. Schrezenmeir, S.H. Knutsen, Effects of orally administered yeast-derived

beta-glucans: A review, Molecular Nutrition & Food Research, 58 (2014) 183-193.

[2] X. Zhang, Y. Zhao, Y. Xu, Y. Pan, F. Chen, A. Kumar, G. Zou, X.-J. Liang, In situ self-assembly of peptides in glucan particles for macrophage-targeted oral delivery, Journal of Materials Chemistry B, 2 (2014) 5882-5890.

[3] F. Garello, R. Stefania, S. Aime, E. Terreno, D. Delli Castelli, Successful Entrapping of Liposomes in Glucan Particles: An Innovative Micron-Sized Carrier to Deliver Water-Soluble Molecules, Molecular Pharmaceutics, 11 (2014) 3760-3765.

[4] M. Aouadi, G.J. Tesz, S.M. Nicoloro, M. Wang, M. Chouinard, E. Soto, G.R. Ostroff, M.P. Czech, Orally delivered siRNA targeting macrophage Map4k4 suppresses systemic inflammation, Nature, 458 (2009) 1180-1184.

[5] E.R. Soto, G.R. Ostroff, Characterization of Multilayered Nanoparticles Encapsulated in Yeast Cell Wall Particles for DNA Delivery, Bioconjugate Chemistry, 19 (2008) 840-848.

[6] G. Shi, L. Rao, H. Yu, H. Xiang, H. Yang, R. Ji, Stabilization and encapsulation of photosensitive resveratrol within yeast cell, Int J Pharm, 349 (2008) 83-93.

[7] H.L. Mai, T.H. Phuong, T.N.T. Bich, H.T.T. Hong, T.H.T. Minh, T.M. Thu, H.T.T. Nhu, N.D. Huu, P.N. Xuan, Q.D. Tuan, Preparation and Antitumor-promoting Activity of Curcumin Encapsulated by 1,3-ß-Glucan Isolated from Vietnam Medicinal Mushroom Hericium erinaceum, Chemistry Letters, 40 (2011) 846-848.

[8] A. Dahan, A. Beig, D. Lindley, J.M. Miller, The solubility–permeability interplay and oral drug formulation design: Two heads are better than one, Advanced Drug Delivery Reviews, 101 (2016) 99-107.

[9] S.P. Chaudhari, A. Gupte, Mesoporous Silica as a Carrier for Amorphous Solid Dispersion, arXiv preprint arXiv:1707.00036, (2017).

[10] J. Brouwers, M.E. Brewster, P. Augustijns, Supersaturating Drug Delivery Systems: The Answer to Solubility-Limited Oral Bioavailability?, Journal of Pharmaceutical Sciences, 98 (2009) 2549-2572.

561107 Multiscale Analysis of Topically Applied Nanoparticulate Adhesives: Biodistribution and Biotransformation

In recent years, metal and metal oxide nanoparticles have received increasing interest in the biomedical world due to their high reactivity and versatility. [1] Despite very promising approaches and proof of concepts, the distribution and complex interaction of nanoparticles with the surrounding tissue are still poorly understood and remains a major obstacle to translation. The rational and meaningful development of nanomedicines is only possible with a good grasp of nanoparticle fate on multiple length scales.

Here we demonstrate the correlative and multiscale analysis of a nanoparticulate tissue adhesive based on hybrid ceria/bioglass nanoparticles. We shine light on the biodistribution and biotransformation of this complex metal oxide system by a multitude of techniques. One week after topical application[2] of the nanoparticles onto the subcutaneous tissue of a rat, we run elemental analysis on the harvested organs to determine that nearly all particles have stayed in the skin. We then identify skin regions with high nanoparticle content using MicroCT. Histological sectioning of said regions gives us direct access to the nanoparticles and the surrounding tissue. We discover that a major part of the nanoparticles have stayed in the subcutis via XRF mapping of the histological skin sections. We then further identify and characterize the surroundings of the nanoparticles by histological staining and electron microscopy. Hence, we are able to show colocalization of nanoparticles and macrophages. We then harness advanced material characterization techniques such as Raman spectroscopy and ToF-SIMS to further classify the surroundings of the nanoparticles. Finally, using high-resolution FIB-SEM, we are able to show that the nanoparticles are tightly confined inside macrophage membranes.

In summary, we were able to harness a plethora of advanced characterization techniques to gain insight on the biodistribution and biotransformation of topically applied complex nanoparticles. This work provides a roadmap for the multiscale analysis of nanomaterials in tissue and thus provides the groundwork for the safe, relevant and rational design of new nanomedicines.

(1) Ramos, A. P.; Cruz, M. A. E.; Tovani, C. B.; Ciancaglini, P. Biomedical Applications of Nanotechnology. *Biophys Rev* **2017**, *9* (2), 79–89. https://doi.org/10.1007/s12551-016-0246-2.

(2) Lese, I.; Graf, D. A.; Tsai, C.; Taddeo, A.; Matter, M. T.; Constantinescu, M. A.; Herrmann, I. K.; Olariu, R. Bioactive Nanoparticle-Based Formulations Increase Survival Area of Perforator Flaps in a Rat Model. *PLOS ONE* **2018**, *13* (11), e0207802. https://doi.org/10.1371/journal.pone.0207802.

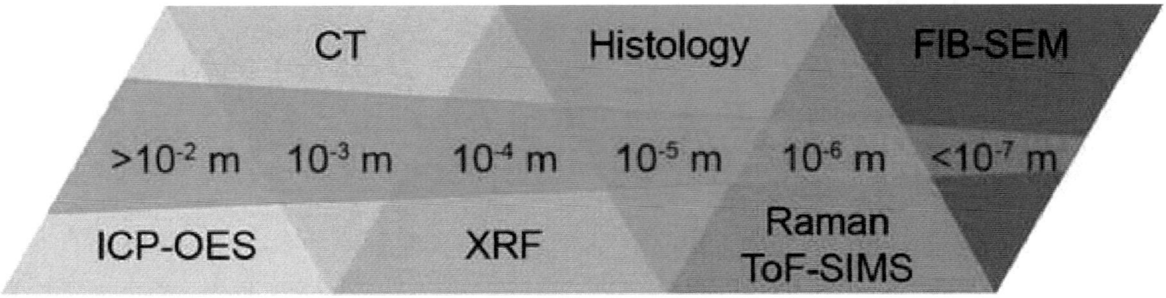
CT
Histology
FIB-SEM
>10-2 m
10-3 m
10-4 m
10-5 m
10-6 m
<10-7 m
ICP-OES
XRF
Raman
ToF-SIMS

587536 Controlled Particle Suspensions for Determination of Intrinsic Dissolution Rate in Simulated Intestinal Fluids at Medium Throughput

The intrinsic dissolution rate (IDR) of active pharmaceutical ingredients (API) is a key property that aids in early drug development. It can guide in the selection of formulation strategies or in the comparison of different polymorphs and salt forms to improve dissolution and thereby drug absorption in the intestine. In industrial settings, it is also an important quality control parameter during API production. A number of small-scale experimental techniques to determine IDR have emerged in the past years that facilitate rapid API screening in biorelevant dissolution media. In this study, we develop a robust method for medium throughput screening of IDR and thereby establish the largest IDR dataset in open literature to date. Dissolution profiles are measured in a small-scale apparatus using in situ fiber optic probes to measure amounts of drug dissolved by UV absorbance spectroscopy. Nineteen compounds with diverse physicochemical properties were studied in both fasted and fed state simulated intestinal fluids. The assays are carried out using either compacted powder discs or stabilized particle suspensions depending on the solubility of the API in the corresponding dissolution medium. IDR measurements were not solely linked to API solubility in either dissolution media. Ionization was weakly correlated to IDR in acids, exhibiting higher IDR compared to bases in both biorelevant media. Multivariate data analysis revealed that IDR strongly depends on compound partitioning into bile salt and phospholipid micelles in the simulated intestinal fluids, a process that in turn is governed by API lipophilicity, hydrophobicity, and ionization. The medium throughput method established here for the determination of IDR in biorelevant media, will be used to establish an extensive IDR database that ultimately can be used to predict IDR in silico based solely on molecular descriptors of the API.

566023 Avoiding Distribution Confusion: Using the Information in Particle Size Distributions

Particle size distributions of solid active pharmaceutical ingredients (APIs) govern the material properties important to bioavailability such as dissolution. To characterize API properties, rich data sets such as Morphologi G3 or surrogate measures such as laser light scattering are used to measure the full population of particles within a sample. However communicating this distribution often relies on histograms of the distribution or simple summary measures of volume weighted quantiles, e.g D10, D50, D90. These measures cannot adequately capture multi modal distributions and can be insufficient for setting material specifications.

By using concepts from nonparametric statistics and information theory, the information loss incurred by using these summary metrics and histogram binning can be quantified. A method is proposed for compressing a high resolution distribution into a summary measure that best captures the underlying particle characteristics. In addition to a theoretical study of information, an example of different summarization approaches applied to a PBPK model for a model API will be presented. The resulting approach of maximizing the value of particle characterization data is a useful methodology for communicating and modeling the behavior of particulate pharmaceutical compounds.

574784 Combining Liquid-Liquid Phase Separation and Mesoporous Carrier to Deliver Drug Product Intermediate

Combining Liquid-Liquid Phase Separation and Mesoporous Carrier to deliver Drug Product Intermediate

Moussa Boukerche, Sean Garner, and Nandkishor K. Nere

Center of Excellence for Isolation and Separation Technologies (CoExIST), Process Research and Development, AbbVie Inc

Liquid-Liquid Phase Separation (LLPS), typically known as oiling out is an undesired phenomenon when developing a robust and scaleable antisolvent crystallization process. LLPS can occur in ternary phase system composed of the molecule, a solvent and an anti-solvent. The solvent and anti-solvent alone are fully miscible but the presence of the solute can force the system to phase split and deliver a binary solution made of a continuous (solute rich) and a dispersed phase (solute lean). This phenomenon will impact the purity and physical properties of the molecule of interest and typically delivers a sticky amorphous phase not suitable for drug product development. In order to circumvent this issue and take advantage of LLPS, a phase diagram determination combined with the use of Mesoporous carrier (Neusilin US2 in this case) allows the preparation of Drug Product Intermediates (DPI) with tunable drug load from 0.1 to 65wt%. The DPI has the physical properties of the carrier and is not impacted by the physical properties of the Drug Substance (DS). This DPI has the potential to be readily suitable for direct compression, reduce the risk of variability of content uniformity, sticking behavior, polymorphic phase transition during DP processing. This approach is different than the Incipient Wetness Impregnation technique that is traditionally used to load mesoporous carrier. This presentation will describe the workflow of this new approach to quickly enable DPI by triggering LLPS purposefully to load mesoporous carrier with a defined drug load.

All authors are employees of AbbVie and may own AbbVie stock. AbbVie sponsored and funded the study; contributed to the design; participated in the collection, analysis, and interpretation of data, and in writing, reviewing, and approval of the final publication.

565588 Optimising the Nucleation of Polypeptides Using Nanotemplates: The Case Study of Insulin

Optimising the nucleation of polypeptides using nanotemplates: The case study of insulin

Frederik J. Link*[a], Jerry Y.Y. Heng[a]

[a]Department of Chemical Engineering, Imperial College London, South Kensington Campus, London SW7 2AZ, UK

*f.link18@imperial.ac.uk

Keywords: insulin, nanotemplates, crystallisation

Bio-active polypeptides have attracted interest in research because of their significant use in medical, biological and pharmaceutical industries. Thereby, the crystalline form offers significant pharmacokinetic advantages with regards to achieving higher bioavailability and better release control [1]. In the case of insulin, which is widely used for treating diabetes, the most common route of administration is by subcutaneous injections of either pure microcrystals or a mixture of microcrystals and amorphous insulin. Due to the slow dissolution of the crystals, the crystalline administration leads to a slow release of insulin into the blood stream [2]. However, one of the main limitation is the lack of controllability of the nucleation mechanism. However, nanotemplates are remarkable in controlling nucleation. Recent studies showed that by using small inorganic particle with a highly porous surface the induction time of nucleation of polypeptides can be significantly accelerated even if the supersaturation is low [3]. Furthermore, it is believed that nucleation is enhanced if the pore diameter has the same order of magnitude as the radius of gyration of the polypeptide [4]. This stems from the suggestion that nucleation in porous surfaces gets enhanced due to local increases in supersaturation which is controlled by the diffusibility of the protein into the pores. However, little is known about this mechanism and the impact of the chemical properties of the surface of the nanotemplates on controlling nucleation. It is believed that porous surfaces trigger nucleation due to the local increase in supersaturation. Therefore, investigating the non-covalent interaction between the surface patches of the nanotemplates and insulin will open up new routes for controlling the crystallisation of insulin.

Insulin has a molecular weight of 5.8 kDa and consist of 2 polypeptide-chains, A and B, with combined 51 amino acids. Furthermore, the two chains are inter-linked by two disulfide bonds and, additionally, chain A is intra-linked with an additional disulfide bond. Citrate buffer with a pH near the isoelectric point of insulin was used with zinc-sulphate as precipitant. Different concentrations were tested to identify the most promising ones. Since insulin is stored in a rhombohedral crystal configuration in combination with zinc-sulphate in the pancreas, zinc-sulphate was selected as the most appropriate precipitate. The impact of nanotemplates on crystallisability is investigated by mean of measuring induction time and crystal yield. Therefore, batch crystallisation experiments will be carried out in which the crystal yield and induction time will be tracked via measuring the concentration of insulin within the solution over time. To achieve this, the concentration of insulin is determined via UV-vis absorption in combination with Bradford's method [5]. Different types of nanotemplates will be investigated which differ in their surface chemistry and porosity, amount and size.

By using nanotemplates, the induction time of crystallising a complex molecule such like insulin can be reduced significantly even at low supersaturation. Although the nanotemplates do not have a notable influence on the crystal quality, multiple configurations can be observed which is due to the local availability of dissolved zinc-sulphate. When in the vicinity of zinc-sulphate insulin crystallises in its rhombohedral form, whereas otherwise it crystallises in a cubic shape.

Further work will be done to characterise the impact of the nanotemplates with regards to understanding the impact of the chemistry of the surface patches and whether the crystallisation can be enhanced solely due to local supersaturation or also because of specific non-covalent interaction between the surface patches and insulin. Additionally, the impact of the seeds on the solubility of insulin has to be investigated to evaluate the impact of the nanotemplates on the overall yield. All these findings should ultimately result in the ability of engineering appropriate nanotemplates to enhance the crystallisability of insulin. Further work will also be done in investigating the potential of scaling up from 1 ml to 10 and 100 ml scale.

References

1. Pechenov, S., et al., *Injectable controlled release formulations incorporating protein crystals.* J Control Release, 2004. **96**(1): p. 149-58.
2. Norrman, M. and G. Schluckebier, *Crystallographic characterization of two novel crystal forms of human insulin induced by chaotropic agents and a shift in pH.* BMC Struct Biol, 2007. **7**: p. 83.
3. Shah, U.V., et al., *Crystallization of Proteins at Ultralow Supersaturations Using Novel Three-Dimensional Nanotemplates.* Crystal Growth & Design, 2012. **12**(4): p. 1772-1777.
4. Shah, U.V., D.R. Williams, and J.Y.Y. Heng, *Selective Crystallization of Proteins Using Engineered Nanonucleants.* Crystal Growth & Design, 2012. **12**(3): p. 1362-1369.
5. M., B.M., *A Rapid and Sensitive Method for the Quantitation of Microgram Quantities of Protein Utilizing the Prinicple of Protein-Dye Binding.* Analytical Biochemistry, 1976. **72**: p. 248-254.

600d - SURFACE-INDUCED NUCLEATION STRATEGIES: SEEKING SYMMETRIES BETWEEN SELF-ASSEMBLY OF HETERONUCLEANTS AND CRYSTALS

Fiora Artusio[1], *Giacomo Ceccone*[2], *Roberto Pisano*[1*]

[1]*Politecnico di Torino, Torino, Italy*
[2]*European Commission, Joint Research Centre, Ispra, Italy*

*Corresponding author, e-mail: roberto.pisano@polito.it

Biography

Fiora Artusio is a Ph.D. candidate in Chemical Engineering at Politecnico di Torino. She is working at the Molecular Engineering (MolE) Lab under the supervision of Prof. Pisano. She got her MSc Degree in 2016 studying photo-activated polymerization for the synthesis of nanocapsules for drug delivery. Her research is now focused on the development of experimental methodologies for the control of pharmaceuticals and biopharmaceuticals crystallization. She has carried out research activity in Paris (LSI, Ecole Polytechnique) and at JRC – European Commission, concentrating on electron microscopy and surface analysis techniques respectively. She is currently working on protein crystallization at Laboratorio de Estudios Cristalograficòs (LEC) in Granada.

Giacomo Ceccone is a Staff Scientist at Joint Research Centre (JRC) – European Commission in Ispra. He received his Degree in Physics from University of Pavia (Italy) in 1984. After post-doctoral fellow at the University of Minnesota (USA) and TASC Laboratory, (Trieste, ITALY) he joined the JRC in 1990 where he is working in the Directorate F – Health, Consumers and Reference Materials. Dr Ceccone's main research interests concern the application of surface analysis techniques (e.g. XPS, ToF-SIMS, Synchrotron Radiation-based spectro-microscopies) to the physico-chemical characterization of surfaces with particular emphasis on nanomaterials and micro/nano-plastics. Dr Ceccone has published more than 150 papers in peer-reviewed journals, 4 book chapters and delivered more than 80 presentations at international conferences.

Roberto Pisano is a Professor of Chemical Engineering at Politecnico di Torino (Italy), where he received his Ph.D. in 2009. Professor Pisano's research focuses on the application of both computational and experimental methods to engineering chemical products and processes, with particular emphasis to pharmaceutical processing and formulation of both small molecules and biologics. He has worked with many pharmaceutical companies in research or consulting. He has published more than 80 papers, 11 book chapters, 1 edited book on freeze-drying of pharmaceuticals and currently has 4 patents issued or pending.

Abstract

The present paper presents the application of surfaces having defined and controlled attributes as heteronucleants for the crystallization of a model pharmaceutical molecule. The synthesis of the substrate was optimized in order to relate surface features to the crystallization outcome. Extremely flat and topographically uniform glass supports bearing amino and thiol head groups were successfully synthesized and characterized by means of contact angle, AFM and XPS analyses. Such surfaces were then used as supports for aspirin (ASA) crystallization in order to investigate their influence on nucleation kinetics. Compared to untreated glass, amino-functionalized glass was dramatically nucleation-active, whereas thiol-functionalized supports strongly repressed ASA heterogeneous nucleation. The promoting or inhibiting action towards the stabilization of ASA nuclei on a functionalized surface and their successive growth into crystals was therefore related to the chemistry of exposed head groups.

Introduction

Nucleation represents the core of a variety of processes occurring in nature, such as ice or cloud formation [1], but it is also responsible of amyloid aggregation leading to Alzhemeir's disease [2] or even creation of virus capsids [3]. The formation of new phase in defined environment as a consequence of self-assembly and ordering processes represents a *fil-rouge* for all these phenomena. More specifically, nucleation can occur homogeneously or heterogeneously, depending on whether aggregation mechanism involves the presence of single or multiple phases. In the latter case, stable nuclei are formed taking advantage of favorable interactions with foreign surfaces which lower the energetic cost for new interface area formation. However, heterogeneous nucleation often occurs in uncontrolled manner because of impurities in the system, leading to significantly different crystallization outcomes. This aspect is pivotal to pharmaceutical and biopharmaceutical industries as nucleation not only affects process kinetics, but also crystalline form, habit and size selection, as well as downstream processes [4].

Recently, the role of external surface features on crystallization of Active Pharmaceutical Ingredients (APIs) is being investigated by relating surface chemistry and morphology to nucleation kinetics and polymorph selection [5–7]. For example, surfaces carrying pores were found to strongly accelerate nucleation kinetics, provided that these were not spherical [8]. Selecting one specific crystalline form of a given substance can also be achieved by simply acting on the heteronucleants, keeping constant all the other crystallization conditions [9–11]. However, it is difficult to isolate and distinguish between chemical and morphological interactions between a surface and a to-be-crystallized molecule.

Spatial confinement of chemical reactions represents a useful strategy for the synthesis of a variety of devices, such as nanocapsules and nanohydrogels [12,13]. Here, by confining the functionalization reaction to an active solid-liquid interface, Self-Assembled Monolayers (SAMs) on glass were synthesized. We present an organic approach to the study of heteronucleants quality attributes in order to relate them to crystallization of small molecule drugs. The complete and diversified characterization of surface properties is vital, as the interplay of complex and different aspects of a surface is a key-aspect when promoting or inhibiting nucleation. By carefully tailoring synthesis conditions, we synthesized high-quality SAMs on amorphous substrates and successively applied them to the crystallization of a pharmaceutical model molecule.

Materials & Methods

Chemicals

Glass coverslips were purchased from Neuvitro (Vancouver, USA). Hydrogen peroxide (30 wt % in water, ACS reagent), sulfuric acid (ACS reagent, 95.0 – 98.9%), anhydrous toluene (99.8%), toluene (ACS reagent, ≥ 99.5%), ethanol (puriss. p.a., ≥ 99.8%), 3-aminopropyltrimethoxysilane (APTMS, 97%), 3-mercaptopropyltrimethoxysilane (MPTMS, 95%) and water (HPLC grade) were purchased from Sigma-Aldrich (Cesano Maderno, Italy).

Synthesis of self-assembled monolayers

Glass coverslips were pre-activated and cleaned in 3:1 *piranha* solution for 1 h and thoroughly rinsed with water afterwards. Then, they were incubated in 5.4 x 10^{-2} M solution of silane in anhydrous toluene. Reaction time was 30 min for APTMS and 15 h for MPMTS. Final rinsing with toluene, toluene-ethanol mixture and ethanol served as last synthesis step.

Surface characterization

Contact angle (CA) analyses were firstly carried out in order to assess the effectiveness of functionalization step and surface uniformity. A MSE DigiDrop (GBX, France) was used to dispense 3 µL HPLC water drops on substrates and calculate sessile drop contact angles. Surface topography was evaluated by means of Atomic Force Microscopy (AFM, Solver NANO, NT-MDT Spectrum Instruments, Russia). Cantilever frequency was 1 Hz, scanned area was 1 x 1 µm and 256 lines per scan were acquired. Finally, chemical analyses were performed with X-Ray Photoelectron Spectroscopy (XPS, AXIS ULTRA, DLD Kratos Analytical, UK). Pass energy was set at 80 keV. Wide spectra were collected from 1100 to 0 eV values of binding energy. The angle between sample surface and analyzer was set at 90°. Calibration for all the spectra involved setting hydrocarbon C1s at 285.0 eV. Data were processed with CasaXPS software (version 2.3.20).

Crystallization studies

We designed and implemented a high-throughput crystallization platform in order to perform a large number of crystallization experiments simultaneously. Aspirin was dissolved in 38% ethanol in water and crystallization was carried out in wells containing SAMs. Supersaturation was set at 1.8 and crystallization temperature was 15 °C. Crystallization outcome was studied according to kinetic and thermodynamic aspects, involving nucleation induction time and crystalline form and habit selection. The first aspect was studied by means of optical microscopy coupled to motorized X-Y stage (M125C, Leica Microsystems, Germany) and *in-house* cooling system, enabling time-lapse multiple-positions acquisition and ensuring maintenance of precise supersaturation conditions until the nuclei formation. Nucleation kinetics was evaluated by calculation of detectable induction time, corresponding to the time when detectable crystals could be evinced inside each crystallizer.

Results

SAMs were built starting from methoxy-silane molecules carrying a propyl chain with a terminal functionality. The latter represented the only difference among our monolayers, being thiol or amino-terminated. The portfolio of elected end groups permits the deep investigation of secondary interactions between surface and solute molecules and their impact on nuclei formation. As a matter of fact, our final goal is to design and develop a versatile benchmark for crystallization studies assessing the extent of specific and non-specific interactions. This is achieved by only changing substrate, while all crystallization conditions, *e.g.* supersaturation, solvent, temperature, are not to be modified.

Proving surface functionalization

Trimethoxysilanes were chosen because of their high reactivity and degree of packing when organized in monolayers. These molecules have been anchored to pre-activated glass substrates by means of condensation reactions between the surface hydroxyl groups and the hydrolyzed methoxy groups of silanes. The synthesis conditions have been optimized in terms of reaction solvent, functionalizing agent, reaction time and temperature in order to ensure the formation of reproducible single layers of organic matter on glass. As a matter of fact, for crystallization studies devoted to the mechanistic understanding of surface effects on nucleation, it is crucial to deal with extremely uniform substrates both as concerns the sample itself, but also *intra* and *inter*-batch homogeneity. This need arises from the intrinsic stochastic nature of nucleation, thus requiring a high number of experiments involving the same experimental conditions to get statistically significant data.

Thorough characterization of the crystallization substrates has been carried out in order to confirm functionalization effectiveness and gain a structured overview of surface properties. First, contact angle analyses confirmed surface functionalization: as regards MPTMS, glass surface turned out to be more hydrophobic after monolayer assembly, whereas, for APTMS, a more hydrophilic surface was obtained, as reported in Table 1. Absence of local deposits of polymerized silanes was confirmed by AFM, which also confirmed the achievement of an extremely regular functionalization. Moreover, some interesting insights into surface roughness were collected. It turned out that the presence of a monolayer on the surface did not alter surface mean roughness, whereas, when uncontrolled functionalization was carried out, major effects on this parameter were detected. More specifically, untreated glass represented an extremely flat surface, having R_q well below 1 nm, *i.e.* 130 pm. Then, covalent bonding of an organic monolayer on top led to comparable average roughness values for all the functionalities. This feature represents a particular appealing aspect of the application of high-quality SAMs to heterogeneous nucleation studies since, as topography is not altered by the functionalization process, the effect of different superficial chemistries may be isolated from morphological interactions between substrate and solute molecules.

The effectivity of functionalization was corroborated by means of XPS. Analyses were carried out for different purposes: the main goal was the confirmation of the presence of characteristic elements of SAMs on glass. Compared to untreated glass, MPTMS and APTMS-functionalized supports led to sulfur and nitrogen characteristic peaks in the XPS spectra, respectively. Further confirmation derived from the comparison of characteristic elements ratios between the ideal and real case. C/S or C/N theoretical ratios for SAMs are equal to 3. We measured 6.6 and 4.4 respectively. Slightly higher elemental ratios were attributed to superficial contamination of the samples, but also denoted that functionalization was likely to occur in monolayers.

The characterization of SAMs led us to confirm the suitability of these surfaces to be used as heteronucleants in crystallization studies.

Table 1. Contact angle, roughness and elemental composition of untreated glass, thiol- and amino-terminated SAMs.

Sample	**CA, °**	**%O**	**%Si**	**%C**	**%S**	**%N**
Untreated glass	49.1 (+/- 1.9)	54.6	21.6	20.4	-	-
Thiol-SAM	65.7 (+/- 1.7)	55.3	21.3	19.1	2.9	-
Amino-SAM	37.1 (+/- 0.6)	30.3	15.0	43.1	-	9.8

Addressing the effect of surface chemistry on crystallization of pharmaceuticals

In this study, self-assembled monolayers (SAMs) have been selected as ideal candidates to probe superficial chemical effects on nuclei formation and successive crystal growth. An amorphous substrate was selected for supporting monolayer assembly in order to exclude epitaxial interaction between the final heteronucleant and solute molecules. In this way, we were able to exclude complex ordering phenomena rising from lattice matching between surface and growing nuclei. In addition, such an approach enables the development of crystallization supports which are structurally inert, but chemically active towards nucleation. Extremely flat surfaces were selected in order to deconvolute, as much as physically possible, the contribution of surface chemistry and topography to nucleation phenomena.

Starting from experimental observation, cumulative probability distribution curves for ASA crystallization were calculated and reported in Figure 1a. Untreated glass was considered as reference for comparing drug nucleation kinetics.

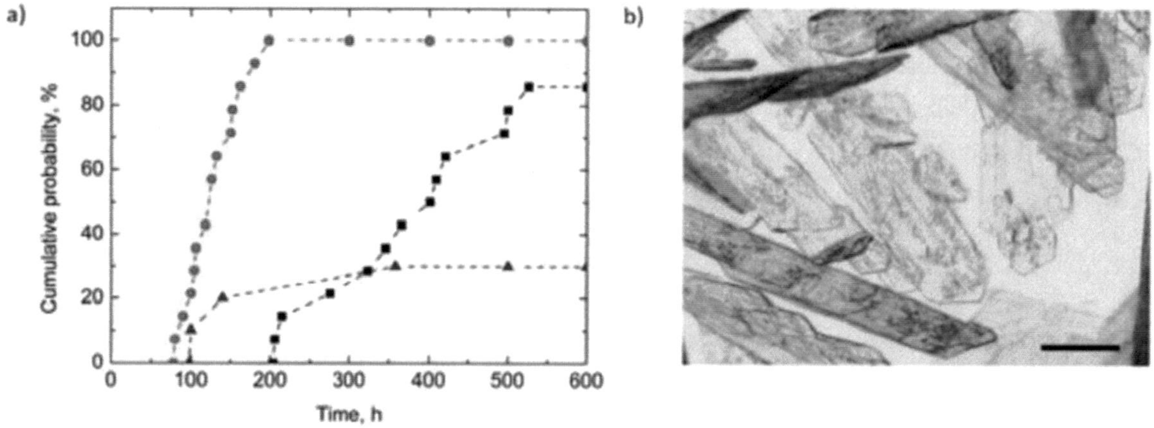

Figure 1. a) Cumulative probability of observing nucleation phenomena for untreated (square, black curve), thiol-SAMs (triangle, blue curve) and amino-SAMs (circle, red curve). b) Optical microscope image of ASA crystals grown on SAMs, scale bare is 100 μm.

Both SAMs were responsible for halving nucleation onset time compared to the untreated substrates. However, the amino-SAMs led to a dramatic boost in nucleation kinetics, as well as the concentration of nucleation events in a shorter time interval. These observations corroborated the nucleation activity of amino surface groups towards ASA. Moreover, crystals were observed in all the wells during the test time, leading to 100% probability of observing nucleation in the system. We related this ability to favorable acid-base interactions between surface and ASA. Conversely, the thiol-SAMs strongly inhibited ASA nucleation, leading to only 30% crystallized wells. This property was attributed to electrostatic repulsion between solute molecules and surface, impeding stable cluster formation in contact with the surface. ASA crystallized according to a plate-like configuration on all the investigated surfaces, as depicted in Figure 1b.

Conclusion

An optimized method for the synthesis of APTMS and MPTMS self-assembled monolayers is here presented, along with a set of characterization techniques for the evaluation of surface quality and features. Such a reference platform is to be linked to the crystallization behavior of heteronucleants. Self-assembly phenomena were therefore exploited at two different levels: firstly, for the synthesis of heteronucleants with defined properties and, then, for the study of the interplay between surfaces and self-assembly of solute molecules in solution leading to crystals. Such surfaces could act as ASA nucleation promotors or inhibitors, thus enabling the tuning of nucleation kinetics.

Abbreviations

AFM, Atomic Force Microscopy; API, Active Pharmaceutical Ingredient; APTMS, 3-aminopropyltrimethoxysilane; ASA; Aspirin; CA, Contact Angle; MPMTS, 3-mercaptopropyltrimethoxysilane; SAM, Self-Assembled Monolayer; XPS, X-Ray Photoelectron Spectroscopy.

Acknowledgements

Alexandre Yuri Kitamukai is kindly acknowledged for the experimental activity on ASA crystallization. Part of this work was carried out thanks to the access to the Nanobiotechnology Laboratory under the Framework for Open Access to the Joint Research Centre Research Infrastructures of the European Commission.

References

[1] J.M. Campbell, F.C. Meldrum, H.K. Christenson, Observing the formation of ice and organic crystals in active sites, Proc. Natl. Acad. Sci. 114 (2017) 810–815. doi:10.1073/pnas.1617717114.

[2] P. Friedhoff, M. Von Bergen, E.M. Mandelkow, P. Davies, E. Mandelkow, A nucleated assembly mechanism of Alzheimer paired helical filaments, Proc. Natl. Acad. Sci. U. S. A. 95 (1998) 15712–15717. doi:10.1073/pnas.95.26.15712.

[3] R. Zandi, P. Van Der Schoot, D. Reguera, W. Kegel, H. Reiss, Classical nucleation theory of virus capsids, Biophys. J. 90 (2006) 1939–1948. doi:10.1529/biophysj.105.072975.

[4] A.S. Myerson, Handbook of Industrial Crystallization, 2nd ed., Butterworth-Heinemann, 2002.

[5] F. Artusio, R. Pisano, Surface-induced crystallization of pharmaceuticals and biopharmaceuticals: A review, Int. J. Pharm. 547 (2018) 190–208. doi:10.1016/j.ijpharm.2018.05.069.

[6] Y. Diao, A.S. Myerson, T.A. Hatton, B.L. Trout, Surface design for controlled crystallization: The role of surface chemistry and nanoscale pores in heterogeneous nucleation, Langmuir. 27 (2011) 5324–5334. doi:10.1021/la104351k.

[7] J.M. Ha, J.H. Wolf, M.A. Hillmyer, M.D. Ward, Polymorph selectivity under nanoscopic confinement, J. Am. Chem. Soc. 126 (2004) 3382–3383. doi:10.1021/ja049724r.

[8] Y. Diao, T. Harada, A.S. Myerson, T. Alan Hatton, B.L. Trout, The role of nanopore shape in surface-induced crystallization, Nat. Mater. 10 (2011) 867–871. doi:10.1038/nmat3117.

[9] J. V. Parambil, S.K. Poornachary, R.B.H. Tan, J.Y.Y. Heng, Template-induced polymorphic selectivity: The effects of surface chemistry and solute concentration on carbamazepine crystallisation, CrystEngComm. 16 (2014) 4927–4930. doi:10.1039/C3CE42622J.

[10] S. Roy, N.R. Goud, A.J. Matzger, Polymorphism in phenobarbital: discovery of a new polymorph and crystal structure of elusive form V, Chem. Commun. 52 (2016) 4389–

4392. doi:10.1039/C6CC00959J.
[11] C.P. Price, A.L. Grzesiak, A.J. Matzger, Crystalline polymorph selection and discovery with polymer heteronuclei, J. Am. Chem. Soc. 127 (2005) 5512–5517. doi:10.1021/ja042561m.
[12] F. Artusio, M. Bazzano, R. Pisano, P.E. Coulon, G. Rizza, T. Schiller, M. Sangermano, Polymeric nanocapsules via interfacial cationic photopolymerization in miniemulsion, Polym. (United Kingdom). 139 (2018) 155–162. doi:10.1016/j.polymer.2018.02.019.
[13] F. Artusio, A. Ferri, V. Gigante, D. Massella, I. Mazzarino, M. Sangermano, A. Barresi, R. Pisano, Synthesis of high payload nanohydrogels for the ecapsulation of hydrophilic molecules via inverse miniemulsion polymerization: caffeine as a case study, Drug Dev. Ind. Pharm. (2019) (in press). doi:10.1080/03639045.2019.1672714.

572187 Olanzapine Crystal Symmetry Originates in Preformed Centrosymmetric Solute Dimers

The symmetry of crystals is a central quality that determines their function and allure. Given the uncorrelated symmetries of a crystal and its constituent molecules, the symmetry breaking that begets a crystal phase has been thought to ensue during nucleation. Here we demonstrate that the one of the two symmetry element of olanzapine crystals, an inversion center, emerges in solute dimers extant in the solution prior to crystallization. Time resolved *in situ* atomic force microscopy monitoring of growth steps reveals that crystals grow nonclassically, exclusively by incorporation of dimers; this growth mode has been speculated and inferred from crystal and solute structures, but never experimentally observed. The growth rate of crystal layers exhibits a quadratic dependence on the solute concentration, characteristic of the second-order kinetics of dimer incorporation. Olanzapine dimers constitute a minority solution component. We show that growth by dimers is preferred not owing to greater activation entropy or faster dissociation of attached solvent molecules. The monomer and dimer adsorption energies on the crystal surface, evaluated in all-atom molecular dynamics simulations, suggest overwhelming accumulation of adsorbed dimers, which expedites their incorporation into growth sites. The proposed outlook on the cooperative creation of crystal symmetry illuminates analyses of structure selection, polymorph transformations, chiral separation, and crystal properties of natural and engineered crystals.

572060 Development of a Hybrid Quality-By-Control Framework for Optimal Crystallization Process Design

Strict regulatory standards are dictated by the FDA, EMEA and WHO for the active pharmaceutical ingredients (API) to meet high level quality criteria such as purity, size and shape of the API crystals or polymorphic structure. One of the main criteria is to achieve size uniformity as well as desired yield in industry. Despite of the efforts put in pharmaceutical industry, intensive fines formation and agglomeration of crystals are still common problems, often having a negative impact on the product crystal size distribution (CSD). A general practice to eliminate fines and agglomerates is to apply temperature cycles for internal fines dissolution and de-agglomeration.[1] For this purpose, quality-by-design (QbD) approaches can be applied for the design of the suitable temperature profile, but this requires intensive experimental investigation to explore the whole design space. Quality-by-Control (QbC) has been introduced more recently as a new novel alternative to QbD, which can be applied to control critical quality attributes of a certain process, whereby the desired operating curve like temperature profile is automatically determined by the feedback control algorithm rather than by excessive open-loop experimentation. In this way QbC can significantly reduce required process development time and material used.[2]

The work proposes a novel hybrid framework combining the model-free and model-based quality-by-control (mfQbC and mbQbC) paradigms for the optimal crystallization design with the aim of minimization of fines and agglomerates in the product. This developed framework describes the mfQbC based rapid design used in conjunction with direct nucleation control (DNC) and supersaturation control (SSC), along with (2) the mbQbC framework using a novel population balance model-based platform for the robust optimization of the operating trajectory of the crystallizer to achieve crystals with desired quality attributes.[3] This framework provides an improved alternative to the current industrial practice, by the combined use of feedback control and mathematical models for rapid design of robust crystallization processes. The implementation of the proposed framework enhanced the prediction of crystallization events and was able to achieve better control of the crystallization processes by providing controlled temperature cycles, for de-agglomeration and fines destruction. The information for parameter estimation and optimization calculations was collected from the experiments and the system diagnostics was done in mfQbC part, which also provided a rapid design of robust operating conditions, that were improved to achieve optimal operation subsequently in the mbQbC approach.

The advantage of the mfQbC is that it requires minimal preliminary system information, hence, it enables quick and efficient crystallization process design. On the other hand, the mf-QbC operation policy might be sub-optimal, which can be overcome by exploiting the predictive capabilities of the process models used by the mbQbC. In the mfQbC part, DNC experiments were run for two purposes: to determine the primary and secondary nucleation metastable zone widths (MSZW) and to obtain the temperature cycle profile that is required for producing the crystals with desired properties. DNC is an efficient approach to eliminate fines due to the application of temperature cycles. SSC was applied after DNC generates the MSZW information to obtain the temperature profile for only growth mechanism. The resulting temperature profile from SSC can also be easily implemented in industrial distributed control systems (DSCs). After mfQbC, using the mbQbC framework, a population balance model (PBM) was developed to simulate primary and secondary nucleation, growth, agglomeration and de-agglomeration mechanisms as well as dissolution. Model parameters were identified using experimental data for a model API crystallization with high agglomeration and secondary nucleation tendency. Using a model-based robust optimization, in which parametric uncertainty obtained from the model-identification step was also included, different temperature profiles were determined considering the dual objective of fine minimization and maximal de-agglomeration. The optimal temperature profiles obtained was validated experimentally to show the higher quality product that can be obtained using the this QbC framework.

References:

1. Wu, Z., Yang, S. & Wu, W. Application of temperature cycling for crystal quality control during

crystallization. *CrystEngComm* **18**, 2222–2238 (2016).
2. Yang, Y., Song, L. & Nagy, Z. K. Automated Direct Nucleation Control in Continuous Mixed Suspension Mixed Product Removal Cooling Crystallization. *Cryst. Growth Des.* **15**, 5839–5848 (2015).
3. Szilágyi, B., Borsos, Á., Pal, K. & Nagy, Z. K. Experimental implementation of a Quality-by-Control (QbC) framework using a mechanistic PBM-based nonlinear model predictive control involving chord length distribution measurement for the batch cooling crystallization of L-ascorbic acid. *Chem. Eng. Sci.* **195**, 335–346 (2019).

574032 Flow-Induced Crystallization in Materials Extrusion Additive Manufacturing

Semicrystalline polymers are often used in additive manufacturing (AM) processes due to the ease of processing in the melt state and the rapid, dramatic increase in mechanical properties during crystallization. Despite the benefits of AM, the uncertainty in the quality of the printed part makes more widespread implementation challenging. Much of the uncertainty in semicrystalline polymer AM stems from the complex relationship between processing conditions and the resulting semicrystalline microstructure. Here, we present a systematic study of printing conditions on the semicrystalline morphology of parts printed from poly(lactic acid) (PLA). The slow crystallization kinetics of PLA generate parts with low crystallinity, however a secondary annealing process at high temperatures (140 °C) generate a space-filled spherulitic texture, with smaller spherulites near the weld zone between the extruded layers. This spherulite size distribution is attributed to a higher nucleation density templated into the part by the temperature and deformation history of the printing process. We show that a combination of materials characterization and process line infrared thermography can be used in theoretical continuum modeling to predict the spatial variation in spherulite size.

570855 Computational Fluid Dynamics Simulation of the Fused Deposition Modeling Process: Effect of Temperature-Dependent Viscosity on the Shape of Extruded Strand

Fused Deposition Modeling (FDM) is a 3D printing technique based on extrusion of polymer melt. In the FDM process, polymer filaments are melted in the liquefier and then extruded layer by layer to build final products. Although the process is fast and cost-effective, the final product may have rough surface due to the deformation of the extruded strain. In this study, the effects of various process parameters on the morphology of the extruded strand have been investigated. Non-isothermal free surface flow simulation of the extrusion process was performed utilizing OpenFOAM, an open source simulation software, along with our development of a user-defined module. Simulation results for the dimensions (including thickness and width of layer) of an extruded strand have been validated by experimental results. The temperature effect on the surface roughness based on these will be also discussed.

565637 Effect of Polypropylene Melt Flow Index on the Material Extrusion Based Additive Manufacturing of Polypropylene/Hydrogenated Resin Blends

Material extrusion (MatEx) based additive manufacturing (AM) process can produce parts with intricate features that are difficult to fabricate using traditional subtractive manufacturing techniques. However, the MatEx process suffers from two major drawbacks: the mechanical properties of the parts produced using MatEx are often inferior to the bulk material properties and only a limited number of materials are currently compatible with this mode of AM. Isotactic polypropylene (PP) is a popular thermoplastic material which undergoes rapid crystallization that leads to significant trapped residual stress when processed using MatEx, resulting in poor geometric tolerance and mechanical performance.

In this work, PP having different melt flow indices (MFIs) were blended with low molecular weight hydrocarbon resins having different degrees of hydrogenation. The influence of the PP MFI on the crystallinity of the pure polymers and PP/hydrogenated resin blends as well as the melting and crystallization temperature was investigated. For example, the addition of the resins lowered the melting temperature by 7^0C while the crystallization temperature of pure PP (MFI=12g/10 min) decreased from 133.3^0C to 118.6^0C in the blends; thereby increasing the time required for the material to crystallize from the melt. Next, the blends were extruded using a single screw extruder to obtain filaments that can be used in the MatEx process. An understanding of the modifications in the crystallization process significantly impact the z-axis adhesion and residual stress state, which directly affect mechanical properties and warpage in the printed parts. Tensile bars were printed in two different orientations to analyze the mechanical performance and study the effect on part isotropy. The maximum tensile strength and modulus of the printed parts was found to decrease by 21% and 15% respectively. However, moderate post-printing annealing achieved mechanical properties that approached injection molded PP parts by improving the interlayer bonding of the parts. We can leverage these results to modify processing conditions to optimize mechanical properties of the parts generated using MatEx.

569034 Engineering Filament Design for Enhanced Printability and Mechanical Properties of Additively Manufactured Plastic Parts

3D printing enables the net shape manufacturing of objects with minimal material waste and low tooling costs, but the functionality and performance of these objects is generally limited. One of the most common 3D printing technologies and the primary consumer-user technology is fused filament fabrication (FFF), which relies on the extrusion of thermoplastic filaments through a heated nozzle. The melted filament is pushed through the nozzle by plunger action of the unheated filament. The object of interest is built up layer-by-layer as the nozzle is translated in the x-y plane. These 3D printed parts tend to suffer from poor mechanical properties due to weak interlayer interfaces that develop during the printing. There is generally an intrinsic trade-off between part accuracy and mechanical properties, where higher temperatures tend to enhance diffusion of polymer chains across the interfaces but this also tends to allow the material to flow which distorts the shape. Here we discuss routes to manipulate these trade-offs through structured core-shell filaments. Routes to produce parts with less warp (deformation) and improved mechanical properties will be discussed. Finally, we will briefly discuss how the shape of the object affects these properties and the importance of the tool path to obtain desired properties.

576383 3D Printing of Interdigitated Dielectric Elastomer Actuators

Among soft actuators, dielectric elastomer actuators exhibit excellent energy density and strain rate. However, typical fabrication methods are based on sequential planar processing, which results in planar devices with low fabrication rate. 3D printing provides the potential to rapidly fabricate versatile designs of actuators with high throughput. Interdigitated DEAs were fabricated by 3D printing vertical arrays of electrodes followed by casting a dielectric. Minimizing print defects such as end beads in the vertical electrodes is important for eliminating premature breakdown. Print defects are reduced by optimizing the ink rheology; the electrode ink is composed of a vinyl-terminated poly(ethylene glycol ethylene sulfide) oligomers with carbon black as the rheological modifier and conductive filler. The print fidelity is optimized by reducing the molecular weight of the oligomer. However, in order to achieve a low elastic modulus in the cured state, a high molecular weight between crosslinks is required. We implemented a strategy of simultaneous chain extension and crosslinking to achieve low-modulus, highly extensible elastomers using low-molecular weight precursors.

The dielectric is a chemically crosslinked oil gel with a dielectric constant of 6. The acrylate crosslinking mechanism allows chemical bonding with the electrodes, providing structural integrity in the device. The oil component in the dielectric fulfills several purposes, including reducing the uncured viscosity to improve infilling in the electrodes, reducing the tangent delta in the cured state, and providing self-healing characteristics. Self-healing occurs because oil molecules damaged by charge leakage can diffuse away from the damaged site, allowing the actuator to be actuated again.

Multinozzle 3D printing enables rapid fabrication of contractile actuators with actuation strains up to 9%. The electrodes can be fabricated at a rate of 0.73 cm^3/min; a device with an area of 7.5x3.9 cm^2 can be fabricated in ~6 min. The use of 3D printing enables versatile designs, including actuators with orthogonal pixels and prestrain-free rotational actuators.

562611 Direct Print Additive Manufacturing of Optical Fiber Interconnects for Board Level Computing Devices

Integrated photonics technology has many compelling advantages for computing and communications applications, including high speed and extremely wide bandwidth operations. Current systems are typically hybrid assemblies of packaged photonic devices where printed circuit boards often serve to route electrical signals and power, and in some cases, have runs of optical fibers. Development of fully integrated photonic systems would allow for higher transmission rates, lower power requirements, improved signal integrity and timing, less heat generation, and improved security of communication signals. Current technology has not been able to overcome how to densely route optical interconnects through tight spaces without losses due to scattering, how to easily connect optical interconnects to devices and finally how to do this in cost effective packages. Here we show that by using a new direct print additive manufacturing (DPAM) process of fused deposition modeling (FDM) of plastic and micro-dispensing of pastes and inks, along with picosecond laser processing, we can 3D print single and multi-mode optical fibers in a controlled manner such that compact, 3-dimensional optical interconnects can be printed along non-lineal paths. We have produced working optical interconnects with fibers core diameters from 77 microns to as small as 12 microns. Our results demonstrate surface roughness in 3D printed 77 micron-diameter optical fibers of less than 450 nm, with optical transmission rates above 46% for proof of concept devices. We anticipate our proof of concept devices to be a starting point in the development of more sophisticated electro-optical computing devices using this new DPAM technique.

568610 Cure-through Correction in Continuous Stereolithographic 3D Printing

Recently, continuous stereolithographic 3D printing (SL) technologies have been developed which significantly increase print speeds. By introducing a *dead zone* adjacent to the window where polymerization does not occur, continuous SL eliminates several time-consuming steps present in traditional SL and enables speeds up to 2000 $mm \cdot h^{-1}$. Two methods have been reported to generate the dead zone: (i) oxygen inhibition, and (ii) dual-wavelength exposure with complementary photoinitiation and photoinhibition.

In continuous SL, the print speed must be limited to ensure high fidelity of the printed part. Accuracy along the vertical axis is affected by the penetration depth of light into the resin. When light penetrates too deep into the resin bath, unwanted curing (known as *cure-through*, *overcure*, *the backside effect*, or *print-through error*) can occur, particularly on the backside of exterior surfaces and into internal voids. Cure-through also contributes to dose heterogeneity, introducing internal stresses which can deform the part and further reduce fidelity. Cure-through has typically been mitigated by adjusting the resin formulation to control penetration depth; however, highly absorbing resins require either higher-intensity light sources or, more commonly, slower print speeds.

Alternatively, high fidelity can be achieved without reduced print speeds by modifying slices to account for expected cure-through. This approach requires both a curing model and a slice correction algorithm. While slice correction has been developed for both external features and internal voids in non-continuous SL, existing models for continuous printing are not tailored to this purpose.

We have developed a dose-based curing model and a slice correction algorithm to precisely tune exposure levels and minimize cure-through artifacts in continuous SL. Our approach differs from previous slice-modification methods in several ways. First, we are considering a continuous system, where cure-through is a more significant and complex problem. Second, we are developing corrections using greyscale, while previous approaches were restricted to white and black. Finally, our correction uses an exact mathematical solution, while previous approaches have used heuristics and optimization.

The mathematical approach will be presented along with theoretical and corresponding experimental results from a dual-wavelength continuous SL printer. Our model allows prediction of the printed part resulting from a given set of slices and analysis of the contributions of each slice to the accumulated optical dose at each point. We have also experimentally investigated the effects of dose heterogeneity to determine optimal dose profiles. The method is demonstrated to yield marked improvement in print fidelity while maintaining high print speeds.

559663 Controllable Preparation of Uniform Magnesium Ethoxide Particles: The Effects of *N*-Hexane and Silicone Oil and Their Synergic Effect

Magnesium ethoxide ($Mg(OEt)_2$) is an important organometallic material, widely used in ceramics, catalyst, pesticides, etc., especially as the carrier for Ziegler-Natta catalysts for the synthesis of polyethylene and polypropylene. The synthesis process of magnesium ethoxide using magnesium powder, ethanol and iodine, involves the complicated solid-liquid reactions as well as particle evolution mechanisms including nucleation, growth, and agglomeration. The control of particle morphology and the agglomeration process is the difficulty in the preparation of magnesium ethoxide particles. In particular, it is still lack of relevant regularity research and understanding towards magnesium ethoxide particles with low size and well dispersity.

In this study, *n*-hexane and low viscosity silicone oil were innovatively added to regulate and control the nucleation process. Regularity study of the growth of magnesium ethoxide revealed that: 1) *n*-hexane could lead to the explosive nucleation and compact the particles; 2) low viscosity silicone oil as the inert component played an effective role in dispersing the reaction system to reduce agglomeration between particles. Through the effective control based on the synergistic effect of *n*-hexane and low viscosity silicone oil, it achieves flexible control of particle size, bulk density and dispersibility of magnesium ethoxide particles. The D50 of particles was controlled from 5 to 30 μm, the RSF (relative span factors) was under 1.0, and bulk density ranged from 0.2-0.7 g/cm^{-3}. The addition of *n*-hexane and low viscosity silicone oil improved the flexibility of preparation, and these strategies provided useful guidance for the customized preparation of micron-sized magnesium ethoxide particles serving the large-scale industrial applications.

561625 Multi-Objective Optimization of Simple Crystallization Systems

Multiple-objective optimization problems in seeded batch crystallization are solved for two chemical systems using the method of moments and a transformation proposed by Raisch and coworkers (Vollmer and Raisch, 2003; Vollmer and Raisch, 2006; Hofmann and Raisch, 2010). Six sets of two objective functions are considered: The nucleated mass and the number of nuclei; the crystal number mean size and the nucleated mass; the weight mean coefficient of variation and the weight mean size; the number mean coefficient of variation and the number of nuclei; the batch time and the nucleated mass; and the batch time and the number mean size.

The results show that if one objective is based on higher moments of the nucleated crystals (e.g. the nucleated mass) while the other is based on lower moments (e.g. the number of nuclei or the number mean size of the crystals), the Pareto-optimal front is relatively wide, indicating significant competition between the two objectives. This finding is consistent with the conclusion in previous work (Ma et al., 2002; Ward et al. 2006; Tseng and Ward, 2017). In these cases, a constant growth rate trajectory may represent a good trade-off between two objectives. By contrast, if both objectives are based on higher moments or both objectives are based on lower moments, the trade-off between the two objectives become less significant and the optimal trajectories for each single objective are similar.

This work identifies and quantifies an inherent trade-off that sometimes occurs between objective functions in batch crystallization process. Understanding this tradeoff and the circumstances under which it arises can help engineers to determine effective operating recipes for such processes. The results are illustrated with case studies based on crystallization of potassium nitrate (Miller and Rawlings, 1994) and Pentaerythritol (Bernardo and Giulietti, 2010).

References

Bernardo, A., Giulietti, M., 2010. Modeling of crystal growth and nucleation rates for pentaerythritol batch crystallization. Chemical Engineering Research and Design 88 (10), 1356-1364.

Hofmann, S., Raisch, J., 2010. Application of optimal control theory to a batch crystallizer using orbital flatness. In 16th Nordic Process Control Workshop, Lund, Sweden 25–27.

Ma, D. L., Tafti, D. K., Braatz, R. D., 2002. Optimal control and simulation of multidimensional crystallization processes. Computers & Chemical Engineering 26 (7-8), 1103-1116.

Miller, S. M., Rawlings, J. B. (1994). Model identification and control strategies for batch cooling crystallizers. AIChE Journal, 40, 1312–1327.

Tseng, Y. T., Ward, J. D. (2017). Comparison of objective functions for batch crystallization using a simple process model and Pontryagin's minimum principle. Computers & Chemical Engineering 99, 271-279.

Vollmer, U., Raisch, J., 2003. Control of batch cooling crystallization processes based on orbital flatness. Int. J. Control, 76, 1635–1643.

Vollmer, U., Raisch, J., 2006. Control of batch crystallization - A system inversion approach. Chem. Eng. Process., 45, 874–885.

Ward, J. D., Mellichamp, D. A., Doherty, M. F. (2006). Choosing an operating policy for seeded batch crystallization. AIChE Journal 52, (6), 2046–2054.

562230 Scalable Fabrication of Multielement Ceramic Oxide Particles By an Aerosol Process

Aerosol processes have long been regarded as scalable methods to fabricate functional particles for the applications in energy storage and conversion. They have distinguished advantages to bridge the gap between the synthesizing at laboratory-scale and manufacturing at commodity-scale, for example, short residence time, facile instrument requirement, and simple operation. Aerosol processes have been successfully applied to produce particles with controllable particle-to-particle composition, morphology, and crystallinity. Unfortunately, those particles generally have a restricted number of elements. This is mainly attributed to the complex particle evolution, involving multiphase chemical reactions in less than seconds. When it comes to multielement powders, element segregation, phase impurity, and a large variety of morphologies can be expected in the product. Therefore, we developed an aerosol process, flame spray pyrolysis (FSP), to make multielement particles. To demonstrate the capability of our FSP process, garnet-type solid-state electrolyte, $Li_7La_3Zr_2O_{12}$ (LLZO), has been produced with a simple post-calcination step to achieve the cubic LLZO (c-LLZO) phase.

The conventional methods to synthesize c-LLZO rely on solid-state reactions starting from micron-sized oxides, repeated ball-milling, pellet-pressing, and calcination. Since the solid-state reactions are restricted by the long-diffusion distance and low material diffusivity, we utilized the FSP process. In FSP, the starting materials are mixed uniformly in precursor solutions to shorten the diffusion length. In the as-synthesized powders, Li, La, and Zr are distributed uniformly. The particles displayed only one crystal phase. Starting from those particles, c-LLZO powders were obtained through the calcination with a temperature below 700 ^{o}C.

571414 Controlling the Internal Structures of Polymeric Microspheres Via the Introduction of a Water-Soluble Organic Solvent

1. Introduction

Polymeric microspheres with various internal structures have attracted increasing interest in modern science and technology due to their attractive characteristics, especially the usefulness of their inner spaces [1–5]. Considerable effort has been made to control the internal structures of polymeric microspheres, including the utilization of microbubbles as templates [6,7] or the template removal method [8], emulsion solvent extraction [9], evaporation [10] and diffusion method [11], the self-assembly method [12,13] and the double emulsion method [14]. However, the reported methods are not versatile as it is not easy to control polymeric microspheres' internal structures by using a particular method. Only a few papers have reported controlling microsphere structures, from the porous to the hollow, by decreasing the polymer concentration [10,15]. Therefore, there is a need to develop a versatile method which can easily control the internal structures of polymeric microspheres. In our previous work [15], we demonstrated a time-saving method with a high yield for producing porous polymeric microspheres in a foam phase, which can be developed as a continuous process for industrial production. In this work, we are proposing a new strategy to control the internal structures of polymeric microspheres based on the reported method [15] by introducing a water-soluble organic solvent. It enables the easy control of the internal structures of polymeric microspheres. The proposed method provides a practical industrial solution to make versatile polymeric microspheres.

2. Materials and Methods

Polymeric microspheres were prepared in a foam phase. Polystyrene (PS, polymer), n-heptane (porogen) and dichloromethane (DCM, an organic volatile solvent), and ethanol (EtOH, a water-soluble organic solvent) were supplied by Xi Long Chemical Co., Ltd., Nanjing, China (AR). Emulsion stabilizer polyvinyl alcohol 1788 (PVA) was supplied by Chengdu Kelon Chemical Regent Co., Ltd., Chengdu, China (AR). In a typical experiment, a mixture of PS, n-heptane and DCM was decanted into a 250 mL three-neck round-bottom flask, where the mixture was admixed with an aqueous solution containing PVA at a concentration of 1.0 wt % and EtOH at various concentrations. Experimental reagents are shown in Table 1. The resulting dispersion was immersed in a water bath at room temperature and thoroughly vortexed at 500 rpm for 15 min. After that, the temperature of a water bath was ramped up at 1 °C/min and kept at 38 °C for 30 min. The bubble phase thus generated, along with PS microspheres, was transferred to a beaker in which a magnetic bar stirred the liquid at 500 rpm. At the same time, hot water at 90 °C was continuously flown into the beaker and interacted thoroughly with the bubbles. The heated water allowed DCM to evaporate and removed n-heptane. An optical microscope (LEICA DM4000 M, Leica Microsystems, Shanghai, China and a scanning electron microscope (SEM, JEOL JSM-6500, Peabody, USA) were used to detect and display the surface morphology and the internal structures of polymeric microspheres. The particle size of PS microspheres was measured by using a laser particle diameter analyzer (Microtrac S3500, York, USA). Each sample was measured three times to calculate the average diameter and the standard error.

Table 1. Preparative conditions of polymeric microspheres. PS: polystyrene; DCM: dichloromethane; PVA: polyvinyl alcohol.

Sample	**PS**	***n*-Heptane**	**DCM**	**PVA**	**Aqueous Phase**	**Temperature Gradient**
	g	**g**	**g**	**g**	**100.0 g**	**°C/min**
E0	10.0	6.0	60.0	1.0	Water	1.0

E1	10.0	6.0	60.0	1.0	5.0 wt % EtOH solution	1.0
E2	10.0	6.0	60.0	1.0	7.0 wt % EtOH solution	1.0
E3	10.0	6.0	60.0	1.0	10.0 wt % EtOH solution	1.0
E4	10.0	6.0	60.0	1.0	14.0 wt % EtOH solution	1.0
E5	10.0	6.0	60.0	1.0	20.0 wt % EtOH solution	1.0
T1	3.0	6.0	60.0	1.0	10.0 wt % EtOH solution	1.0
T2	6.0	6.0	60.0	1.0	10.0 wt % EtOH solution	1.0
T3	10.0	6.0	60.0	1.0	10.0 wt % EtOH solution	1.0
T4	15.0	6.0	60.0	1.0	10.0 wt % EtOH solution	1.0
P1	3.0	6.0	60.0	1.0	Water	1.0
P2	6.0	6.0	60.0	1.0	Water	1.0
P3	10.0	6.0	60.0	1.0	Water	1.0
P4	15.0	6.0	60.0	1.0	Water	1.0

3. Results and Discussion

3.1. The Effect of Introducing a Water-Soluble Organic Solvent on the Internal Structures of Polymeric Microspheres

The effect of introducing a water-soluble organic solvent (EtOH) to the continuous phase on the evolution of the internal structures of polymeric microspheres was first validated, as shown in Figure 1. Without EtOH, PS microspheres were in possession of porous structures, as shown in Figure 1a, which was the same as the results reported previously [15]. After introducing EtOH, the internal structures of PS microspheres evolved from porous to hollow. When EtOH concentration was low (5.0 wt %), both porous microspheres and hollow microspheres with a porous shell were observed (Figure 1b). However, the dimension of the pores was much larger than that prepared without EtOH. This was regarded as a transition point, as hollow structures were prepared above the concentration, as seen in Figure 1c-e. A further increase in EtOH concentration (20.0 wt %) led to the collapse of the structures. The result indicates that introducing a water-soluble organic solvent enables the control of the internal structures of polymeric microspheres.

Figure 1. Optical microscope images and SEM images of PS microspheres prepared under varying experimental conditions: concentrations of EtOH at (**a**) 0, (**b**) 5.0 wt %, (**c**) 7.0 wt %, (**d**) 10.0 wt %, (**e**) 14.0 wt %, (**f**) 20.0 wt %.

3.2. The Mechanism Resulting in Controlling the Internal Structures of Polymeric Microspheres

According to the above analysis, a probable mechanism to control the internal structures of polymeric microspheres has been deduced as shown in Figure 2. The solubility parameters of polystyrene, n-heptane and dichloromethane, ethanol, and water are 8.7–9.1, 7.45, 9.7, 12.7, 23.2 ($cal^{1/2} \cdot cm^{-3/2}$) respectively [16] (pp. 52–54). EtOH molecules are more likely to diffuse towards the oil droplets (Figure 2I–III) during the suspension process (Figure 2(1–2)), which contain PS, n-heptane and DCM. Meanwhile, DCM molecules on the surface of the oil droplets diffuse towards EtOH molecules. EtOH and DCM are miscible and they may form a miscible organic solvents layer at the interface, as shown in Figure 2III. By elevating the temperature (Figure 2(3)), DCM molecules evaporate from the organic solvents layer, and then the organic solvents layer is inclined to extract DCM molecules form the out surface of an oil droplet as supplements, which drive the phase separation between PS and n-heptane (Figure 3IV). More importantly, the presence of the organic solvents layer acts as a barrier, and DCM may not evaporate completely instantaneously. PS is in a softening state with residual DCM, so both PS and n-heptane are flowing and they can easy mingle to form a few large pores or even one single extremely large pore in the inner of an un-solidified microsphere (Figure 2VI–VIII). When these un-solidified microspheres are delivered to the foam phase (Figure 2(3)), the organic solvents layer vanishes quickly and the residual DCM evaporates completely. Solidified PS microspheres (Figure 2IX–XI) are obtained after using hot water to remove n-heptane, PVA and the foam phase (Figure 2(4)). No miscible organic solvents layer as a barrier exists when only water is used as the continuous phase; the evaporation of DCM is too quick, which drives a quick phase separation between PS and n-heptane to form porous structures. When EtOH concentration is low, the barrier presents itself; however, the effect is weak. Therefore, porous structures are present, but the pore number decreases while the dimension increases, as illustrated in Figure 2IX. A further increase in EtOH concentration increases the thickness of the miscible organic solvents layer and the barrier effect increases. The softening PS and n-heptane have more time for mingling to form a few single large pores (Figure 2X) or even one single large pore (Figure 2IX) at a proper EtOH concentration. When EtOH concentration is extremely high, the organic solvents layer is thick enough to dissolve the spherical structure. No microsphere can be prepared. It is worthwhile to note that, after introducing EtOH, only a few or even one singe pore is observed on the microsphere surface. This is because the un-solidified microsphere is surrounded by the miscible organic solvents layer and the layer can dissolve polymers nearby, creating a smooth surface. After they are delivered to the foam phase, the residual DCM flows rush out from the most vulnerable areas to produce a few or even only one singe pore on the surface.

Figure 2. (a) Schematic diagrams of the process to prepare polymeric microspheres; (b) the mechanism to control the inner structures of polymeric microspheres by introducing a water-soluble organic solvent.

4. Conclusions

In this study, we demonstrated a new strategy to control the inner structures of polymeric microspheres. This enables the versatile control of the polymeric microspheres' internal structures, from porous to hollow, only by introducing a water-soluble organic solvent to our previously reported method.

587386 Quantitative analysis of heat transfer in a rotary drum

Heat transfer in granular processing and handling is commonly encountered in material manufacturing. Granular processing consumes roughly ten percent of the world's energy [1]. Approximately 1–3% of the world's energy is utilized in cement industries alone [2]. Since most of these processes are energy-intensive, the handling of granular materials is tightly connected to a high-energy demand. The flow properties of granules alone are governed by their intrinsic material properties (e.g., cohesion, the coefficient of friction and yield stress) and particle properties (e.g., size distribution, morphology, porosity, and roughness), which makes predicting the heat transfer behavior of dynamic granular processes even more complex. Over the last five decades, there has been a continued interest in understanding the role of system parameters and the mechanisms of heat transfer between granular media and the boundary surfaces. The thermal behavior of the granules may change dramatically during processing, introducing variability into the process and affecting final product quality and characteristics. It is, therefore, important to understand the heat transfer mechanism in process specific granular systems.

This work focuses on understanding the heat transfer mechanism in the granular bed inside a rotary drum, one of the most commonly used process equipment. Experiments are performed using silica particles to investigate the granular flow and heat transfer mechanism inside a stainless-steel rotary drum. All the modes of heat transfer are quantified under varying operation conditions to establish a strong understanding of the heat transport. For this, both experiments and CFD-DEM (using MFIX-DEM, an open source multi-solver suite) simulation techniques are used to analyze the thermal behavior. The experimental setup is designed to handle up to 1000^0C, making it possible to study all modes of heat transfer via conduction, convection, and radiation. Omega thermocouples and a mid-wave infrared (MWIR) camera were used to record the temperature profile. The digital image analysis (DIA) and the particle image velocimetry (PIV) are combined with IR thermography (DIA/PIV/IR technique [3]) to obtain combined quantitative hydrodynamic and thermal data sets.

Reference:

- Duran, Jacques. "Introduction."Sands, Powders, and Grains. Springer New York, 2000. 1-18.
- Engin, T., Ari, V., 2005. Energy auditing and recovery for dry type cement rotary kiln systems – a case study. Energy Convers. Manag. 46, 551–562.
- Patil, A. V, Peters, E. A. J. F., Sutkar, V. S., Deen, N. G., & Kuipers, J. A. M. (2015). A study of heat transfer in fluidized beds using an integrated DIA / PIV / IR technique. Chemical Engineering Journal, 259, 90–106.

575999 Experimental and Numerical Study of a Radial Multi-Zone Vortex Chamber Spray Dryer

High-G operation in vortex chambers allows intensifying interfacial mass, heat and momentum transfer [1]. In previous studies, application to particle drying [2,3] and fine particle coating [4] was demonstrated. Combining high-G intensified gas-solids contact, gas-solids separation and solids segregation was also considered [5]. In the present work, a variant called the radial multi-zone dryer (RMD) [6] is experimentally and numerically studied focusing on application in spray drying, characterized by relatively dilute operation. Two distinct temperature zones are created without physical barrier, a radially central zone where hot air with a temperature up to 350°C is injected and a peripheral zone where air injected through vortex chambers at a temperature of around 100°C.

The proposed multi-zone operation allows significant process intensification while preventing degradation of the product. Liquid droplets are injected in the hot central zone, co- or counter-current with the hot air. Fast initial drying is achieved in the hot central zone while burning the product is prevented by rapid evacuation of the produced particles to the colder periphery under the action of the centrifugal force generated by the vortex chamber(s). Typical residence time of the particles in the hot zone can be limited to a few milliseconds. Drying is continued in the colder periphery where high-G operation intensifies interfacial mass, heat and momentum transfer and ensures efficient gas-solids separation.

The experimental set-up that was used is flexible and allows studying various multi-zone concepts, varying the dimension/design of the different zones and the type of spray nozzle. Extension chambers allowing to minimize air consumption and to optimize gas-solids separation can be added. Tests were carried out with total air flow rates of maximum 1000 Nm^3/h, liquid flow rates of up to 6 g/s and hot air temperatures of maximum 350°C. Both hollow-cone and full-cone nozzles were tested. Three types of experiments were carried out: (i) in the absence of liquid injection, (ii) injecting water, and (iii) injecting milk. The axial and radial temperature profiles in the device were measured by means of 16 thermocouples. Comparison of the profiles in the three types of experiments and visual observations allowed gaining understanding in the complex flow pattern of the gas and droplets and the extent of evaporation in specific regions. Deposition of powder on the walls and in the gas and solids outlets was also carefully analyzed. Protection of the nozzle from fouling required a new protective sleeve design.

To gain further insight in the flow pattern, CFD simulations were carried out, adopting the RANS-approach and a coarse-grained discrete particle model to track the motion of the droplets. Evaporation was accounted for and the behavior of droplets of different size studied.

Experimental studies and CFD simulations of spray drying in a radial multi-zone spray dryer show that the technology is feasible and promising for intensified two-step drying. Efficient separation of a hot central zone and cold periphery and an extremely short residence time of the droplets/particles in the central hot zone can be achieved. As a result, using 350°C hot air, high-quality milk powder could be produced. A full cone nozzle was shown to be more efficient and stable than a hollow cone nozzle, especially with counter-current injection of hot air.

References

1. Axel de Broqueville, Juray De Wilde, Numerical investigation of gas-solid heat transfer in rotating fluidized beds in a static geometry, In Chemical Engineering Science, Volume 64, Issue 6, 2009, Pages 1232-1248, ISSN 0009-2509.
2. Eliaers, Philippe ; De Wilde, Juray. Drying of Biomass Particles: Experimental Study and Comparison of the Performance of a Conventional Fluidized Bed and a Rotating Fluidized Bed in a Static Geometry. In: Drying Technology, Vol. 31, no.2, p. 236-245 (2013)
3. Philippe Eliaers, Jnyana Ranjan Pati, Subhajit Dutta, Juray De Wilde, Modeling and simulation of biomass drying in vortex chambers, In Chemical Engineering Science, Volume 123, 2015, Pages 648-664, ISSN 0009-2509,

4. Philippe Eliaers, Axel de Broqueville, Albert Poortinga, Tom van Hengstum, Juray De Wilde, High-G, low-temperature coating of cohesive particles in a vortex chamber, In Powder Technology, Volume 258, 2014, Pages 242-251, ISSN 0032-5910
5. Juray De Wilde, George Richards, Sofiane Benyahia, Qualitative numerical study of simultaneous high-G-intensified gas–solids contact, separation and segregation in a bi-disperse rotating fluidized bed in a vortex chamber, Advanced Powder Technology, Volume 27, Issue 4, July 2016, Pages 1453-1463, ISSN 0921-8831
6. Axel de Broqueville, Juray De Wilde, Thomas Tourneur, Device for treating particles in a rotating fluidized bed, WO/2018/203745, November 2018

Acknowledgements

The authors would like to acknowledge the technical support of Luc Wautier, the support of the Dtuch Institute for Sustainable Process Technology (ISPT) and the financial support by the Dutch Rijksdienst voor Ondernemend Nederland (RVO).

567697 Spatially-Averaged Models for Moderately Dense Gas-Particle Flows Including a Thermal Energy Balance

Moderately dense gas-particle flows, like fluidized beds or risers, are used for a wide range of applications, from food drying and polymer production to the direct reduction of iron ore. While for small-scale fluidized beds, simulation is quite feasible using averaged equations of motion on the microscale (kinetic-theory based two-fluid model), the occurring physical phenomena cannot be fully numerically analyzed in large scale reactors yet, due to a large range of involved scales. In fluidized bed reactors of several meters in height, the smallest stable structures are usually only several particle diameters wide [1]. By employing coarse enough grids to make a simulation of full-scale reactors possible, mesoscale structures and, thus, their contributions to the gas-solid drag, intra- and interphase heat transfer between the particles and the gas, as well as to other macroscale flow properties, are not resolved. However, the unresolved heterogeneous structures have a significant influence on the macroscale flow properties. The velocity fluctuations around particle clusters give rise to turbulence, i.e. cluster induced turbulence [5,6]. Not accounting for this turbulence leads, for example, to an overestimation the gas-particle heat transfer.

A model accounting for the unresolved terms can be derived by spatially averaging the kinetic theory based two-fluid model equations [4]. The heat transfer coefficient devided by the solids volume fraction can be approximated by its zeroth order Taylor series expansion about the filtered variables. This approximation is used to close the unresolved part of the filtered heat transfer using a concept similar to the drift velocity [2,3]. This so-called drift temperature is the difference between the gas-phase temperature and the gas-phase temperature as seen by the particles. A transport equation for the variance of the temperature of the gas-phase is derived in order to obtain a closure model for the drift temperature.

An a-priori analysis shows that the proposed closure model for the unresolved heat transfer fits well with the predictions obtained by filtering highly resolved fine grid simulation data. In addition, closure models for the other unresolved terms in the filtered thermal energy equation and in the transport equation for the variance of the temperature are validated against the filtered fine-grid simulation data and numerical experiments.

References

[1] Agrawal K, Loezos PN, Syamlal M, Sundaresan S. The role of meso-scale structures in rapid gas-solid flows. *J. Fluid Mech.* 2001;445:151–185.

[2] Ozel A, Gu Y, Milioli CC, Kolehmainen J, Sundaresan, S. Towards filtered drag force model for non-cohesive and cohesive particle-gas flows. *Phys. Fluids* 2017;29:103308.

[3] Schneiderbauer S, Saeedipour M. Approximate deconvolution model for the simulation of turbulent gas-solid flows: An a priori analysis. *Phys. Fluids* 2018;30:023301.

[4] Schneiderbauer S. A spatially-averaged two-fluid model for dense large-scale gas-solid flows. *AIChE J.* 2017;63(8):3544-3562.

[5] Fox RO. On multiphase turbulence models for collisional fluid-particle flows. J. *Fluid Mech.* 2014;742:368-424.

[6] Capecelatro J, Desjardins O, Fox RO. Numerical study of collisional particle dynamics in cluster-induced turbulence. *J. Fluid Mech.* 2014;747:R2.

568819 Validating Optimal Drag Corrections for the Simulation of Sand Biomass Co-Fluidization

The bubbling fluidized bed reactor is widely used in the fast pyrolysis of biomass to generate syngas and bio-oil. In this process, the biomass is co-fluidized with sands. Computational Fluid Dynamic (CFD) simulation of this reactor is challenging due to the large number of sand particles and lack of accurate drag corrections. In this research, the tracked number of sand particles is reduced by using the coarse grained discrete element method (CGDEM) where many sand particles are lumped into a numerical parcel. The drag corrections proposed by Gidaspow, Syamlal-O'Brien, and Ganser are evaluated using the experimental measured pressure drops at six different locations. Among all the tested four different superficial gas velocities and four biomass mass fractions, the Syamlal-O'Brien drag corrections shows superiority over others. The accuracy was also improved by using Syamlal-O'Brien drag corrections for sand parcels and using the non-spherical Ganser drag corrections for biomass particles.

562271 Particle-Scale Investigations of Reactive Fluidized Beds Based on CFD-DEM Coupling Methods

Fluidized beds have been widely used in the industrial applications of solid fuel energy conversion. Though efforts have been taken to investigate reactive fluidized beds under various conditions, there is still a lack of thorough understanding on the interplays among the gas-solid hydrodynamics, heat transfer, and chemical reactions. It is noted that most previous numerical studies on the reactive fluidized beds were based on the continuum-based simulations, in which it is difficult to arrive the particle-scale information. In comparison, the simulation based on the computational fluid dynamics (CFD) - discrete element method (DEM) can give predictions with higher accuracy and resolution, which facilitates in-depth analysis on the fluidized beds.

A computational framework of CFD-DEM for the dense reactive gas-solid flow has been developed. The models are first validated against a series of experiments, which shows good agreement between the predictions and measurements. In this study, three typical reactive applications operated under the bubbling fluidization regime are investigated, namely the (i) biomass fast pyrolysis, (ii) coal gasification and (iii) coal combustion. The results show that different fluidized bed applications show distinct patterns of particle entrainment, gas-solid mixing and heat transfer. The effect of the gas-solid mixing on the performance of the reactor is emphasized. The chemical reactions in turn affects the gas-solid hydrodynamics by interphase mass transfer.

561881 A Structural Model of Sub-Grid Drift Velocity for Fluidized Gas-Particle Flows

The drift velocity, defined as the correlation between the local fluid velocity and the local solid volume fraction, is a sub-grid quantity representing the local inhomogeneity of the meso-scale structure for fluidized gas-particle flows. It is shown that the drift velocity correlates well with the filtered drag force through analysis of numerous numerical results. However, the drift velocity is not available in coarse-grid simulations and a closure is required. In analogy to the modelling of the sub-grid scale (SGS) scalar flux in single-phase turbulent flows, several models for the drift velocity have been developed in the literature. This work aims to propose a new structural model for the drift velocity. This model considers the effect of the adjusted slip velocity, the summation of slip velocity and drift velocity, at scales above the filter size Δ. Denoting the scale larger than the filter size as $\gamma\Delta$, it is found that the magnitude of the drift velocity at the scale of Δ decreases with the increase of the ratio between the adjusted slip velocity and the slip velocity at the scale of $\gamma\Delta$. The database used is generated through fully resolved two-fluid model (TFM) simulations in a 3D periodic fluidized system with Geldart A particles in different volume fractions. The TFM results are filtered by volume averaging and used to construct the closure of drift velocity. Furthermore, a *priori* assessment of the present model is presented.The comparisons between the present model and the models proposed by Parmentier et al. *(AIChE J.* 2017;63:3544–3562) and Ozel et al. (*Int J Multiphase Flow*. 2013;55:43–63) are made. Results show that this new structural model gives better performance than the previous models in terms of higher Pearson correlation coefficients as well as narrower probability distribution function of the relative error between the predicted filtered drag force and the exact filtered drag force directly obtained from the database. Besides, the new model requires less computational costs than previous models due to that the necessity of dynamic adjustment of the model constant is diminished. It is concluded that this work provides an accurate and efficient model to estimate the drift velocity in coarse grid simulation for the drag force modeling.

572737 A Cascading Approach to Develop a Filtered Drag Force Model for Large-Scale Gas-Particle Flows

Gas-particle flows in fluidized beds are inherently unstable with multiscale structures. These systems can be studied with Two-Fluid Model (TFM) which solves continuum equations of motion for both fluid and particle phases. When a small mesh size up to a few particle diameters is used, TFM can resolve the fine multiscale structures such as bubbles and clusters fairly accurately. However, simulations with small mesh grid size become prohibitively expensive for larger scale systems, leading to the development of filtered Two-fluid model (fTFM). fTFM requires accurate filtered drag model to account for the significant sub-grid contribution to homogeneous drag force, which originates from inhomogenieties within the filtering volume[1-4].

Recently we have developed a neural-network-based filtered drag model from a dense fluidized bed case[5]. Compared to previous modeling attempts for filtered drag force model, we included an additional marker, gas-phase pressure gradient, which is identified through theoretical derivation and budget analysis. Validation tests through *a priori* analysis show high prediction accuracy, and *a posteriori* analysis with coarse grid size up to 27dp show good agreement between fine- and coarse-grid simulations. In the present study, we further investigate how this neural-network-based filtered drag model can be improved by incorporating filter size and Froude number, to make this model more generally applicable, and efficient for fTFM simulations.

Most industrial devices are very large and grid sizes used to simulate flows in these units are invariably much larger than the range considered in our neural-network based filtered drag model. In fact, most computationally developed coarse models are based on rather small fine grid simulations, and they essentially extrapolate filter size effect to these larger systems, whose accuracy is not established. In the present study, we set out to examine how drag correction changes with filter size for larger filter sizes. We approach this problem through *cascading*, where one performs one or more additional rounds of filtering of the fTFM. Cascading also allows us to examine how the dominant physics changes with scale.

Briefly, the fine-grid simulations are first used to develop the closures for an fTFM. This fTFM is then solved with coarser grids, whose size is in the range of the first round of filtering, to generate computational data which is further filtered to get drag corrections which are now valid for even larger filter sizes. This procedure is repeated until we reach a mesh grid size of desired level of coarsening for most large industrial systems. We first validate this approach rigorously and then examine how drag correction changes with filter size for large filter sizes.

[1] Agrawal, K., Loezos, P. N., Syamlal, M., & Sundaresan, S. (2001). The role of meso-scale structures in rapid gas–solid flows. *Journal of Fluid Mechanics*, *445*, 151-185.

[2] Igci, Y., Andrews, A. T., Sundaresan, S., Pannala, S., & O'Brien, T. (2008). Filtered two-fluid models for fluidized gas-particle suspensions. *AIChE Journal*, *54*(6), 1431-1448.

[3] Ozel, A., Fede, P., & Simonin, O. (2013). Development of filtered Euler–Euler two-phase model for circulating fluidised bed: high resolution simulation, formulation and a priori analyses. *International Journal of Multiphase Flow*, *55*, 43-63.

[4] Parmentier, J., Simonin O., & Delsart,O. (2012) A functional subgrid drift velocity model for filtered drag prediction in dense fluidized bed." *AIChE Journal* 58.4: 1084-1098.

[5] Jiang, Y., Kolehmainen, J., Gu, Y., Kevrekidis, Y. G., Ozel, A., Sundaresan, S. (2019) Neural-network-based filtered drag model for gas-particle flows. *Powder Technology*, 346, 403-413

576779 Comparative Analysis of Drag Model Predictions for Fluidized Bed Applications

A stream of liquid or gas can be used to suspend granular materials into a fluidized state. Such fluidized beds of particles are commonplace in industrial operation because of their efficient mass, momentum and energy transfer, for example, fluidized beds are commonly used for catalyzing reactions, mixing, and heat transfer operations. Such industrial processes could benefit from the physical insights and predictions generated via models of fluid and solid-particle flows. The drag force is the momentum exchange between the fluid and solid particles that causes fluidization, and therefore correctly modeling the drag force is particularly important to accurate fluidized bed predictions. Currently, a gamut of drag models exist. Prior works have shown that the predicted fluidized bed behavior depends on the drag model selected. Additionally, while individual reports generally identify an accurate drag formulation for a specific experimental comparison, a single drag model demonstrating consistent accuracy across these previous studies does not exist. Hence, it is not clear which drag model is appropriate for a given application, a priori. Furthermore, a link between the drag model formulation and the macro-scale prediction of fluidized bed behavior is currently lacking. Here, simulations of fluidized bed behavior are presented for varied system conditions and drag model formulations. The aim of this work is to develop an understanding of expected variability in macro-scale predictions of fluidization due to permutations in the model inputs. The uncertainty in fluidized bed prediction resulting from changing the drag model formulation is quantified and explored across a wide parameter space is presented.

560354 Numerical Modeling on Gas-Solid Two-Phase Flow Including Thin Plates

Multiphase flows including granular media is widely encountered in the chemical engineering. Because the multiphase flows are regarded to be extremely complicated phenomena, introduction of numerical simulation techniques is desired for the better understanding. The Discrete Element Method coupled with Computational Fluid Dynamics method, so-called DEM-CFD method[1], has been often employed in various multiphase flow systems in industries so far. In the DEM-CFD method, the Local volume average technique[2] is introduced to evaluate the void fraction in each CFD grid. Although the advantages on the representation of the macroscopic behavior of multiphase flows has been proven in the previous studies[3]–[5], it is of great difficulty to calculate the multiphase flows including the thin plates.

To solve this problem, the new discretization technique in the DEM-CFD method is developed, namely the Refined grid model. In the Refined grid model, two different scales of grids are flexibly used: one is for the calculation of void fraction, drag force and buoyancy; the other is for the computation of the fluid equations. Using two kinds of grids, the multiphase flows including thin plates can be calculated in the staggered grids. In the current study, the adequacy of the Refined grid model is proven through the validation tests in the fluidized bed system. Besides, applicability of a scaling law model such as the coarse graining DEM to the Refined grid model is shown for the first time. Consequently, introduction of the numerical simulation techniques is drastically accelerated.

Acknowledgement

This study was financially supported by JSPS KAKENHI (17H02825), JSPS KAKENHI (17KK0110) and by Grant-in-Aid for JSPS Research Fellow from the Japan Society for the Promotion of Science (18J12267).

References

[1] Y. Tsuji, T. Kawaguchi, and T. Tanaka, "Discrete particle simulation of two-dimensional fluidized bed," *Powder Technol.*, vol. 77, no. 77, pp. 79–87, 1993.

[2] T. B. Anderson and R. O. Y. Jackson, "A Fluid Mechanical Description of Fluidized Beds," *Ind. Eng. Chem. Fundam.*, vol. 6, no. 4, pp. 527–539, 1967.

[3] H. Yao, Y. Mori, K. Takabatake, X. Sun, and M. Sakai, "Numerical investigation on the influence of air flow in a die filling process," *J. Taiwan Inst. Chem. Eng.*, Dec. 2017.

[4] Y. Mori, C.-Y. Wu, and M. Sakai, "Validation study on a scaling law model of the DEM in industrial gas-solid flows," *Powder Technol.*, vol. 343, pp. 101–112, Feb. 2019.

[5] K. Takabatake, Y. Mori, J. G. Khinast, and M. Sakai, "Numerical investigation of a coarse-grain discrete element method in solid mixing in a spouted bed," *Chem. Eng. J.*, vol. 346, pp. 416–426, Aug. 2018.

574640 Compressible Nutsche Cake Filtration – Retarding Effect of Residual Base-Cake

Production scale filtration of fine needle morphology crystalline solids is often challenging, since needles tend to give rise to compressible, slower filtering cakes. In this study, a slurry comprised of needles that has a difficult Nutsche filtration at production scale was studied in a laboratory setting to better explore the factors that affect this filtration. Laboratory filtration was accomplished on a 1L "pocket" Nutsche filter. The filtration was characterized by measuring the inherent cake resistance and media resistance under different conditions. Increasing the driving pressure showed limited improvement in filtration rate, indicating a moderately high level of compressibility. Filtration at colder temperatures showed a retardation from both increased viscosity and a higher inherent cake resistance. However, while these filtration experiments were insightful, scale-up calculations were under-predicting the filtration times at production scale. Further investigation focused on base-cakes, which are thin 1 – 2 cm heels left behind on many production filters, including the agitated Nutsche used in this process. A thin residual base-cake increased filtration times 2X and then nearly 3X in successive filtrations as the base-cake resistance increased. Incorporating these base-cake resistances into the scale-up calculations resulted in predictions that nearly match the production scale results. Removal of the base-cake in this application holds significant potential to increase productivity of the filtration unit operation.

574639 Free Surface Profiles in Centrifuge Filtration at Low Speeds – Theoretical and Computational Solutions

The speed of rotation of the centrifuge plays an important role in the filtration rate of any centrifuge based filtration systems. As the filtration proceeds, a wet cake starts to form in the centrifuge. The speed of rotation also plays a very important role in the compressibility of this wet cake. For very compressible cakes, the centrifuging is done at low speeds of rotation to minimize compression of the cake. If the speed is too low, it could lead to loss of slurry from the bottom of the centrifuge. Hence it is important to understand the effect of speed of rotation of the centrifuge on the liquid free surface profile. In this paper, theoretical solutions are presented for various scenarios of centrifuge operation. These results are compared with computational fluid dynamics (CFD) based predictions for the free surface profiles. This analysis helps in determining the optimum centrifuge speeds for filtration of very compressible cakes.

560916 Developing Hollow Fiber Air Filters for Filtration of Ultrafine Particles

Air pollution is an important issue in recent decades. Among various air pollutants, particulate matters (PM) are of particular concerns due to their adverse health effects to human being. To remove these airborne particles, researchers have developed a variety of filters, such as high efficiency particulate air (HEPA) and electrospun air filters. However, most filtrations show a limited filtration efficiency for removing ultrafine particles (particles <0.3 mm).

Rather than normal flat sheet filters, the hollow fiber is another mature technique which has been applied to nanofiltration and gas separation. The particles separated by hollow fibers are within nanometer or even angstrom sizes. Moreover, hollow fibers are self-supported and have a large volume to area ratio. Therefore, we believe hollow fibers would be a great candidate as air filters to remove PMs, especially for ultrafine particles. The objectives of this study are to (1) fabricate suitable hollow fiber air filters by studying their membrane formation mechanisms; (2) investigate the effects of dope additives to fabricate fibers with high gas flux and filtration efficiency; (3) examine the operation mode of air filtration using hollow fiber filters to obtain the best separation performance.

In this study, we have developed various hollow fiber air filters from different polymer materials, such as polyvinylidene difluoride (PVDF), polyacrylonitrile (PAN), and polyethersulfone (PES). The experimental results showed that our hollow fibers were able to remove > 99.999% particles down to 30-50 nm. To further enhance the gas permeance and lower the pressure drop of hollow fibers, we introduced high molecular weight polyethylene glycol (PEG) and Pluronic F127 in polymer dopes. Moreover, the dope concentrations were decreased. As a result, the N_2 permeance of hollow fiber was increased by 5 time. In addition, we also developed another approach to increase the hollow fiber gas permeance by adding silica particles in dopes and washing out from the resultant hollow fiber subsequently. Finally, to ease the module fabrication process, the mechanical properties of hollow fibers were also considered. Comparing to PVDF and PES hollow fibers, PAN hollow fiber possessed the most balanced filtration performance in terms of N_2 permeance, mechanical properties, and filtration efficiency. In summary, hollow fiber air filters demonstrate an excellent filtration efficiency of ultrafine particles which show a great potential to be a new generational air filter.

572094 Cyclone Design and Performance at Restriced Space

Cyclone separators are widely used in industry for separating solid particles from gases. The standard cyclone has gas and particles exiting the device at opposite sides and is either equipped with a tangential inlet or – mostly for the usage in multiclones - with an axial inlet. A competitive but less explored alternative to the standard cyclone is the uniflow cyclone. Like a pipe, a uniflow cyclone has gas and particles passing through it in only one direction. Like in the standard cyclone the vortex flow is generated either by swirl vane inserts or by a tangential inlet at the entrance of the device. In contrast to standard cyclones clean gas and separated particles leave the uniflow cyclone at the same side. Compared to standard cyclones, uniflow cyclones are much more compact. This design reduces their investment costs and makes them particularly interesting for applications with limited space. It also allows a simple and cost-effective implementation in piping systems. Even though the principle of uniflow cyclones has long been known, not much literature has been published on how to design and to calculate them. In recent years comprehensive experimental and theoretical studies of uniflow cyclones have strongly improved their understanding and led to approved design criteria and calculation methods similar to those which have already been achieved for standard reverse flow cyclones.

Often space for a dedusting unit in an industrial plant is limited. If saving of space is of high priority or if there is even only a limited space available for purifying a given gas volume flow the question arises whether under the restriction of restricted space uniflow cyclones may be preferable over standard cyclones. This question is addressed by applying well proven calculation programs for standard cyclones according to [1,2] and for uniflow cyclones as described in [3-5]. The results indicate in agreement with experiments:

For separating a specified gas solids mixture at a specified pressure drop
- a uniflow cyclone requires significantly less space (diameter and length) than a standard reverse flow cyclone but is generally also slightly less efficient, at least in its present state design.
- the efficiencies of the standard cyclone and of the uniflow cyclone approach each other with increasing gas volume flow, i.e. decreasing cyclone size, and furthermore with increasing pressure drop, particle size and/or particle density.
- a uniflow cyclone can be more efficient than a standard cyclone if space for dedusting the gas solids flow is limited and the available pressure drop is low. In other words, uniflow cyclones can achieve higher separation efficiencies per volume than standard cyclones for low pressure drops.

A deeper understanding of those results is obtained by analyzing the principal differences between standard cyclones and uniflow cyclones regarding their performance data. This analysis includes amongst others the quantitative relationship between the gas and particle velocities and the performance data of both cyclone types. Its implication on erosion potentials in both cyclone types is discussed as well.

The results described above are transferable to multicyclone systems, i.e. systems of many parallel cyclone cells within a common housing having a common solids hopper for solids discharge. Principally increasing the number of parallel cyclone cells and decreasing at the same time their size improves the efficiency of a multiclone without changing the base area and without affecting the pressure drop, provided a uniform distribution of the gas and the solids feed into each single cyclone cell can be achieved and bypass flows through the solids discharge openings from one cyclone cell to the other can be avoided. In many cases multicyclones use swirl tube cells. Analogously, multicyclones can be made from uniflow cyclone cells. Often these devices are used in space-limited applications, e.g. as third-stage separators in fluid catalytic cracking (FCC) processes [6-8]. Multiclones using uniflow cyclone cells at ambient conditions are applied e.g. in combustion engines of agricultural machines, locomotives, construction machines and others for pre-cleaning the suction air of the engine before the final filter [9,10]. Industrial mobile cleaners are a further area of application.

[1] E. Muschelknautz, V. Greif, Cyclones and other gas-solids separators, in Circulating Fluidized Beds (Eds: J. R. Grace, A. A. Avidan, T. M. Knowlton), Blackie Acad. & Profess., London (1997) 181-213.
[2] VDI-Wärmeatlas Atlas, Zyklone zum Abscheiden fester Partikel aus Gasen. 12th German ed., 2018.
[3] U. Muschelknautz, Analytical approach for calculating the separation efficiency of uniflow cyclones, Chem. Eng. Technol. 35 (12) (2012) 2099-2106.
[4] U. Muschelknautz, Performance comparison of uniflow cyclones and standard cyclones, Proceedings of 12th International conference on fluidized bed technology, Krakow, 421-428, 2017.
[5] U. Muschelknautz, Design criteria of multi cyclones in a limited space. Proceedings of 8th World Congress on Particle Technology, Orlando, USA, 2018.
[6] W. Peng, A. C. Hoffmann, H. Dries, Separation Characteristics of Swirl-Tube Dust Separators, AIChE J. 50 (1) (2004) 87-96.
[7] M. Kraxner, U. Muschelknautz, S. B. R. Karri, R. Cocco, T. M. Knowlton, Applicability of a Uniflow Cyclone as a Third Stage Separator in the FCC-Process. AIChE – American Institute of Chemical Engineers – Annual Meeting, Minneapolis / Minnesota, USA, 2011.
[8] S. Krishnamurthy, W. Koves, Apparatuses and methods for gas-solid separations using cyclones. Uop Llc, Patent US-8419835 B2, 2013.
[9] V. Greif, Multizyklon-Blöcke zur Staubabscheidung vor Verbrennungsmotoren – Auslegung und Betriebserfahrungen. VDI Berichte 1511 (1999), p. 215-226.
[10] M. Kraxner, Empirical evaluation of design criteria for uniflow cyclones in multicyclone boxes, PhD Thesis, TU Munich, Germany, 2013.

564589 Self-Cleaning Electrostatic Oil Mist Collector By Using Corona Discharge and Surface Dielectric Barrier Discharge

Oil mist emitted during cooking is a major source of PM in the urban area. As its high efficiency and low pressure drop compared to other methods, ESP (Electrostatic Precipitator) is the most commonly used in the removal of oil mist. However, by long time operation with ESP, oil mist would be deposited on the surface of electrode. Particle deposition on electrode would reduce the particle removal efficiency of ESP. Therefore, ESP requires regular cleaning to maintain its initial performance. As cooking oil fume is very sticky when it's deposited, it is very hard to be removed. To solve this problem, non-thermal plasma can be used to remove oil mist deposition on its electrode. In this study, we developed self-cleaning ESP using SDBD (Surface Dielectric Barrier Discharge) plasma. SDBD plasma is one of non-thermal plasma generated by high voltage AC or pulsed DC power. In the plasma region, there are lots of active species (radicals, electrons). These active species can decompose oil mist deposited at the surface of electrode.

Similar to conventional ESP, our device uses corona discharge for particle charging and DC electric field for particle collection. After long time operating in particle collection, the electrode needs cleaning because of the oil mist deposition. In this condition, by modifying the power circuit, SDBD plasma is generated at the surface of its electrode. By SDBD plasma generated at the surface, oil mist deposited on the electrode is decomposed to form gas phase products (CO, CO_2).

By increasing the applied voltage from 0 to 5 kV, particle collection efficiency also increased from 21% to 94.4%. After particle collection, our device works in electrode cleaning mode by using SDBD plasma. After operating SDBD plasma on the contaminated electrode, CO and CO_2 were detected at the outlet. Simultaneously, oil mist deposited at the surface was removed after operating SDBD plasma. From this result, it can be concluded that oil mist at the surface of the electrode was converted to CO and CO_2 by SDBD plasma.

563442 Optimization of Classifier Height and Nozzle Angle in a Jet Mill

Our experimental study investigates the optimum height of a classifier and the optimum angle of a nozzle inside the grinding chamber for a jet mill. In these experiments, first, we prepare a mixture of different sized particles. In case 1, this mixture is prepared using two different sized particles, and in case 2 using three different sized particles. Then, the mixture is placed inside the milling chamber and air pressure is applied through the grinding nozzle. The outlet from the chamber is collected using the classifier, which is placed at the center of the milling chamber and whose height is *h* from the base of (inside) the chamber. Finally, output from the classifier is analyzed and its efficiency is found out for coarser particles, i.e. mass percentage of the coarse particles left inside the chamber, and for finer particles, i.e. mass percentage of finer particles in the output.

From our experiments, we found that the optimum height of a classifier (hc) and the optimum angle of a nozzle, for both the cases, as 10mm and 45°-60°, respectively. It is observed that with increasing $h(<hc)$, the classifier efficiency increases for coarser particles and reaches its maximum (100%) at h=10mm. During this, the classifier efficiency decreases for finer particles. With Further increase in $h(>hc)$, we do not observe any variation (increase) in the efficiency of classifier for coarser particles, but certainly decreases for the finer particles. Therefore, the optimum height of the classifier has to be 10 mm, where the efficiency for coarser particles first became 100%. It is also observed that for nozzle angles of greater or lesser than 45°, we see the decrease or increase in efficiency of the classifier for finer particles. Therefore, at the above-mentioned optimum values, the classifier produces the final product with the smaller particle sand narrow distribution, which is expected from a jet mill.

567094 Enhancement of Fluorite Flotation Process with the Combination of in-Situ Modification By Ball-Milling and Gas-Liquid Microdispersion

Fluorite is the most important upstream raw material of fluorine industry. As main and the most economical source of fluorine, high grade fluorite (CaF_2>97%) is of great demand in order to obtain high performance fluorine-containing solar cell film, ionic superconducting materials and other kinds of fluoropolymers. However, complex association is quite common in fluorine ore, effective fluorite content of which could only reach 7 % to 20 % in the nature. As the most universal separation method, higher request has been put forward to froth flotation to get fluorite concentrate. Facing challenges due to the similar Ca site of fluorite and calcium carbonate and the easiness of quartz to be activated by the existence of calcium, countless improvements have been contributed to froth flotation during decades, mainly focused on mixture of novel additives, effective equipment and so on. However, rarely did the investigators focus on the problem of high energy consumption caused by times of scavenging and mechanical agitation. What is more, it is still lacks a green method to separate the fine and complex paragenetic particles without expensive additives and high emission.

In this study, we realized the separation of fine fluorite particles with the help of micro-bubbles and in-situ modification during the ball-milling. Instead of traditional water milling, we directly added oleic acid and Na_2CO_3 into the ball-milling jar, which was found to be the optimal feeding and modification mode by contrast. Firstly, owing to the regulation of pH caused by Na_2CO_3, an appropriate zeta potential was achieved, effectively preventing the aggregation of the fine particles. Secondly, due to the mechanical force of the milling, adsorption of oleic acid onto the surface of the particles was intensified, leading to a superior dispersion of the initial suspension under minimum amount of surfactant. In addition, it was proved by the size analyzer that the selectivity and accuracy of the adsorption between surfactant, fluorite and quartz particles was remarkably enhanced. Furthermore, a controlled and high efficient gas-liquid microdispersion technology was introduced to intensify the froth flotation process in terms of productivity and selectivity. Within the limited space less than 8 cubic millimeter, due to sufficient dispersion of membrane and strong shear of continuous phase of suspension, we obtained gas bubbles swarm with average diameter of 1 mm. Compared with traditional flotation where bubbles are usually in the size of centimeter level, the contact area of the gas-liquid-solid system in our work was expanded by thousand fold, which maximized the interaction between hydrophobic fluorite and the bubbles, decreased the concentration of the depressant and reduced gas consumption at the same time. Thanks to the combination of the approaches mentioned above, the grade of fluorite could reach 98 % in one time flotation just within a palm-sized blocked swiftly, which notably increased the flotation efficiency. Besides, the density and size of the bubble group could be precisely controlled by the gas flux along with mechanical equilibrium of each single bubble, which brings good adaptability to various particles with different size, concentration or surface characteristics.

In a word, with the coupling of in-situ modification of ball-milling and membrane dispersion flotation, instead of utilizing other complex combination of additives, we obtained a satisfactory grade and recovery only with the most fundamental and cheapest long-chain fatty acids oleic acid as collector and sodium silicate as depressant. Owing to the accurate interaction between tiny particles and additives, usage of the additive was minimized and the flotation could be once completed without times of selection, which cut down the consumption of water treatment, energy and fixed investments to a great extent. Moreover, it is worth noting that the simplified process is of huge potential to be applied to the separation of different kinds of tiny solid compounds effectively through the adjustment of micro bubbles flexibly.

578884 A Course at Purdue University on Particle, Powder, and Compact Characterization

Purdue's Center for Particulate Products and Processes offers a dual-level course entitled "Particles, Powders, and Compact Characterization" every other spring. The goal of this course is to familiarize students with the properties and methods used to characterize the mechanical behavior of particles, powders, and compacts, with the intention of using these properties for process and performance design. Topics covered in the course include sampling, particle size and size distributions, particle shape, particle density, particle surface area, powder bulk density, powder compressibility, powder flow, compact elastic modulus, Poisson's ratio, yield and fracture strength, viscoelasticity, and variation with porosity. Students work in small teams in the laboratory to measure sample properties in order to gain hands-on experience using standard laboratory equipment. In addition, students also develop small computational projects related to characterization. Students who have enrolled in the course have been from aerospace, mechanical, chemical, agricultural and biological, and materials engineering, as well as industrial and physical pharmacy, reflecting the inter-disciplinary nature of the course topics. Students successfully completing the course are able to: (1) define and describe the significant properties of particles, powders, and compacts, (2) describe and demonstrate techniques used to measure these properties, and (3) demonstrate how these properties are useful in product and manufacturing performance.

567948 Non-Linear Dependence of Macroscopic Powder Properties on Powder Composition

Non-linear Dependence of Macroscopic Powder Properties on Powder Composition

Sara Fathollahi[1,2], Eva Faulhammer[1], Benjamin J. Glasser[3], Johannes G. Khinast[1,2,*]

1: Research Center Pharmaceutical Engineering (RCPE) GmbH, 8010 Graz, Austria

2: Graz University of Technology, Institute of Process and Particle Engineering, 8010 Graz, Austria

3: Department of Chemical and Biochemical Engineering, Rutgers University, Piscataway, NJ 08854, USA

Corresponding author at: Graz University of Technology, Institute of Process and Particle Engineering, 8010 Graz, Austria.

E-mail address: khinast@tugraz.at (J.G. Khinast)

Keywords: Powder mixtures, Macroscopic properties, API, Lactose, FT4 powder rheometer

Abstract

Powder processing steps, such as feeding and blending, are critical in many industries, including the pharmaceutical industry. For example, within a continuous tablet manufacturing environment, powder blending impact quality of drugs. Powder feeding is highly relevant in continuous processes. For the design of such operations the powder properties should be known, including the particle size distribution, flowability, compressibility, electrostatic chargeability and tendency to segregate. Typically, these properties are measured for pure components. Although it is known that mixtures of pure materials do not follow a linear dependence of the macroscopic powder properties on the mixture fraction, few studies exist that quantify these effects. This, however, is important for the design and optimization of powder processing operations, such as the ones utilized in a continuous manufacturing plant.

In summary, the development of solid dosage forms and the associated manufacturing processes requires a good understanding of the relationship between powder composition and the properties of the powder [1, 2, 3]. Thus, in our work binary mixtures of powders at different composition were prepared in a Turbula mixer. Then, macroscopic powder properties were measured, including bulk and tapped density (BD and TD), powder compressibility, permeability and flowability.

In our presentation we show that there is a non-linear dependence of the relevant properties and composition and propose a model for describing these dependencies. This is highly relevant for designing processes, especially in the context of continuous pharmaceutical manufacturing.

List of References

1. Forte G, Clark PJ, Yan Z, Stitt EH, Marigo M. Using a Freeman FT4 rheometer and Electrical Capacitance Tomography to assess powder blending. Powder Technol [Internet]. Elsevier B.V.; 2018; Available from: http://linkinghub.elsevier.com/retrieve/pii/S0032591017309701
2. Worku ZA, Kumar D, Gomes JV, He Y, Glennon B, Ramisetty KA, et al. Modelling and understanding powder flow properties and compactability of selected active pharmaceutical ingredients, excipients and physical mixtures from critical material properties. Int J Pharm [Internet]. Elsevier B.V.; 2017;531(1):191–204. Available from: http://dx.doi.org/10.1016/j.ijpharm.2017.08.063
3. Leaper MC, Ali K, Ingham AJ. Comparing the Dynamic Flow Properties and Compaction Properties

of Pharmaceutical Powder Mixtures. Chem Eng Technol. 2018;41(1):102–7.

572744 Comparative Study of Sugar Powder Consistency Using an FT4 Powder Rheometer and Drag Force Flow (DFF) Sensor

The rheological properties of seven common sugar powders were characterized using an FT4 Powder Rheometer® and a Lenterra Drag Force Flow (DFF) sensor. The data generated by both the FT4 Powder Rheometer and DFF sensor reflect mechanical resistance of the powders to forced flow. During dynamic flow FT4 analysis, a precise volume of powder is subjected to the flow pattern of a rotating blade that moves downward through the powder bed. Measurements of the torque and force applied to the blade are combined to calculate Basic Flowability Energy (BFE), quantifying the powder's resistance to the blade motion. [1]

The recently developed DFF sensing technology provides instantaneous readings of the force exerted by particulate flow to a narrow pin immersed within process equipment. The output of the DFF sensor quantifies the drag force, which is a function of material properties such as particle size, morphology and adhesion [2]. The DFF sensor is primarily intended for real-time, in-line process monitoring where the probe is installed within processing equipment. The technology has shown to be a valuable Process Analytical Technology (PAT) tool for High Shear Wet Granulation (HSWG) [3]. For this study, an DFF test station was assembled, where a probe was installed into a laboratory scale mixer with a bottom-mounted impeller.

DFF sensor measurements have previously been compared to FT4 BFE data during HSWG process cycles [4]. The study confirmed that both the in-line DFF sensor and at-line FT4 similarly responded to changes in wet mass densification during the granulation cycle for the majority of blends. This research also highlighted some differences between the measurement trends for formulations with higher binder content. These differences are deemed to be related to wet mass attrition during the delay between the DFF and FT4 measurements, respectively. While DFF sensor measured the flow force in real time inside the granulator, it took between 7 to 12 minutes to discharge the powder from the granulator and run an FT4 test. Even this relatively small time delay may have caused the wet mass properties to change between analysis techniques.

576559 Material Property Database and Its Applications in Predictive Batch and Continuous Unit Operation Modeling

This work includes GSK's efforts in collaboration with Rutgers to build a database with relevant material properties. An important objective of a database is not just to compile the material data but link it to different applications. Towards that end, a bottom up approach was identified, where the unit operations of interest were listed. The required predictive capabilities and the desired output for each of these unit operations were identified. The material attributes that impacted the performance and the output of these unit operations were identified and were decided to be included in the database. Materials were characterized for their flow (characterized using two different shear cells), density, permeability, basic flow energy, wall friction and particle size. A Brookfield Powder flow tester and an FT4 rheometer were used to characterize the flow of powders. The database enabled comparing the output from two different shear cells as well as two different attachments (sample holders) for the same shear cell. The database was linked to a variety of applications; predicting feeder variability, demonstrating mixing rules, surrogate identification and identifying redundant material properties.

The comparison of the Brookfield powder flow data obtained from the shear cell test using two different sample holders demonstrated that the data were not interchangeable. The discrepancy in the data between the two cells was more pronounced for good flowing materials. A characteristic line was shown to exist between the measurements from the two cells, allowing the data to be used interchangeably. A comparison between the repeatability of the measurements from FT4 and Brookfield was also conducted and the reproducibility of each of the measurements was determined. An important goal for such a database is to identify surrogates for the desired unit operation as well understand the redundancy in several material properties. The ability to merge different databases was also investigated. Merging different databases, compiled within the same or different organizations, can help expand the given database rapidly and cost effectively, saving resources and the need to repeat testing for some common excipients. It can also help in standardizing the methodology for data collection. A similar approach will be demonstrated where two different databases containing similar material properties were analyzed and approaches will be demonstrated to combine them, if feasible.

556664 Wide Range Particle Sizing for Submicron By Scattering Light Image Analysis

Particle size is one of the important influential factors for product quality such as pharmaceuticals, ceramics, magnets, and other devices using powder, paste or slurry. Monitoring and control of particle size is a great challenge in these products. Particle size monitoring technology, such as Focus Beam Reflectance Measurement (RBRM) and Particle Vision and Measurement (PVM), can measure precise particle size distribution with single particle detection. However, these methods cannot measure submicron and smaller particles. To solve this issue, we developed a new method which could measure submicron particles by analyzing scattering light with microscope. In this method, a size of individual particle is estimated based on small-angle scattering light intensity acquired from microscope image. We studied the limitations of measurable range of particle size for different materials by experiments and simulations using light scattering theory. As a result, this method can measure precise particle size distribution of materials including submicron particles, and the lower limit of measurable particle size is 200 nm in case of alumina and depends on refractive index of the material.

578831 Polymer and Metalic Powders Flowability & Electrical Charges Characterisation for Additive Manufacturing

Granular materials and fine powders are widely used in industrial applications. To control and to optimize processing methods, these materials have to be precisely characterized. The characterization methods are related either to the properties of the grains (granulometry, morphology, chemical composition ...) and to the behaviour of the bulk powder (flowability, density, blend stability, electrostatic properties ...). However, concerning the physical behaviour of bulk powder, most of the techniques used in R&D or in quality control laboratories are based on old measurement techniques. During the last decade, we have updated these techniques to meet the present requirements of R&D laboratories and production departments. For this purpose, the measurement processes have been automatized and rigorous initialization methods have been developed to obtain reproducible and interpretable results.

In this paper, two new powders characterisation techniques will be presented: a high-resolution tapped density analyser, the GranuPack and a patented triboelectric effect analyser called the GranuCharge™. To show the possibilities of these instruments, two steel powders (from Carpenter and LPW) and three polyamides 12 powders were selected (from Duraform and Arkema). The Duraform polyamide 12 powder was also used to study the impact of a Selective Laser Melting Process on the modification of the surface properties of this sample.

578830 Characterizing the Effect of Powder Flow Aid Additives

We show the effect of the addition of a small fraction of flow aid agents on the flow of cohesive powders. In particular, we will focus on mesoporous silica (MPS) grains playing a role on both capillary and electrostatic forces. We will compare the effect of MPS grains with the addition of magnesium stearate and other classical flow aid agents. The effect on the flow is quantified with the rotating drum method (GranuDrum instrument). According to both grain and pore size of MPS, the cohesiveness decrease can be immediate (for small MPS grains having small pore size) or on the longer term (for larger MPS grains having higher pore size). The effect on the powder electrostatic properties is analysed with a recently developed powder tribo-electrometer called GranuCharge. The quantity of electrostatic charges created during a flow in contact with stainless steel is decreased by the presence of MPS grains. We show that this decrease is induced by a modification of electric resistivity. The MPS grains are adsorbing the humidity contained in the powder and are forming a network having a low resistivity, improving the diffusion of electrostatic charges.

568090 On the Feasibility of Utilising Compaction Equation Parameters As Material Key Attribute Descriptors

In pharmaceutical manufacturing processes such as pharmaceutical tableting, an understanding of a materials flow behaviour is imperative, as it dictates product quality in terms of content and weight uniformity [1]. Various compendial methods exist in measuring powder flow properties such as bulk and tapped density, angle of repose, Carr's compressibility index and Hausner ratio [2]. Other methods such as dynamic flow and shear measurements can be undertaken using a rheometer [2]. However, the accuracy of these methodologies is subject to performance conditions, which may not be representative of the process in question [1]. Additionally, the knowledge of material specific key attributes such as a materials susceptibility to lubricant poisoning also allows for the adequate selection and pairing of materials during formulation development. The use of compaction equations as a means of analysing the pharmaceutical compression process of single- and multi-component powders and the physical correlation of parameters to powder properties has the advantage of requiring significantly less material and time. A combination of parameters from these equations has been shown to provide a useful indication of critical material properties during direct compression[3]. In recent work, it was observed that a combination of parameters from these equations correlated well to derived compressibility values obtained from a rheometer [3]. In the current work, selected parameters are compared to compendial methodologies in order to establish the feasibility of utilising these parameters as indicators for critical material attributes such as the flow property and lubricant sensitivity of materials.

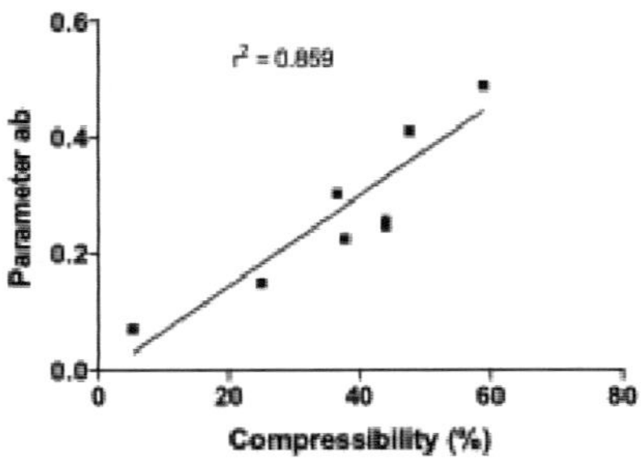

Figure 1: Correlation of compaction parameter ab to measured compressibility values

References

1. Prescott, J.K. and R.A. Barnum, *On powder flowability.* Pharmaceutical technology, 2000. **24**(10): p. 60-85.

2. Shah, R.B., M.A. Tawakkul, and M.A. Khan, *Comparative evaluation of flow for pharmaceutical powders and granules.* AAPS PharmSciTech, 2008. **9**(1): p. 250-258.

3. Solomon, S., et al., *Particle Engineering of Excipients: A Mechanistic Investigation into the Compaction Properties of Lignin and [Co]-Spray Dried Lignin.* International Journal of Pharmaceutics, 2019.

574650 The Effect of Ambient Gas Species on Tribocharging

The Effect of Ambient Gas Species on Tribocharging

X. Liu[1], J. Kolehmainen[1], A. Park[2], S. Sundaresan[1]

[1] *Department of Chemical and Biological Engineering, Princeton University, Princeton, New Jersey 08540, USA*

[2] *Department of Chemical Engineering, Columbia University, New York, New York 10027, USA*

Tribocharging refers to the phenomenon that materials pick up static charges upon contact with other materials [1]. Ambient gas is known to limit the maximum charge acquired by tribocharging on a surface through dielectric breakdown [2-4], but most studies mainly concern tribocharging in air or nitrogen [2]. In this study, we seek to investigate the effect of ambient gas species on tribocharging through vibrated bed experiments in nitrogen and argon.

We first conduct vibrated bed experiments with soda lime glass particles and polyethylene particles of different sizes in an acrylic bed. In each experiment, particles of one chosen narrow size range and one mass loading are shaken for a long time in the container and the total charge on particles is then measured by a Faraday cup. We conducted these experiments in dry nitrogen and dry argon and observed that the charge level measured in nitrogen is higher than that in argon.

We then improve our current charge transfer model used in discrete element method (DEM) simulation [5-6] by including the effect of dielectric breakdown. Although the breakdown electric field strength varies with the separation distance between particles [4], as a simplification, we only consider dielectric breakdown between contacting particles. We set up complimentary DEM simulation and found lowering the dielectric breakdown field strength results in lower charge level. This suggests that the difference in charge level measured in nitrogen and argon could stem from the differences in the dielectric field strength.

The results of these experiments and analyses will be described in this presentation.

[1] Lacks, D., Mohan, S. Contact electrification of insulating materials. J. *Appl. Phys.* 2011; 44(45).

[2] Fotovat, F., Bi, X. T., Grace, J. R. Electrostatics in gas-solid fluidized beds: a review. *Chem Eng Sci.* 2017; 173: 303-334.

[3] Matsuyama, T., Yamamoto, H. Charge-relaxation process dominates contact charging of a particle in atmospheric condition: II. The general model. *J Phys D. Appl Phys. 1997*; 30(15): 2170.

[4] Matsuyama, T. A discussion on maximum charge held by a single particle due to gas discharge limitation. *AIP Conf. Proc.* 2018; 1927: 020001.

[5] Laurentie, J. C., Traore, P., Dascalescu, L. Discrete element modeling of triboelectric charging of insulating materials in vibrated granular beds. *J Electrostat* 2013; 71(6): 951-957.

[6] Kolehmainen, J., Ozel, A., Boyce, C. M., Sundaresan, S. Triboelectric charging of monodisperse particles in fluidized beds. *AIChE J* 2017; 63(6): 1872–1891.

562011 Ddpm-DEM Simulation of Biomass Gasification in a Bubbling Fluidized Bed Reactor

Biomass gasification is a complex chemical process in which biomass is converted into syngas, a combination of carbon dioxide, carbon monoxide, hydrogen, and methane. The syngas could be then used as a fuel in internal combustion engines, gas turbines, or fuel cells for the production of heat, mechanical energy, or power, or as a feedstock for the synthesis of liquid fuels and chemicals. Among the various gasification reactors, the fluidized bed (FB) reactor presents good prospects due to its high rates of heat and mass transfer, good temperature control, and its excellent mixing properties. In a typical FB reactor, fuel feed, together with inert bed material which acts as heat capacitance for the fuel, are fluidized by the gasifying agents, such as air, steam, pure oxygen or their combination. There are many chemical processes within a real biomass FB reactor, such as mixing, segregation, collision, particle heat-up, drying, pyrolysis, volatile matter combustion, and char reaction. Moreover, their scales are greatly separated, which results in detailed study of the entire gasification process being a challenging task. Computational fluid dynamics (CFD) models have become more and more popular in recognizing the dense gas-solid flow dynamics and chemical reactions in FB reactors.

Generally, there are two approaches for modeling gas-particle interaction numerically in dense beds: the Eulerian two fluid model (TFM) approach and the Eulerian–Lagrangian approach. In the TFM approach, solid and gas phases are both treated as interpenetrating continua, and the kinetic theory of granular flows (KTGF) approach is used to calculate particle interaction by solving an additional equation for solids fluctuating energy or granular temperature. The main disadvantage of the TFM approach is the long calculation time required for evaluating the time-averaged solution. The consideration of particle size distribution can significantly increase this time, as an additional dispersed phase for each particle size must be used. In the Eulerian–Lagrangian approach, the gas phase is treated as a continuum while the solid phase is solved by tracking a large number of particles through the flow field calculated from the Navier–Stokes equations for the gas phase. All the forces applied on the particles are calculated to find their new positions and velocities. The solid phase can exchange momentum, mass, and energy with the gas phase. When the particle phase is solved by discrete element method(DEM), the Eulerian-Lagrangian approach is also called DDPM-DEM model. It can offer detailed microscopic information at the particle level, such as particle trajectory, particle-particle and particle-fluid interaction, and transient forces acting on each particle, which is extremely difficult, even impossible to obtain by E-E approach.

A crucial point when using DDPM-DEM is the computing power for simulation comparing with E-E. Therefore, the DDPM-DEM simulations are often restricted to the order of 10^4 particles and are mostly restricted to 2D or quasi-3D solutions. If chemical reactions are added, computation is more and more complicated and expensive. To date most of the DDPM-DEM studies performed have been focused on the hydrodynamics of the isothermal fluidized bed and there have been few works on the simulation of dense gas-solid flow coupling with chemical reactions.

From all the above mentioned models the E-L approach potentially offers the most accurate description not only of the particle motion, but also of chemical reactions and heat and mass transfer between the dispersed phase and the gas phase at the individual particle scale. Although one can find quite many publications of inert particle simulations using the DEM, only few applications of the DEM for particle systems involving heat and mass transfer and chemical reactions are available in the literature.

The aim of this study is to develop a comprehensive CFD-DEM model capable of describing dense, thermal, and reactive multiphase flows like biomass gasification in a fluidized bed reactor. Based on the model validations against lab-scale bubbling fluidized bed experiments, this study evaluates the effect of different operating parameters on the gasification reactions. In particular, the relationship between gas-solid mixing and chemical reactions is investigated.

564706 Kinetic Modeling of Low-Temperature PM Combustion with Relative Velocity in the Presence of Water Vapor

Particulate matter (PM) is a well-known air pollutant, and mainly soot emitted from fossil fuel combustion. The diameter of PM has decreased in recent years with improvement of combustion technology. Especially, the concentration of $PM_{2.5}$ (diameter less than 2.5 μm) increases. While bag filter, diesel particulate filter, and electrostatic precipitator are conventionally used to remove PM from exhaust gas of combustor, collection efficiency of these devices is not sufficiently high for fine PM. In our research group, fluidized bed is used as a PM removal device to collect fine PM by adhesion force, which is dominant force for PM diameter of less than several tens μm. This device collects PM by depositing PM on bed particle, and collection efficiency is 100% for $PM_{2.5}$ due to the collection principle of this device. However, the amount of PM deposition increases through continuous PM collection and PM aggregation is formed on bed particle. PM aggregation becomes large with time. Large PM aggregation easily separates from bed particle by getting large flow resistance, and collection efficiency decreases. It is necessary to take counter measure for maintaining high collection efficiency. PM is mainly composed of combustible substances, and fluidized bed is well-known as a low-temperature combustor. Thus, this device can burn PM at low-temperature. PM combustion inhibits the formation of PM aggregation, and collection efficiency easily maintains initial value. It is observed that the fluidized bed type PM removal device is continuously regenerated at the bed temperature of 400 ‹C while maintaining the collection efficiency of 100%, in contrast to the conventional continuous regeneration devices that require temperatures of 600 to 650 ‹C for PM combustion [1].

Water vapor is generated by fossil fuel combustion. Exhaust gas of combustor includes water vapor at concentration of 5-15 vol.%, but the effect of water vapor on continuous regeneration is not investigated in our previous work [1]. It is reported that water vapor promotes PM combustion at above several hundred degrees. Moreover, the combustion rate of PM increases with relative velocity between gas and solid on PM collection and combustion experiments by means of this device [1]. In order to further develop this device as a continuous regenerating type PM removal device, it is necessary to understand the transport phenomena and PM combustion characteristics in this device. The effects of water vapor and relative velocity on carbon combustion were investigated at low temperatures, and the kinetic model was constructed to take the relative velocity into account. The kinetic model was applied to numerical simulation of a fluidized bed type PM removal device to confirm the validity of this model and to investigate transport phenomena, PM adhesion characteristics, and PM combustion behavior in the presence of water vapor.

PM combustion rate in the presence of water vapor should be measured with considering relative velocity between gas and PM. A new thermogravimetric apparatus was built to create collision between gas and PM. Figure 1 shows a schematic diagram of the experimental apparatus. PM is maintained in thermogravimetric apparatus by hanging on electric balancer. Water vapor is added at concentration of 0-15 vol.% in this experimental apparatus by saturating air with water vapor. Air including water vapor was collided with PM to represent contact between gas and PM in the fluidized bed type PM removal device. Relative velocity between the gas and PM was expressed as the superficial velocity. Carbon black was used as a PM on collection and combustion experiments by means of fluidized bed. However, it is difficult to be placed carbon black in this thermogravimetry apparatus while direct collision with air. Carbon block was used to collide with gas in this thermogravimetric apparatus, and it has almost same combustion rate per unit Brunauer-Emmett-Teller (BET) surface area as carbon black. Combustion rate was obtained from measuring weight change of carbon block by electric balance.

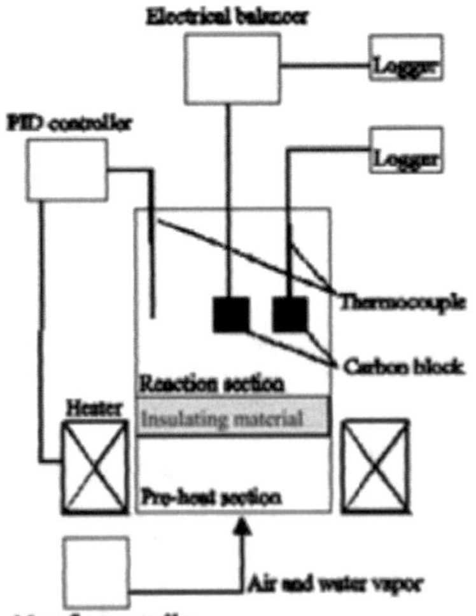

Figure 1 Schematic diagram of thermogravimetric apparatus.

Figure 2 shows the plot of combustion rate versus the relative velocity between the carbon block and the gas at temperature of 400 ‹C, water vapor of 0 and 10 vol.%. Combustion rate at water vapor 10 vol.% is larger than that at water vapor 0 vol.%. The promoting effect of water vapor is detected from this result. It is reported that complex groups (e.g., carboxyl groups) are generated with water vapor even at several hundred degrees, and increase with the water vapor concentration. Carboxyl groups transfer into CO_2 and OH at temperatures ranging from 100-400 ‹C. It is considered that this is the main reaction for the promotion of PM combustion by water vapor at bed temperature of 250-400 ‹C. Moreover, combustion rate increases with increasing gas-PM relative velocity. The carbon combustion rate depends on oxidant concentration at carbon surface. Oxidant concentration decreases by PM combustion and increases by oxidant mass transfer. When the consumption of oxidant is larger than the supply of oxidant, the oxidant concentration decreases and the combustion rate becomes small. High combustion rate is obtained owing to supplying more oxidant to the carbon surface by a rise of relative velocity. Especially, the dependence on relative velocity at water vapor of 10 vol.% is larger than that at water vapor of 0 vol.%. Oxidant consumption increases due to the promoting effect of water vapor on PM combustion. It is necessary to supply more oxidant to maintain PM combustion rate. Thus, the effect of relative velocity becomes large as water vapor.

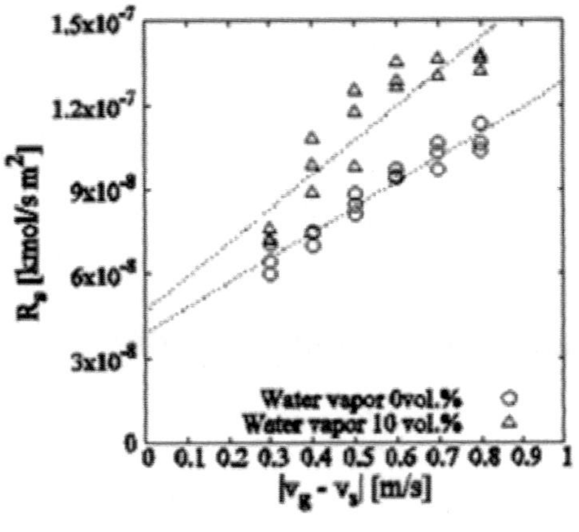

Figure 2 Dependence of combustion rateon relative velocity.

Numerical analysis of fluidized bed type PM removal device is performed under the same condition as those for our previous experimental work. The Euler-Euler method was used for the simulation of fluidized bed type device. PM adhesion is represented by adhesion model. Adhesion model is constructed in our previous study to represent adhesion behavior [2]. It describes the relationship between the amount of the PM deposition and collection efficiency. Collection efficiency is calculated as a function of the amount of PM deposition and decreases with increasing amount of PM deposition. Kinetic model is constructed in the thermogravimetric experimental work, and kinetic parameters are determined by fitting. Kinetic model of PM combustion was incorporated into the simulation code for the fluidized bed type PM removal device. The effect of relative velocity was taken into account as a mass transfer term in this kinetic model [3]. PM combustion amount is calculated from the kinetic model. The amount of PM deposition decreases with increasing PM combustion amount. Figure 3 shows time histories of collection efficiency at superficial velocity of 0.4 m/s, PM diameter of 100-125 μm, PM concentration of 100 mg/m^3, bed temperature of 20, 400 ‹C, water vapor 0 and 10 vol.%. PM diameter and concentration are severe conditions to clearly observe the effects of PM combustion and water vapor on continuous regeneration of this device. The numerical and experimental collection efficiencies decrease with time at all conditions by formation and separation of PM aggregation on bed particles. Numerically obtained collection efficiency is good agreement with experimental data at ambient temperature. Thus, if numerical results agree with experimental results at continuous regenerating mode, the validity of kinetic model is confirmed. Collection efficiency at 400 ‹C is larger than that at ambient temperature, because PM combustion decreases the amount of PM deposition and inhibits formation of PM aggregates on the bed particle. Furthermore, collection efficiency increases in the presence of water vapor. The amount of PM deposition at water vapor 10 vol.% are maintained smaller than that at water vapor 0 vol.% due to promoting effect of water vapor on PM combustion. The differences between numerical and experimental obtained collection efficiencies are at most 1.6%. The numerically obtained collection efficiency shows a reasonably good agreement with the experimental collection efficiency. As a result, the validity of the kinetic model was confirmed.

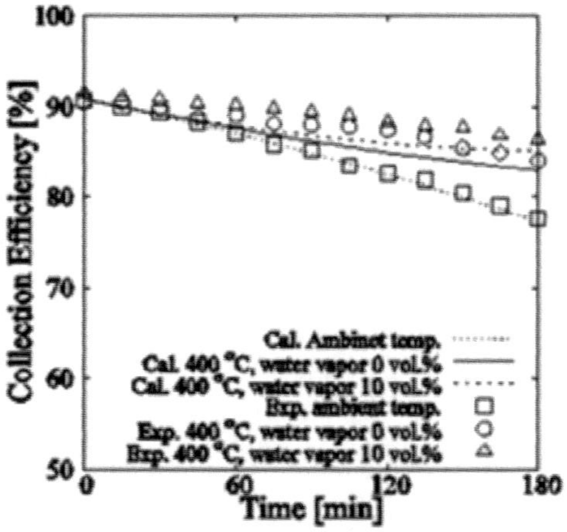

Figure 3 Comparison of numerical and experimental obtained collection efficiencies.

[1] Yamamoto, T., Yokoo, K., Kusu, A., & Tatebayashi, J. Highly efficient particulate matter removal by a fluidized-bed-type device operated in continuous regeneration mode. *Powder Technology* 2018, 323, 86-94.

[2] Yamamoto, T., Tsuboi, T., & Tatebayashi, J. A numerical simulation of PM adhesion characteristics in a fluidized bed type PM removal device by a finite volume Eulerian–Eulerian method. *Powder Technology* 2016, 288, 26-34.

[3] Guedea, I., Díez, L.I., Pallarés, J., & Romeo, L.M. On the modeling of oxy-coal combustion in a fluidized bed. *Chemical Engineering Journal* 2013, 228, 179-191.

567752 Particle Attrition By Chemical Reaction during the Redox Process of Chemical Looping Combustion

Due to the large carbon footprint of existing fossil fuel based powered plants, it is necessary to find potential alternatives to existing gas scrubber technology. One suggestion is by using high temperature, dry sorbent technologies like chemical looping combustion (CLC). CLC provides increased power efficiencies while additionally having lower energy penalties towards carbon capture than traditional combustion and gasification technologies that include carbon capture. CLC uses metal oxide particles termed “oxygen carriers” in a cyclic redox mode typically using two reactors to separate the combustion of carbonaceous fuel and power generation/ oxygen regeneration step. CLC is typically operated in a circulating fluidized bed (CFB) reactor where the oxygen carriers are first reacted with the carbonaceous fuel than are transported to the air reactor to regenerate the oxygen in the reduced carrier. Chemical looping combustion faces some uncertainty due to economic feasibility. This arises partially from the solids makeup rate as a result of particle attrition encountered during operation. Particle attrition is a concern for chemical looping combustion due to the mechanical, thermal, and chemical stresses that an oxygen carrier experiences. Mechanical attrition typically occurs in two primary mechanisms: wear and impact. Thermal stresses resulting from the high temperature and potential temperature swing experienced in the separate fuel and air reactors can lead to oxygen carrier degradation by increasing the amount of internal stresses and affect material properties such as hardness and fracture toughness. Finally, chemical stress due to the redox process in chemical looping can lead to degradation of the oxygen carrier directly or indirectly by increasing the likelihood of mechanical attrition through increased internal stresses. While some studies have observed degradation due to undesired chemical reactions such as carbon formation from methane decomposition, the effect of chemical reactions have not been effectively studied for the particle attrition involved in chemical looping combustion.

This study aims to investigate and explain the effect of chemical reactions on a hematite ore oxygen carrier. This is done by utilizing a conical jet cup that can operate similar to a fluidized bed reactor by allowing for high temperature and under redox conditions. To focus on the chemical reactions involved in chemical looping, the bed is operated in a fixed bed mode targeting the transition from hematite (Fe_2O_3) to magnetite (Fe_3O_4). Effects of fuel (H_2 and CH_4), temperature (800-900°C), and cycles (25-50 cycles) are investigated. A second set of experiments will utilize the tangential jet of the jet cup to determine the impact on mechanical attrition during the
indicate that for the hematite ore undergoing the redox process cause minor fines to be elutriated during the reduction process. The jet cup indicates that the material becomes weakened due to the chemical reaction as the oxygen carrier undergoing a 25 redox cycle experiences nearly 3 times more mechanical attrition than the unreacted oxygen carrier.

573755 Intensification of Ethylene Production By Chemical Looping Oxidative Dehydrogenation through Exergy Loss Minimization

The production of ethylene is one of the most energy and carbon intensive processes among commodity chemicals. Recently, we reported a chemical looping approach to oxidative dehydrogenation of ethane (CL-ODH), where a metal oxide redox catalyst is used to supply oxygen to convert ethane to ethylene and water. The redox catalyst is subsequently circulated into a separate reactor were the lattice oxygen is regenerated in air. In addition to experimental data showing excellent ethylene yield and selectivity, our process modeling of the CL-ODH systems project up to 81% fuel savings with respect to ethane steam cracking, which is the standard industrial process for ethylene production. These projected large energy savings are surprising given the high thermal efficiencies reported for steam cracking reactors/furnaces (~95%) owing to decades of industrial research and optimization to improve cracking plant efficiency. The net cracking reaction, in which ethane decomposes into ethylene and hydrogen, is very endothermic and requires high temperatures (>1000 °C). However, a large amount of energy is recovered in the form of high temperature steam, for power generation, and hydrogen, for use as fuel-gas or as a chemical feedstock. The high thermal efficiencies of cracking, the loss of hydrogen and the practical reactor limitations of conventional ethane and oxygen co-feed ODH has led to some industrial skepticism about the ability of ethane-ODH to intensify ethylene production. However, our AspenPlus modeling studies using experimental redox catalyst results indicate that significant energy savings and CO_2 emission reductions can be achieved through CL-ODH even when hydrogen is fully credited.

We present experimental and modeling data of a robust redox catalyst for use in circulating fluidized beds that supports the viability of the CL-ODH system. We also present extensive process modeling of exergy (exploitable energy) loss in cracking and CL-ODH process to identify the sources of the energy savings in CL-ODH. We show that CL-ODH can reduce the exergy loss of ethylene production by over 4 GJ/tonne High Value Products and reduce CO_2 emissions by up to 87%. We identify multiple locations where CL-ODH reduces the exergy destruction when compared with the high (1st law) efficiency steam crackers. It is shown that the in-situ combustion of hydrogen and inherent heat integration of circulating fluidized ODH reactors offer significant exergy advantages over cracking furnaces.

567186 Synthesis and Characterization of Aerosol Particles and Reactions for the Solar Driven Hybrid-Sulfur-Ammonia Water-Splitting Cycle

Thermochemical water-splitting cycles are some of the most promising and renewable methods for hydrogen production. The Hybrid-Sulfur-Ammonia (HySA) cycle is one such cycle that ideally utilises the whole spectrum of solar irradiation by combining one photochemical with three thermochemical steps. All steps have been studied in the past, but there are still open challenges and opportunities to improve the reported solar-H_2 efficiency of around 25%. One of the advantages of the HySA cycle is the fluid-only thermochemical steps, i.e. no crystallisation and solidification of the employed sulfate-pyrosulfate molten salts. However, to achieve this requirement, large quantities of the pyrosulfate salt, e.g. 10:1 for potassium, are required in order to maintain the binary solid-solution in the liquid state. Consequently, such large quantities lead to large process streams, excess heating energy, and massive storage and reactor units.

Earlier our group presented the concept of an approach that addresses this issue by conducting all thermochemical steps with aerosol particles, i.e. solid-gas states instead of liquid-gas states. Also, aerosol phase reactors could utilize direct solar heating with the potential for higher solar to chemical energy efficiency compared to the indirect ones. Currently, there are multiple studies on particle reactors such as moving and fluidization beds but very few on aerosol reactors. On the other hand, aerosol phase processes are very similar to solid fuel combustion and gasification, and free-board reactions. Therefore, in this study, we investigated the potential and efficiency of the aerosol phase reactors in the HySA cycle experimentally and as an extension in other similar thermochemical processes. Particles were synthesized via evaporation, crystallization, and drying of precursor aqueous solutions at various potassium sulfate, potassium pyrosulfate, and ammonium sulfate ratios. Ammonium sulfate is a product of earlier steps that participates in the thermochemical steps. The exact mechanism, onset temperatures, and reaction energies were defined at bench-scale with a combination of thermogravimetric analysis (TGA), differential scanning calorimetry (DSC), Fourier transformation infrared spectroscopy (FTIR), and mass spectroscopy (MS). For a selected number of particle compositions and later reference, reactions were studied at the "bulk" phase (fixed bed) in a high-temperature tube reactor while evolved gases were continuously monitored via an MS. Finally, the same were studied in an aerosol phase reactor, where particles were fed using a powder feeder in a countercurrent flow configuration. Overall, reaction kinetic rates in the aerosol reactor were similar with the bench scale experiments, but conversion efficiency was lower although significantly higher than in the bulk phase. As a next step, the solar to chemical energy efficiency will be investigated utilizing a high-flux solar simulator and the presented aerosol reactor.

572627 Evaluating Different Dry Coating Processes for Preparing Fine-Grade High Functional Microcrystalline Cellulose

Previous studies (1, 2) showed that dry particle coating can be used to develop new excipients through a batch process using LabRAM (a high intensity mixer) and a continuous process using fluid energy mill (FEM). In addition, another device that is continuous in its operation can also be used, called a conical mill or comil (3). Although these processes share the same mechanism of dry coating that small guest particles get coated onto the big host particles via high intensity mechanical forces, there are subtle differences which could mean that the properties of the excipients prepared from these methods would be different from each other. Therefore, this work aims to evaluate the relative effectiveness of different dry coating methods for developing fine grade high functional excipients. In order to achieve a head-to-head comparison, the starting material in all cases was Avicel PH-102 milled down using FEM to ~30 µm size. The milled Avicel PH-102 was dry coated with 1.0 wt% A200 using both LabRAM and Comil. In parallel, Avicel PH-102 was also milled and simultaneously dry coated with 1.0 wt% A200 to obtained dry coated fine excipient of ~30 µm size. The properties of the excipients prepared via three different methods were compared, and the ability to improve the processability of blends with a challenging API (model BCS II drug, Ibuprofen) at 60% drug loading was also investigated. The results show that all the three dry coating techniques allow significant property enhancement of fine grade MCC. Moreover, all the excipients prepared from the three processing methods demonstrated significant processability enhancement as compared to other commercial grade MCC based excipients. In addition, the prepared excipients from FEM showed the best overall performance on improving the properties of Ibuprofen blends and tablets, followed by Comil and then LabRAM. Overall conclusions are: (1) LabRAM processing had the best overall coating quality, achieved through a simple batch process, (2) Comil processing was deemed to be a promising continuous dry coating process but required an intensive pre-blending process, especially for cohesive materials, to achieve good processability, and (3) FEM processing was found to be a most feasible continuous dry coating process for both cohesive and good flowing materials. This study therefore highlights the importance of selecting a robust method for dry particle coating that will deliver the desired functionality in tableting.

1. Chen L, Ding X, He Z, Huang Z, Kunnath KT, Zheng K, et al. Surface engineered excipients: I. improved functional properties of fine grade microcrystalline cellulose. International Journal of Pharmaceutics. 2018;536(1):127-37.
2. Chen L, Ding X, He Z, Fan S, Kunnath KT, Zheng K, et al. Surface engineered excipients: II. Simultaneous milling and dry coating for preparation of fine-grade microcrystalline cellulose with enhanced properties. International Journal of Pharmaceutics. 2018;546(1):125-36.
3. Chattoraj S, Shi L, Sun CC. Profoundly improving flow properties of a cohesive cellulose powder by surface coating with nano-silica through comilling. Journal of Pharmaceutical Sciences. 2011;100(11):4943-52.

572604 Characterization and Mitigation of Stick-Slip in Pharmaceutical Powders

Consistent powder flow is an important property for process control of pharmaceutical unit operations like milling and die filling of tablet presses. In literature, stick-slip is identified as one of the possible mechanisms for intermittent powder flow. In large material hoppers, stick-slip can also result in silo vibrations commonly referred to as silo quaking and honking. These silo vibrations can have major impact on the structural integrity of the silo as well as the surrounding environment. In literature, stick-slip is well documented for stiff particles of high sphericity. However, the literature on characterizing stick-slipfor pharmaceutical powders and providing potential mitigation strategies is terse. The objectives of the present work are to:

- Develop a method for measurement and prediction of stick-slip using bulk solid properties
- Discuss some of the common pitfalls of using ring shear tester data for characterization of stick-slip without additional processing
- Provide strategies to mitigate the stick-slip and
- Present experimental results showing the effectiveness of these mitigation strategies

562492 Effect of Particle Properties and Formulation Composition on Sticking Behavior during Pharmaceutical Tabletting

Tablets comprise the majority of pharmaceutical products for oral solid dosage forms in part due to their low cost of manufacturing, convenience of dosing to patients, and high stability. Tablets are typically produced by compacting powders or granules in a die using a rotary press. The formulation intended for tabletting must have adequate flow properties to consistently fill a tabletting die from a hopper and must also have sufficient compression properties to form a tablet with the mechanical strength to undergo subsequent coating, packaging, and handling. While adequate flowability and tabletability are critical to attain during development for formulations intended for tableting, some drug substances may also have a propensity to adhere to the tableting punch. So-called sticking can result in cosmetic defects or can prevent manufacturing altogether for an otherwise acceptable formulation.

This work investigates the effect of particle properties and formulation composition on sticking behavior during tabletting. Various active pharmaceutical ingredients (API) were blended with commercial grades of microcrystalline cellulose (MCC) ranging in size from 20-110 μm. Effects of particle size and surface area were systematically studied. Inter-particle adhesion was also studied by using a dry-coating technique to modify particle surface properties. This study concludes that both particle size/surface area and inter-particle adhesion force significantly affect sticking behavior. Tablets with higher tensile strength (i.e. higher inter-particle adhesion) comprised of API with lower surface area relative to the excipient surface area resulted in less sticking. This study also found that the difference between the tablet tensile strength of the blend, s_{Blend}, and that of the API, s_{API}, normalized by the surface area fraction of API in the blend, $f_{SA,API}$, correlates well with the rate of mass adhered to the punch. This quantity, referred to as the stinking index (SI), is shown below.

$$SI=(1/f_{SA,API})*(s_{Blend} - s_{API})$$

The sticking index can be used to assess effects of particle properties and formulation composition on sticking. In particular, the sticking index shows that the use of fine particle size/high surface area MCC (d_{50} = 20 μm) in place of coarser grades of MCC significantly reduce sticking and may offer an effective and practical method of addressing sticking during formulation development.

All authors are employees of AbbVie. The design, study conduct, and financial support for this research was provided by AbbVie. AbbVie participated in the interpretation of data, review, and approval of the publication.

565550 Impact of Particles Needle Shape, a Critical Material Attributes (CMAs), on Miniaturized Pharmaceutical Units Operation

The US Food and Drug Administration (FDA) encourages the adoption of Quality by design (QbD) principles in drug product development and manufacturing[1]. FDA emphasizes quality must be built into the product. QbD elements include the critical quality attributes (CQAs) of the drug product, identification of critical material attributes (CMAs) and critical process parameters (CPPs), and linking CMAs and CPPs to CQAs[1]. A CMA is a physical, chemical, biological, or microbiological property or characteristic of an input material that should be within an appropriate limit, range, or distribution to ensure the desired quality of that drug substance, excipient, or in-process material. Typically materials move from one unit operations to another unit operations. CMAs can significantly impact on pharmaceutical unit operations, process consistency, and product quality attributes. Hence, material properties need to be tested and CMAs need to be defined and controlled. Drug substances physical properties such as particle morphology (shape) has a great impact when design and developing a robust drug product having intended CQAs.

At MIT we developed a miniaturized, portable, re-configurable and automated system to manufacture drug product from drug crystals[2, 3]. The compact, portable system, roughly the size of a household oven, [72.4 cm (length) ×53.3 cm (width) ×134.6 cm (height)] is built to manufacture on-demand pharmaceutical direct compressible tablets on a scale of hundreds to thousands per day. Its design combines off-the-shelf devices with custom-designed automated mechanisms. The whole unit consists of: upper process streams start with powder feeding and end with blending and lower process streams include powder dispensing thru to making tablets. These two process streams operate independently under high-level software control, allowing for smooth integration between the two. Several commercial APIs having different dosage strength were manufactured and tested to meet U.S. Pharmacopeia standards. In the case of needle-shaped Ciprofloxacin HCl (CIPRO), processing difficulties were observed in different unit operations: feeding, dispensing even with the few grams of CIPRO.

Typically several strategies have been considered to handle needle shaped particles: alteration to crystallization process using crystal habit modifying agents, manipulation of solvents, and mechanical manipulation (e.g. milling). Each approach has its own drawbacks. Given the limitations, milling and micronization are still, at present, an attractive approach routinely used to obtain consistent particle size and shape modification during downstream formulation processing. However, the shape alteration of a few grams of materials by using conventional equipment is not typical.

In this research work, we emphasize on how particle shape, especially needle-shaped, a critical material attribute, impacted on different unit operations of the system even at a small scale and discussed different grinding options were attempted to improve the processability of needle-shaped particles. Finally, we showed the MIT team designed grinder performed well for small scale CIPRO powder, the powder can easily be processed through the system, and the manufactured oral dosage tablets showed a comparable performance of the commercially marketed CIPRO tablet.

References:

1. Lawrence XY, Amidon G, Khan MA, Hoag SW, Polli J, Raju G, et al. Understanding pharmaceutical quality by design. The AAPS journal. 2014;16(4):771-83.
2. Azad MA, Osorio JG, Brancazio D, Hammersmith G, Klee DM, Rapp K, et al. A compact, portable, re-configurable, and automated system for on-demand pharmaceutical tablet manufacturing. International journal of pharmaceutics. 2018;539(1-2):157-64.
3. Myerson A, Azad M, Hammersmith G, Brancazio D, Osorio Caicedo J. Systems and methods for

the fabrication of tablets, including pharmaceutical tablets. U.S. Provisional Patent Application US 2018/0271791 A1, Published on September 27, 2018.

572722 Coarsening Approach in Modeling Gas-Particle Flows in Dry Powder Inhaler

Coarsening approach in modeling gas-particle flows in dry powder inhaler

X. Liu[1], J. Kolehmainen[1], S. Sundaresan[1]

[1] *Department of Chemical and Biological Engineering, Princeton University, Princeton, New Jersey 08540, USA*

Dry powder inhaler (DPI) is commonly used to treat respiratory diseases by delivering active pharmaceutical ingredient (API) particles to the airways and/or lungs of the patient [1]. Because of the strong cohesive interaction between small API particles, larger carrier particles are utilized to agglomerate with API particles so as to improve fluidization, and then deagglomerate to release API particles for drug delivery [2-3]. Computational fluid dynamics coupled with discrete element method (CFD-DEM) has the advantage of capturing relevant particle-fluid and particle-particle interactions in this system but is limited by the simulation size [4]. For example, a typical DPI simulation requires tracking a few hundred thousand carrier particles and about ten million API particles [4]. Rigorous coarsening of the simulation to speed up the computations would rapidly expand the use of simulations to guide the development of DPI formulations and more efficient drug delivery.

With this in mind, the present study explores a coarsening approach, which tracks all the carrier particles, but only a subset of representative API particles. Such coarse-graining has been used in other contexts (for example, see [5]). The merits and limitations of this coarse-graining approach will be discussed in this presentation.

[1] Islam, N., Gladki, E. Dry powder inhalers (DPIs)—a review of device reliability and innovation. *Int. J. Pharm.* 2008; 360(1-2): 1.

[2] Yang, J., Wu, C. Y., Adams, M. Numerical modelling of agglomeration and deagglomeration in dry powder inhalers: A review. *Curr. Pharm. Des.* 2015; 21(40): 5915.

[3] Newman, S. P., Chan, H. K. In vitro/in vivo comparisons in pulmonary drug delivery. *J Aerosol Med Pulm Drug Deliv*. 2008; 21(1): 77.

[4] van Wachem, B., Thalberg, K., Remmelgas, J., Niklasson-Björn, I. Simulation of dry powder inhalers: Combining micro-scale, meso-scale and macro-scale modeling. *AIChE J.* 2017; 63(2): 501.

[5] Derksen, J. J., Sundaresan, S., Van Den Akker, H. E. A. Simulation of mass-loading effects in gas–solid cyclone separators. *Powder Technol*. 2006; 163(1-2): 59.

577170 Surface Engineering of Inhaled Pharmaceutical Particles By Atomic and Molecular Layer Deposition

Active pharmaceutical ingredients (APIs) and pharmaceutical excipients are typically small solid organic particles. The morphology and surface characteristics of both APIs and excipients play a crucial role in both their manufacturing and administration into the human body. On the one hand, the physical properties of pharmaceutical powders, e.g., particle size distribution, particle shape, density, and cohesiveness, as well as extrinsic factors such as moisture, temperature and triboelectricity strongly affect their bulk behavior (e.g., flowability, dispersibility and compactibility), and thus the manufacturing of dosage forms. For instance, poor powder flow can lead to weight variation and poor content uniformity during tabletting and capsule filling operations or poor dispersibility from dry powder inhalers. On the other hand, the size and shape as well as the crystal structure of pharmaceutical particles have a significant impact on their administration and bioavailability, and thus on the uptake by the human body. For example, particle size is crucial for pulmonary drug delivery, where particles with an aerodynamic diameter smaller than 5 μm are required to reach the targeted action sites of the deep lung. To achieve the desired inhalable particle size, the most commonly used method is micronization which delivers particles down to the micrometer or, in some cases, nanometer size. However, micronized particles are highly cohesive and moreover might present amorphous regions that crystallize over time in an uncontrolled manner. A growing amount of new inhaled APIs includes amorphous drugs, which are often very moisture-sensitive, as well as expensive and less potent drugs, which require improved flowability and aerosolization efficiency to meet the drug load requirements and to minimize their cost. Moreover, there is no commercially viable extended-release technique for inhaled drug particles. Therefore, there is an unmet need for novel solutions to provide surface modification of inhaled pharmaceutical powders that lead to improved processability and stabilization of solid state forms (e.g., amorphous, metastable, polymorphs, hydrates) and, most importantly, clinical benefits, e.g., controlled release of API from dosage forms.

Conventionally, improved flow properties, enhanced stability and tailored dissolution behavior have been achieved by placing a coating over the pharmaceutical particles. Nevertheless, due to the irregular shape and inhomogeneous surfaces of pharmaceutical particles, current technologies, based on both wet and dry synthesis, offer limited control over the coating thickness and poor drug loading efficiency, thus providing unsatisfactory performance of the final pharmaceutical formulation. Atomic layer deposition (ALD) is an established technique for the synthesis of thin films for various applications ranging from semiconductors to energy storage devices. Recently, it has been gaining attention in the pharmaceutical field to modify the single particle and bulk powder properties such as the drug release. Compared to the conventional methods of drug particle coating, ALD has a number of advantages: control over the amount of deposited material, conformality, and its solventless nature. A typical ALD process relies on sequential surface reactions of two gaseous precursors, which chemisorb on the solid substrate in alternate pulses, separated by inert gas purging. In doing so, ALD lends itself to the growth of multiple inorganic films, typically metal oxides, with exquisite thickness control at the sub-nanometer level.

ALD on pharmaceutical powders has been demonstrated by using rotary and fluidized bed reactors (FBRs) [1-3]. Despite being convenient systems for coating powders at the lab scale, rotary reactors are not suitable to manufacture the quantities of products required at the industrial scale. Instead, FBRs are inherently scalable, particularly if operated at atmospheric pressure. Moreover, FBRs are well-established technologies in the pharmaceutical industry. Therefore, ALD in atmospheric-pressure FBRs constitutes a scalable and cost-effective route to engineer pharmaceutical particles.

In this work, we demonstrate the use of ALD as a route to modify the properties of inhaled pharmaceutical particles, namely (i) dispersibility, (ii) stability to moisture and (iii) dissolution rate. We deposit ultrathin oxide ceramic films, namely Al_2O_3, TiO_2 and SiO_2, as well as hybrid inorganic-organic films on both API (e.g., budesonide and salbutamol) and excipient (i.e., lactose) particles, both crystalline and amorphous. The ALD process is carried out at ambient conditions in a fluidized bed reactor for a range of cycles from 10 to 50, using TMA/O_3, $TiCl_4$/H_2O and $SiCl_4$/H_2O as precursors for Al_2O_3, TiO_2 and SiO_2 ALD, respectively. The deposition strongly depends on the surface crystal structure of the particles. Time-of-flight secondary ion mass spectrometry and transmission electron microscopy reveal the deposition of uniform and conformal nanofilms on crystal surfaces, whereas uniform but non-conformal nanofilms are observed on amorphous surfaces [4]. The dispersion properties are evaluated both in the liquid and dry state. The ALD-coated particles exhibit considerably higher dispersibility in both water and ethanol solutions, thus suggesting higher bioavailability, than the uncoated ones. In-vitro aerosolization testing by the next generation impactor shows improved fine powder delivery (<5 µm, i.e., particle size range relevant for inhalation) and greatly reduced powder retention in the inhaler for the ALD-coated particles. The dispersion and aerosolization properties are retained even upon different conditions of temperature (25-40 °C) and relative humidity (60-75 %) over 3 weeks. Finally, in-vitro dissolution tests and cell absorption studies reveal more sustained release with increasing film thickness.

Nanoengineering of APIs for inhaled drug delivery is a novel application of ALD. Being a dry process at near ambient conditions, ALD provides an almost non-invasive platform to engineer surfaces of pharmaceutical particles, regardless of their shape, size and rugosity. Stabilization of sensitive particles and high energy solid forms with a conformal thin film is an attractive prospect for both drug product manufacturing and storage. Moreover, nanoscale films on inhaled particles can provide less cohesive powders with improved aerosolization properties, and thereby improved lung deposition, as well as at the same time powders with a tailored dissolution rate. This concept opens up exciting opportunities to produce more complex materials for targeted and controlled drug delivery.

References

1. Zhang, D., M.J. Quayle, G. Petersson, J.R. van Ommen, and S. Folestad, *Atomic Scale Surface Engineering of Micro- to Nano- Sized Pharmaceutical Particles for Drug Delivery Applications.* Nanoscale, 2017.
2. Kääriäinen, T.O., M. Kemell, M. Vehkamäki, M.-L. Kääriäinen, A. Correia, H.A. Santos, L.M. Bimbo, J. Hirvonen, P. Hoppu, S.M. George, D.C. Cameron, M. Ritala, and M. Leskelä, *Surface modification of acetaminophen particles by atomic layer deposition.* International Journal of Pharmaceutics, 2017. **525**(1): p. 160-174.
3. Hellrup, J., M. Rooth, A. Johansson, and D. Mahlin, *Production and characterization of aluminium oxide nanoshells on spray dried lactose.* International Journal of Pharmaceutics, 2017. **529**(1–2): p. 116-122.
4. Zhang, D., D. La Zara, M.J. Quayle, G. Petersson, J.R. van Ommen, and S. Folestad, *Nanoengineering of Crystal and Amorphous Surfaces of Pharmaceutical Particles for Biomedical Applications.* ACS Applied Bio Materials, 2019.

572162 Evaluating the Potential Benefits of Cohesion

Powders are often categorized as being free-flowing or cohesive, with a common perception that highly cohesive materials will experience more processing challenges whereas free-flowing powders will perform well. For example, Shear Cell testing generates a flow function coefficient which categorizes powders into discrete groups and primarily assesses their ability to discharge from hoppers and silos, where higher cohesion will typically inhibit flow. However, hopper discharge is often only the first stage of a manufacturing processes, with subsequent unit operations applying a range of stress conditions and flow regimes. Under these conditions less cohesive powders may not always deliver the best results.

Simple classifications do not describe the breadth of behavior that powders can exhibit, nor identify how this can change depending on the process environment. In order to optimize productivity and achieve high quality final products, powders must be selected based upon their compatibility with a process, rather than upon the generalization that cohesive powders exhibit poor performance.

This presentation demonstrates how a degree of inter-particular bonding and friction can be necessary for optimal performance in certain applications. Powder rheology is used to evaluate flow properties under a range of stress and strain conditions, and demonstrates how this data can be applied to better understand and predict performance in a number of unit operations. The potential benefits of greater absolute cohesion within coating, filling and dosing applications are discussed, for example to promote the formation of robust coatings and stable agglomerates.

568246 Linking Moisture Induced Variations in Pharmaceutical Primary Powders to Their Tableting Performance

A drug is only a small proportion of a pharmaceutical tablet. Other excipients, for example binders and disintegrants, make up the bulk volume of the dosage form and therefore the properties they possess are important in determining the final quality and compaction attributes of a tablet. Pharmaceutical formulations are often wet granulated, for example via twin screw granulation (TSG), prior to compaction to ensure homogeneity and to improve their flowability. The tableting properties of pharmaceutical formulations can be characterised by their compressibility and compactibility. Compaction consists of compression and consolidation of a particulate-gas system due to an applied force and such mechanisms have become an important part of pharmaceutical development.

We have observed that tablet tensile strength varies depending on the relative humidity (RH) upon storage of compressed tablets or granules produced via TSG. This work has led to further investigation of pharmaceutical primary powders and how tablet properties change with moisture content. The aim of the current work is to investigate the influence of material properties, for example moisture content and solubility on final tablet attributes. Ungranulated powders of varying solubility, hygroscopicity and molecular weight were chosen for testing purposes. Powders were equilibrated at differing relative humidity levels before using characterisation techniques to determine tablet compactibility, porosity, and tensile strength. Tabletability describes the strength a tablet possesses after compaction under a given pressure. The tabletability profiles in Figure 1 show tablet strength as a function of compaction pressure for MCC and starch.

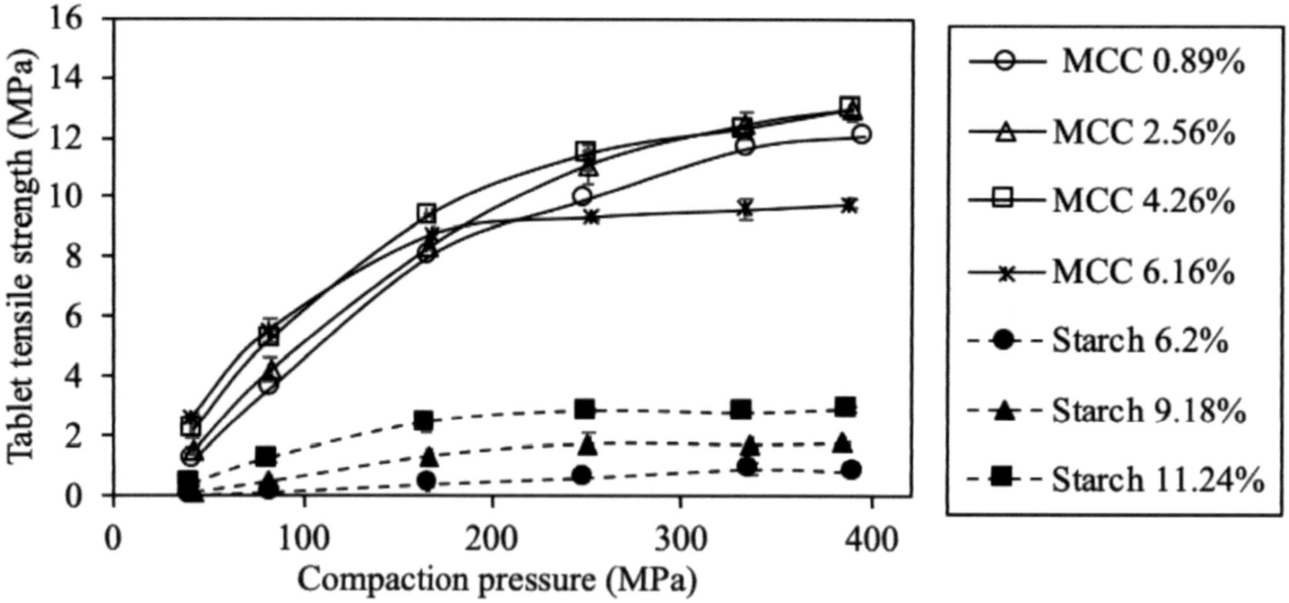

Figure 1. Tabletability profiles for MCC and starch.

Due to the hygroscopic nature of microcrystalline cellulose (MCC) and starch each powder increases in moisture content upon storage at increasing relative humidity. Tablet strength increases with increasing compaction pressure in all cases. A loss in tabletability is seen with the highest moisture content (6.16%) at compaction pressures over 180 MPa for MCC, below this compaction pressure the higher moisture content produces the strongest tablets. Where water aids compaction at low compression, it hinders tableting at high compaction pressure. A tabletability profile for lactose (Pharmatose 200M) shows no significant change in moisture content or tablet strength across the range of relative humidity the powders are stored at.

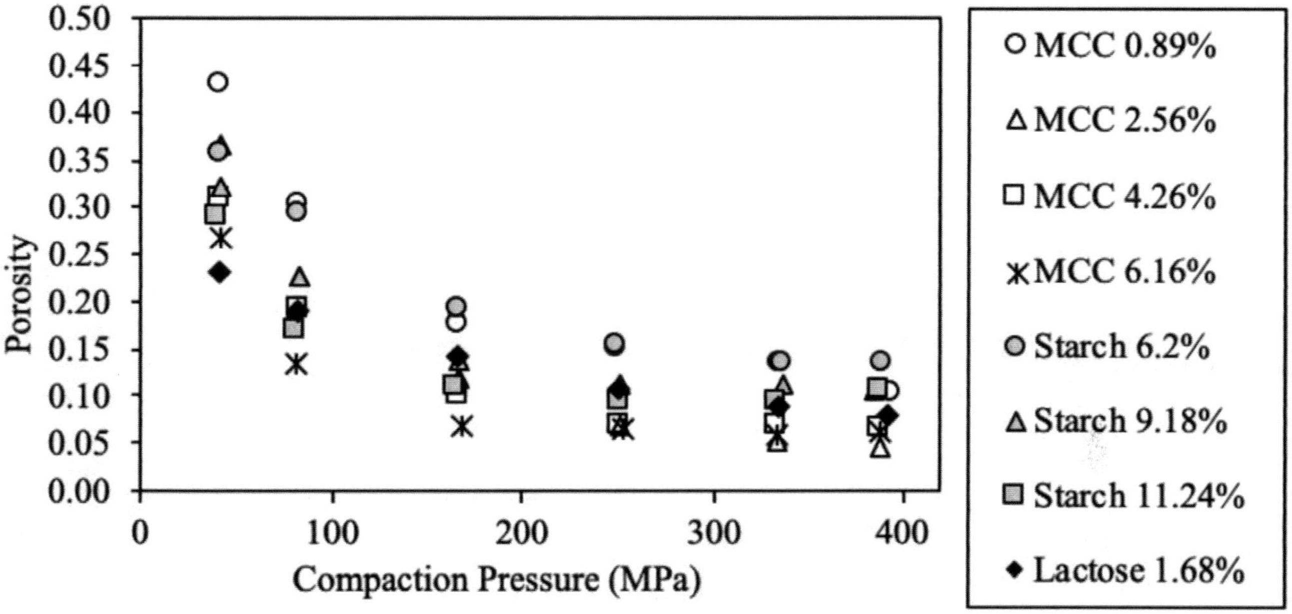

Figure 2. Compressibility profiles for MCC, starch and lactose.

Compressibility describes how the porosity of a powder changes under pressure and is shown for each material in Figure 2. With increasing pressure, the tablet density increases, and tablet porosity approaches zero. MCC is the most plastic material, followed by lactose and finally starch [2]. Porosity approaches zero faster the more plastic a powder is because they eliminate pores via plastic deformation more effectively. The increasing moisture content in MCC and starch also plasticizes the material further, leading to a faster reduction in porosity with increasing moisture content.

By selecting a range of powders with differing properties it is hoped a link can be made between structure, moisture content and tablet tensile strength that can be applied across multiple materials in the future, along with valuable information that can be applied to optimise the TSG process.

References

[1] C. C. Sun. Mechanism of moisture induced variations in true density and compaction properties of microcrystalline cellulose, International Journal of Pharmaceutics, 346 (2008) 93-101.

[3] D. H. Choi et. al. Material properties and compressibility usingHeckel and Kawakita equation with commonly used pharmaceutical excipients. Journal pd Pharmaceutical Investigation, 40(4) (2010) 237-244.

571318 Coarse-Grained DEM for Large-Scale Fluidized Bed Applications

Fluidized beds are used in a wide range of applications including those in the chemical and process industry. To ensure the efficient and safe operation of these unit operations, Computational Fluid Dynamics (CFD) is often employed during the design and optimization of the units. Yet, when it comes to numerical simulation of industrial scale units, classical methods like the Kinetic Theory of Granular Flow (RANS-KTGF or KTGF) and the Discrete Element Method (DEM) are limited by their hardware requirements. In the case of KTGF it is the high number of cells needed to resolve all important flow structures (i.e., grid resolution), whereas DEM is limited by the particle count that can be tracked efficiently. To avoid this, in the KTGF framework, sub-grid or filtering models like the EMMS-drag model can be used. However, this increases the complexity of the model by adding several unknown closures.

A promising DEM framework-based approach is the so-called Coarse-Grained DEM (CG-DEM). This approach tries to circumvent the hardware-related limitations of DEM by combining several individual particles into one representative parcel or cloud. As in DEM, the parcels trajectories and collisions are tracked individually and interaction forces (i.e., gravity, drag) are scaled to account for the artificially reduced number of interactions.

This current study will show the capabilities of CG-DEM in the field of large-scale fluidized bed applications. Particles of different Geldart Group classification are investigated. The impact of drag, collision force scaling, and the effect of increasing the number of particles per parcel on the accuracy of the numerical results will be compared and discussed. The results will be compared with experimental data, taken from a twelve-inch semi-circular pilot-scale unit.

574097 Acquiring DEM Validation Data on an Industrial Scale

Modeling gas-solid flows using discrete element method (DEM) simulations offers promising predictive capabilities for industrial operations. With DEM, Newton's equations of motion are solved for each particle in the simulation and therefore all collisions are tracked. Furthermore, unlike kinetic-theory-based continuum models, enduring particle-particle contacts can be resolved in DEM such that dense flow of granular material can be simulated. However, the expense of tracking each individual particle frequently prevents the use of DEM simulations at industrial scale. As a result, increasing the computational speed of DEM simulations is an area of active research. An important prerequisite to the reliable use of DEM is validation at the relevant time and length scales. Although DEM hydrodynamic predictions have been compared to numerous small-scale experiments, DEM predictions on an industrially-relevant scale have been prohibitive due to the computational overhead. In this effort, experiments were designed for purposes of DEM validation in an industrially-relevant system. The system considered here is a stripper unit, which involves the co-current flow of solids moving downwards and fluidizing gas moving upwards. The purpose of the stripper unit in an industrial operation is to remove product gases from catalyst particles entrained from a reactor system - e.g., removing the product gases from the interstitial spaces between catalysts. For the purposes of this study, experiments were performed in the industrially-sized system with large, relatively monodisperse particles and conditions controlled in order to limit the number of particles in the experimental section to approximately 10^9 particles. A detailed explanation of the experimental apparatus will be provided alongside the results. Additionally, the measurements taken to characterize inputs into DEM simulations will be discussed.

576013 Evaluating Mixing Performance of a Resonant Acoustic Mixer Both Experimentally and Using a Coupled Lbm-DEM Technique

Bladed mixers are widely used in many industrial processes that require mixing of a liquid phase and a solid phase (pharmaceutical, polymers, cement, propellant, etc.). Resonant acoustic mixing (RAM) is a non-contact mixing method that uses low frequency vibration to promote mixing and is a relatively new alternative to mix complex blends. RAM mixers are touted as being able to enhance mass transport, require a shorter time to mix, and apply less shear to the materials than bladed mixers, all valuable characteristics for mixing materials that are sensitive to shear and/or highly viscous. Nevertheless, the operation of this class of device is not well understood and little formal literature research is available. The primary purpose of this project is to evaluate the mixing performance of a highly-loaded solid phase within a viscous liquid using a RAM device. Our methodology involves both experimental and computational examination of this device. First, we develop a code that mimics the RAM and captures the fluid flow using the lattice Boltzmann method (LBM) in three dimensions. This model is coupled with the discrete element method (DEM) in order to elucidate the impact of particle loading on the operation of this mixer. Experimentally the effect of the size of the particles in a bimodal mixture and that of the viscosity of the fluid and mixing parameters (mixing time and acceleration) on the mixing performance was examined via the uniformity of the resultant blend. In order to determine the uniformity of the blend the relative standard deviation and the mean mass fraction were used. The acceleration, bimodal distribution size and viscosity have an important effect in the mixing performance experimentally and the same trend was observed using the model.

562281 Numerical Verification and Validation of the Particle Tracking Technique in Predicting Particle Velocity and Solids Distribution in a Travelling Fluidized Bed with CFD-DEM

Numerical verification and validation of the particle tracking technique in predicting particle velocity and solids distribution in a travelling fluidized bed with CFD-DEM

Yupeng Xu[1,8], Tingwen Li[2], Sina Tebianian[3], Jamal Chaouki[4], Thomas W. Leadbeater[7], Rouzben Jafari[4], David J. Parker[5], Jonathan Seville[5], Naoko Ellis[6], John R. Grace[6]

[1] National Energy Technology Laboratory, U.S. Department of Energy, Morgantown, WV, USA

[2] SABIC Corporate Research and Development

[3] IFP Energies Nouvelles, Process Design and Modeling division, Lyon, France

[4] Département de génie chimique, Ecole Polytechnique, Montréal, QC Canada H3T 1J4.

[5] Positron Imaging Centre, University of Birmingham, Birmingham, B15 2TT, UK

[6] Department of Chemical & Biological Engineering, University of British Columbia, Vancouver, Canada V6T1Z3

[7] Department of Physics, University of Cape Town, South Africa

[8] West Virginia University Research Corporation, Morgantown, WV 26506, USA

Abstract

Gas-solid fluidized beds are widely used in energy and chemical industry applications due to their good mixing and heat transfer ability. However, the design and scale-up of the reactor and improvement of reactor performance are still hindered by the lack of fundamental knowledge of physical phenomena in these systems due to the opaque and abrasive nature of the solid particles, turbulence, and multiple scales of motion and flow structure. Several non-invasive techniques such as radioactive particle tracking (RPT) and positron emission particle tracking (PEPT) have been deployed to measure the average particle velocity and the mean solid distribution in recent years in a "travelling fluidized bed". These techniques require the use of tracer particles which differ in size, density and/or shape from the bulk particles of interest. In determining averaged velocities based on such techniques, a choice must be made between averaging velocities of particles crossing a virtual plane over a period of time ("face-average" approach) and those passing through a defined volume over time (the "volume-average" approach). Moreover, different methods have been developed to calibrate the particle tracking measurements in order to quantify the mean solid concentration in fluidized beds.

In the current research, CFD-DEM simulations were conducted to study the gas-solid flows in the traveling fluidized bed and the behavior of different particles, including bulk sand particles and tracer particles for radioactive particle tracking and positron emission particle tracking. The general flow hydrodynamics of numerical simulation are compared with measured values obtained using different experimental techniques. As shown in Figure 1, the formation, movement and burst of a typical square nosed slug for a superficial gas velocity of 0.4 m/s are well captured.

Analyses were carried out to derive particle velocity and mean solid concentrations from the tracer particle data. The different averaging techniques for particle velocity and different calibration approaches for solids volume fraction are also compared. It is shown that both the particle velocity and the mean solid concentration can be measured reliably based on representative tracer particles. As an example, comparison of the cross-section averaged solids volume fraction along the bed height between simulation and different experimental techniques are shown in Figure 2 for superficial velocities of 0.40 and 0.60 m/s.

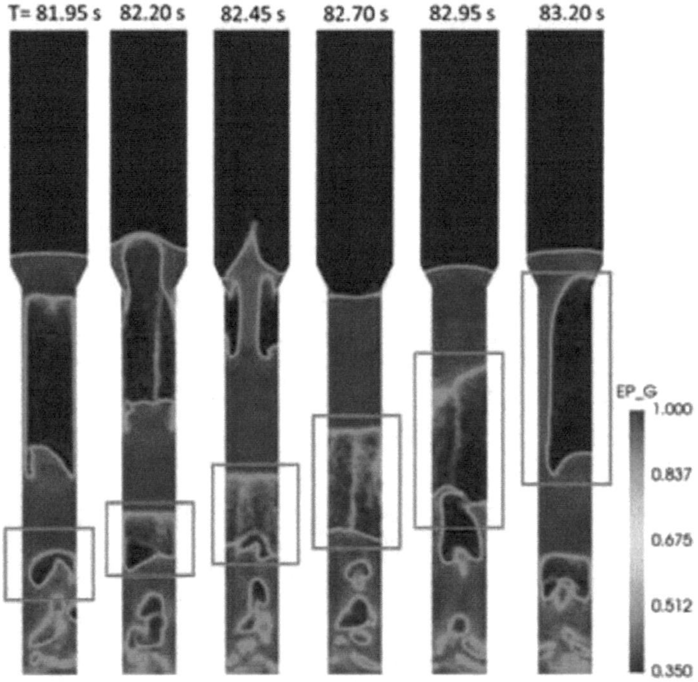

Figure 1. Simulation results of the formation, movement and burst of a typical square-nosed slug for a superficial gas velocity of 0.4 m/s.

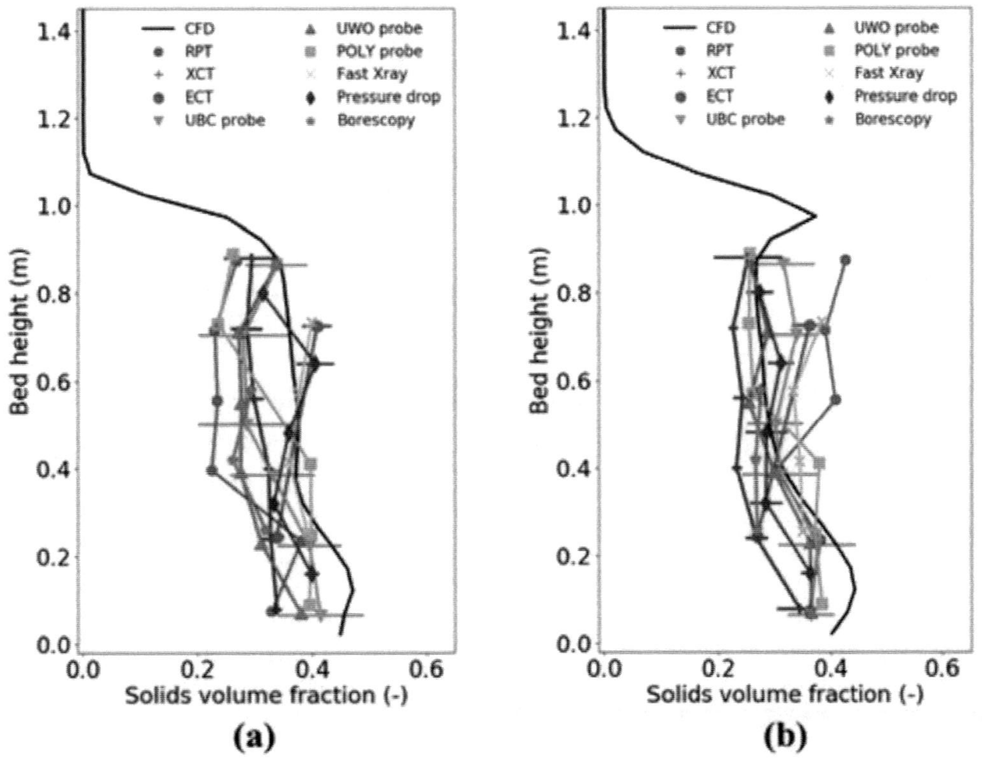

Figure 2. Comparison of cross section averaged solids volume fraction along the bed height between simulation and different experimental techniques at (a) U_g=0.40 m/s, (b) U_g=0.60 m/s.

567748 Cluster Formation Mechanism in Boundless Down-Flow Gas-Particle Systems

Abstract submitted to the Particle Technology Forum in the AIChE Annual Meeting 2019. Orlando, USA.

Cluster formation mechanism in boundless down-flow gas-particle systems

Noel Gomez[a,b], Alejandro Molina[b], Guy B. Marin[a], and Kevin M. Van Geem[a]

[a] Laboratory for Chemical Technology. Ghent University. Technologiepark 125-9052 Ghent, Belgium.

[b] Universidad Nacional de Colombia – Sede Medellín. Facultad de Minas, Calle 80 No 65-223, Medellín, Colombia

Key-words: Particle clustering, downer, computational fluid dynamics, particle technology.

Several Large-Eddy simulations (LES) coupled with the discrete element method (DEM) were performed in 3D boxes with periodic boundary conditions to study the dynamics of particle clusters. The investigation is based on a parametric analysis varying the Stokes number (Stk) and Reynolds number (Re). These dimensionless groups characterize the hydrodynamic conditions of the system.

Particle clusters do not allow an ideal contact between phases in the reactor and, therefore, the limitations for heat and mass transfer increase, and as such, may reduce the efficiency of the reactions (low-conversion zones). In reactors with downward flowing fluid, so-called downers, the positive effect of gravity in the stream-wise direction prevents particle back-mixing; however, clustering is not entirely prevented. It has been proven the strong effect of walls on the formation of clusters in downers (Gomez et al.), boosted by the low turbulence and velocity in this zone, and by the fact that particles lose energy when they hit the walls. It is also possible to find agglomerations in regions far from the wall-influenced zones of the reactors. This fact suggests that there are other mechanisms of cluster formation that allow the presence of such structures in the reactor.

One way to study these phenomena is to use computational fluid dynamics (CFD) under a Euler-Lagrange approach, which allows the assessment of the gas and particle dynamics independently, the local interactions between the two phases, and collisions between particles. For that aim, 3D boxes with periodic boundary conditions were used to analyze the dynamics of particles in down-flow systems. The simulations covered a range of solid/gas mass ratio (CTO) from 5 to 30, which corresponds to a mean particle volume fractions from 0.001 to 0.03 over the simulated volume, typical values for gas-solid reactive flows in downers. The Smagorinsky LES model was implemented to calculate the subgrid-scale turbulent values over the computational cells in which the volume was divided. The LES model takes into account the main fluctuations in the velocity fields of the gas phase and the influence of these fluctuations on the particle motion.

It was found that clusters are formed even at Stk lower than one, value up-to which it is presumed that particles follow the gas streamlines and do not form clusters. Once the cluster is formed, it flows as one single entity for a fraction of a second, during which its 'cluster Stokes number' is normally above one, mainly because of its larger characteristic length (3 to 6 times the particle diameter) than that of a single particle. This behavior promotes the formation of groups of clusters that flow independently and are not regularly seen in wall-influenced regions (see Figure 1).

The analysis was also carried out under reactive conditions, using the catalytic cracking of gasoil as an example. Stk increased with conversion because the gasoil cracked into lighter molecules, leading to a decrease in the gas density and viscosity. In consequence, bigger and long-lasting clusters were formed.

Figure 1. Gas velocity: 1 m/s. CTO: 15. Stk: 0.5.

Bibliography

Gomez and Molina, 2019, Chem. Eng. Technol. 10.1002/ceat.201800463

568028 Modeling of Complex Liquid-Solid Flow of Particle Swelling in Slurry Loop Reactors

A swelling-dependent two-fluid model (STFM) is developed for the liquid-solid flows of swelling particles in polyethylene reactors. The model integrates the two-fluid model (TFM) with a species transport equation (STE) to account for the diffusion of alkane molecules from the liquid bulk to the amorphous region of particles, and a population balance equation (PBE) to consider the aggregation of swelling particles. Simulations show that only the TFM fails to capture the main features of swelling systems. By contrast, the STFM captures the gradual increase of power consumption due to particle swelling and aggregation, which agrees with the experiments in a stirred tank. The STFM predicts also the slug formation and a sharp increase of power consumption in a slurry loop reactor as well as the solid accumulation behind pump. The difference of model prediction for stirred tanks and loop reactors suggests the potential of reactor optimization by enhancing local mixing while still keeping high solid concentration for productivity.

571192 Development of Mucous-Penetrating Levodopa-Loaded Microparticles for Treatment of Parkinson's Disease

Drug loaded microparticles for pulmonary drug delivery applications have gained considerable interests, owing to their ability to provide sustained drug delivery for the treatment of various diseases. The treatment of Parkinson's Disease (PD) by administrating Levodopa (L-dopa) via the pulmonary route provides distinct advantages over the oral formulations. The enhanced stability of the drug increases the period of contact with the local mucosa coupled with the presence of large absorptive areas of the lungs allows a low dosage of drug and reduces the potential side effects. However, the presence of a mucous layer on the pulmonary membranes also poses a major hurdle for delivering drug molecules to the underlying epithelium, especially in the elderly patients, who are the largest fraction of the population suffering from PD. The mucous build-up can entrap the drug molecules and causes delayed release or even degradation.

This work aims to formulation microparticles loaded with L-dopa and study the effect of the presence of the mucous on drug delivery. Characterization of microparticles are performed by Scanning Electron Microscopy (SEM), Anderson Cascade Impactor (ACI), Fourier Transform Infrared (FT-IR) spectroscopy, X-ray Diffraction Spectra (XRD), Thermogravimetric Analysis (TGA), Differential Scanning Calorimetry (DSC). Tapped Density and Drug Dissolution Study are also conducted. The stability of L-dopa and the mucus-penetrating capabilities will be studied in saline solution, mucin aqueous solution, artificial mucous and simulated interstitial lung fluids.

577271 Roles of SDS in Griseofulvin Release from Nanocomposites and Amorphous Solid Dispersions: A Comparative Study

It has been reported that 40% of the marketed products and up to 70% of the new chemical entities (NCE) are poorly water-soluble [1]. Intestinal absorption of these drugs turns out to be rate-limiting due to the poor aqueous solubility, which resulted in low bioavailability. Among different approaches to enhance bioavailability, two platform approaches have gained prevalence both in the scientific literature and marketed products: nanoparticle-based dosage forms (nanocomposites) and amorphous solid dispersions (ASDs) [2]. In the nanoparticle-based dosage form, dissolution rate is improved due to the dramatic increase in the surface area of the drug crystals via size reduction down to the nanoscale [2, 3]. On the other hand, amorphous form of the drug can offer significant enhancement in the extent–rate of dissolution due to its higher free energy and higher apparent solubility. Although both drug nanocomposites and ASDs have been studied for decades, a scientifically fair head-to-head comparison of these two forms of the drug having identical formulation is largely missing in the literature.

While the impact of surfactants on drug release from drug nanocomposites is well-understood [4], the same cannot be said for ASDs as there are conflicting reports in the literature. Not only did the use of surfactant in ASD improve the dissolution performance, it also enhanced the physical stability [5,6]. Other studies also reported improved dissolution profiles of ASDs containing surfactants compared to ASDs without surfactants [7,8]. On the contrary, Liu et al. [9] reported that the presence of SDS with PVP-VA significantly improved the apparent solubility of a poorly water-soluble drug, sorafenib, but reduced the recrystallization inhibition ability of PVP-VA resulting in poor dissolution performance. Medarević et al. [10] have also observed a negative impact of a surfactant (poloxamer 188) on the storage stability of carbamazepine–Soluplus ASD, even though the surfactant increased the dissolution performance significantly. Therefore, it is fair to state that the use of surfactant can be beneficial/detrimental depending on the specific drug–polymer–surfactant system and requires further investigation. More importantly, none of the aforementioned studies elucidated the roles of surfactants on the wettability enhancement of the hydrophobic drug. Also, impact of surfactant in the release rate of a poorly water-soluble drug from nanocomposites vs. ASDs with identical formulation has not been examined.

Here, we aim to examine the roles of surfactants in drug release from nanocomposites vs. ASDs with identical formulation, which are prepared by spray drying as platform technology using drug suspension-based (W) and solution-based (S) feeds, respectively. Nanocomposites were prepared by wet media milling of griseofulvin (GF) suspensions followed by spray drying. Hydroxypropyl cellulose (HPC) and Soluplus® (Sol) were used as stabilizers–matrix formers with or without sodium dodecyl sulfate (SDS, surfactant). ASDs were produced by spray-drying the identical drug–polymer formulation dissolved in acetone–deionized water. Three different drug:polymer (HPC/Sol) ratios, i.e., 1:1, 1:3, and 1:5, were used for both drug nanocomposite and ASD production. Laser diffraction, XRPD, redispersion test, modified Washburn method for drug wettability [11], drug desupersaturation experiments, and *in vitro* USP II dissolution testing was used for the characterization of the spray-dried powders.

XRPD results indicate that for both polymers, spray-drying of the drug suspension-based (W) feed led to drug nanocrystals dispersed in the polymeric matrix (nanocomposite), whereas that of the drug solution-based (S) feed led to ASD. In the dissolution tests, all formulations with SDS exhibited faster GF release than those w/o SDS due to the higher hydrophilicity of the former, which was quantified by the wetting effectiveness factor calculated via the modified Washburn method. This factor ($\cos\theta_S/\cos\theta_W$) [11] for HPC, HPC–SDS, Sol, and Sol–SDS solutions (with respect to the deionized water as basis) was calculated to be 20.9, 42.1, 2.65, and 4.65, respectively, showing the wettability enhancement by the SDS. For 8.9 mg GF dose (non-supersaturating condition), immediate release (>80% in 20 min) of GF was observed from the nanocomposites due to the inclusion of SDS in the formulation, whereas release rate was significantly reduced w/o SDS. These results can be explained by the lower wetting effectiveness and incomplete redispersion of the drug nanoparticles w/o SDS in the formulation. Similarly, presence of SDS in the ASD formulation significantly improved the release rate of GF from both HPC and Sol matrices. However, HPC matrix alone was able to release GF significantly faster than Sol matrix alone from ASD formulation, due to the slower erosion of the Sol matrix compared to the HPC matrix. Interestingly, under non-supersaturating conditions, nanocomposites released GF faster than ASDs due to the faster erosion of the polymeric matrices and complete redispersion of the drug nanoparticles when both formulations had SDS.

For 100 mg GF dose (supersaturating condition), the dissolution tests on ASDs with HPC–SDS or HPC alone showed relatively low supersaturation ratio (as percentages), i.e., up to ~44%, whereas notable supersaturation was generated by Sol–SDS (up to 475%) and Sol alone (up to 249%) at 210 min. This could be explained by the fast recrystallization of GF in the dissolution medium and/or ASD matrix when HPC was used as a matrix polymer, and Sol appeared to be an effective recrystallization inhibitor. The solubility parameter difference for the GF–polymer pairs were estimated as follows: 7.2 and 11.8 $MPa^{0.5}$, respectively, for GF–Sol and GF–HPC. The poor miscibility of GF with HPC could explain the observed lower extent of supersaturation and faster recrystallization as compared with Sol. The drug desupersaturation results show that SDS did not play significant role in drug supersaturation maintenance. This finding seems to suggest that the slower GF release without SDS could be attributed mainly to the lower wettability, which results in slower erosion of the matrix. Nanocomposites with SDS exhibited faster GF release than the those w/o SDS. Interestingly, Sol–SDS nanocomposites also supersaturated the dissolution medium. Finally, ASD with GF–Sol provided higher GF supersaturation than the GF–Sol nanocomposite due to the miscibility of GF–Sol and recrystallization inhibition provided by Sol in ASD, whereas both ASD and nanocomposites with GF–HPC performed similarly with slight supersaturation.

Overall, this study demonstrates how to produce both nanocomposites and ASD of the same drug with identical formulation using spray dryer as a tool and to perform a comparative assessment of their dissolution performance. Presence of SDS had significant positive impact on the GF release from ASDs and nanocomposites, but does not have significant impact on the supersaturation maintenance. This study also suggests that the drug–polymer miscibility is required to achieve the advantages of ASD under supersaturating condition, whereas nanocomposites could be competitive and may outperform ASDs in formulating potent (low dose) drugs.

References:

[1] Duarte, Í., Santos, J.L., Pinto, J.F., Temtem, M., 2015. Screening methodologies for the development of spray-dried amorphous solid dispersions. Pharm. Res. 32, 222-237.

[2] Brough, C., Williams, R., 2013. Amorphous solid dispersions and nano-crystal technologies for poorly water-soluble drug delivery. Int. J. Pharm. 453(1), 157–166.

[3] Kesisoglou, F., Panmai, S., Wu, Y., 2007. Nanosizing—oral formulation development and biopharmaceutical evaluation. Adv. Drug Delivery Rev. 59, 631–644.

[4] Bhakay, A., Rahman, M., Davé, R.N., Bilgili, E., 2018. Bioavailability enhancement of poorly-water soluble drugs via nanocomposites: formulation–processing aspects and challenges. Pharmaceutics, Special Issue: Dissolution Enhancement of Poorly Soluble Drugs, Guest Editors: E. Bilgili, R.N. Dave, 10(3), 86

[5] Goddeeris, C., Willems, T., Houthoofd, K., Martens, J., Van den Mooter, G., 2008. Dissolution enhancement of the anti-HIV drug UC 781 by formulation in a ternary solid dispersion with TPGS 1000 and Eudragit E100. Eur. J. Pharm. Biopharm. 70, 861–868.

[6] Joshi, H.N., Tejwani, R.W., Davidovich, M., Sahasrabudhe, V.P., Jemal, M., Bathala, M.S., Varia, S.A., Serajuddin, A.T., 2004. Bioavailability enhancement of a poorly water-soluble drug by solid dispersion in polyethylene glycol–polysorbate 80 mixture. Int. J. Pharm. 269, 251–258.

[7] Li, B., Konecke, S., Harich, K., Wegiel, L., Taylor, L.S., Edgar, K.J., 2013. Solid dispersion of quercetin in cellulose derivative matrices influences both solubility and stability. Carbohydr. Polym. 92, 2033–2040.

[8] Ghebremeskel, A.N., Vemavarapu, C., Lodaya, M., 2007. Use of surfactants as plasticizers in preparing

solid dispersions of poorly soluble API: selection of polymer–surfactant combinations using solubility parameters and testing the processability. Int. J. Pharm. 328, 119–129.

[9] Liu, C., Chen, Z., Chen, Y., Lu, J., Li, Y., Wang, S., Wu, G., Qian, F., 2016. Improving oral bioavailability of sorafenib by optimizing the "Spring" and "Parachute" based on molecular interaction mechanisms. Mol. Pharmaceutics 13, 599–608.

[10] Medarević, D.P., Kachrimanis, K., Mitrić, M., Djuriš, J., Djurić, Z., Ibrić, S., 2016. Dissolution rate enhancement and physicochemical characterization of carbamazepine-poloxamer solid dispersions. Pharm. Dev. Technol. 21, 268–276.

[11] Li, M., Ioannidis, N., Gogos, C., Bilgili, E., 2017. A comparative assessment of nanocomposites vs. amorphous solid dispersions prepared via nanoextrusion for drug dissolution enhancement. Eur. J. Pharm. Biopharm. 119, 68–80.

578784 Layer-Wise Agglomeration of Urea Granules

Urea has been widely used as a nitrogen fertilizer due to its high nitrogen content. However, because of its high water-solubility, the nitrate content is lost as surface runoff. In this study, different binders were used to alter the internal structure of the urea granule to control the dissolution behavior and reduce the rate of dissolution. Urea, with binders, was granulated in layers. The core of the granule was manufactured by granulating technical urea powder in a drum granulator, with corn starch as the binder. A second layer was added, by drum granulation, to the core in order to obtain a nutrient release pattern that matches the crop nitrate demand. Corn starch, PEG 4000, and corn starch hydrogel were used as binders for the second layer. The density, thermal properties, fracture pressure, internal porosity, and dissolution rate of the two-layer granules were measured using standard methods. Dissolution rate of the double layer granules were slower than the market urea granules. The mechanical fracture pressure of the double layer granules with hydrogel binder was the highest among these three types of double layer granules. Among these three types of granules, the dissolution behavior of the granule with starch hydrogel on the second layer was identical to the crop nitrate demand curve. Overall, the double layer granule with corn starch binder in the core and hydrogel as the binder in the second layer had a better physical and dissolution properties. Though the pattern of dissolution of double layer granule was similar to the crop nitrate demand curve, but, a soil based study is needed to verify the nitrate release characteristics in the future.

576291 Impact of Dry Coating on Dissolution Profile of Micronized BCS Class II Drugs: Influence of Coating Material Hydrophobicity

Impact of dry coating on dissolution profile of micronized BCS class II drugs: Influence of coating material hydrophobicity

Sangah Kim, Liang Chen, Kuriakose Kunnath, Ecevit Bilgili, Rajesh Dave*

BCS Class II drugs exhibit poor aqueous solubility, which leads to slow and poor absorption through the gastrointestinal tract and in turn inadequate bioavailability. One way to overcome this issue is the micronization of drug particles, thus enhancing their surface area and in turn the dissolution rate[1]. However, such fine particles tend to agglomerate during the processing and storage, which negates the benefits of micronization[2-5]. Simultaneous micronization–dry coating and dry coating of micronized drug particles have been shown to be effective for reducing the extent of agglomeration and increasing the surface area of drug particles[4,5]. While previous investigations focused on the impact of dry coating on drug particle agglomeration and dissolution enhancement, the impact of coating material on the surface hydrophobicity of drug particles has not been examined. Hence, the current study aims to fill this gap by examining the impact of hydrophobicity of the applied dry-coating material on the dissolution rate of selected BCS class II drugs. Griseofulvin (d_{50} of 10 μm) and Ibuprofen (d_{50} of 20 μm) were selected as model drugs. Each drug was dry coated with either hydrophilic (A200) or hydrophobic (R972P) fumed silica via LabRAM, and theoretical calculations were used to select the amount of coating material for achieving 0-100% surface area coverage (SAC). De-ionized water was used as the dissolution medium to investigate the effect of dry coating. USP IV dissolution test with USP <711> protocol was conducted on each formulation. Collected dissolution data was then fitted by the Peppas model to estimate its parameters, i.e., k and n (R^2 values close to 1). Relative hydrophobicity of different formulations was evaluated by using the modified Washburn method with pre-saturated de-ionized water as the wetting liquid. Using the slope of the square of liquid mass penetrated vs. time curve, we calculated the wetting enhancement factor[7]: $\cos\theta_c/\cos\theta_{uc}$ where $\cos\theta_c$ and $\cos\theta_{uc}$ stand for the cosine of the advancing contact angle of the coated and uncoated drug powder, respectively. This factor is greater than 1 when the coating enhances wettability of the drug powder.

Compiled results from dissolution study and wettability testing show that the hydrophobicity of the applied coating material significantly affects the wettability of the drug powder and thus the dissolution profiles. In general, for both drugs, hydrophobic coating material led to slower drug release than the hydrophilic coating material. The enhancement with coating depended on the type and SAC. On the other hand, the impact of dry coating on the wettability and dissolution of different drugs was markedly different. Surface wetting rate was the dominant factor dictating the dissolution rate of dry coated Griseofulvin. The higher the wettability enhancement of Griseofulvin, as quantified by the higher wetting effectiveness factor $\cos\theta_c/\cos\theta_{uc}$, the faster the dissolution (higher k of the Peppas model). When coated with R972P, micronized GF exhibited slower dissolution, and this impact got stronger with an increase in SAC of R972P. However, for Ibuprofen, the analysis revealed a more convoluted relationship of dissolution profiles with the SAC and type of coating material. Surprisingly, the 50% SAC with R972P coating led to wettability enhancement and faster drug dissolution despite the hydrophobicity of the coating material. We hypothesize that the impact of both agglomerate size and wettability after dry coating are important factors that govern the dissolution rate of Ibuprofen. This hypothesis was tested with QICPIC, a dynamic particle image analyzer, which does not cause significant dispersion of agglomerates during the measurement, thus yielding information about the inherent agglomerate particle sizes in the samples. Overall, this study has demonstrated the importance of coating material hydrophobicity and SAC, through their impact on wettability and extent of drug particle agglomeration, in the dissolution of coated, micronized drug particles. The fundamental understanding of the mechanisms controlling the drug release from coated, micronized drug particles will ultimately enable rational design of solid oral dosages using micronized BCS Class II particles.

References

1. Yang, J., Sliva, A., Banerjee, A., Dave, R. N., & Pfeffer, R. (2005). Dry particle coating for improving the flowability of cohesive powders. *Powder Technology, 158*(1-3), 21-33. doi:10.1016/j.powtec.2005.04.032
2. Chen, Y., Yang, J., Dave, R. N., & Pfeffer, R. (2008). Fluidization of coated group C powders. *AIChE Journal, 54*(1), 104-121. doi:10.1002/aic.11368
3. de Villiers, M. M., & Van der Watt, J. G. (1994). The Measurement of Mixture Homogeneity and Dissolution to Predict the Degree of Drug Agglomerate Breakdown Achieved Through Powder Mixing. *Pharmaceutical Research: An Official Journal of the American Association of Pharmaceutical Scientists, 11*(11), 1557-1561. doi:10.1023/A:1018997418322
4. Dressman, J. B., & Fleisher, D. (1986). Mixing-tank model for predicting dissolution rate control of oral absorption. *Journal of Pharmaceutical Sciences, 75*(2), 109-116. doi:10.1002/jps.2600750202
5. Huang, Z., Scicolone, J. V., Gurumuthy, L., & Dave, R. N. (2015). Flow and bulk density enhancements of pharmaceutical powders using a conical screen mill: A continuous dry coating device. *Chemical Engineering Science, 125*, 209-224. doi:10.1016/j.ces.2014.05.038
6. Huang, Z., Xiong, W., Kunnath, K., Bhaumik, S., & Dave, R. N. (2017). Improving blend content uniformity via dry particle coating of micronized drug powders. *European Journal of Pharmaceutical Sciences, 104*, 344-355. doi:10.1016/j.ejps.2017.04.012
7. Li, M., Ioannidis, N., Gogos, C., & Bilgili, E. (2017). A comparative assessment of nanocomposites vs. amorphous solid dispersions prepared via nanoextrusion for drug dissolution enhancement. *European Journal of Pharmaceutics and Biopharmaceutics, 119*, 68-80. doi:10.1016/j.ejpb.2017.06.003

570279 3D Printing of Tuneable Coloured Agglomerates - Strain Distribution Study of a Single Agglomerate Design

The current study presents a novel and reliable method for producing 3D printed agglomerates with different colour distributions and material properties with two-fold aims: providing feasible and accurate control on compression of multiple copies of the same 3D printed agglomerate under different compression angles, and better tracking of individual particle positions after agglomerate breakage.

Multi-coloured agglomerates in cubic tetrahedral and random sphere shapes (used in a previous study [1]) were printed with both rigid and soft bonds using a Statasys Connex 3 using standard vendor polymers and colours to indicate different regions of the agglomerate. The surface and structural properties of the agglomerate, including surface roughness and printing accuracy, were analysed. The agglomerate breakage behaviours under quasi-static compression in an Instron were analysed as a function of bond strength (by varying the polymer used to print the bonds), loading rate (0.02-1 mm/s) and loading directions. The strain distribution was then plotted for the random sphere agglomerate structure as a function of these variables. In addition, agglomerate structures with designed internal macro-voids in different positions and sizes were also created and their breakage behaviour analysed an effort to better understand parameters governing the mechanical properties of agglomerates with cavities and voids. The results showed than a single agglomerate has a distribution of strengths, depending on the orientation and strain rate. This is the first time that a wide distribution of "strength" for a single agglomerate has been demonstrated, as the 3D printing technique allows multiple identical copies of the same agglomerate to be tested under systematically controlled conditions. Similarly, the removal of particles and bonds in key positons also showed effects on the agglomerate strength. Overall the results demonstrate a new technique to understand agglomerate deformation and breakage, including due to defects and cavities, which is an inevitable process during particle handling in industry but poorly understood at present.

Reference

[1] Ge, R., et al., *3D printed agglomerates for granule breakage tests.* Powder Technology, 2017. **306**: p. 103-112.

572635 Interaction of Eugenol and Novel Lignin Grafted Silica Nanoparticles with Synthetic Lipid Bilayers

Lignin oligomers have broadly been recognized as a potential source of renewable aromatics for advanced materials, with a goal of developing functionalized surfaces with desirable selectivity (for separations), antimicrobial, and antifouling properties. Surface modification with lignin components also allows the quantification of interactions between lignin oligomers with solutes and biomolecules for the development of structure-surface property relationships. The potential to translate the properties of small-lignin derived molecules to surfaces has been limited due to the lack of well defined lignin oligomers. With the availability of lignin oligomers of known chemistry synthesized through a "building up" approach, this work develops and validates a generalizable approach to the functionalization of surfaces with lignin oligomers using versatile click chemistry. Here, allyl-modified guaiacyl β-O-4 guaiacyl (GG) lignin dimer is synthesized and attached to mesoporous silica nanoparticles (MSNPs) via click chemistry. The ability of GG lignin-dimer functionalized particles to interact with and disrupt synthetic lipid bilayers is compared to MSNPs functionalized with allyl group containing eugenol (l-hydroxy-2-methoxy-4-allylbenzene), a known natural antimicrobial. Uniform MSNPs were synthesized, and post-synthesis thiol attachment (SH) using (3 mercaptopropyl) trimethoxysilane (MPTMS) was quantified by Ellman's test. The resultant SH-MSNPs were conjugated with GG lignin dimers and eugenol separately by a thiol-ene reaction under ultraviolet light in presence of a photo initiator. Thermogravimetric analysis (TGA) was used to determine that approximately 0.22 mmol of eugenol and 0.16 mmol of the GG dimer were attached to every 1 gram of the MSNPs. The interaction of the functionalized MSNPs with a model phospholipid bilayer of 1,2-Dipalmitoyl-sn-Glycero-3-Phosphocholine (DPPC) (representing cell membranes) supported on gold surface was investigated using a Quartz Crystal Microbalance with Dissipation (QCM-D). While eugenol-grafted particles in PBS (0.5 mg/mL) were found to be benign and nondestructive to the structure of the bilayer interpreted from slight changes in the mass on the QCM-D sensor after exposure of the bilayer to the particles, dimer attached particles with the same concentration caused considerable lipid mass loss associated with disruption of some of the membrane, suggesting higher biocidal activity. This suggests the potential for the functionalization of nanoparticles with specific lignin oligomers to be a path to novel bio-inspired materials and provide a usage for lignin building blocks recovered from lignin deconstruction.

573898 Design of All-Solid-State Lithium-Ion Battery Based on Particle Coating Technology

As a promising candidate of next generation secondary battery with both high safety and high energy density, all-solid-state lithium-ion secondary batteries (ASS-LIBs) have attracted much attention. In the ASS-LIBs, inorganic solid electrolytes (SEs) are used instead of the organic liquid electrolytes. SEs are non-flammable materials and stable under higher voltage and broad temperature range. Due to these features, the ASS-LIBs are highly expected to be a next generation secondary battery for EVs. In the ASSLIBs, transfer of lithium ions during charging and discharging only occurs at the interfacial contact surface between an electrode particle (active material) and SE. Therefore, development of particle processing technologies to construct the solid-solid interfacial contacts is important issue.

We here propose to use a dry coating process to produce composite particle of electrode active material and solid electrolytes. In the dry coating process, lager electrode active material (core particles) were directly coated with SEs (guest particles) by mechanical forces without using any solvents and binders. As a result, FESEM and EDX images showed that the active material were almost fully covered with SEs without breakage of the active material, indicating that the solid-solid interfacial contacts between active material and SEs were constructed within a primary composite particle. A performance of ASS-LIB-test-cell exhibited remarkable improvement when the dry coated particles were used. It was concluded that the dry coating process is very effective to produce composite particle for electrode of ASS-LIBs.

555830 Impact of Dry Coating to Improve Bulk Properties of Needle-like Morphology

It has been demonstrated that the powder properties of API including flow-ability, bulk density, and compaction can be significantly improved via dry coating. These results have been explained theoretically in terms of surface coverage area and surface energy. However, this approach has not been demonstrated on high aspect ratio needle-shaped APIs. The primary purpose of this work is to conduct a systematic study on the impact of dry coating on the powder properties of needle-shaped APIs.

570548 The Calibration of DEM Input Parameters for Biomass Using a Cubical Triaxial Tester

This talk will present the experiments and procedures used to calibrate DEM properties for bulk milled corn stover over a range of loading conditions. Cubical Triaxial Tester (CTT) experiments were performed at Penn State University to obtain the loading, unloading, and shear behavior of the corn stover under conditions similar to those found in a compression screw feeder; which is where the calibrated bulk parameters will be used. A DEM simulation of the CTT experiments was developed using a nonlinear normal force model and a linear spring sliding friction tangential force model, and implemented in the commercial DEM package Rocky. This simulation was also coupled with a Python optimization algorithm which simultaneously minimizes the error between the DEM simulation predictions and experiments for the entire range of test conditions.

576251 Understanding Particle Scale Effect of Mixture Segregation

One of the challenges in pharmaceutical, food, and chemical product development is the ability to relate particle scale properties to behaviors in complex processes. A key issue in processing powders in these industries is mixture segregation, which is responsible for nearly one third of all processing issues with these types of powders. The trouble with solving segregation effects by product design is that the effects are driven by various mechanisms that can act at the same time. Each mechanism, or cause, of segregation is due to a characteristic of the particles that make up a mixture. For example, sifting segregation is induced by differences in particle size. Angle of repose segregation is induced by differences in friction between particles. Rebound segregation is due to differences in coefficients of restitution. Air entrainment segregation is facilitated by particle size differences and true particle densities. Experiments can be constructed to isolate each mechanism, but it is difficult to combine the effects of these individual mechanisms to predict overall segregation because one must determine how much of each mechanism is active in the mixture. This issue is the focus of this paper. Key particle scale parameters are selected for each mechanism of segregation, and these key parameters were measured for each mixture. The segregation intensity of multiple mixtures was then measured and this was combined through proposed segregation models to relate segregation to particle scale properties. Segregation flux velocity functions were determined and then used to estimate the expected variance each mechanism has on the total segregation in the system. This allowed separation of each mechanism on a component-by-component basis, so that the formulator knows what to change about the various components in the system so as to mitigate the segregation of the mixture.

564118 Scale up of Heat Transfer for Dry Granular Material in a Bladed Mixer

Drying is an important manufacturing step in the production of active pharmaceutical ingredients (APIs) following the crystallization and filtration of the material. Several approaches exist to carry out the drying procedure, but a common method used by the pharmaceutical industry is agitated drying. During the process, a wet bed of API is heated in a jacketed cylindrical vessel while being mixed by a rotating impeller until the moisture content is reduced to a desired level. The complexity of the process stems from the fact that heat transfer, mass transfer, and changes in physicochemical properties can occur simultaneously throughout drying. Complications often plague the procedure, including issues such as lengthy drying times, over-drying, nonuniform drying, agglomeration, attrition, and form changes. These circumstances make agitated drying a complicated process to understand and control. When considering scale up, these challenges are coupled with the difficulties typically associated with transferring knowledge from lab scale to pilot or manufacturing scale. As a result, it can be difficult to design an appropriate drying protocol that optimizes heat transfer and can be translated from scale to scale while minimizing the risk for adverse conditions.

In this work, we decouple the problem and focus on studying the heat transfer aspect of agitated drying using dry glass beads. More specifically, we investigate the influence of bed fill height and geometric scaling on heat transfer by studying both the average and standard deviation of the bed temperature. Our approach consists of employing a combination of experiments and discrete element method (DEM) modeling techniques to analyze how heat transfer scales with material fill level and equipment sizing. We carry out a design of experiments and compute heating times as well as heat transfer coefficients for the different conditions. We find that fill height and geometric scaling influence the flow and compressibility of the bed and therefore creates a balance between conduction and granular convection as the dominant mode of heat transfer. We leverage the discrete nature of DEM modeling to evaluate features such as the coordination number and normal overlap between particles. We find that the relationship between fill height, geometric scaling, and heating time is complex and depends on the geometry area available for heat transfer and the particle contacts. Finally, we relate simulation findings to experimental results for validation. This work provides better fundamental insights into how heat transfer scales in a bladed mixer and could greatly aid in the development of drying protocols.

572350 Investigating the Influence of Processing Temperature on Powder Flowability

Numerous unit operations operate at elevated temperatures and/or require powders to be heated in preparation for processing, for example drying operations are common across numerous industries . However, predicting powder behaviour at elevated temperatures is challenging due to the complex network of factors that influence powder flow properties. Powders are bulk assemblies which consist of intricately variable particles with varying quantities of entrained gases and liquids, all with different heat capacities. These different constituent phases, as well as their interactions with each other, may be highly sensitive to variations in ambient temperature and to different extents. It is therefore evident that there is a significant demand to understand the rheological properties of powders at temperatures which are more representative of the environmental conditions within a process.

A strong understanding of the thermal deformation temperatures of powders is required to support the optimisation of in-process operating temperature. At elevated temperatures, powders may undergo thermal deformation, where irreversible changes to properties such as particle shape and surface texture may occur, in addition to particles beginning to melt and fuse together. Particle deformation can have a significant impact on the rheological properties of the powder bulk. For example, in Additive Manufacturing (AM) applications, many powders are heated prior to deposition in order to lower the temperature gradient for sintering. Changes in the flowability of powders as a result of being exposed to high temperatures can therefore have a significant impact on the performance of recycled powder and the quality of final products.

In addition to the impact of temperature on the structure of solid particles, it is also important to consider that processing at elevated temperatures can also affect liquid within the powder bed. Heating powders, for example during drying applications, can begin to evaporate liquid bridges and significantly reduce the level of capillary bonding.

The temperature can also impact on the interaction between the solid particles and gases within the powder bed, for example, gas adsorption (on to solid surfaces) is reduced at high temperatures, changing the chemical properties of the particle surfaces. Furthermore, elevated temperatures can also activate chemical reactions between surface groups, or between surface groups and entrained gases and liquids, again changing the surface chemical properties and ultimately the powder flow behaviour.

This presentation will highlight findings from a study investigating the influence of temperature on the dynamic flow properties of powders. In this study, a selection of powders was equilibrated at a range of temperatures and their rheological properties evaluated using an FT4 Powder Rheometer® (Freeman Technology Ltd, UK). Comparison with behaviour at ambient temperature was undertaken to assess the impact of exposure to elevated temperatures and to demonstrate the importance of characterising powders under conditions that are relevant to the process. As temperature-induced changes also tend to be non-reversible, the hysteresis behaviour of a range of powders were also investigated by cycling the temperature regime, with the results clearly demonstrating the importance of fully understanding the thermal history of the powder been processed.

565001 Investigating the Moisture Content Impact on the Flow Behaviour of a Wet Glass Beads Powder

Granular materials are well known to exhibit a broad range of intricate behaviours and might vary in size and shape; in addition, moisture can be present. In the chemical and process industries, where those materials are widely handled and produced [1], studying the flow behaviour of such materials is of critical importance. Indeed, understanding the impact of internal (e.g., particles size and shape) and external (e.g., applied stresses, moisture content) physical properties on the granular flow behaviour will help industrial practitioners handle and produce particulates in an efficient and less costly way.

In this work, we investigate the impact of moisture (at various concentrations) on the flow behaviour of a powder of wet glass beads using the shear cell of the automated powder rheometer (FT4, Freeman Technology). Shear testers are well known for the relative simplicity of their procedures used to evaluate the stresses needed to generate shear; the shear can lead to either compaction or dilation states of the studied material, under a given applied normal stress. The experimental procedure of the FT4 shear test is as follow: the bladed rotational shear head moves downwards to apply vertical consolidation stress (normal stress σ) on the flat powder sample placed in a steady vessel. Once the desired normal stress is reached, the shear head starts to rotate inducing a shear stress (τ) that increases until the powder bed fails. This procedure is usually repeated at different consolidation stresses in order to build up the yield loci: $\sigma=\mu\tau+c$, where m is the angle of internal friction and c is the so-called cohesive strength. The Mohr circle analysis applied on the yield loci allows the evaluation of the flow function coefficient FF which provides information on the flow behaviour of the studied powder, where high FF corresponds to a free flowing powder (Jenike [2]).

In this study, a powder made of monodispersed (500 mm diameter) glass beads is wetted at different concentrations ω (from 0.1% to 5%) and tested with the FT4 shear cell. The FT4 automated conditioning blade is used to prepare each wetted sample using the same automated mixing procedure. This method allows also an in-situ test, where the shear test is performed directly on the mixed sample. The angle of internal friction and the cohesive strength and the flow function coefficient were then evaluated at different initial consolidation stresses and water contents. The first results showed that, at a fixed water concentration, the angle of internal friction is independent of the initial consolidation stress. The values of flow function showed that, according to Jenike classification [2], a wet powder with a fixed w can be classified as free flowing (FF>10) if the powder is highly consolidated or as cohesive (2<FF<4) at low consolidation stresses. This might be explained by the fact that the liquid bridges created by the water content can be broken at high consolidation stresses. In addition, for a fixed initial consolidation stress, the angle of internal friction appears to be independent of the water content with a value equal to the one obtained in a dry case. Furthermore, it appears that the cohesive strength c increases when increasing the water content and reaches an asymptotic value. However, it should be noted that even if this final observation is in good agreement with the one found in the literature [3], it was obtained using one initial consolidation stress. Therefore, further measurements, including different consolidation stresses, will be performed in order to confirm the first measurements and observations.

References

[1] F. J. Muzzio, A. Alexander, C. Goodridge, E. Shen, and T. Shinbrot, "Solids mixing part A: fundamentals of solids mixing," *Handb. Ind. Mix. Sci. Pract.*, pp. 887–985, 2004.

[2] A. W. Jenike, *Storage and flow of solids*, Rev. ed. Salt Lake City, Utah : University of Utah, 1970.

[3] V. Richefeu, M. S. El Youssoufi, and F. Radjai, "Shear strength properties of wet granular materials," *Phys. Rev. E Stat. Nonlin. Soft Matter Phys.*, vol. 73, p. 051304, May 2006.

AUTHOR INDEX

Abbasian, Javad 347
Abdullahi, Hassan 137
Ade, Nilesh 52
Adepu, Manogna 197, 393
Afarinmade, Azeem 20
Aglave, Ravindra 442
Ahmad, Stephanie 452
Ahsan, Syed 348
Akabua, Eyram 268
Akhavan, Ali 211
Akroyd, Jethro 171
Alaie, Seyed 115
Alayoubi, Alaadin 282, 283, 290
Albadarin, Ahmad B. 2, 419
Alberini, Federico 57
Albrecht, Jacob 365
Almgren, Ann 220
Ambrose, Kingsly 411, 455
Amichetti, Marissa 432
Amini, Negin 92, 227, 289, 458
Amini, Shahriar 131
An, Keju 212
Anderegg, David 329
Anderson, Carl A. 199
Anderson, Justin 265
Andino, Jean M. 212
Anthis, Alexandre 355
Antic, Anthony 289
Arastoopour, Hamid 347
Arevalo, Faustin 452
Arjmandi-Tash, Omid 173
Armstrong, Madeleine 139
Artusio, Fiora 369
Asare, Shardrack O. 459
Asfari, Sarah 282, 283, 290
Ashok, Anchu 102
Ashraf, Muhammad 282, 283, 290
Askarishahi, Maryam 204
Austin, David 60
Azad, Mohammad 434
Baba, Tomoya 195
Bagaria, Pranav 50, 51
Bailey, Maximilian R. 313
Bairq Sr., Zain Ali Saleh 266
Baran, Oleh 442
Bari, Rozana 260, 304
Barkley, Stuart J. 256
Barrett, Ross 46
Barzegar, Mohammad 16
Baser, Deven 186
Bassous, Nicole 192
Batchelor, Hannah K. 440
Bayham, Samuel 45, 428
Beaudoin, Stephen P. 253, 291
Beetham, Sarah 236
Behdani, Behrouz 381
Belekar, Viraj Vilas 36
Bell, John 220
Benedetti, Antonio 415
Benítez, Laura Andrade 273
Benz, Dominik 313
Bergström, Christel A. S. 364
Bhethanabotla, Venkat R. 385
Bhosale, Rahul 430
Bhujbal, Sayali 452
Bi, Huichao 96
Bi, Xiaotao 84
Bilgili, Ecevit 55, 452, 456
Bird, David 340
Biserni, Stefano 273
Biswas, Pratim 37
Blake, Diane A. 20
Boje, Astrid 171
Borghard, William G. 142, 200
Bortner, Michael J. 329, 382
Bose, Arijit 20
Boukerche, Moussa 366
Boyce, Christopher M. 12
Braatz, Richard D. 127
Bradfish, Kacie 291
Brady, Sarah 327
Brancazio, David 434
Brauer, Nicolás Torres 217
Breault, Ron 43
Breault, Ronald W. 45, 346, 428
Brenza, Timothy M. 356
Brindle, Joseph 264
Brockbank, Katrina 198
Brunsteiner, Michael 270
Buckley, David 141
Bull, Sarah 180, 307
Bullard, Joseph W. 34, 147
Bunchatheeravate, Pongpumin 34, 147
Burcham, Christopher L. 137
Burns, Mark A. 386
Buttini, Francesca 273
Campo-Deano, Laura 134
Capece, Maxx 433
Capecelatro, Jesse 236
Capellades, Gerard 434
Carmichael, Christopher 466
Carmody, Alan 114
Ceccone, Giacomo 369
Champ, Theodore 180, 307
Chan, Gabriela Ruphuy 358
Chan, Steven H. 461
Chang, Ji Woong 265
Chaouki, Jamal 445
Chappelow, Chris 403
Chase, Emily 459

AUTHOR INDEX

Chaudhari, Purvali 50, 52
Chaudhuri, Bodhisattwa 199, 287, 348
Chen, Jie 140
Chen, Liang 431, 456
Chen, Xiao 205, 399
Chen, Xiaoshuang 167
Chen, Yanjie 455
Chen, Yu-Yen 188
Chen, Zhifeng 265
Cheng, Guanzhi 191
Cheng, Zhuo 186
Chew, Jia Wei 15
Chintersingh-Dinnall, K.-Lee A. 250
Chortos, Alex 384
Christau, Stephanie 357
Chung, Neal Tai-Shung 405
Clancy, Don 415
Clark, Rosemary 211
Clarke, David R. 384
Clarke, James 275
Clayton, Jamie 465
Cliff, Alexander 196
Cloete, Jan Hendrik 131
Cloete, Schalk 131
Cocco, Ray 15, 109, 129, 345, 401, 442, 443
Codan, Lorenzo 168
Coelho, Alexander 452
Coleman, Sarah 181
Contreras, Lourdes 114, 350
Coppens, Marc-Olivier 6, 211
Coultas-McKenney, Caralyn A. 291
Crownover, Robin 341
Cruz, Celia N. 282, 283, 290
Curtis, Jennifer S. 34, 147
Curtis, Jennifer Sinclair 206, 235
Dabbagh, Firas 39
Daftary, Vivek 141
Dahl, Steven R. 443
Dam-Johansen, Kim 96
Dartevelle, Emma M. 382
Das, Arit 382
Dattani, Rajeev 439
Davari, Seyyed Ali 251
Davé, Rajesh 269, 431, 456, 461
Davies, Clive E. 16, 196
Daw, C. Stuart 213
De Beer, Martin 386
De Beer, Thomas 169, 276, 278, 280
De Broqueville, Axel 394
De Lasa, Hugo 217
De Wilde, Juray 221, 394
Dean, Joshua 257
Deng, Wei 191
Deng, Zhekai 145
Denton, Aric 260, 304
Dhakate, Mahesh 409
Dhondt, Jens 278
Diemer Jr., R. Bertrum 80, 174
Dillard, David 329
Dixit, Mudit 88
Dobosz, Kerianne M. 38
Dong, Juyao 181
Dong, Quanxiao 191
Doshi, Pankaj 114, 276
Dou, Jian 187
Doukkali, Mouhcine 249
Dowling, Denis 327
Dreizin, Edward 248, 303, 306, 343
Dreizin, Edward L 250, 305
Dreizin, Edward L. 247
Dsouza, Peter V. 42
Duan, Fan 205
Duan, Yifei 144
Dumesic, James 262
Duron-Torres, Sergio M. 101
Duscher, Gerd 251
Duzik, Adam 330
Elias, George 238
Ellis, Naoko 84, 445
Ellwood, Kevin 115
Emady, Heather N. 197, 393
Engstrom, Joshua 141, 461
Ennis, Benjamin 176, 177, 344
Ennis, Brandon 176, 344
Ennis, Bryan J. 176, 177, 344
Erdemir, Deniz 141
Eren, Ayse 378
Escalante-Gacia, Ismailia L. 101
Fakhrabadi, Ehsan Akbari 14
Fan, Jianren 398
Fan, L.-S. 82
Fan, Liang-Shih 185, 186, 188, 189
Fan, Yi 145
Fathima, Navaira 430
Fathollahi, Sara 412
Faulhammer, Eva 412
Feng, Jie 139
Ferguson, Steven 327
Fernández, Verónica Daza 273
Ferreira, Ana 365
Finney, Charles E. A. 213
Folestad, Staffan 313, 437
Fondren, Zachary 260, 304
Forgber, Thomas 114, 350
Francia, Victor 6
Francqui, Filip 175, 417, 418
Frawley, P. J. 81
Freeman, Tim 198, 414, 439, 465
Freireich, Ben 109, 129, 345, 401, 442, 443
Fukuoka, Morio 196

AUTHOR INDEX

Fullard, Luke........16, 196
Furar, John........115
Fushimi, Chihiro........86
Gabbott, Ian........440
Gadikota, Greeshma........48
Gaffney, Anne........224
Galindo-Rosales, Francisco J.........134
Galinsky, Nathan........428
Gamble, John........275, 365
Gao, Hongxia........266
Gao, Jinghui........21
Gao, Jinsen........352
Gao, Xi........35, 397
Gao, Yijie........348
Garcia, Caterina Bartomeu........192
García-Batancourt, M. Luisa........101
Gardner, John M.........329
Garner, Sean........366
Gasteyer, Nathaniel........462
Gerken, Lukas R. H.........355
Ghoroi, Chinmay........252
Gilchrist, James F.........21, 309
Gilmer, Eric L.........329
Girard, Kevin........327
Glasser, Benjamin........13, 412
Glasser, Benjamin J.........464
Glotzer, Sharon C.........100
Goetze, Josh........185
Golberg, Jordan........256
Göltl, Florian........262
Goméz, Alejandra De La Rosa........101
Gómez, Noel........448
Gong, Xun........194
Gonzalez, Marcial........238, 462
Good, David........365
Gordiichuk, Pavlo........181
Goswami, Partha S.........203, 210
Gottfried, Jennifer........251
Goudeli, Eirini........3, 89, 183
Gounaris, Chrysanthos E.........267
Grace, John R.........84, 445
Grafton, Miles C. E.........16
Greenwood, Richard........275
Grillo, Fabio........4
Groven, Lori........342
Guan, Guoqing........86
Guenther, Chris........216
Guildenbecher, Daniel........49
Gump, Christopher........312
Gunda, Harini........252
Guo, Mengqing........185, 188
Guo, Yu........206, 235
Habibi, Nahal........357
Hagan, Matthew........339
Hajiesmaili, Ehsan........384
Halota, Madeline........464
Hammersmith, Gregory........434
Hanley, Thomas R.........22
Hanuš, Jaroslav........358
Hapgood, Karen........289, 458
Hapgood, Karen P.........92
Harbottle, David........60
Harding, Matthew........327
Hardy, Baptiste........221
Haribal, Vasudev Pralhad........429
Haritos, Victoria........227
Harper, Daniel........98
Harrison, Andrew........263
Hartig, Julia........9, 207
Hartmanshenn, Clara........464
Hasan, Mohammad J.........311
Hassanpour, Ali........60
Hastings, Daniel........303, 343
Hauser, Thomas........220
He, Jibao........20
He, Xiaoqing........310
He, Yanping........390
Heindel, Theodore J.........36, 146
Heng, Jerry Y. Y.........367
Henson, Michael A.........38
Hermans, Andre........284
Hernandez, Juan D. Frías........101
Herrmann, Inge K.........94, 355, 362
Hill, Chanel........180, 307
Hill, Justin J.........330
Hill, Priscilla J.........91
Hittel, Benjamin........263
Hlabangana, Nkosikhona........1
Ho, Raimundo........140
Hogan Jr., Christopher J.........167
Hoops, Jordan A.........356
Hou, Dianhao........385
Hrenya, Christine........443
Hrenya, Christine M.........208, 220
Hsieh, Daniel........141
Hu, Chenshu........398
Huang, Ci........306
Huang, Zhengliang........19
Hunter, Timothy N.........60
Hwang, In Sik........422
Hwang, Jungho........408, 422
Ingram, Andrew........57, 275, 440
Irizarry, Roberto........168
Isenberg, Natalie M.........267
Ismail, Hamza........2
Issangya, Allan........345
Jacob, Karl........145
Jafari, Rouzben........445
Jajcevic, Dalibor........114, 350
Jasim, Ahmed........310

AUTHOR INDEX

Jasuja, Kabeer....252
Jharimune, Suprita....265
Jiang, Ming....399
Jiang, Peng....97
Jiang, Yiyang....206
Jiang, Yundi....400
Jibrin, Naseem....176, 177, 344
Jin, Baitang....261
Johansen Sr., Stein Tore....239
Johanson, Kerry....63, 148, 463
John, Vijay T.....20
Johnson, Ashley E....311
Johnson, J. Karl....98
Jones, A-Andrew....192
Jones, John....275
Jones, Ryan P.....143
Joseph, Eugene G.....382
Joshi, Jyeshtharaj B.....409
Jurtz, Nico....442
Kacica, Clayton....37
Kaewpetch, Thitiporn....309
Kakosimos, Konstantinos E.....430
Kalyva, Agni E.....430
Kamath, Anusha....37
Kang, Myung Soo....408
Kape, Adrian....114
Kapoor, Yash....284
Karim, Ayman M.....88
Karri, Reddy....109, 112
Karri, S. B. Reddy....129, 345
Keevend, Kerda....355
Kelesidis, Georgios A.....3, 179
Kenvin, Jeffery....198
Khakhar, Devang V.....409
Khan, M. Arif....459
Khan, Mohammed....131
Khinast, Johannes G.....114, 350, 412, 464
Kim, Na Yoon....192
Kim, Sangah....431, 456
Kim, Shinhyuk....127
Kishida, Masahiro....423
Kitchens, Christopher L.....99
Klan, Jonathan....98
Klee, David....434
Klinzing, Gerard....284
Knobbe, Tom....17
Knowlton, T. M.....112, 129, 345
Knutson, Barbara L.....459
Kobler, Mirjam....273
Kodam, Madhusudhan....403, 404
Koh, Yung P.....260
Kolehmainen, Jari....400, 420, 436
Kolis, Stanley P.....52
Konstandopoulos, Athanasios G.....430
Kopcha, Michael....290
Kopp, Nathaniel D.....141
Kortshagen, Uwe R.....167
Kotter, Lance....342
Kotula, Anthony P.....380
Kovar, Thomas M.....12
Kozawa, Daichi....181
Kraft, Markus....171
Kraume, Matthias....442
Krzystowczyk, Emily....187
Kulatilaka, Waruna....49, 50
Kumar, Anand....102
Kumar, Ashish....276
Kumar, Prashant....140
Kumar, Rohit....237, 432
Kunnath, Kuriakose....269, 431, 456, 461
Kwon, Dong Eui....193
La Zara, Damiano....4, 313, 437
Lahann, Joerg....357
Lakerveld, Richard....325
Lamarche, Casey Q.....401, 443
Lamprakou, Zoi....96
Lan, Xingying....352
Lan, Zhou....387
Lane, Gregory....141
Lau, Raymond....451
Lawrence, Adam....255
Leadbeater, Thomas....445
Lee, Jay H.....127
Lee, Kai....276
Lee, Youn-Woo....193
Lese, Ioana....362
Leu, Ming....381
Leverant, Calen....97
Lew, Tedrick Thomas Salim....181, 194
Lewis, Jennifer A.....384
Li, Boyang....38
Li, Fanxing....187, 429
Li, He....229
Li, Jingzhe....415
Li, Ping....230
Li, Shiguang....261
Li, Si W....287
Li, Tianyi....13
Li, Tingwen....213, 445
Li, Wenhui....88
Li, Yanjie....235
Liang, Xinhua....261
Liang, Yujia....389
Liang, Zhiwu....266
Libera, Joseph A.....389
Liberatore, Matthew....14
Lim, Jim C.....84
Lin, Junjie....398
Lin, Sen....230
Lindrud, Mark....141

AUTHOR INDEX

Link, Frederik J. 367
Litster, James D. 173
Liu, Jing 200
Liu, Meishen 48
Liu, Peiyuan 208, 220
Liu, Qingzhao 235
Liu, Sitong 354
Liu, Tianyu 100
Liu, Wanjiao 115
Liu, Xiaoyu 420, 436
Liu, Xufung 224
Liu, Yan 185
Liu, Yancong 352
Liu, Yifan 191
Liu, Yifei 262
Liu, Yong 216
Liu, Yu 238
Liu, Yubai 201
Liu, Zheng 232
Liu, Zhi-Hua 228
Loh, Kai Chee 225
Lopez, Ramon E. 444
Lopez-Rodriguez, Rafael 327
Lu, Huilin 315
Lu, Liqiang 130, 397
Lueptow, Richard M. 143, 144, 145
Lumay, Geoffroy 175, 417, 418
Luo, Kun 398
Lvov, Yuri 20
Lynn, Bert C. 459
Ma, Anson 287
Madlmeir, Stefan 114
Maffia, Gennaro J. 224
Malekzadeh, Mohammad 93
Malsbury, Adriene 141
Manser, Kimberly 284
Mantanus, Jérôme 276
Mantri, Rao 461
Mao, Chen 223
Marin, Guy B. 448
Mashuga, Chad 49, 50, 51, 52
Mason, Leah 381
Masuyama, Tomoyuki 460
Matsune, Hideki 423
Matter, Martin T. 94, 355, 362
Mavrikakis, Manos 262
Mazumder, Shanta 226
McCarthy, Joseph J. 444
McGough, Patrick 403
McGowan, Thomas F. 111
McKenna, Gregory B. 260, 304
McLaren, Christopher 12
Medina, David 192
Meesters, Gabrie 17
Melocchi, Alice 285
Mercandelli, Alberto 273
Meyer, Jonathan 276
Mi, Gujie 192
Michael, James B. 256
Misawa, Tomonari 416
Mishra, Ipsita 208
Mitchell, Christopher 464
Mlinar, Laurie 140
Monington, Laura 439, 465
Moon, James 357
Moradipour, Mahsa 459
Moraes, Ângela Maria 190
Morton, David 92, 289, 458
Morud, John 239
Mozaffari, Saeed 88
Mpourmpakis, Giannis 88, 267
Mudryy, Ruslan S. 292
Mukherjee, Dibyendu 251
Muliadi, Ariel 223
Müller, Christoph R. 12
Mursalat, Mehnaz 248, 303
Muschelknautz, Ulrich 129, 406
Musgrave, Charles B. 180, 307
Mushonga, Natsai 1
Musser, Jordan 220
Muzzio, Fernando J. 13, 415
Myerson, Allan 434
Nadeem, Humair 146
Nadgouda, Sourabh 186, 188
Nadimpalli, N. Kumar Varma 184
Nagy, Zoltan K. 378
Nakai, Masahiro 133
Nakamura, Hideya 182, 195, 460
Neal, Luke 429
Nere, Nandkishor 140, 366
Nguyen, Duy 20
Nigra, Michael M. 264
Nikitczuk, Whitney 141
Nissley, Rebecca 284
Nofsinger, Rebecca 284
Nopens, Ingmar 169, 276
Nordstrom, Fredrik 90
Norris, Contessa 291
Nott, Prabhu R 42
O'Brien, Robert C. 180, 307
Ochoa, Alejandro Molina 448
Ochyl, Lukasz 357
Ocone, Raffaella 46, 466
O'Connor, Thomas 282, 283
Ohsaki, Shuji 182, 195, 460
Ojo, Olakunle 20
Olaleye, Akeem 18
Omarova, Marzhana 20
O'Toole, Rebecca 312
Ottino, Julio M. 143, 144, 145

AUTHOR INDEX

Ozel, Ali 400
Pai, Dhananjay A. 411, 455
Pan, Hao-Jen 388
Pan, Jianhua 189
Papageorgiou, Charles D. 378, 464
Pardikar, Kunal S. 223, 234
Parietti, Federico 285
Park, Ah-Hyung Alissa 8, 420
Park, Geonhwan 193
Park, Hee Jeong 193
Park, Joontaek 381
Park, Minkyung 194
Parker, David J. 445
Parker, Nathaniel 55
Parviz, Dorsa 194
Passalacqua, Alberto 36
Patil, Amit 239
Paudel, Amrit 270, 273
Paul, Abhishek 238
Paul, Mahasweta 451
Pawar, Pallavi 415
Peeters, Michiel 276
Pellejero, Lorena Baranda 192
Penn, Alexander 12
Perez, Lisa M. 52
Perez-Viramontes, Nicté J. 101
Peterman, Amanda 224
Petersson, Gunilla 437
Petrie, Frankie A. 311
Petrouskie, Brandon 115
Pettinau, Alberto 214
Piepke, Tobias 46
Pinto, Joana T. 270
Pirenne, Thomas 175
Pirker, Stefan 39
Pisano, Roberto 369
Pitt, Kate 173
Pittman, Zachariah 99
Poortinga, Albert 394
Porcu, Andrea 214
Porfirio, Tiago 134
Portal, Ricardo J. F. 35
Portier, Christoph 276
Pothier, Christina 432
Pozarlik, Artur 394
Pratsinis, Sotiris E. 94, 179
Priestley, Rodney D. 139
Pritchard, Zachary D. 386
Probst, Susana Ecenarro 273
Procopio, Adam 284
Prud'Homme, Robert K. 139
Pukkella, Arjun Kumar 184
Puri, Virendra 462
Qin, Lang 185, 186
Quayle, Mike J. 313, 437
Quon, Justin L. 378, 464
Rad, Farnaz Esmaeili 347
Radl, Stefan 10, 204
Rahman, Mahbubur 55, 452
Rahman, Umair Jamil Ur 394
Raikes, Michelle 90
Raj, Rishi 312
Raman, Ritesh 203, 210
Ramirez, Emilio 213
Rankin, Stephen E. 459
Rapp, Kersten 434
Rauchenzauner, Stefanie 396
Redding, Gabe P. 16
Reding, Nicholas 54, 178
Reeves, Ryan D. 330
Remiro, Paula De Freitas Rosa 190
Reynolds, Gavin K. 440
Reza, M. Toufiq 226
Rhodes, Martin 211
Ribeiro, Erick L. 251
Rim, Guanhe 8
Rimmer, Sophie 57
Rinaldi, Carlos 354
Rioux, Robert M. 265
Roberts, Zane 253
Rock, Reza 115
Rodriguez, Natasha V. 199
Rogers, William A. 35, 130, 214, 397
Roginski, Andrew 291
Rohani, Parham 93
Rosa, Paulo De Tarso Vieira E 190
Rowan, Steven 43, 45, 346
Roy, Manish 338
Roy, Somesh 183
Ruffley, Jonathan 98
Runkana, Venkataramana 184
Ryckaert, Alexander 276, 278
Saha, Pretom 226
Saini, Ankit 49, 50
Sakai, Mikio 402
Sakurai, Risa 182, 460
Šalamúnová, Petra 358
Salar-Behzadi, Sharareh 350
Salehi, Mohammad-Sadegh 204
Salon, Ivan 358
Sannidhi, Abhinav 22
Sansare, Sameera 287
Saranteas, Kostas 38
Sarkar, Avik 114, 350
Sastri, Bhima 214
Savliwala, Shehaab 354
Schiffman, Jessica D. 38
Schlegel, Andrea 94
Schlegel, Jessica 284
Schneiderbauer, Simon 39, 41, 396

AUTHOR INDEX

Schoell, Jochen 168
Schoenitz, Mirko 247, 248, 250, 303, 305, 306, 343
Schweizer, Christian 49, 50, 51
Scott, Douglas 139
Scott, Timothy F. 386
Semiã£o, Viriato 134
Sen, Koyel 199, 287
Seto, Takafumi 167
Seville, Jonathan P. K. 445
Shahnam, Mehrdad 130, 213, 214, 397
Shang, Zeyu 261
Shen, Yangyang 142, 200
Shen, Ye 85
Sheverev, Valery 414
Shi, Hongbin 308
Shi, Xiaogang 352
Shiflett, Mark B. 54, 178
Shin, Jaeuk 408
Shinebaum, Rachael 440
Shirazian, Saeed 2
Shrey, Fnu 51
Silmore, Kevin S. 194
Silverman, Roy 113
Simon, Sindee L. 260, 304
Singh, Mehakpreet 2
Sinha, Kushal 36, 140, 146
Siochi, Emilie J. 329
Sippel, Travis R. 255, 256
Sittisomwong, Poom 37
Skliar, Mikhail 338
Smith, Derrick 284
Smith, Kurt 33
Smith, Rachel M. 173
Solomon, Michael J. 100
Solomon, Samuel 419
Son, Steven F. 253
Soundarrajan, Narasimhan 45
Sousa, Ricardo 133
Stapaniuk, Vadim 414
Starsich, Fabian 94
Stauffer, Fanny 276
Steendam, René R. E. 81
Štepánek, František 358
Stepanov, Victor 260, 304
Stickel, Jonathan J. 14
Strano, Michael 181, 194
Subramaniam, Shankar 36, 146
Subramanian, Sivakumar 184
Sun, Feilong 437
Sun, Jingyuan 19
Sun, Shuying 230
Sun, Yalun 250
Sun, Yanlin 390
Sundaram, Shyam 44
Sundaresan, Sankaran 8, 400, 420, 436
Swaminathan, Shrikant 237, 432
Sweere, Anton 394
Swihart, Mark T. 93
Syamlal, Madhava 220
Szilagyi, Botond 378
Takabatake, Kazuya 402
Tamblyn, Richard 57
Tanaka, Jun 416
Tang, Christina 263
Tanksale, Akshat 227
Tao, Jing 141
Tao, Yi 13
Tarabokija, James 452
Tarlochan, Faris 102
Tatebayashi, June 423
Tausendschön, Josef 10
Taylor, Michael G. 267
Tebianian, Sina 445
Teleki, Alexandra 364
Tew, Jonathan D. 173
Thakur, Subhash 237, 432
Theuerkauf, Jörg 145
Thomas, Kiran Mathew 325
Thornton, Tony 198
Tipton, Roger B. 385
Tobyn, Mike 275, 365
Todd, Paul W. 22
Tomas, Jan 358
Tomassone, M. Silvina 33, 142, 200
Tong, Andrew 188, 189
Tong, Yen Wah 225
Tourneur, Thomas 394
Tripathi, Naveen M. 175
Trogrlic, Martina 114, 350
Trout, James 20
Tseng, Yu-Ti 388
Tsutsumi, Atsushi 87
Tyrrell, Rory 81
Umbanhowar, Paul B. 143, 144, 145
Ureña-Benavides, Esteban E. 311
Valente, Pedro 133
Valluri, Siva Kumar 247
Van De Steene, Shana 280
Van Den Akker, Harry E. A. 18
Van Geem, Kevin M. 448
Van Hauwermeiren, Daan 169, 276
Van Ommen, J. Ruud 4, 209, 211, 313, 437
Vansaders, Bryan 100
Vasquez, Erick S. 311
Vasudevan, Ashvin Kumar 305
Vekilov, Peter 377
Vervaet, Chris 280
Veser, Götz 268
Vetter, Thomas 137
Vicente, Joã£o 134

AUTHOR INDEX

Vigh, Tamas....276
Vogt, Bryan D....383
Vuong, Tien....263
Walker, Gavin....2, 18
Walton, Kenneth....338
Wang, Allison....434
Wang, Chi-Hwa....85, 229
Wang, Dawei....189
Wang, Dongyi....8
Wang, Guixin....233
Wang, Jingdai....19
Wang, Liang-Yi....405
Wang, Shunxing....233
Wang, Tao....308
Wang, Xijun....187
Wang, Yujun....201
Ward, Jeffrey D....388
Wassgren, Carl....206, 235, 411, 455, 462
Wassgren, Carl R....223, 234, 238
Watano, Satoru....182, 195, 460
Weber, Justin....43, 346
Webster, Thomas J....192
Wei, Fei....83
Wei, Liping....85
Wei, Yun....233
Weiglein, Emma....291
Weimer, Alan W....9, 180, 207, 307, 312
Weinell, Claus Erik....96
Weller, Thomas M....385
Wemmers, Anton....394
Wendt, Lynn M....224
Whitaker, Darren....2
White, Tommi....310
Wickham, Jason....253
Winckelmans, Grégoire....221
Winn, Michael....177, 344
Wu, Kaiqiao....6
Wu, Yingya....352
Wutscher, Thomas....270, 273
Xie, Delong....390
Xie, Yongjiang....191
Xing, Lei....230
Xing, Yangchuan....310
Xu, Dandan....235
Xu, Dikai....189
Xu, Fei....233
Xu, Yupeng....214, 445
Yahia, Lyes Ait Ali....46, 466
Yamamoto, Tsuyoshi....423
Yang, Jiao....200
Yang, Jiecheng....34, 147
Yang, Jingsi....43, 45, 346
Yang, Ning....202, 450
Yang, Shiliang....15
Yang, Yahui....268
Yang, Yao....19
Yang, Yongrong....19
Yangcheng, Lu....387
Yao, Zhiyi....85
Yazdanpanah, Nima....282, 283
Ye, Lu....410
Yi, Hojae....462
Yin, John....414, 439
Yin, Xiangyu....267
Yokoo, Kento....423
Yonamoto, Yoshiki....416
Yoon, Bola....312
Yoon, Tae Jun....193
Young, Daniel....433
Yu, Jia....35, 130
Yu, Jianguo....230
Yu, Kyungseok....193
Yu, Liya E....405
Yu, Zhaosheng....206
Yuan, Joshua....228
Yule, Robert E....415
Yusoff, Siti N. Mohd....173
Zeevi, Michael....263
Zellnitz, Sarah....270, 273
Zenz, Jonathon....110
Zhang, Di....437
Zhang, Donghui....20
Zhang, Fuweng....209, 437
Zhang, Haitao....38
Zhang, Jingxin....225
Zhang, Jun....92, 289, 458
Zhang, Lifeng....262
Zhang, Peng....19
Zhang, Yitao....188, 189
Zhao, Ya....15
Zheng, Kai....132, 431
Zheng, Xinguo....191
Zhou, Qiang....205, 399
Zhou, Rongtao....202, 450
Zhu, Keke....256
Zhu, Linhua....390
Zidan, Ahmed....282, 283, 290
Zlochower, Isaac....53
Zou, Jiangdong....233